Memory Mechanisms in Health and Disease

Mechanistic Basis of Memory

Memory Mechanisms in Health and Disease

Mechanistic Basis of Memory

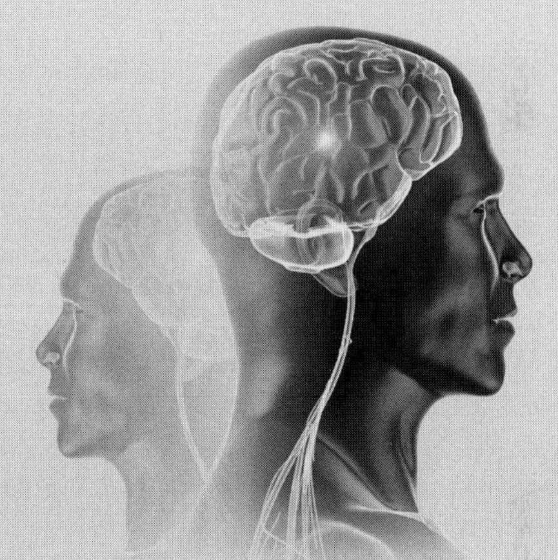

Editor

Karl Peter Giese

King's College London, UK

NEW JERSEY • LONDON • SINGAPORE • BEIJING • SHANGHAI • HONG KONG • TAIPEI • CHENNAI

Published by

World Scientific Publishing Co. Pte. Ltd.
5 Toh Tuck Link, Singapore 596224
USA office: 27 Warren Street, Suite 401-402, Hackensack, NJ 07601
UK office: 57 Shelton Street, Covent Garden, London WC2H 9HE

British Library Cataloguing-in-Publication Data
A catalogue record for this book is available from the British Library.

MEMORY MECHANISMS IN HEALTH AND DISEASE
Mechanistic Basis of Memory

ISBN-13 978-981-4366-69-4
ISBN-10 981-4366-69-2

Typeset by Stallion Press
Email: enquiries@stallionpress.com

Printed by FuIsland Offset Printing (S) Pte Ltd Singapore

Contents

Contributing Authors

Cristina M. Alberini
Department of Neuroscience
Department of Psychiatry
Mount Sinai School of Medicine
New York, NY 10029, USA
E-mail: cristina.alberini@mssm.edu

Giorgia Albieri
MRC Centre for Neurodegeneration Research
King's College London
De Crespigny Park, London SE5 8AF, UK

Dhananjay Bambah-Mukku
Department of Neuroscience
Mount Sinai School of Medicine
New York, NY 10029, USA

Dillon Y. Chen
Department of Neuroscience
Mount Sinai School of Medicine
New York, NY 10029, USA

Gerald T. Finnerty
MRC Centre for Neurodegeneration Research
King's College London
De Crespigny Park, London SE5 8AF, UK
E-mail: gerald.finnerty@kcl.ac.uk

Andre Fischer
Laboratory for Aging and Cognitive Diseases
European Neuroscience Institute
Grisebach Str. 5, D-37077 Goettingen, Germany
E-mail: a.fischer@eni-g.de

Paul W. Frankland
Neurosciences & Mental Health
The Hospital for Sick Children
555 University Ave., 6018 McMaster Building
Toronto, Ontario, M5G 1X8, Canada
E-mail: paul.frankland@sickkids.ca

Katsuo Furukubo-Tokunaga
Department of Life and Environmental Sciences
University of Tsukuba, Japan
E-mail: furubuko-tokunaga.gm@u.tsukuba.ac.jp

Karl Peter Giese
Department of Neuroscience, Institute of Psychiatry
King's College London
125 Coldharbour Lane, London, SE5 9NU, UK
E-mail: karl.giese@kcl.ac.uk

Frank Hirth
Department of Neuroscience
MRC Centre for Neurodegeneration Research
Institute of Psychiatry
King's College London
De Crespigny Park, London SE5 8AF, UK

Sheena A. Josselyn
Program in Neurosciences & Mental Health
Hospital for Sick Children
555 University Ave.
Department of Physiology
Institute of Medical Sciences
University of Toronto
Toronto, ON, M5G 1X8, Canada
E-mail: sheena.josselyn@sickkids.ca

Satoshi Kida
Department of Bioscience
Faculty of Applied Bioscience, Tokyo University of Agriculture
1-1-1 Sakuragaoka, Setagaya-ku, Tokyo 156-8502, Japan
E-mail: kida@nodai.ac.jp

Kimberly Lackenby
Department of Neuroscience
Institute of Psychiatry,King's College London
125 Coldharbour Lane, London, SE5 9NU, UK

Zoe N. Ludlow
Department of Neuroscience
MRC Centre for Neurodegeneration Research
Institute of Psychiatry
King's College London
De Crespigny Park, London SE5 8AF, UK

Thomas J. McHugh
Laboratory for Circuit and Behavioural Physiology
RIKEN Brain Science Institute
2-1 Hirosawa, Wako City, Saitama, 351-0198 Japan
E-mail: tjmchugh@brain.riken.jp

Keiko Mizuno
Department of Neuroscience
Institute of Psychiatry, King's College London
125 Coldharbour Lane, London, SE5 9NU, UK
E-mail: keizo.mizuno@kcl.ac.uk

Masuo Ohno
Center for Dementia Research
Nathan Kline Institute
New York University School of Medicine
Orangeburg, New York, USA
E-mail: mohno@nki.rfmh.org

Athanasia Papoutsi
Institute of Molecular Biology and Biotechnology (IMBB)
Foundation for Research and Technology (FORTH)
Department of Biology
University of Crete
Heraklion, Crete, Greece

Panayiota Poirazi
Institute of Molecular Biology and Biotechnology (IMBB)
Foundation for Research and Technology (FORTH)
Heraklion, Crete, Greece
E-mail: poirazi@imbb.forth.gr

Victor I. Popov
Institute of Cell Biophysics
Russian Academy of Sciences
Pushchino, Russia

P. Joel Ross
Program in Neurosciences & Mental Health
Hospital for Sick Children
555 University Ave.
Toronto, ON, M5G 1X8, Canada

Derya Sargin
Program in Neurosciences & Mental Health
Hospital for Sick Children
555 University Ave.
Toronto, ON, M5G 1X8, Canada

Melanie J. Sekeres
Program in Neurosciences & Mental Health
Hospital for Sick Children
555 University Ave.
Department of Physiology
University of Toronto
Toronto, ON, M5G 1X8, Canada

Kyriaki Sidiropoulou
Institute of Molecular Biology and Biotechnology (IMBB)
Foundation for Research and Technology (FORTH)
Department of Biology
University of Crete
Heraklion, Crete, Greece

Michael G. Stewart
Department of Life Sciences
The Open University
Milton Keynes, UK
E-mail: m.g.stewart@open.ac.uk

Scellig S.D. Stone
Neurosciences & Mental Health
The Hospital for Sick Children
555 University Ave., 6018 McMaster Building
Toronto, Ontario, M5G 1X8, Canada

Shahid H. Zaman
Cambridge Intellectual & Developmental Disabilities Research Group
Department of Developmental Psychiatry
University of Cambridge
Cambridge CB2 8AH, UK
E-mail: shz10medschl.cam.ac.uk

Introduction

Karl Peter Giese*

Learning and memory are of fundamental importance for the survival of animals. Thus, it should be of no surprise that highly evolved animals, such as mammals, have very sophisticated memory mechanisms. Biomedical research in the last 40 years has provided many insights into memory mechanisms in normal, healthy conditions. More recent research has also begun to address how memory mechanisms are affected in memory disorders, such as Alzheimer's disease. However, much more research is needed to develop a satisfactory understanding of memory mechanisms in health and disease. It is hoped that such mechanistic insights will ultimately lead to the development of "memory pills" to treat memory dysfunction. This book provides an overview of current highlights of mechanistic studies of memory and it addresses the clinical application of contemporary memory research. The book is suitable not only for final-year BSc Neuroscience, MSc Neuroscience and PhD students, but it also provides critical reviews for postdocs and PIs.

Learning and memory are properties of a living brain and they are defined operationally at the behavioural level. Consequently, learning and memory cannot be studied in cell culture or brain slices. Such reduced preparations are suitable for the investigation of cellular mechanisms, which may or may not be relevant for learning and memory. Candidate mechanisms must be studied in behaving animals. Various behavioural tasks have been designed to study learning and memory. For example, the hidden-platform version of the Morris water maze (Morris, 1982, Nature *297*, 681–683) and contextual fear conditioning (see Chapter 14) are widely used for studies of mechanisms in the hippocampus. Most learning and memory tasks are well explained in basic textbooks, so they are not described in detail in this book.

For the mechanistic analysis of learning and memory, it is important to distinguish between processes. The following processes can be distinguished:

* King's College London, Institute of Psychiatry, Department of Neuroscience, 125 Coldharbour Lane, London, SE5 9NU, UK.

memory formation, memory storage, and memory retrieval. Memory formation consists of acquisition and consolidation. Furthermore, it is useful to distinguish between memory formation after one training trial and memory formation after repeated training trials, as the underlying mechanisms seem to differ (e.g. Irvine *et al.*, 2006, Trends Neurosci *29*, 459–465). Additionally, there are retrieval-induced processes, such as memory reconsolidation and memory extinction. The current mechanistic understanding is most advanced for memory formation,whilst not much is known about mechanisms underlying memory retrieval. Accordingly, in this book, we cover mainly mechanisms underlying memory formation (Chapters 1–8, 13, 14), but mechanisms of memory storage (Chapter 9) and retrieval-induced processes (Chapter 10) are also considered.

In mammals, there are independent memory systems that require distinct brain areas. For example, memories for places require the hippocampus whilst motor skill memories are not affected by hippocampal loss. Thus, to develop an understanding of memory mechanisms, it is important to know what brain regions are relevant. The identification of relevant brain regions involves studies of engagement (e.g. using imaging approaches) and inactivation/lesion experiments to establish that the area of interest is required for memory formation and/or storage and/or retrieval. This anatomical knowledge helps to define the neuronal circuitry which forms and stores particular memories. However, molecular and cellular studies are needed to address how these memories are formed and stored. Basic textbooks sufficiently describe distinct memory systems and their underlying anatomy so that this information is not presented much in this book here. However, Chapters 11 and 12 address contemporary ideas about memory circuits.

The most widely investigated cellular mechanisms underlying memory formation is long-term potentiation (LTP), a long-lasting enhancement of transmission at a given synapse. Hence, the first chapter in this book is dedicated to the role of LTP in memory formation. This chapter aims to provide critical insights into studies of the role of LTP. Further chapters in this book are dedicated to other cellular mechanisms of memory that are less widely discussed, but that are still very important. These mechanisms include synaptogenesis (Chapter 2), rewiring of the connections between neurons (Chapter 8), changes in neuronal excitability (Chapter 3), and adult neurogenesis (Chapter 4). We discuss extensively molecular mechanisms underlying memory consolidation, some of which are sex-dependent (Chapters 5–7). The molecular basis of memory storage is considered in Chapter 9.

Learning and memory are not only impaired in Alzheimer's disease, but are also affected in autism, bipolar disorder, depression, mental retardation, post-traumatic stress disorder and schizophrenia. The nature of these impairments is unclear. However, there are some insights into memory dysfunction in Alzheimer's disease. This research is discussed in Chapter 13; Chapter 14 illustrates how these insights might be useful to develop treatments for patients with Alzheimer's disease. Finally, Chapter 15 explains in more detail what impact knowledge of memory mechanisms is expected to have for psychiatry.

I want to thank the following scientists who gave invaluable feedback that helped to shape this book: Drs. Ted Abel (Penn University, USA), Sam Cooke (MIT, USA), John Disterhoft (Northwestern University, USA), Diane Hanger (King's College London, UK), Jonathan Lee (University of Birmingham, UK), Anna Need (Duke University, USA), Matthew Nolan (Edinburgh University, UK), Marco Peters (Darts Neuroscience, USA), Claire Rampon (University of Toulouse, France), Emilie Rissman (University of Virginia, USA), Todd Sacktor (SUNY Downstate Medical Center, USA), Miao-Kun Sun (Blanchette Rockefeller Neuroscience Institute, USA), Jeffrey Vernon (University College London, UK), and Marcelo Wood (University of California Irvine, USA).

Long-term Potentiation and Memory

1

Karl Peter Giese*

1. Introduction

How are memories formed and stored? This has been a fundamental question in neuroscience and decades of research have established that there is no simple answer. Distinct types of memories engage and require different brain regions. Circuits, cell types, cellular plasticity and molecular signalling vary across brain region. Therefore, it is unlikely that one particular molecular or cellular mechanism accounts for memory in general. Nonetheless, long-lasting enhancement of chemical synaptic transmission is believed by many researchers to be *the* cellular basis of memory. The experimental model of this long-lasting synaptic plasticity is long-term potentiation (LTP). While it is unlikely that LTP-like plasticity is the only cellular basis of memory, it has emerged that it is likely to be *a* memory mechanism. This conclusion results from a large number of studies that investigated the properties of LTP, its underlying molecular mechanisms and the correlation of LTP alterations with learning and memory. A recent Medline search showed that over 6,300 papers have been published on LTP. This chapter aims to summarize highlights of these accounts.

2. Key properties of LTP

LTP is an activity-induced, long-lasting increase in chemical synaptic transmission (Bliss and Collingridge, 1993). In brain slices, LTP can last for several hours and in the intact brain, it can last several months (Abraham *et al.*, 2002). Before the discovery of LTP, electrophysiologists had not identified neuronal plasticity that lasted for more than a few minutes (Lømo, 2003). It was acknowledged that such short-lasting plasticity could not support the storage of long-term memory,

---

* King's College London, Institute of Psychiatry, Department of Neuroscience, 125 Coldharbour Lane, London, SE5 9NU, UK.

which may last a lifetime. Due to its longevity, LTP is thought to model a mechanism of long-term memory. Additionally, LTP can be induced in many brain regions that are relevant for particular forms of long-term memory. These brain regions include the amygdala, hippocampus, neocortex, and even the cerebellum, a brain structure important for motor memory in which the converse process of long-term depression (LTD) has long been championed as a storage device.

Most LTP experiments have used transverse hippocampal slices to study LTP at CA1 synapses. This preparation has two major attributes: First, the ease with which pharmacological treatments can be both washed in and washed away, allowing for interrogation of the contribution of molecular targets to the induction, expression and maintenance phases of LTP and, second, the simple, monolayered anatomy of the hippocampus, allowing for analysis of synaptic events using extracellular field recording techniques (Skrede and Westgaard, 1971). To induce LTP, an electrical tetanus (a high-frequency barrage of electrical pulses) is normally applied to an afferent fibre pathway (in this case, the Schaffer collaterals) via a stimulating electrode while recording synaptic events in a postsynaptic population of neurons (in this case, CA1 pyramidal cells). Depending on the frequency and pattern of tetanisation distinct types of CA1 LTP can be induced (Table 1.1). A standard tetanus (e.g. 100 Hz, 1 s) induces "early" LTP (E-LTP) that declines within 3 hours. The induction of this LTP type requires activation of NMDA receptor channels that allow for calcium entry at the postsynapse (Bliss and Collingridge, 1993). A stronger tetanus (200 Hz) not only induces NMDA receptor-dependent LTP, it also evokes voltage-gated calcium channel (VGCC)-dependent LTP (Grover and Teyler, 1990). Furthermore, multiple tetani, such as four 100-Hz tetani, induce "late" LTP (L-LTP) that lasts for at least 8 hours in hippocampal slices, which is about the life time of this tissue preparation (Frey *et al.*, 1988). L-LTP induction is NMDA receptor-dependent and the late phase of L-LTP requires protein synthesis. However, depending on the stimulation protocol, distinct types of L-LTP are induced; only some of which require gene transcription (Huang and Kandel, 2005). Taken together, it is not possible to generalize LTP. However, in the context of memory, NMDA receptor-dependent long-lasting potentiation is primarily considered (Martin *et al.*, 2000).

In addition to its longevity, LTP has further properties that make it an intuitive memory mechanism: (1) It is input specific, meaning that only synapses that experience tetanisation induce LTP; (2) It is co-operative, meaning that LTP induction requires tetanisation above a particular threshold. Therefore, only particular inputs can generate LTP; (3) It is reversible; (4) It is associative, meaning that a subthreshold tetanus can only induce LTP if it coincides with a stronger, LTP-inducing tetanus at neighbouring synapses. This associativity is reminiscent

Table 1.1. Distinct types of hippocampal CA1 LTP and how they correlate with learning and memory.

Stimulation for induction	Characteristics	Correlation with learning and memory
100-Hz tetanus	Requires NMDA receptor activation, alphaCaMKII autophosphorylation (Cooke *et al.*, 2006) and GluR1 (Zamanillo *et al.*, 1999). This LTP type is most commonly studied.	Specific block correlates with impaired spatial working memory (Reisel *et al.*, 2002).
Multiple 100-Hz tetani	Late LTP phase requires gene transcription and protein synthesis (Frey and Morris, 1997).	Sex-specific deficit of late phase correlates with sex-dependent impairment in spatial reference memory (Mizuno *et al.*, 2007). This LTP is not essential for contextual fear memory after a massed training session (Irvine *et al.*, 2011).
Theta bursts	Requires NMDA receptor activation, alphaCaMKII autophosphorylation (Giese *et al.*, 1998), but is GluR1-independent (Hoffman *et al.*, 2002). The late phase requires local translation but not gene transcription and share mechanisms of BDNF-induced LTP (Huang and Kandel, 2005).	This type of LTP appears sufficient for spatial reference memory (Hoffman *et al.*, 2002; Reisel *et al.*, 2002).
200-Hz tetanus	Voltage-gated calcium channel-dependent (Grover and Tyler, 1990).	This LTP type does not seem to be induced in the brain after training in a memory task, as the training-induced increase in synaptic transmission if fully dependent on NMDA receptor activation (Gruart *et al.*, 2006; Whitlock *et al.*, 2006).

of Pavlovian conditioning, in which a behaviourally neutral conditioned stimulus (e.g. tone) is paired with an unconditioned stimulus (e.g. shock). Moreover, the conversion of E-LTP into L-LTP can be associative, if the induction of E-LTP (resulting from one tetanus) occurs around the same time as the induction of L-LTP (resulting from multiple tetani) at another synapse of the same dendritic branch (Frey and Morris, 1997; Govindarajan *et al.*, 2011).

3. Training-induced occurrence of LTP

Until 1997, it was not known whether LTP-like plasticity is induced under physiological conditions, in particular whether it occurs in the brain after training in a behavioural memory task. Up to this time, it was conceivable that the LTP-inducing tetanisations provided by the stimulation electrodes do not mimic firing patterns that occur in the brain of the behaving animal. However, in 1997, it was shown that tone fear conditioning induces an LTP-like synaptic potentiation in the amygdala (Rogan *et al.*, 1997). Later in the hippocampus, an NMDA receptor-dependent LTP-like synaptic potentiation was detected after training in memory tasks (Gruart *et al.*, 2006; Whitlock *et al.*, 2006). The first of these studies used hippocampus-dependent trace eyeblink conditioning and it detected an NMDA receptor-dependent increase of excitatory transmission at CA1 synapses (Gruart *et al.*, 2006). Learning in this task could be prevented by LTP saturation using tetanic stimulation. The other study detected an NMDA receptor-dependent increase of excitatory transmission at CA1 synapses after training in the passive avoidance task (Whitlock *et al.*, 2006). This increase in synaptic transmission occluded further induction of LTP. Later the same group also found an induction of LTP-like synaptic potentiation in neocortex after visual experience (Cooke *et al.*, 2010). An experience-dependent LTP-like synaptic potentiation has also been detected in the motor cortex (Rioult-Pedotti *et al.*, 2000). Taken together, these studies suggest that an LTP-like long-lasting synaptic potentiation can be induced by behavioural experience.

4. LTP and Hebb's postulate

In his book the *Organization of Behaviour*, 1949, Donald Hebb wrote a theoretical contribution as to how a memory mechanism might operate. Hebb's postulate that "when neuron A repeatedly or persistently fires neuron B, then some metabolic or growth changes are induced so neuron A fires neuron B more efficiently" proposes that the communication between neurons is enhanced to form a memory. Many publications mention that LTP follows Hebb's postulate. However, strictly speaking, this is not entirely correct. Hebb's postulate refers to the firing of neurons and never discusses synapses. The term "LTP" very specifically describes the process of enhancing excitatory synaptic transmission and an increase of transmission at a single synapse is unlikely to result in a change of neuronal firing. Multiple synaptic inputs are usually needed to evoke action potentials. Nevertheless, Hebb's postulate captures the essence of why LTP is an attractive mnemonic device. The induction of L-LTP can be clustered on the same dendritic branch and such clustering likely contributes to increased neuronal firing (Govindarajan *et al.*, 2006, 2011). Additionally, the induction of LTP is associated with an E–S potentiation (E = excitatory postsynaptic potential; S = spike) (Bliss and Collingridge, 1993). E–S potentiation affects neuronal firing without changing synaptic strength. Instead, changes in intrinsic cell excitability lead to more effective induction of action potential firing for a given excitatory input. The mechanisms underlying E–S potentiation are not known and whether or not E–S potentiation relates to memory is understudied.

5. Molecular mechanisms underlying LTP

As already mentioned, depending on the tetanisation, different forms of LTP can be induced at a given synapse. Furthermore, very different mechanisms can underlie LTP induction, expression and maintenance at distinct synapses. For example, mossy fibre LTP in hippocampal area CA3 is induced at the presynapse, whereas CA1 LTP is induced at the postsynapse (Bliss and Collingridge, 1993). It is beyond the scope of this chapter to discuss the molecular mechanisms underlying all known types of LTP. Instead the chapter will focus on key molecular mechanisms underlying LTP at CA1 synapses (Figure 1.1). It is commonly accepted that this type of LTP requires posttranslational modification and trafficking of synaptic proteins. In the later LTP stages, structural changes at the existing synapses occur, but growth of new synapses is not involved (see Chapter 2).

The induction of CA1 LTP requires activation of the NMDA receptor at the postsynapse (with the exception of VGCC-dependent LTP that is induced with a very strong tetanus). NMDA receptor activation leads to postsynaptic entry of calcium that acts as second messenger to induce molecular signalling (Bliss and Collingridge, 1993). A key molecule that becomes activated by this calcium entry is the calcium/calmodulin-dependent protein kinase II (CaMKII) (Lisman *et al.*, 2002). CaMKII is an abundant, multifunctional serine/threonine kinase that

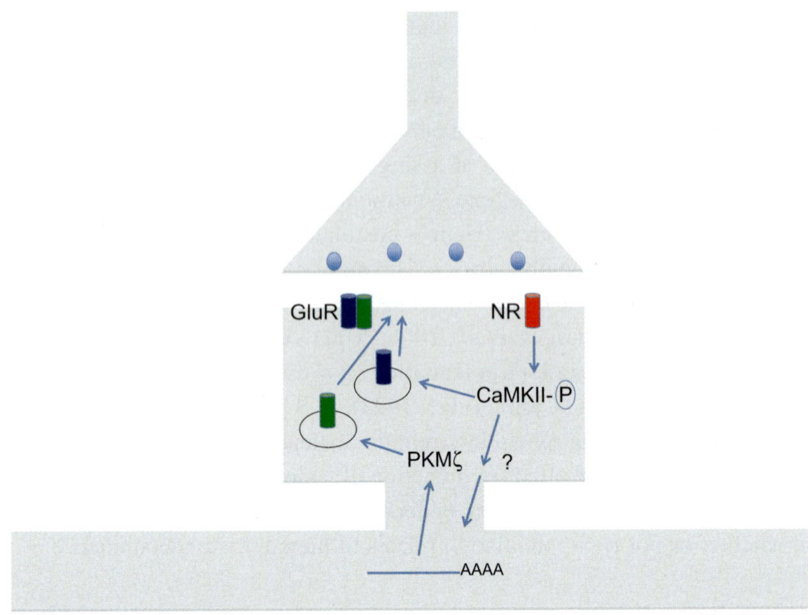

Figure 1.1. Simplified schematic of key processes underlying 100-Hz–induced hippocampal CA1 LTP. The induction of this LTP requires calcium influx through the NMDA receptor (NR). This leads to autophosphorylation of alphaCaMKII that is essential for the LTP induction. The autophosphorylated CaMKII is likely to contribute to trafficking of GluR1 subunits (blue), which leads to increased GluR expression in the synaptic membrane. Autophosphorylated CaMKII may also regulate the locate translation of PKMzeta that is required for the maintenance of LTP enabling GluR2 trafficking and counteracting a homeostatic endocytosis of GluR2 subunits.

switches its activity state from calcium-dependence to -independence upon autophosphorylation. CaMKII autophosphorylation is essential for the induction of NMDA receptor-dependent LTP, as demonstrated with various induction protocols (Giese *et al.*, 1998; Yasuda *et al.*, 2003; Cooke *et al.*, 2006; Irvine *et al.*, 2011). The function of CaMKII autophosphorylation is likely to prolong molecular signalling to trigger LTP induction (Irvine *et al.*, 2006).

A further critical step for the establishment of LTP is AMPA receptor trafficking that increases AMPA receptor density and consequently enhances glutamatergic transmission (Malinow, 2003). AMPA receptor subunits are localized in the plasma membrane and in internal vesicles. LTP induction leads to the fusion of internal vesicles, resulting in the delivery of additional AMPA receptor subunits to the postsynaptic membrane. Increasing the density of the AMPA receptors at the postsynapse leads to enhanced glutamatergic transmission. In adult CA1 neurons, the AMPA receptor subunit GluR1 traffics after LTP induction and this requires CaMKII activity (Malinow, 2003). However, in immature hippocampal neurons, trafficking of the GluR4 subunit is of relevance, which is not CaMKII dependent (Malinow, 2003). Additionally, in immature hippocampus, LTP can also be expressed presynaptically (Ward *et al.*, 2006). In agreement with this, in immature hippocampus CA1 LTP does not require CaMKII autophosphorylation (Yasuda *et al.*, 2003). In adult hippocampus, additional calcium-dependent signalling, such as participation of protein kinases A (PKA) and C (PKC), is involved in trafficking of GluR1 subunits (Kessels and Malinow, 2009). It is interesting to note that GluR1 trafficking is not required for some types of CA1 LTP in adult hippocampus (Frey *et al.*, 2009) and that trafficking of the GluR2 subunit is also important for CA1 LTP (Sacktor, 2011). A key regulator of GluR2 trafficking is an atypical isoform of PKC, PKMzeta. PKMzeta maintains GluR2 trafficking and thereby maintains LTP. Interestingly, PKMzeta protein is locally synthesized from mRNA in dendrites. Thus, PKMzeta is required for the establishment and maintenance of protein synthesis-dependent L-LTP.

The most widely studied form of CA1 L-LTP is dependent on gene transcription in the nucleus. In this case, a signal is transmitted from the synapse to the nucleus to induce transcription. Translocation of importin subunits appears to be critical for this (Thompson *et al.*, 2004). In the nucleus, the transcription factor cAMP-responsive element-binding protein (CREB) is essential for the induction of transcription-dependent LTP (Josselyn and Nguyen, 2005). Various kinases can activate CREB including calcium/calmodulin-dependent kinase IV (CaMKIV) and mitogen-activated protein kinase (MAPK). It is not clear how many genes are transcribed after LTP induction. LTP-induced gene transcriptions appear to occur in cascades as the initial wave of transcription involves immediate-early transcription factors, which can trigger further waves of transcription. Most of the newly synthesized mRNAs are translated into plasticity-related proteins (PrPs) in the soma of the neuron. Thereafter, the PrPs diffuse into the dendritic tree and are targeted to the activated synapse. The activated synapse has a "tag" so that the diffusing PrPs can be captured selectively by this synapse (Frey and Morris, 1997). Tagging involves a change of diffusion into the spine through the spine neck (Okada *et al.*, 2009) and CaMKII signalling is required for tagging (Redondo and Morris, 2011). It remains unclear what the identity of the critical PrPs is and what functions they carry out at potentiated synapses. However, in current models, it seems that PrPs do not induce local synaptogenesis, but, instead, seem to strengthen the existing, potentiated synapse. Accordingly, models of memory storage currently posit that only existing synapses need to be modified in a "permanent" manner (see below).

It remains unclear how many signalling processes are involved in the induction, expression and maintenance of LTP. More than 100 different knockout mice have LTP deficits. In the vast majority of the cases, LTP is reduced but not completely blocked. Conceptually, these knockout studies have not tested whether a particular signalling molecule is a core component or just a modulator of LTP (Lisman *et al.*, 2003).

6. Is LTP required for memory?

To establish that the experimental model LTP reflects a true memory mechanism, the impact of selective manipulations of LTP must be assessed experimentally at the behavioural level (learning and memory are properties of a behaving animal). The first study of this kind was performed in the 1980s by Richard Morris (Morris *et al.*, 1986) and it has been followed by hundreds of experiments. Taken together, these studies strongly suggest that LTP-like plasticity is important for memory, although there is also evidence that alternative plasticity processes can be critical (e.g. Giese *et al.*, 2001; Irvine *et al.*, 2011).

Attempts have been made to block LTP at the electrophysiological level, using strong tetanisation to saturate LTP at most synapses. However, these saturation experiments have not been conclusive, possibly because not all synapses could be saturated by electrical stimulation. In combination with lesioning of one of the two hippocampi, LTP saturation impairs spatial memory formation (Moser *et al.*, 1998), suggesting that LTP-like plasticity is required for spatial memory formation.

More conclusive findings have resulted from studies blocking LTP at the molecular level. For example, LTP can be inhibited pharmacologically, although not many specific blockers are available. Accordingly, Richard Morris used the NMDA receptor blocker AP-5 to inhibit LTP induction (Morris *et al.*, 1986). He showed that infusion of AP-5 into rat hippocampi impairs spatial memory formation. Therefore, this study suggested that NMDA receptor-dependent LTP-like plasticity is required for spatial memory formation. However, the use of a very high dose of AP-5 affected the performance of the animals in the water maze that was used to assess spatial learning. Therefore, Morris *et al.* repeated the experiment with different doses of AP-5 (Davis *et al.*, 1992). They found that a lower dose of AP-5 fully blocked the induction of CA1 LTP, but only partially impaired spatial memory formation. A further limitation of the AP-5 studies is that blocking the NMDA receptor not only blocks LTP, it also impairs other plasticities, such as LTD and synaptogenesis. The latter is not believed to contribute to LTP. Instead, LTP refers explicitly to the strengthening of existing synapses.

In the early 1990s, Alcino Silva established that genetic modification of mice is a useful tool to investigate the function of LTP-like plasticity in learning and memory. In Susumu Tonegawa's lab, he knocked out the gene encoding α-isoform of CaMKII and showed that this reduces NMDA receptor-dependent CA1 LTP and impairs spatial memory formation (Silva *et al.*, 1992a, b). These experiments represent the beginning of a molecular biological era to study the role of LTP-like plasticity in learning and memory. While Silva's study was of fundamental importance, it had its limitations. For example, it was noted that the alphaCaMKII knockout mice have seizures (Butler *et al.*, 1995), which in principle may impair spatial memory formation. Furthermore, it emerged that CA1 LTP is reduced, but still present, in the alphaCaMKII knockout mice and that the remaining plasticity was supported by a compensatory translocation of beta-CaMKII (Elgersma *et al.*, 2002). Later, both of these confounds were overcome in alphaCaMKII-T286A knock-in mice, which cannot autophosphorylate at threonine 286 (Giese *et al.*, 1998), thus preventing the activity of the kinase in absence of calcium. These latter mutants have a complete block of NMDA receptor-dependent CA1 LTP (Giese *et al.*, 1998; Yasuda *et al.*, 2003; Cooke *et al.*, 2006; Irvine *et al.*, 2011), they have no seizures, and they are impaired in spatial memory formation (Need and Giese, 2003). Shortly after Silva's study, Seth Grant and Eric Kandel published that genetic knockout of another kinase, fyn, impaired CA1 LTP and caused deficits in spatial memory formation (Grant *et al.*, 1992). However, it emerged that these mutants can form spatial memory when helped to overcome a swimming deficit in the water maze (Huerta *et al.*, 1996). The fyn knock-out mice have obvious developmental abnormalities in the hippocampus; the number of neurons is drastically increased due to a lack of apoptosis during development. This observation inspired some researchers to claim that knockout studies are generally confounded by developmental abnormalities offering a "trivial" explanation for impaired memory formation. However, it is now known that at least some knock-outs have impaired memory formation due to deficits that are unrelated to developmental abnormalities. For example, $K_v$beta1.1 knock-out mice have impaired memory formation at a young age, but memory formation is improved in old age, indicating that deficit in plasticity, and not a developmental abnormality, accounted for the impairment at a young age (Need *et al.*, 2003). Furthermore, deficits in memory formation can be pharmacologically rescued in various mouse disease models, suggesting that impaired plasticity, and not abnormal development, accounts for the behavioural deficits in these models (Ehninger *et al.*, 2008).

Knock-out mutations have also been restricted to particular brain regions or cell types in order to improve the analysis of LTP manipulation and its impact on memory formation. Region-restricted knock-outs can be generated with the Cre/loxP recombination system. In transgenic mice, Cre recombinase is expressed under control of a region-specific promoter and when expressed, Cre deletes an essential region from a target gene that is surrounded by loxP recombination sites. There are only a few currently known brain region-specific promoters available, so not many region-restricted knockouts have been produced. Nonetheless, the essential NMDA receptor subunit NR1 was knocked out specifically in hippocampal area CA1 (Tsien *et al.*, 1996). These mutants not only fail to exhibit NMDA receptor-dependent CA1 LTP, but they are also impaired in memory formation in various hippocampus-dependent tasks (Tsien *et al.*, 1996; Rampon *et al.*, 2000). These results are in agreement with the findings obtained by Richard Morris using the pharmacological approach to block the NMDA receptors (Morris *et al.*, 1986; Davis *et al.*, 1992) and they suggest an important role for mechanisms of NMDA receptor-dependent CA1 LTP in memory formation. Interestingly, environmental enrichment of the CA1-NR1 knock-out mice rescues the impairments in non-spatial memory tasks (Rampon *et al.*, 2000). The rescue has been linked with synaptogenesis in CA1. Thus, in addition to mechanisms of CA1 LTP, synaptogenesis may enable memory formation in non-spatial tasks.

7. A specific role for CA1 LTP

In most mutant mouse studies, a reduction or increase of NMDA receptor-dependent CA1 LTP correlates with altered hippocampus-dependent learning (Silva, 2003). This strongly suggests that NMDA receptor-dependent LTP-like plasticity in CA1 has an important role in learning. However, some studies found conditions that allowed for hippocampus-dependent memory formation despite impaired CA1 LTP (e.g. Rampon *et al.*, 2000; Reisel *et al.*, 2002; Irvine *et al.*, 2011). Thus, it is emerging that NMDA receptor-dependent LTP-like plasticity in CA1 may have a specific, rather than general, role in memory formation. Analyses of the following two mutant mouse lines make this case (see also, Table 1.1).

GluR1 knock-out mice lack NMDA receptor-dependent CA1 LTP after a standard tetanisation (100 Hz, 1 s) and these mice have normal memory formation in the spatial reference memory version of the water maze, where the hidden platform is in a fixed location (Zamanillo *et al.*, 1999). Importantly, lesion experiments showed that this type of spatial memory formation is hippocampus dependent in the mutants (Reisel *et al.*, 2002). Further behavioural analyses revealed that the GluR1 knock-out mice are impaired in hippocampus-dependent spatial working memory tasks, such as memory formation in the T maze (Reisel *et al.*, 2002). These results initially suggested that NMDA receptor-dependent LTP-like plasticity in CA1 has a specific role in hippocampal memory formation, being essential for spatial working, but not reference, memory. However, follow-up LTP investigations revealed that theta burst stimulation leads to a slow, gradual development of NMDA receptor-dependent CA1 LTP that requires CaMKII signalling (Hoffman *et al.*, 2002; Frey *et al.*, 2009; Romberg *et al.*, 2009). These findings are in agreement with the notion that distinct mechanisms of CA1 LTP may have different roles in memory formation. Mechanisms of GluR1-independent LTP appear to be sufficient for spatial reference memory formation, whilst mechanisms of GluR1-dependent LTP appear to be essential for spatial working memory. Interestingly, GluR1-independent LTP develops slowly and it might be not fast enough to reflect mechanisms of working memory. Both of these CA1 LTP types (GluR1-dependent and -independent) require the activation of the NMDA receptor and it remains to be seen whether both types of plasticity are induced by behavioural experience.

In contrast with GluR1 knockout mice, autophosphorylation-deficient alphaCaMKII-T286A knockin mice (T286A mutants) do not have NMDA receptor-dependent CA1 LTP (Giese *et al.*, 1998; Yasuda *et al.*, 2003; Cooke *et al.*, 2006; Irvine *et al.*, 2011). Standard tetanisation, theta burst stimulation and multi-tetanisation cannot induce this type of LTP in the T286A mutants. In contrast, NMDA receptor-independent CA1 LTP can be induced in the T286A mutants (Giese *et al.*, 1998), but such plasticity does not seem to be induced by training of rodents in a memory task (Gruart *et al.*, 2006; Whitlock *et al.*, 2006). The complete block of NMDA receptor-dependent CA1 LTP in the T286A mutants in combination with contextual fear conditioning studies have suggested that NMDA receptor-dependent LTP-like plasticity has a specific role in memory formation. Contextual fear conditioning is a hippocampus-dependent task, in which a mouse is conditioned to a neutral environment (context). After a single training trial, the T286A mutants cannot form contextual fear memory (Irvine *et al.*, 2005, 2011). However, after five training trials, the T286A mutants form a contextual fear memory. This contextual fear memory in the T286A mutants is hippocampus dependent (Irvine *et al.*, 2011). Thus, mechanisms of NMDA receptor-dependent CA1 LTP appear important for one-trial memory formation but they are not essential for memory formation after repeated training (massed training session).

8. LTP and long-term memory storage

After memory formation is completed, memory needs to be stored. Do the mechanisms underlying memory formation also contribute to long-term memory storage? Some molecular mechanisms underlying memory formation are transient and therefore, they cannot contribute to memory storage that persists for longer time. However, at the cellular level, LTP-like plasticity may contribute to both formation and storage of memory. If LTP contributes to memory storage, then blocking of LTP maintenance should impair memory.

LTP occurs at existing synapses. Thus, LTP maintenance must involve molecular signalling processes that are immune to protein turnover. Every protein has a lifetime in the range of hours to days that is much shorter than long-term memory or LTP, lasting several months or longer. Theoretically, molecular feed-forward systems were proposed to underlie memory storage (Crick, 1984; Lisman, 1985). CaMKII was the first protein to be championed as a storage device (Lisman *et al.*, 2002). This is because the autophosphorylation of CaMKII switches the kinase into an ON state. Furthermore, CaMKII is a holoenzyme that consists of 12 subunits and the autophosphorylation is an intersubunit reaction (Hanson *et al.*, 1992). Thus, the turnover of one subunit may lead to the incorporation of a newly synthesized, non-autophosphorylated subunit that can be switched ON by a neighbouring subunit in the ON state, thereby enabling immunity to molecular turnover. While the model is very elegant, it cannot be supported by experimental evidence. Specific blockade of CaMKII after LTP induction does not impair the maintenance of LTP (Buard *et al.*, 2010) and loss of CaMKII autophosphorylation does not correlate with failure of long-term memory storage (Irvine *et al.*, 2005, 2006; Buard *et al.*, 2010).

Instead, the constitutively active kinase PKMzeta is now recognized as being critical for LTP maintenance (Sacktor, 2011). mRNA encoding the PKMzeta isoform is localized in dendrites where it is translated on demand. Under basal conditions, the mRNA translation is inhibited by the protein Pin1. However, this translational block is relieved when signalling is induced at a nearby synapse. Once translated, PKMzeta phosphorylates Pin1 and blocks the translational repression of the PKMzeta mRNA. Thus, a positive feedback loop of local translation assures that PKMzeta activity outlasts protein turnover. At the activated synapse, the continued PKMzeta activity leads to enhanced trafficking of the AMPA receptor subunit GluR2. GluR2 trafficking needs to be continuously increased to overcome a homeostatic mechanism that opposes an increase of GluR2 subunits in the postsynaptic membrane. A synthetic peptide termed ZIP (PKMzeta inhibitory peptide) can be used to specifically block PKMzeta and this peptide inhibits the maintenance of CA1 LTP when applied after induction (Pastalkova *et al.*, 2006). Additionally, ZIP blocks spatial memory when delivered locally in the hippocampus several days or weeks after training. In fact, many different memories can be blocked by ZIP application after training

(Serrano *et al.*, 2008; Shema *et al.*, 2009). Erasure of memory storage correlates with impaired GluR2 trafficking, providing a strong link between a mechanism of LTP maintenance and memory storage (Migues *et al.*, 2010). However, hippocampal infusions of ZIP do not block contextual fear memory (Serrano *et al.*, 2008; Kwapis *et al.*, 2009). Since contextual fear memory storage requires the hippocampus, this suggests that LTP maintenance mechanisms are not important for this type of memory storage. In summary, LTP maintenance appears to be important for storage of most, but not all memories. A more comprehensive discussion of mechanisms underlying memory storage can be found in Chapter 9.

9. LTP and sex differences

The vast majority of LTP studies have been done with male rodents. A possible reason for this sex bias may be the idea that studies using females will have larger variances than studies with males, due to changes in hippocampal synapse number during the oestrous cycle. However, such a difference in variance has not been observed when comparing data from males and females randomized for the oestrous cycle. It has become clear from these studies that the signalling mechanisms underlying CA1 LTP differ between males and females (Mizuno and Giese, 2010). Interestingly, sex-specific impairments in CA1 LTP correlate with sex-specific deficits in hippocampus-dependent memory formation. A more comprehensive discussion of sex differences in memory formation can be found in Chapter 7.

10. Conclusion

The discovery of LTP opened a new, exciting subfield within neuroscience that has produced a wealth of experimental findings. LTP models an endogenous strengthening of existing synapses that is very likely to be a memory mechanism. Furthermore, it seems likely that particular types of LTP-like plasticity have specific roles in learning and memory, such as enabling one-trial memory formation. Interestingly, LTP is impaired in mouse models of diseases including Alzheimer's disease, which results in major memory deficits in humans (Chapter 13). Thus, developing a full mechanistic understanding of LTP promises to greatly aid the development of treatments for memory deficits (see also, Chapters 14 and 15). However, it has also emerged that LTP is not the only correlate of memory. Other types of plasticity, such as *de novo* synaptogenesis (Chapter 2), modulation of neuronal excitability (Chapter 3), neurogenesis in adult brain (Chapter 4), rewiring of neuronal connections (Chapter 8), and LTD may also contribute to learning and memory.

References

Abraham, W.C., Logan, B., Greenwood, J.M., and Dragunow, M. (2002). Induction and experience-dependent consolidation of stable long-term potentiation lasting months in the hippocampus. *J Neurosci* **22**, 9626–9634.

Bliss, T.V. and Collingridge, G.L. (1993). A synaptic model of memory: Long-term potentiation in the hippocampus. *Nature* **361**, 31–39.

Buard, I., Coultrap, S.J., Freund, R.K., Lee, Y.S., Dell'Acqua, M.L., Silva, A.J., and Bayer, K.U. (2010). CaMKII "autonomy" is required for initiating but not for maintaining neuronal long-term information storage. *J Neurosci* **30**, 8214–8220.

Butler, L.S., Silva, A.J., Abeliovich, A., Watanabe, Y., Tonegawa, S., and McNamara, J.O. (1995). Limbic epilepsy in transgenic mice carrying a Ca^{2+}/calmodulin-dependent kinase II alpha-subunit mutation. *Proc Natl Acad Sci U S A* **92**, 6852–6855.

Cooke, S.F. and Bear, M.F. (2010). Visual experience induces long-term potentiation in the primary visual cortex. *J Neurosci* **30**, 16304–16313.

Cooke, S.F., Wu, J., Plattner, F., Errington, M., Rowan, M., Peters, M., Hirano, A., Bradshaw, K.D., Anwyl, R., Bliss, T.V., and Giese, K.P. (2006). Autophosphorylation of alphaCaMKII is not a general requirement for NMDA receptor-dependent LTP in the adult mouse. *J Physiol* **574**, 805–818.

Crick, F. (1984). Memory and molecular turnover. *Nature* **312**, 101.

Davis, S., Butcher, S.P., and Morris, R.G. (1992). The NMDA receptor antagonist D-2-amino-5-phosphonopentanoate (D-AP5) impairs spatial learning and LTP *in vivo* at intracerebral concentrations comparable to those that block LTP *in vitro*. *J Neurosci* **12**, 21–34.

Ehninger, D., Li, W., Fox, K., Stryker, M.P., and Silva, A.J. (2008). Reversing neurodevelopmental disorders in adults. *Neuron* **60**, 950–960.

Elgersma, Y., Fedorov, N.B., Ikonen, S., Choi, E.S., Elgersma, M., Carvalho, O.M., Giese, K.P., and Silva, A.J. (2002). Inhibitory autophosphorylation of CaMKII controls PSD association, plasticity, and learning. *Neuron* **36**, 493–505.

Frey, M.C., Sprengel, R., and Nevian, T. (2009). Activity pattern-dependent long-term potentiation in neocortex and hippocampus of GluA1 (GluR-A) subunit-deficient mice. *J Neurosci* **29**, 5587–5596.

Frey, U., Krug, M., Reymann, K.G., and Matthies, H. (1988). Anisomycin, an inhibitor of protein synthesis, blocks late phases of LTP phenomena in the hippocampal CA1 region *in vitro*. *Brain Res* **452**, 57–65.

Frey, U. and Morris, R.G. (1997). Synaptic tagging and long-term potentiation. *Nature* **385**, 533–536.

Giese, K.P., Fedorov, N.B., Filipkowski, R.K., and Silva, A.J. (1998). Autophosphorylation at Thr286 of the alpha calcium-calmodulin kinase II in LTP and learning. *Science* **279**, 870–873.

Giese, K.P., Peters, M., and Vernon, J. (2001). Modulation of excitability as a learning and memory mechanism: A molecular genetic perspective. *Physiol Behav* **73**, 803–810.

Govindarajan, A., Kelleher, R.J., and Tonegawa, S. (2006). A clustered plasticity model of long-term memory engrams. *Nat Rev Neurosci* **7**, 575–583.

Govindarajan, A., Israely, I., Huang, S.Y., and Tonegawa, S. (2011). The dendritic branch is the preferred integrative unit for protein synthesis-dependent LTP. *Neuron* **69**, 132–146.

Grant, S.G., O'Dell, T.J., Karl, K.A., Stein, P.L., Soriano, P., and Kandel, E.R. (1992). Impaired long-term potentiation, spatial learning, and hippocampal development in fyn mutant mice. *Science* **258**, 1903–1910.

Grover, L.M. and Teyler, T.J. (1990). Two components of long-term potentiation induced by different patterns of afferent activation. *Nature* **347**, 477–479.

Gruart, A., Munoz, M.D., and Delgado-Garcia, J.M. (2006). Involvement of the CA3-CA1 synapse in the acquisition of associative learning in behaving mice. *J Neurosci* **26**, 1077–1087.

Hanson, P.I. and Schulman, H. (1992). Neuronal Ca^{2+}/calmodulin-dependent protein kinases. *Annu Rev Biochem* **61**, 559–601.

Hoffman, D.A., Sprengel, R., and Sakmann, B. (2002). Molecular dissection of hippocampal theta-burst pairing potentiation. *Proc Natl Acad Sci USA* **99**, 7740–7745.

Huang, Y.Y. and Kandel, E.R. (2005). Theta frequency stimulation induces a local form of late phase LTP in the CA1 region of the hippocampus. *Learn Mem* **12**, 587–593.

Huerta, P.T., Scearce, K.A., Farris, S.M., Empson, R.M., and Prusky, G.T. (1996). Preservation of spatial learning in fyn tyrosine kinase knockout mice. *Neuroreport* **7**, 1685–1689.

Irvine, E.E., Vernon, J., and Giese, K.P. (2005). AlphaCaMKII autophosphorylation contributes to rapid learning but is not necessary for memory. *Nat Neurosci* **8**, 411–412.

Irvine, E.E., von Hertzen, L.S., Plattner, F., and Giese, K.P. (2006). AlphaCaMKII autophosphorylation: A fast track to memory. *Trends Neurosci* **29**, 459–465.

Irvine, E.E., Danhiez, A., Radwanska, K., Nassim, C., Lucchesi, W., Godaux, E., Ris, L., and Giese, K.P. (2011). Properties of contextual memory formed in the absence of alphaCaMKII autophosphorylation. *Mol Brain* **4**, 8.

Josselyn, S.A. and Nguyen, P.V. (2005). CREB, synapses and memory disorders: Past progress and future challenges. *Curr Drug Targets CNS Neurol Disord* **4**, 481–497.

Kessels, H.W. and Malinow, R. (2009). Synaptic AMPA receptor plasticity and behavior. *Neuron* **61**, 340–350.

Kwapis, J.L., Jarome, T.J., Lonergan, M.E., and Helmstetter, F.J. (2009). Protein kinase Mzeta maintains fear memory in the amygdala but not in the hippocampus. *Behav Neurosci* **123**, 844–850.

Lisman, J.E. (1985). A mechanism for memory storage insensitive to molecular turnover: A bistable autophosphorylating kinase. *Proc Natl Acad Sci USA* **82**, 3055–3057.

Lisman, J., Schulman, H., and Cline, H. (2002). The molecular basis of CaMKII function in synaptic and behavioural memory. *Nat Rev Neurosci* **3**, 175–190.

Lisman, J., Lichtman, J.W., and Sanes, J.R. (2003). LTP: Perils and progress. *Nat Rev Neurosci* **4**, 926–929.

Lømo, T. (2003). The discovery of long-term potentiation. *Philos Trans R Soc Lond B Biol Sci* **358**, 617–620.

Malinow, R. (2003). AMPA receptor trafficking and long-term potentiation. *Philos Trans R Soc Lond B Biol Sci* **358**, 707–714.

Martin, S.J., Grimwood, P.D., and Morris, R.G. (2000). Synaptic plasticity and memory: An evaluation of the hypothesis. *Annu Rev Neurosci* **23**, 649–711.

Migues, P.V., Hardt, O., Wu, D.C., Gamache, K., Sacktor, T.C., Wang, Y.T., and Nader, K. PKMzeta maintains memories by regulating GluR2-dependent AMPA receptor trafficking. *Nat Neurosci* **13**, 630–634.

Mizuno, K. and Giese, K.P. Towards a molecular understanding of sex differences in memory formation. *Trends Neurosci* **33**, 285–291.

Morris, R.G., Anderson, E., Lynch, G.S., and Baudry, M. (1986). Selective impairment of learning and blockade of long-term potentiation by an N-methyl-D-aspartate receptor antagonist, AP5. *Nature* **319**, 774–776.

Moser, E.I., Krobert, K.A., Moser, M.B., and Morris, R.G. (1998). Impaired spatial learning after saturation of long-term potentiation. *Science* **281**, 2038–2042.

Need, A.C. and Giese, K.P. (2003). Handling and environmental enrichment do not rescue learning and memory impairments in alphaCamKII(T286A) mutant mice. *Genes Brain Behav* **2**, 132–139.

Need, A.C., Irvine, E.E., and Giese, K.P. (2003). Learning and memory impairments in Kv beta 1.1-null mutants are rescued by environmental enrichment or ageing. *Eur J Neurosci* **18**, 1640–1644.

Okada, D., Ozawa, F., and Inokuchi, K. (2009). Input-specific spine entry of soma-derived Vesl-1S protein conforms to synaptic tagging. *Science* **324**, 904–909.

Pastalkova, E., Serrano, P., Pinkhasova, D., Wallace, E., Fenton, A.A., and Sacktor, T.C. (2006). Storage of spatial information by the maintenance mechanism of LTP. *Science* **313**, 1141–1144.

Rampon, C., Tang, Y.P., Goodhouse, J., Shimizu, E., Kyin, M., and Tsien, J.Z. (2000). Enrichment induces structural changes and recovery from nonspatial memory deficits in CA1 NMDAR1-knockout mice. *Nat Neurosci* **3**, 238–244.

Redondo, R.L. and Morris, R.G. (2010). Making memories last: The synaptic tagging and capture hypothesis. *Nat Rev Neurosci* **12**, 17–30.

Reisel, D., Bannerman, D.M., Schmitt, W.B., Deacon, R.M., Flint, J., Borchardt, T., Seeburg, P.H., and Rawlins, J.N. (2002). Spatial memory dissociations in mice lacking GluR1. *Nat Neurosci* **5**, 868–873.

Rioult-Pedotti, M.S., Friedman, D., and Donoghue, J.P. (2000). Learning-induced LTP in neocortex. *Science* **290**, 533–536.

Rogan, M.T., Staubli, U.V., and LeDoux, J.E. (1997). Fear conditioning induces associative long-term potentiation in the amygdala. *Nature* **390**, 604–607.

Romberg, C., Raffel, J., Martin, L., Sprengel, R., Seeburg, P.H., Rawlins, J.N., Bannerman, D.M., and Paulsen, O. (2009). Induction and expression of GluA1 (GluR-A)-independent LTP in the hippocampus. *Eur J Neurosci* **29**, 1141–1152.

Sacktor, T.C. (2011). How does PKMzeta maintain long-term memory? *Nat Rev Neurosci* **12**, 9–15.

Serrano, P., Friedman, E.L., Kenney, J., Taubenfeld, S.M., Zimmerman, J.M., Hanna, J., Alberini, C., Kelley, A.E., Maren, S., Rudy, J.W., *et al.* (2008). PKMzeta maintains spatial, instrumental, and classically conditioned long-term memories. *PLoS Biol* **6**, 2698–2706.

Shema, R., Hazvi, S., Sacktor, T.C., and Dudai, Y. (2009). Boundary conditions for the maintenance of memory by PKMzeta in neocortex. *Learn Mem* **16**, 122–128.

Silva, A.J., Paylor, R., Wehner, J.M., and Tonegawa, S. (1992a). Impaired spatial learning in alpha-calcium-calmodulin kinase II mutant mice. *Science* **257**, 206–211.

Silva, A.J., Stevens, C.F., Tonegawa, S., and Wang, Y. (1992b). Deficient hippocampal long-term potentiation in alpha-calcium-calmodulin kinase II mutant mice. *Science* **257**, 201–206.

Silva, A.J. (2003). Molecular and cellular cognitive studies of the role of synaptic plasticity in memory. *J Neurobiol* **54**, 224–237.

Skrede, K.K. and Westgaard, R.H. (1971). The transverse hippocampal slice: A well-defined cortical structure maintained *in vitro*. *Brain Res* **35**, 589–593.

Thompson, K.R., Otis, K.O., Chen, D.Y., Zhao, Y., O'Dell, T.J., and Martin, K.C. (2004). Synapse to nucleus signaling during long-term synaptic plasticity; a role for the classical active nuclear import pathway. *Neuron* **44**, 997–1009.

Tsien, J.Z., Huerta, P.T., and Tonegawa, S. (1996). The essential role of hippocampal CA1 NMDA receptor-dependent synaptic plasticity in spatial memory. *Cell* **87**, 1327–1338.

Ward, B., McGuinness, L., Akerman, C.J., Fine, A., Bliss, T.V.P., and Emptage, N.J. (2006). State-dependent mechanisms of LTP expression revealed by optical quantal analysis. *Neuron* **52**, 649–661.

Whitlock, J.R., Heynen, A.J., Shuler, M.G., and Bear, M.F. (2006). Learning induces long-term potentiation in the hippocampus. *Science* **313**, 1093–1097.

Yasuda, H., Barth, A.L., Stellwagen, D., and Malenka, R.C. (2003). A developmental switch in the signaling cascades for LTP induction. *Nat Neurosci* **6**, 15–16.

Zamanillo, D., Sprengel, R., Hvalby, O., Jensen, V., Burnashev, N., Rozov, A., Kaiser, K.M., Koster, H.J., Borchardt, T., Worley, P., *et al.* (1999). Importance of AMPA receptors for hippocampal synaptic plasticity but not for spatial learning. *Science* **284**, 1805–1811.

Structural Synaptic and Dendritic Spine Plasticity in the Hippocampus

2

Michael G. Stewart* and Victor I. Popov†

1. Introduction

The connections between neurons consist of synapses, a term coined by Charles Sherrington in 1893 to denote the point at which the nervous impulse is transmitted from one nerve cell to another. Each synapse has an axonal varicosity, a bouton with a presynaptic element making contact with a postsynaptic density (PSD) which contains receptors, scaffold and signalling proteins (Gu and Zeng, 2009). PSDs are found typically on dendritic spines which are the small twig-like protrusions from dendrites. Spines were first identified by Camillo Golgi in 1873 using potassium dichromate and silver nitrate, and later detailed by Santiago Ramon y Cajal in 1893 (DeFelipe, 2006).

A synapse with spine is shown schematically in Figure 2.1A. Synapses can be identified at the electron microscope (EM) level (Figures 2.1A, Ba) by the presence of a PSD and presynaptic vesicles close to the presynaptic membrane, which is referred to as the active zone of the synapse. Excitatory synapses (Figure 2.1A, Ba–c) have an asymmetrical synaptic junction with a prominent PSD, and synaptic vesicles of spheroid character (in the hippocampus normally containing glutamate), whereas inhibitory synapses (Figures 2.1C–E) are typically GABA-ergic, have a synaptic junction that is symmetrical in form and possess ellipsoidal synaptic vesicles. The synapse shown in Figure 2.1A has a perforated PSD and is an example of a synaptic bouton making a single contact with a dendritic spine.

The physiological basis of learning at the level of neural circuitry is believed to result from a change in the efficacy of synaptic transmission. While such changes will result from a series of events at the molecular level, ultimately they will most likely be expressed as structural plasticity, reflected in both strengthening and growth of existing synapses and/or development of new synapses.

* Dept of Life Sciences, The Open University, Milton Keynes, UK.

† Institute of Cell Biophysics, Russian Academy of Sciences, Pushchino, Russia (deceased 2011).

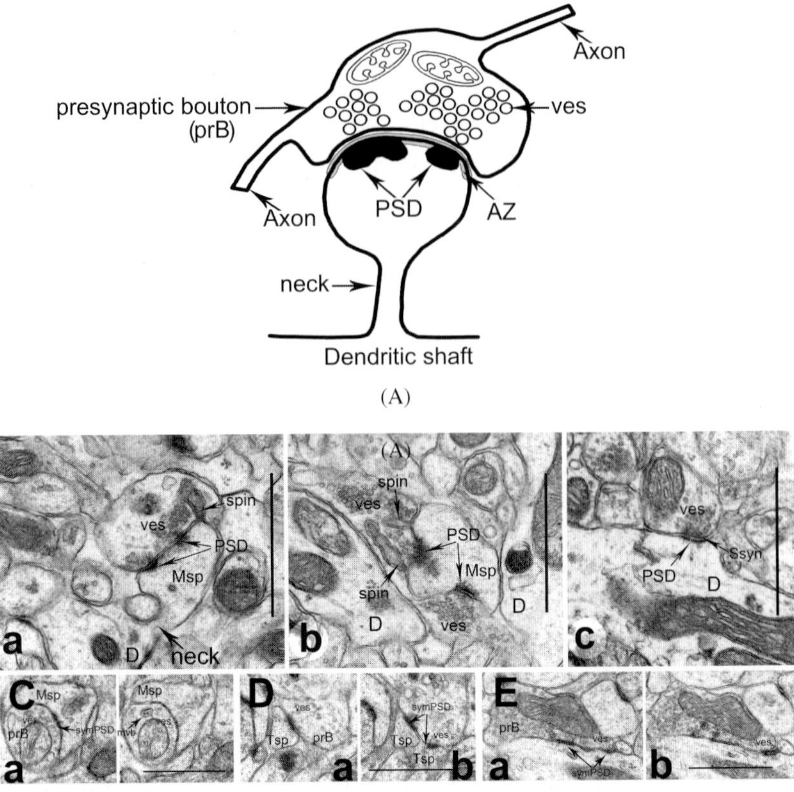

(B)

Figure 2.1. (A) presynaptic bouton (prB) contacts a large spine head (mushroom in type). The postsynaptic density (PSD) is shown with arrows and there is a perforation between the two parts of this PSD. The synaptic apposition zone (AZ) is greater in size than the PSD and its limits are shown on the diagram. There is a thin spine neck which leads to a dendritic shaft (adapted from Popov *et al.*, 2008).

B(a–c) Electron micrographs with examples of three asymmetric synapses with a prominent PSD and spheroid vesicles. Spinules are located both (Ba) outside the PSD and (Bb) within the PSD and they penetrate into each presynaptic bouton in Figures Ba–b. Both synapses contact a dendritic spine of the mushroom form (Msp),

(Bc) Electron micrographs with example of a shaft symmetric synapse (Ssyn). D: dendrite; ves: presynaptic vesicles.

(C–E) Symmetric synapses for both (a–b) mushroom (Msp) and (Da–b) thin (Tsp) dendritic spines (Figures Ca–b and Da–b, respectively). For symmetric synapses, presynaptic (ves) are ellipsoidal in shape. Figures Ea–b show symmetric shaft synapse, where symPSD is symmetric PSD.

All scale bars are 1 μm.

To understand how this might occur, it is necessary to appreciate the structural relationship between synapses and dendritic spines (DeRoo *et al.*, 2009) because the term synaptic plasticity, restricted to the synapse *per se*, tells only a partial aspect of the story of how synaptic efficacy might change.

The structural association between PSDs and dendritic spines can be seen clearly in 3-dimensional (3D) reconstructions, as in Figures 2.2A and B which is from a 10 μm dendritic segment of a granule cell in the dentate gyrus. The PSD can be simple and unperforated (Figures 2.2A, 2Ca), or perforated (Figures 2.1A, Cb). The PSD can also be found in a U-shaped form (Figure 2.2BCc) or segmented, i.e. in separate parts (Figure 2.2Cd). Spines are a major site for excitatory synapses in the central nervous system while inhibitory synapses more commonly occur directly on dendritic shafts. Spines play primarily connective, electrical and biochemical roles in neuronal physiology (DeFelipe, 2006; Harris, 1999; Harris and Stevens, 1989; Hering and Sheng, 2001; Lisman, 1989; Matsuzaki *et al.*, 2001; 2004; Popov *et al.*, 2004; 2005; Segal, 2005; Svoboda *et al.*, 1996; Yuste and Bonhoeffer, 2001).

The relationship between a spine and synapse is not necessarily one-to-one (Fiala *et al.*, 1998; Harris, 1999; Geinisman *et al.*, 2001; Toni *et al.*, 1999, 2001; Trommald and Hulleberg, 1997; Popov and Stewart, 2009). There is great variability in the morphological characteristics of spine types, including their size, the shape of the spine head and neck, and the size and distribution of the PSD (or densities where more than one synapse makes input to a spine). The different types of spines may actually be a continuum of spine shapes that can change form depending on physiological circumstances, from the small thin spines, to the large so-called mushroom spines (see below). Intrinsic fluctuations in their volume and hence shape can explain synapse maintenance over long periods. Kasai *et al.* (2010a) suggested that spontaneous generation, selection and strengthening of spines underlies learning and memory. Their "fifteen" rules of spine structural plasticity, or "spine learning rules", quantitatively describes the creation, enlargement, and shrinkage of spines with different volumes. The spine forms are essentially similar to those shown in Figures 3.2A–H, which is a classification scheme for synapses based subjectively according to their morphology and the nature of their postsynaptic contact (Peters and Kaiserman-Abramof, 1970). Of the categories presented diagrammatically in Figures 2.3A–H: three spine types are classed as either "mushroom", "thin", or "stubby", (thin and mushroom are also shown in the 3D reconstructions in Figure 2.2A, and mushroom in more detail in Figure 2.2B.). A fourth category comprises synapses where the presynaptic bouton contacts the dendrite directly and is termed a "shaft" synapse (Figures 2.3A, 1Bc, 2Ea–b). A spine is classified as *mushroom* (*Msp,* Figures 2.2A, B, Da-d, 2.3E–F) if its head is appreciably wider than the width of the neck; thin (*Tsp,* Figures

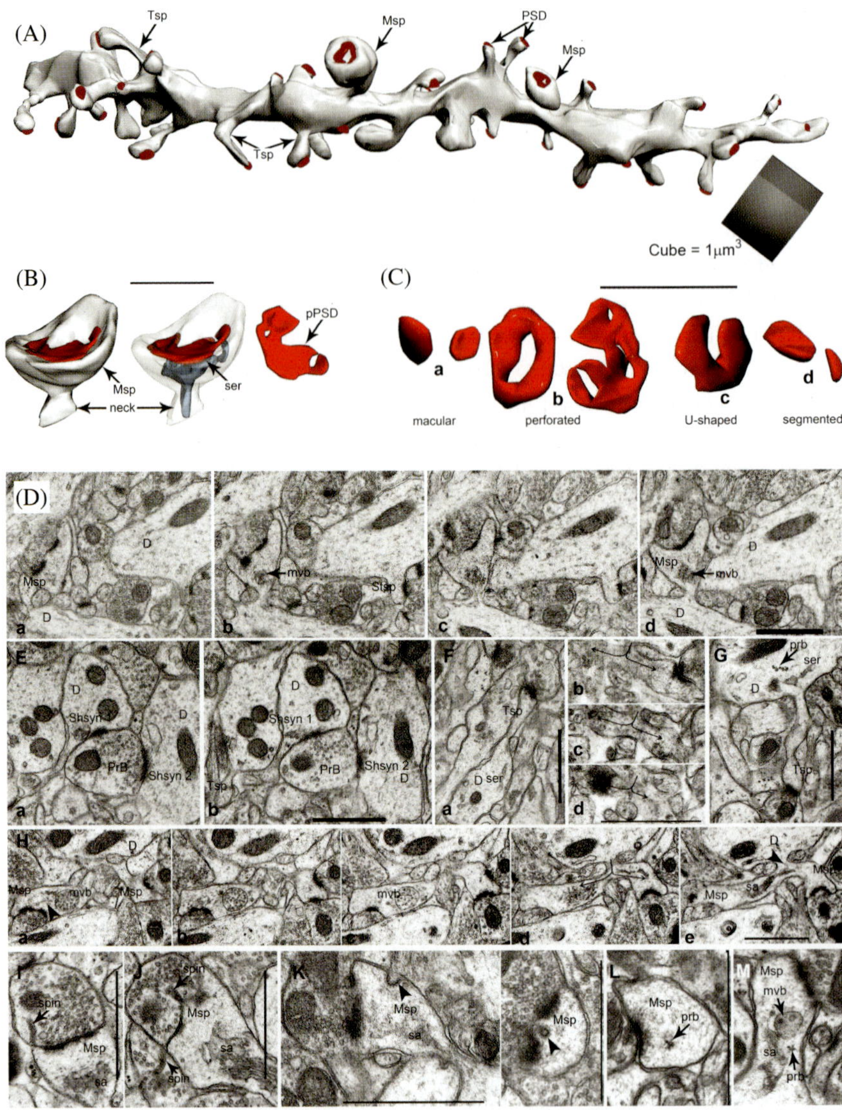

Figure 2.2. *(*A) 3-Dimensional (3D) reconstructions of a dendritic segment (~10 μm) showing a number of spine types: Msp, mushroom spines; Tps thin spines (see also Figure 3).

(B) 3D reconstructions of a large spine mushroom type from ~70 serial images together with spine neck. The PSD is shown at the spine head and to the right, there is a 3D semi-transparent version of the mushroom spine; the perforated PSD is next to it. Note the presence of the smooth endoplasmic reticulum (ser).

2.2A, 2.3C, D), if its length is greater than the width of its neck and head, and *stubby (Ssp)* (Figures 2.2Db, 2.3B) (sometimes termed sessile), if the width of the neck is similar to its length. Generally, the volume of a thin spine is approximately 10 times less than that of an *Msp*. Figure 2.3 shows how thin, and mushroom spines can change in shape) and note also that Figure 2.3D and G show branched forms of thin and mushroom spines, and in the 3–D reconstruction, a branched mushroom spine is presented in Figure 2.4Ca. A branched dendritic spine can also form synapses with the same bouton (Toni *et al.*, 1999; Popov and Stewart, 2009).

How structural aspects of synapses and spines are regulated and which molecules account for these morphological modifications is still poorly understood, despite the fact that such information might provide important clues as to the molecular mechanisms. A direct connection between functional efficacy of synapses and the type of dendritic spine is unclear, though suppression of synapse efficacy is accompanied with reversible retraction of thin spines and CA3 thorny excrescences during hibernation arousal (Popov *et al.*, 1992; 2007). Recovery from stress reverses a volume decrease in thorny excrescences in CA3 while water

---

Figure 2.2. (***Continued***) (C) The main types of PSDs are shown: (a) macular (unperforated PSD); (b) perforated; (c) U-shaped; (d) segmented.

(Da–d) Four serial images showing both mushroom (Msp) and stubby (Stsp) dendritic spines (D: dendritic shaft).

(Ea–b) Two serial images of shaft (Shsyn) synapses (Shsyn 1 and Shsyn 2).

D: dendritic shaft; mvb: multivesicular body; PrB: presynaptic bouton axonal varicosity).

(F) Examples of single thin spine (Tsp) (a) and branched thin spine (b–d).

(G) Thin spine in base of which cisterns of smooth endoplasmic reticulum (ser) and free polyribosomes (prb).

(H) Branched mushroom spine (Msp) on five serial images. Arrowhead shows base of a mushroom spine originating from dendrite (D).

(I) Spinule (spin) on periphery of mushroom spine head. Abundant spine apparatus (sa).

(J) Spinule (spin) within perforated PSD of mushroom spine head. Abundant spine apparatus (sa).

(K) Clathrin-coated omega-like profile on mushroom spine (Msp) membrane.

(L) Clathrin-coated vesicle in mushroom spine (Msp).

(M) Free polyribosomes (prb) within mushroom spine (Msp).

All scale bars are 1 μm.

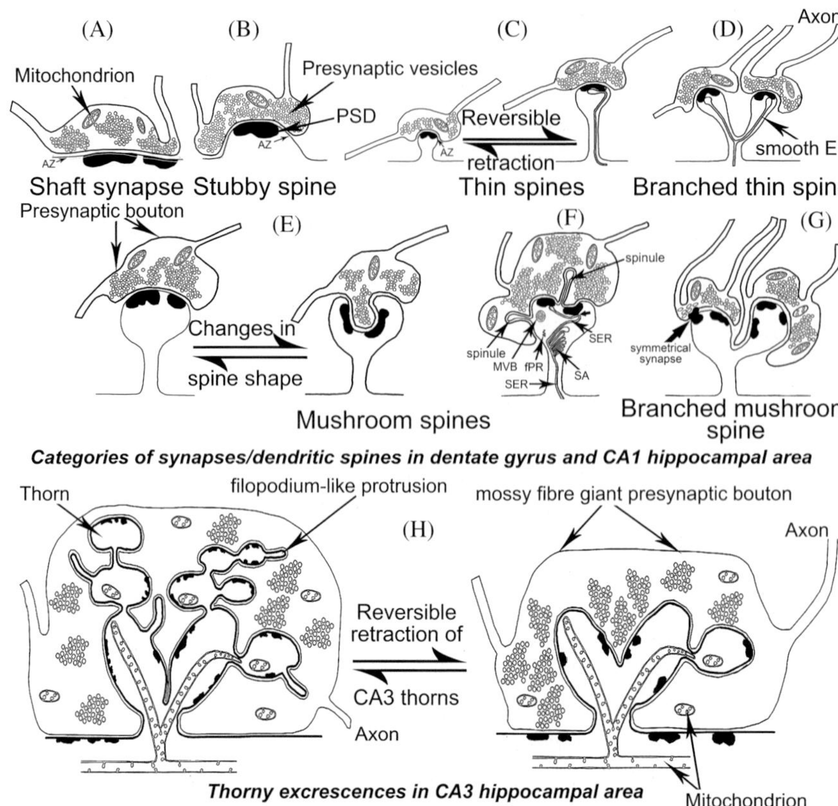

Figure 2.3. (A–G) Classification of spines and synapses in the medial perforant path terminal zone of the dentate gyrus. Figure 2.3A–G show a diagrammatic representation of the four spine categories: shaft, stubby, thin, and mushroom. Note the reversible retraction of thin and mushroom spines, which allows shape and size change. Note also that thin and mushroom spines can be found in branched form (AZ: apposition zone).

(H) Synaptic contact between thorny excrescence and its thorns with mossy fibre giant bouton/varicosity on proximal portions of apical dendrites of CA3 hippocampal neurons represent thorny excrescences, which in contrast to dendritic spines from dentate gyrus and CA1 area, contain mitochondria and tubulin microtubules. Also, thorny excrescences contain all the organelles typical of mushroom dendritic spines. Dendrites of mossy cells (giant interneurons) in hilus also contain thorny excrescences. Schematically thorny excrescences represent stalks covered with thorns. Most likely, the thorns have filopodial origins similar to that of dendritic spines. (H: right side) One form of thorny excrescence plasticity shows retraction of thorns (Popov *et al.*, 1992; Stewart *et al.*, 2005).

maze training increases the surface area of the PSDs (Stewart *et al.*, 2005b). Mushroom spines have been called "memory spines", typically exhibiting enhanced PSD complexity, and/or dimensions, after physiological stimulation (Bourne and Harris 2007; 2008). The PSDs of these spines will likely contain a higher concentration of glutamate receptors, and more smooth endoplasmic reticulum and endosomes (Bourne and Harris, 2008). Figure 2.3 shows that thin spines may be a transition phase becoming mushroom on stimulation, or shrinking back to the dendrite when activity is reduced. Stubby spines (Figure 2.3B) and shaft synapses (Figure 2.3A) would thus represent a stage pre-thin spine development.

Some synaptic boutons make contact with two or more spines, and are termed multiple synapse boutons (MSBs), this is shown in Figures 2.2Ea, b, and 2.4A, B where a presynaptic bouton (prb) makes contact with two mushroom spines (Figures 2.3Aa, b), and a mushroom and thin spine (Figure 2.3Ba, b). Figure 2.4Cb shows a 3D reconstruction of an axon contacting dendritic spines with the synaptic boutons from the enpassant axon making multisynaptic contacts with spines. The most notable examples of multi-synapse boutons are shown making contact on dendritic thorny excrescences in CA3 (Figures 2.3 H and 2.5Aa–c). Formation of multi-synapse boutons has been suggested to be connected with activation of synapses in long-term potentiation (LTP) (Toni *et al.*, 1999, 2001; Trommald and Hulleberg, 1997), learning (Geinisman *et al.*, 2001; Jones, 1999), synaptogenesis (Fiala *et al.*, 1998) and after maintenance of living hippocampal slices *in vitro* (Kirov *et al.*, 1999).

Recent publications have shown dendritic spines in CA1 and the dentate gyrus (DG) that may be contacted by two or more axonal boutons (Nikonenko *et al.*, 2003; Genoud *et al.*, 2004; Stewart *et al.*, 2005; Popov and Stewart, 2009), termed multi-innervated spines (MIS), as in the diagrammatic example of the branched thin and mushroom spines (Figures 2.3D, G), and in the 3D reconstructions in Figures 2.4D and E.

2. Methods to study structural synaptic and dendritic plasticity

The ability to study connections between synapses and spines has been enhanced greatly by optical advances in confocal microscopy and more recently stimulated emission depletion microscopy (STED) (Wilt *et al.*, 2009; Nagerl *et al.*, 2008; 2010), which has enabled quantification of factors such as Ca^{2+} entry into spines. These techniques applied to living tissue (normally acute hippocampal slices) also permit measurements of changes in spine shape and spine number (De Simoni *et al.*, 2004; Engert and Bonhoeffer, 1999; Matsuzaki *et al.*, 2001; Svoboda *et al.*, 1996; Yuste

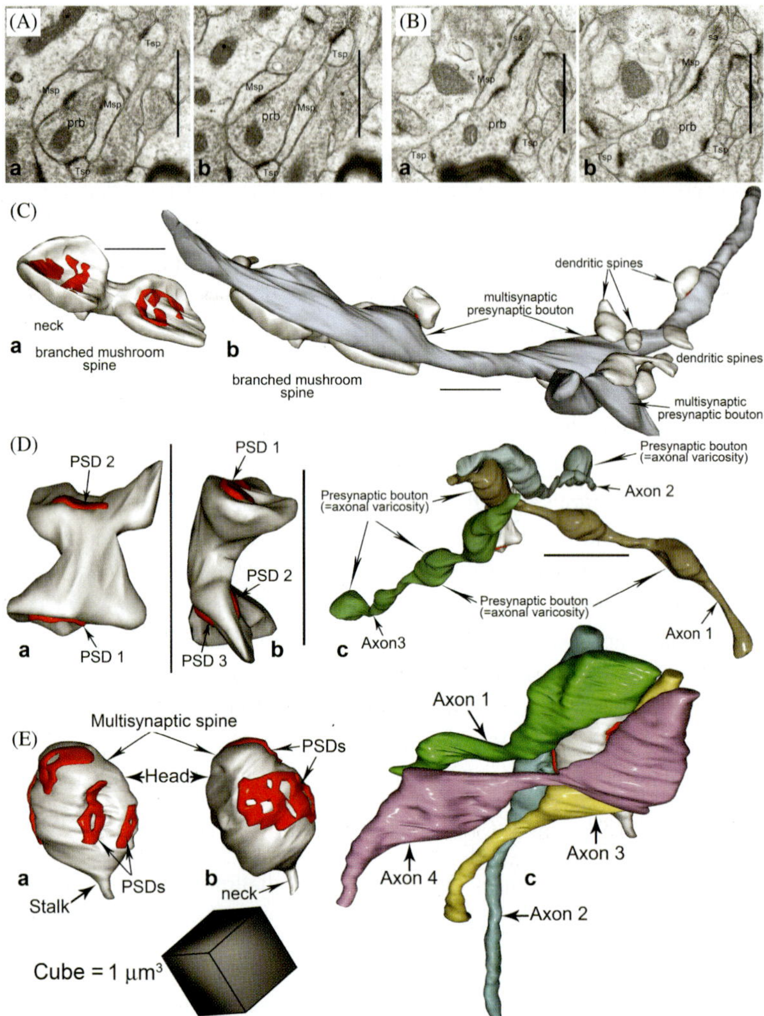

Figure 2.4. (A–B) Serial sections (Aa, b) showing multi-synaptic presynaptic boutons (prb) contacting with mushroom (Msp) and (Ba, b) thin (Tsp) dendritic spines originating from different dendrites.

(Ca) Multi-synaptic or *en passant* axon contacting with branched mushroom spines and other dendritic spines with the formation of multi-synaptic presynaptic bouton. (Cb) Axonal segment with multi-synaptic boutons; also shown are dendritic spines.

(D) Multi-synaptic dendritic spine contacting with three axons (a–b). (Dc) Three axonal segments contacting with multi-synaptic dendritic spines.

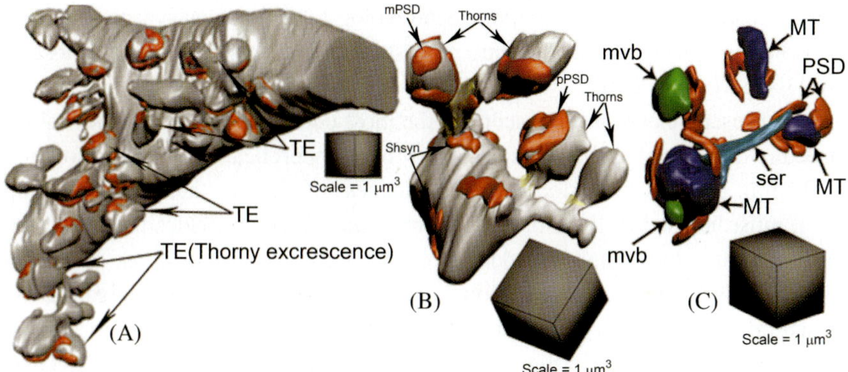

Figure 2.5. Ultrastructure of CA3 thorny excrescences which are contacted by giant boutons from mossy fibres.

(A) 3D reconstructed proximal portion of CA3 apical dendrite covered with thorny excrescences. Red colour shows PSDs.

(B) 3D reconstruction shows thorns, macular (mPSD), and perforated (pPSD) postsynaptic densities. The yellow colour shows filopodium-like protrusions originating from thorny excrescences stalks. Such filopodium-like protrusions produce the thorns.

(C) 3D illustration of inner structure of the thorn showing mitochondrion (MT), multivesicular body (mvb), cisterns of smooth endoplasmic reticulum (ser), and PSDs (red colour).

and Bonhoeffer, 2001), but unfortunately, not as yet with any precision, alterations in synaptic bouton or PSD parameters. The restrictions for estimation of synaptic parameters using light and 2-photon techniques have been addressed previously (Harris, 1994; Moser *et al.*, 1997; Bourne and Harris 2008). This arises because even the latest 2-photon or STED microscopy methods have resolution limitations (0.1 μm for 2-photon microscopy and ~ 60–80 nm and depth up to 120 μm for STED), and with the present level of technological development, are really best applied to tissue *ex vivo,* or *in vivo* very close to a "window" created on the surface of the brain (Golshani and Portera-Cailliau, 2008). Nonetheless, nanoscopy techniques like

---

Figure 2.4. (*Continued*) (E) multisynaptic dendritic spine contacting with four axons with different colours (a–b). Ec shows axons contacting with the same dendritic spine. PSDs are labeled with red colour. All scale bars are 1μm.

STED are likely to be central to future neurobiological research on synapse function, providing new insights into dynamic processes inside living cells at the nanoscale level (Nagerl *et al.*, 2010).

At present, electron microscopy is the most convincing means to view ultrastructure of synapses and spines because it has a theoretical maximum resolution of 0.2 nm. Its main limitation is (i) the thickness of sections that can be cut, and (ii) it must be used in fixed tissue. Electron microscopy provides the resolution necessary to study the fine details of synapses and spines, as shown in Figures 2.1–2.5. Considerable effort has thus been devoted to investigating the structural correlates of synapse function, and the relationship with spines, both *in vitro* and *in vivo*/*in situ*. Earlier methods of synapse morphometry at the electron microscope level extrapolated structural and quantitative information from 2D planar sections of brain tissue to produce 3D estimates, though only the disector method (Sterio, 1986) can be guaranteed to give accurate estimations from 2D images. To fully elucidate the complex structure and relationships that may exist between spines and their synaptic contacts, a detailed visualization at ultrastructural level of both pre- and postsynaptic portions is required. True 3D measures can be obtained from the approach of Harris and Fiala (http://synapses.clm.utexas.edu) to make 3D reconstructions from serial ultrathin sections (using anything up to 150 sections) (Harris, 1999; Harris and Stevens, 1989; Ostroff *et al* 2002; Bourne and Harris, 2007, 2011; Bourne *et al.* 2007; Stewart *et al.*, 2004, 2005, 2008; Medvedev *et al.*, 2010a,b).

3. Structural plasticity of spines

Dendritic spines are highly plastic during development, especially as seen in organotypic culture and in acute slices. Rapid activity-triggered plasticity of spines may relate directly to cognitive processes (Yasumatsu *et al.*, 2008; Kasai *et al.*, 2010b). Confocal studies on living hippocampal slices show the dynamic and often transient nature of spine changes, pointing to the formation of new spines after tetanic hippocampal stimulation (Engert and Bonhoeffer, 1999; Korkotian and Segal, 2001). However, whether dendritic spines move *in vivo* to the extent seen in culture is doubtful (Edwards, 1998). They are more stable *in vivo*, at least in adulthood, and time lapse imaging of dendritic spines in the cerebral cortex suggests they may last throughout life (Bhatt *et al.*, 2009).

Spine morphology influences synaptic input. The length of the spine neck could affect the efficiency of the molecular machinery involved in the modification of glutamate receptors, or the availability of cytoskeletal elements associated with changes in glutamate-mediated function (van Rossum and Hanisch, 1999). Alteration of spine neck length acts as a "fine tuning" mechanism for continuous adjustment of synaptic modification. A constricted neck partitions signals in the spine head, conveys synapse specificity, and protects the parent dendrite from excitotoxicity (Bourne and Harris, 2008). Segal (2010) suggests that a major function of dendritic spines is to protect neurons from synaptic activity-induced rises in intracellular calcium concentrations; hence spine density increases with enhanced synaptic activity, including LTP generation, and shrinks or is replaced by filopodia when synaptic activity is low, as is the case with the deafferented neuron. When spines are lost, neurons are less able to cope with the large synaptic inputs impinging on their dendritic shafts, and these inputs may lead to their eventual death. A strong correlation is seen between spine morphology and the functional properties of excitatory synapses (Bourne and Harris, 2008). Large spines possess larger PSDs, and are more stable over time, whereas in contrast, spines with small heads are less stable (Bourne and Harris, 2008).

4. Synaptic plasticity, LTP, and learning

Determining structural changes associated with learning is a key objective for studies of synaptic plasticity. However, the reality is that determining the nature of structural synaptic plasticity following learning *per se* would be an ambitious task, given the variety of paradigms, and conditions under which learning is studied (Marrone, 2007). Hence, the focus of many studies is on an easily elicitable form of synaptic plasticity in hippocampus, LTP of the perforant path (Bliss and Lømo, 1973; Bliss and Collingridge, 1993), which has become the dominant putative model of the synaptic basis of memory formation (Abraham *et al.*, 2002), although evidence of a direct link between LTP and memory remains unclear. A brief reference here will also be given to additional examples of structural synapse and spine alterations in the counterpart of LTP, long-term depression (LTD), and in other models of plasticity including, stress, kindling and the effects of gonadal hormones on dendritic spine densities in the adult female rat.

An extensive meta-analysis on ultrastructural plasticity associated with hippocampal-dependent learning was undertaken by Marrone (2007), and this has shown a diversity of morphological changes associated with different learning paradigms. As Marrone's study shows, it is difficult to be definitive about the nature of synaptic and dendritic changes overall associated with learning as most morphological changes are paradigm specific. Thus, a direct comparison with

LTP induced structural changes is difficult and will not be attempted in this chapter. However, a summary of learning-related structural alterations from Marrone's very thorough analysis will be provided and generalised findings will be described.

5. What is the structural basis for synaptic plasticity?

The structural basis of synaptic plasticity in a paradigm such as LTP remains the subject of debate (Muller *et al.*, 2000; Sorra and Harris, 1998; Rusakov *et al.*, 1997; Dhanrajan *et al.*, 2003, Popov *et al.*, 2004; Stewart *et al.*, 2005; Medvedev *et al.*, 2010a). There is a lack of agreement on precisely how and indeed whether synapse and dendritic spine morphology and numbers changes as a result of LTP, in part reflecting the differing ways in which LTP is induced — in particular whether *in vivo* or *in vitro*, and if *in vivo* whether in anaesthetised or awake rats which have previously been implanted with electrodes under anaesthesia. Segal (2010) addressed this point, highlighting the factors that contribute to the difficulty in generalizing rules which govern the behaviour of spines. These are: the different types of hippocampal preparations, ages, and imaging conditions used. He points out that we cannot expect to generalize *a priori* from mature, highly spiny neurons recorded in an acute slice or in a cultured slice. Thus, one cannot definitively predict how synapse and spine density, or size, will change as a result of LTP, unless the conditions under which the experiment is performed are described fully. Structural plasticity alterations are, in essence, protocol dependent. In addition, a major issue in studies of structural synaptic and dendritic plasticity using LTP is whether the synapses examined are actually activated, since it is not possible to be certain that *all* synapses have been physiologically stimulated, even in potentiated tissue because hippocampal axons are mainly *en passant* and one axon may synapse with many dendrites of different neurons (Popov and Stewart, 2009).

An ideal solution to ensure that only activated spines and synapses are studied is to use a combined approach of 2-photon microscopy followed by electron microscopy, as shown by Knott *et al.* (2009). Axons and dendrites of GFP-labelled neurons, which have been imaged previously in a live mouse or in slice preparations using 2-photon laser microscopy, are later analysed in fixed tissue using electron microscopy. This allows morphological reconstruction of the synapses both on the imaged neurons, as well as those in the surrounding neuropil. One can thus be sure that an activated synapse and spine is being studied, but for large-scale morphometric studies, this approach would be challenging. A method to ensure that a very large proportion of the synapses examined have been potentiated is the induction of chemical LTP via the addition of tetraethylammonium

(TEA) to a modified CSF containing an elevated concentration of Ca^{2+} and no Mg^{+} (Aniksztejn and Ben-Ari, 1991; Hosokawa *et al.*, 1995; Otmakhov *et al.*, 2004; Stewart *et al.*, 2005).

Following chemical potentiation, in CA1 slices, 3D reconstructions of both dendritic spines and PSDs in CA1 stratum radiatum were made on serial ultrathin sections (Stewart *et al.*, 2005). While there were no changes in synapse or spine density 1 hour after chemical LTP induction, there were alterations in the PSDs. The proportion of synapses with an unperforated PSD (termed macular) decreased (50%) whilst the number of synapses with simple perforated and complex PSDs (non-macular) increased significantly (17%), without significant changes in volume and surface area of the PSD. In addition, the surface area of mushroom spines increased significantly (13%) whilst there were no volume differences in either mushroom or thin spines, or in the surface area of thin spines.

Another method to ensure that only activated synapses are examined was developed by Toni *et al.* (2001), based upon 3D electron microscope reconstruction studies in tetanically stimulated organotypic slices. In this case, the activated synapses were identified via a calcium precipitation method, 1–2 h after LTP. This technique led to the proposal that there was a splitting of the PSD of existing spines with the eventual formation of two separate boutons resulting in the creation of a new immature spine. However, Sorra *et al.* (1998) examined the idea that long-term enhancement of synaptic transmission between neurons results from the enlargement, perforation, and splitting of synapses and dendritic spines. Following a 3D reconstruction study in hippocampal area CA1, they suggested the data showed morphological characteristics inconsistent with synapse and spine splitting. Instead, Sorra *et al.* (1998) suggested perforated synapses and spinules (see description of spinules below) are transient components of synaptic activation, and branched spines appear from synapses forming in close proximity to one another.

6. Learning-related structural changes

The meta-analyses by Marrone (2007) referred to above suggested a reliable profile of morphological change in the hippocampus following learning experiences. Dendritic complexity, including increases in branching and spine density, increases in overall dendritic length, and the size of perforated postsynaptic densities showed consistent increases following training. The density of synapses increases and the nature of perforated synapses show alterations, depending on the duration of training and/or different cell layers of the hippocampal formation. However, it was concluded that only a few morphological parameters typically measured in anatomical studies of plasticity showed significant changes following training. Further, it was suggested that while synaptogenesis was correlated with enduring neocortical plasticity, structural changes were correlated with more transient hippocampal plasticity.

7. Pre- or postsynaptic location of structural plasticity changes?

There is evidence for structural remodelling following LTP both at pre- and postsynaptic levels and Malenka and Nicoll (1999) have summarized the importance of the role played by postsynaptic receptors. It seems likely that the remodelling process is correlated with a change in synaptic strength following stimulation (Nikonenko *et al.*, 2002). Changes in presynaptic function during long-term synaptic plasticity in slices were visualized by Zakharenko *et al.* (2001) and Yuste and Bonhoeffer, (2001) using 2-photon microscopy. Emptage *et al.* (2003) applied optical quantal analysis to reveal a presynaptic component of LTP at hippocampal Schaffer-associational synapses. Synaptic vesicle redistribution appears to be a feature of increased synaptic activity (see below). Synapse enlargement following LTP is indicated by an increase in the presence of polyribosomes in spines (Harris, 2003). Postsynaptic mechanisms of LTP are reflected morphologically in dendritic spine remodelling (Kirov and Harris, 1999; Toni *et al.*, 1999; 2001). Synaptic boutons were smaller after LTD, (Medvedev *et al.*, 2010a) indicating a presynaptic effect.

8. Time scale over which structural changes occur

Short-term structural changes

The time scale over which LTP is studied is a factor in the nature of changes. Morphological changes that occur at shorter times post LTP (<5 h) are most likely due to the rearrangement of existing structures since protein synthesis cannot occur, or at least be completed, in such short time scales. In hippocampal slices, spines have been shown to emerge rapidly (within minutes) from the dendrites, or withdraw depending upon stimulation (Engert and Bonhoeffer, 1999; Yuste and Bonhoeffer, 2001). Electron microscopic studies following LTP suggest stubby spines may become thin, and transform to large mushroom spines (Stewart *et al.*, 2005). These alterations in the structure of dendritic spines and synapse number can occur as early as 30 min post-LTP induction (Desmond and Levy, 1986a, b; 1988; 1990). However, Sorra and Harris (1998) used 3D synaptic reconstructions to show that LTP-induced *in slices* did not result in the overall formation of new synapses at 2-h post-tetanus in hippocampal area CA1.

Synaptic vesicles

Short-term synaptic enhancement is likely to be reflected in an increase in the probability of release of available quanta, and an increase in the number of release sites capable of releasing a quantum (Zuker and Regehr, 2002). However, fixing (preserving) vesicles for visualisation via electron microscopy would not be capable of capturing such rapid movements or the process of fusion of vesicles with the synaptic active zones. More appropriately, fluorescence imaging of synaptic vesicle trafficking has been used in combination with electrophysiological detection of short-term synaptic plasticity (Kavalali, 2007); a 15-s time course has been proposed for synaptic vesicle recycling (Chung and Kavalali, 2009). A recent observation by Matz *et al.* (2010) suggests that the available pool and release probability of synaptic vesicles in hippocampal neuronal cultures may be affected by the size of the active zone at individual synapses and can undergo frequent short-term alterations suggesting a novel form of synapse-specific plasticity.

Nonetheless, some features of vesicle plasticity can be visualised in the electron microscope. Redistribution of synaptic vesicles in the synaptic bouton towards the synaptic active zone is believed to be a characteristic of LTP (Applegate *et al.*, 1987). An electron microscope study found significant increases in vesicles attached to the active zone membrane, and in the percentage of vesicles adjacent to the active zone. Quantification of synaptic vesicles 10 and 20 min after the induction of LTP in the Schaffer-commissural system of the hippocampus demonstrated an overall vesicle density decrease (Applegate and Landfield, 1988). Similarly, changes in synaptic vesicle distribution were shown in an electron microscope study by Lushnikova *et al.* (2008) using organotypic hippocampal slice cultures. Changes in both synaptic vesicle number and their spatial distribution occurred in spine synapses in CA1 at 30 and 60 min following LTP. There was a decrease in the total synaptic vesicle number as well as depletion of the readily releasable synaptic vesicles pool and spatial clustering of synaptic vesicles.

Long-term structural changes

Longer-term structural changes in spines and synapses in hippocampal dentate gyrus and CA1 determined *in vivo* can last for days (Geinisman *et al.*, 1991, 1994; Stewart *et al.*, 2000, Popov *et al.*, 2004; Medvedev *et al.*, 2010a) and will involve biosynthetic processes leading to the formation of new synapses or spines. LTP thus involves activation of protein synthesis on free polyribosomes located in dendrites (Steward and Schuman, 2001; Ostroff *et al.*, 2002; Goldin and Segal, 2003;

Bradshaw *et al.*, 2003). The rapid increase in the rate of protein synthesis can be inhibited by anisomycin, but there may also be accelerated protein degradation via the proteasome system (Karprova *et al.*, 2006). Anisomycin blocks longer-term changes in LTP (>5 h) (Frey *et al.*, 1988, Mochida *et al.*, 2001) whereas the early phase of LTP discussed above is independent of protein synthesis.

Popov *et al.* (2004) made 3D reconstructions of the dentate gyrus 6 h after tissue was stimulated *in vivo*, at a time when protein synthesis could have initiated long-term changes. However, synapse density did not change in potentiated compared to control tissue, but there was an increase in volume and area of unperforated (macular) and perforated (segmented) PSDs. There was also considerable restructuring of pre-existing synapses, with shaft and stubby spines apparently transforming to thin dendritic spines, and mushroom spines changing only in shape and volume. At 24-h post-stimulation Medvedev *et al.* (2010a), also using an *in vivo* approach, but in awake rats (i.e. previously implanted with stimulating and recording electrodes in the entorhinal cortex and dentate gyrus, respectively), showed synapse density did not change in dentate gyrus following LTP. However, there was a significant increase in the proportion of mushroom spines, whereas LTD increased the proportion of thin spines, and both LTP and LTD decreased stubby spine number.

9. Synapse pruning — Masking of plasticity

Increases in synaptic number may be masked by synaptic pruning. An apparent lack of increase in overall synaptic number even at later times post-LTP might actually be the result of both increases and decreases following tetanic stimulation. If new synapses are formed, then spine pruning may occur to keep the synaptic number constant. It is not clear whether spine pruning is an active process associated with an increase in synaptic activity or a passive process caused by lack of afferent activation of the pruned spine. A recent study by Bourne and Harris (2010) showed that in hippocampal slices of adult rats, synaptogenesis began as early as 5 min after theta burst stimulation (TBS), indicated by the fact that polyribosomes were elevated in large spines in CA1 (reflecting local protein synthesis), and at 2 h only those spines with further enlarged synapses contained free polyribosomes but not rough ER. There was an elevation in asymmetric shaft synapses and stubby spines at 5 min and more nonsynaptic filopodia at 30 min but by 2 h the smallest spines were, in contrast, markedly reduced in number. Thus, synapse loss was compensated by enlargement of the remaining excitatory synapses so the summed synaptic surface area per length of dendritic segment was constant across time and conditions.

10. Spinules at the PSD

One of the most notable illustrations of the remodelling of the postsynaptic membrane associated with both structural short and long-term plasticity *in vivo* (Popov *et al.*, 2004), and in hippocampal slices (Toni *et al.*, 2001), are the increases in spinules (spine head protrusions which protrude into synaptic boutons, (Figures 2.1Ba 2.1Bb; 2.2I, 2.3F). This increases very significantly after LTP (and interestingly LTD, Medvedev *et al.*, 2010a) and can be completely blocked by an *N*-methyl-d-aspartate (NMDA) receptor antagonist, CPP, (3-[(R)-2-carboxypiperazin-4-yl]-propyl-1-phosphonic acid), which also blocks LTP (and LTD). An increase in spinule formation is not restricted to LTP, after partial rapid kindling in rats, in hippocampus "giant" spinules originating from mushroom spine heads, and protruding into presynaptic bouton, are usually found near thé perforation in the PSD or at the edge of the PSD (Kraev *et al.*, 2009, Figures 2.1Ba,b). Spinules may remodel the postsynaptic membrane, (Applegate and Landfield, 1988; Geinisman *et al.*, 1993; Toni *et al.*, 1999), or represent an increased motility of spines (Matus, 2000; 2005). Spacek and Harris (2004) suggested that trans-endocytosis of spinules removes excess pre- and postsynaptic membrane after activation, resulting in transient or segmented synapses on mushroom spines.

11. What are the cellular processes involved in spine and synapse remodelling?

Dendritic spines

Dendritic spines have an extensive actin cytoskeleton (Matus *et al.*, 2000; Frost *et al.*, 2010), which enables spines to rapidly change their shapes (in slices at least) in response to stimuli or during hibernation even *in vivo* (Popov *et al.*, 1992). Changes in the shape and size of dendritic spines are correlated with the strength of excitatory synaptic connections and heavily depend on remodelling of the underlying actin cytoskeleton (Hotulainen and Hoogenraad, 2010). Actin has two forms present in spines: polymerised filaments (F-actin) or unpolymerised globular subunits (G-actin). The rapid motility of spines depends on dynamic F-actin fibres (Fischer *et al.*, 1998) and, the application of actin depolymerisation drugs prevents the formation of stable LTP in hippocampal slices (Kim and Lisman, 1999). The stability of the actin cytoskeleton is disrupted by intense glutamate receptor activity following induction of LTP. Mechanisms regulating the spine actin cytoskeleton thus contribute to the persistence of LTP (Fukazawa, 2003). Honkura *et al.* (2008) suggested that the organisation of a dynamic pool of

F-actin plays a key role in spine enlargement. Larger spines have proportionately more glutamate receptors, since the receptors dock to the ends of actin filaments. The resulting increased responsiveness to glutamate in turn helps LTP. The spine often releases this "enlargement pool" of actin into the dendritic shaft, but the pool needs to be physically confined by a spine neck for the enlargement to be long-lasting. Ca^{2+}/calmodulin-dependent protein kinase II regulates this confinement.

In addition, an immediate early gene activity-regulated cytoskeletal protein, Arc (Arg3.1), controls consolidation of LTP through the regulation of local actin polymerization in the dentate gyrus *in vivo* (Messaoudi *et al.*, 2007). Arc mRNA is localized to activated synaptic sites in an NMDA receptor-dependent manner where the newly translated protein plays a crucial role in learning and memory-related molecular processes (Steward and Worley, 2001). Rodriguez *et al.* (2005), using an immunocytochemical method, demonstrated an elevated LTP-associated increase in the expression of Arc protein in the middle molecular layer of the dentate gyrus (2 h after potentiation), with preferential targeting to dendrites and dendritic spines.

Arc has been suggested as a potential "master regulator" of protein synthesis-dependent forms of synaptic plasticity (Bramham *et al.*, 2010). Synaptic tagging (Frey and Morris, 1997; Morris, 2006) and capture of plasticity-related proteins is believed to play a key role in determining if synaptic contacts are strengthened and hence whether or not processes such as LTP are persistent. Synaptic tags capture the products of the synthesis of plasticity proteins set in train by events before, during, or even after an event to be remembered. Tag-protein interactions thus stabilize synaptic potentiation and, by implication, memory. However, the role tags actually play in the process of structural synaptic plasticity is uncertain but are most certainly a crucial factor in modifications at the spine and synapse.

Synaptic boutons

As with dendritic spines, these have an extensive actin cytoskeleton and microtubules which provide the underlying molecular basis for morphological plasticity (but note that dendritic spines in both CA1 and dentate gyrus do not contain microtubules, in contrast to CA3 thorny excrescences). However, its functional significance to the mechanisms of synaptic transmission and synaptic plasticity is unclear. Dillon and Goda (2005), proposed a plausible scheme in which synaptic bouton and spine shape might change with stimulation. Synaptic stimulation increases F-actin in the pre- and postsynaptic terminals and presynaptic F-actin is assembled and/or recruited from synaptic and axonal pools of actin. Actin concentrates at the active synaptic zone while the spine head undergoes a lateral expansion and contracts toward the dendrite shaft in a process coordinated with the remodelling of presynaptic actin.

12. Neural cell adhesion molecule (NCAM)

Cells expressing polysialylated forms of NCAM have a marked increased capacity for structural plasticity (Muller *et al.*, 1996). At the synaptic apposition zone, a key factor in the process of structural synapse remodelling after LTP, and synapse formation, is the neural cell adhesion molecule (NCAM) particularly polysialic acid which is associated with NCAM, as the polysialated neural cell adhesion molecule (PSA-NCAM) (Dityatev *et al.*, 2004). When heterogenotypic cultures containing wild type and NCAM-deficient neurons were used to study mechanisms underlying NCAM-driven preferential synapse formation, it was demonstrated that the polysialylated neural cell adhesion molecule promotes remodelling and formation of hippocampal synapses after induction of LTP. PSA-NCAM is also involved in synaptic plasticity in memory acquisition and consolidation in a fear-conditioning paradigm (Senkov *et al.*, 2006). The precise way in which cell adhesion molecules contribute to synaptic plasticity is unclear but is likely to be connected with the matching of growing structures in pre- and postsynaptic components (Muller *et al.*, 2010).

Synaptic apposition zone curvature changes

Aside from either changes in density and/or enlargement and increases in volume of synaptic boutons following tetanisation involving the actin cytoskeleton, curvature changes of the contact zones between a synapse and spine — the synaptic apposition zone (AZ), and PSD, are amongst the most notable examples of structural plasticity following induction of LTP. Curvature has been used as a criterion for ultrastructural estimation of activity in dendritic spines or synapses (Marrone and Petit, 2002; Marrone *et al.*, 2004, Popov *et al.*, 2004). Curvature of synaptic contacts is described as being either convex or concave (Desmond and Levy, 1983; Markus and Petit, 1989, 2002; Popov *et al.*, 2004). A synapse whose presynaptic element protrudes into the postsynaptic element is referred to as concave, and a synapse whose postsynaptic element protrudes into the presynaptic element is classified as convex. Marrone and Petit (2002) and Connor *et al.*, (2006) demonstrated that LTP was associated with an increased number of excitatory perforated and concave-shaped synapses. Increased numbers of perforated concave synapses were also found to be significantly correlated with the degree of potentiation in the LTP animals. The latter estimations were based upon on single measurements on 2D sections.

Using three dimensional (3D) reconstructions of electron microscope images of ultrathin serial sections (Medvedev *et al.*, 2010b), it was shown that LTP caused a flattening of the curvature of synaptic AZs and PSDs in mushroom and thin dendritic spines while an NMDA receptor antagonist CPP (3-[(R) -2-Carboxypiperazin-4-yl]-propyl-1-phosphonic acid) prevented these changes in the curvature of mushroom and thin spine PSDs and apposition zones. It was suggested that these changes were correlated with a decrease in contact area between the presynaptic bouton and spine head membrane, due to the intensive processes during AMPA receptor insertion into the PSD and receptor recycling (Matsuzaki *et al.*, 2001; Ganeshina *et al.*, 2004a, 2004b; Nicholson *et al.*, 2004, 2006; Nimchinsky *et al.*, 2004).

13. Alterations in synaptic and dendritic organelles

The importance of the cytoskeleton and the role of actin filaments were discussed above in relation to their involvement in shape changes in synapses and dendritic spines. However, underlying these changes are alterations in the synthetic machinery necessary for structural changes and it is possible to visualise some of these at the electron microscope level. Large mushroom spines possess an abundance of smooth endoplasmic reticulum and free polyribosomes (but not rough ER as membrane-associated polyribosomes) (Spacek and Harris, 1997; Ostroff *et al.*, 2002), and a multitude of molecular components (Bourne and Harris, 2008), which may be influenced by experiential factors. In addition, smooth endoplasmic reticulum in dendrites can be shown in 3D reconstructions to continue as the spine apparatus (SA) (Figure 2.1Bii), which forms a lamellar type structure, penetrating into the large mushroom spines reaching even the PSD (Spacek and Harris, 1997). Mice lacking spine apparatus (synaptopodin (SP)-deficient mice) show deficits in LTP and impaired spatial learning supporting the concept of involvement of the spine apparatus in synaptic plasticity (Jedlicka *et al.*, 2008; 2009). Spines containing SP generate twice as large responses to flash photolysis of caged glutamate than SP-negative ones; SP appears to plays an important role in the calcium store-associated ability of neurons to undergo long-term plasticity (Segal *et al.*, 2010).

De Roo *et al.* (2008) demonstrated in organotypic hippocampal slices that spine enlargement and stabilization following LTP were both NMDA- and protein synthesis-dependent, thus enabling synaptic remodelling. The effect of stimulation on large spines most likely involves a sequence of intracellular biochemical events which includes changes in protein synthesis machinery (redistribution of polyribosomes) and in cytoskeletal spine structure due to actin polymerization

(Matus, 2000). This leads to an increase in spine size coupled to increased expression and insertion of AMPA receptor into the PSDs at the spine head (Ganeshina *et al.*, 2004a;b), which in turn would produce larger postsynaptic currents and contribute to the expression of LTP (Matus, 2005). One of the most notable changes is in the smooth endoplasmic reticulum (SER), which can be shown to locate into the large mushroom spines, not only in LTP but also in recovery from stress (Stewart *et al.*, 2005,), and when rats are treated with a cell adhesion molecule mimetic (Popov *et al.*, 2008).

Endosomes

Endosomes can be identified at the electron microscope level primarily in a form termed as multivesicular bodies (MVBs or endosome-like structures) (Spacek and Harris, 1997; Cooney *et al.*, 2002; Stewart *et al.*, 2005). Endosomes are involved with plasticity after synaptic stimulation and play a major role in the endocytotic route sending proteins and lipids to multiple destinations including the cell surface, Golgi complex, and lysosomes (Cooney *et al.*, 2002; Murk *et al.*, 2003; Popov *et al.*, 2008). Ehlers (2000) demonstrated AMPA receptor sorting occurs in early endosomes and is regulated by synaptic activity and activation of AMPA and NMDA receptors, and Blanpied *et al.* (2002) showed that in dendritic spines, endocytotic zones lie lateral to the PSD where they develop and persist independent of synaptic activity, similar to the PSD itself. In a model of stress involving restraint, Stewart *et al.* (2005) showed that in the hippocampus, CA3 thorns contain cisterns of smooth and rough ER and MVBs originating from Golgi stacks, and are localised in both the neuronal body and the base of CA3 apical dendrites. There was an increase in endosome-like structures following water maze training and a decrease following restraint stress which is likely to reflect altered synaptic activity since endosomes are major sorting stations in the endocytotic route sending proteins and lipids to multiple destinations including the cell surface, Golgi complex, and lysosomes (Murk *et al.*, 2003). The downregulation of endosomes retards recycling of GluR1 to the cell surface after NMDA stimulation. Clathrin-coated pits invaginate to form coated vesicles that become large vesicles (Stewart *et al.*, 2005) after the loss of the coat. Large vesicles merge into tubular endosomes and MVB-tubular complexes. Numbers of endosome-like structures increase significantly in TEs, after water maze training and decrease after stress; they are also altered in dendritic segments but are observed to change significantly only after water maze training, increasing by almost 30% (Stewart *et al.*, 2005). A recent study by Sharma *et al.* (2010) suggested signalling endosomes could trigger synapse assembly with data that showed NGF-TrkA signalling endosomes travel from distal axons to cell bodies and dendrites where they promote PSD clustering.

14. Is structural synapse and spine plasticity durable?

Rapid transient alterations in dendritic spine morphology can be demonstrated in hippocampal slices and organotypic cultures, normally in studies of short-term duration, as discussed above. At the electron microscope level, an investigation of structural plasticity in LTP shows it to be long lasting, in experimental periods from 24 h to 3 days (Geinisman *et al.*, 1991; 1994; Weeks *et al.*, 1999; 2000; 2001; Popov *et al.*, 2004, Medvedev *et al.*, 2010a;b).

Hours-long stress paradigms with rodents *in vivo* impairs learning and memory via mechanisms that disrupt the integrity of hippocampal dendritic spines, which is contributed to by corticotropin-releasing hormone (Chen *et al.*, 2010). However, longer-term structural plasticity in spines and synaptic parameters can be rapidly reversed in other *in vivo* experimental paradigms. After 21 days of chronic restraint stress (Stewart *et al.*, 2005), 3D reconstructions were made in stratum lucidum of CA3 thorny excrescences (complex club-like structures originating from the base of dendrites of CA3 pyramidal cells containing a thin stalk covered by small protrusions or thorns). The 3D reconstructions demonstrated that retraction of thorny excrescences had occurred after chronic restraint stress, but this was reversed 24 h after water maze training. After 24 h of water maze training alone, thorny excrescences increased in volume, and the number of thorns per thorny excrescence increased. PSD surface area was unaffected by restraint stress but water maze training increased both number and area of PSDs per thorny excrescence. The proportion of perforated PSDs almost doubled after water maze training.

Fluctuation of gonadal hormones during the female sex cycle of adult rats modulates dendritic spine densities of primary cortical pyramidal Chen *et al.* (2009), as it does in several brain regions including CA1 where fluctuations in dendritic spine number can be observed (Cooke and Wooley, 2005). Moreover, differences in hippocampus-dependent memory formation may result from sex differences in structural synaptic plasticity that are present *before* memory formation (Mizuno and Giese, 2010). There is evidence that the mechanisms of synaptic plasticity show sex-related differences in synaptic signalling and gene transcription (Mizuno *et al.*, 2006).

In a study on the effects of hibernation on CA1 in the Siberian ground squirrel (Popov *et al.*, 2007), a 3D electron microscope study demonstrated reversible retraction of dendritic spines in CA1 when the ground squirrel was subjected to hypothermia-normothermia *in vivo*. However, a similar reversible retraction of dendritic spines was also shown in a non-hibernating mammal, rat, cooled to 19°C, and then brought back to normothermia. Whether LTP-associated structural plasticity is similarly reversible remains to be ascertained, largely at least, at the electron microscope level, because of the massive volume of work involved.

15. Conclusion

One of the best-known models used to study structural synaptic plasticity is LTP of the perforant path in the hippocampus. Although used as a potential model of learning, it cannot be said to show definitive changes that occur during learning *per se*; only studies of structural plasticity after use of a behavioural paradigm would do so. Moreover, even with LTP, it is difficult to generalise the nature of alterations in synapse and dendritic spine morphology as a result of physiological stimulation, unless the conditions under which the experiments are performed are described fully. In short, structural alterations are *protocol dependent* and this relates particularly to whether LTP is induced *in vivo* or *in vitro*; and if *in vivo*, whether in anaesthetised or awake rats which have previously been implanted with electrodes under anaesthesia. Moreover, one cannot assume that structural changes that are induced in an isolated hippocampal slice will occur similarly in an intact animal, where the cells are subject to interaction with entire physiological systems. Nonetheless, as far as can be discerned from data relating to experiments in hippocampal slices (whether acute or organotypic), LTP leads to rapid, dynamic, and often transient, increases in spine parameters, including spine size and number, a process which can be actively viewed in 2-photon systems. Short-term strengthening of a synapse and changes of less than a few hours involve the rearrangement of existing structures but longer-term changes likely involve synaptic tagging, require protein synthesis, and are inhibited by anisomycin.

Alterations in synapse bouton parameters in hippocampal slices are more difficult to ascertain, because of resolution limitations at the 2-photon level; but increases in synapse density appear to parallel increases in spine numbers in some *in vitro* (and *ex vivo*) systems.

In contrast, in the whole animal following hippocampal LTP, large increases in synapse number or dendritic spine density are less prominent than increases in sizes of synapses and spine, and relative proportions of spine types. However, dynamic structural plasticity can be seen in whole animals in response to hibernation, gonadal hormonal alterations in females, kindling and induced stress, with reversible retraction of spines, and alterations in spine sizes.

In learning-related studies of plasticity, only a relatively small number of morphological parameters have been recorded as showing significant changes following training. While synaptogenesis is correlated with enduring neocortical plasticity, structural changes appear correlated with more transient hippocampal plasticity. However, extensive morphological studies in cortical regions, following learning paradigms, have not been carried out to anything like the same extent as following hippocampal LTP studies.

References

Abraham, W.C., Christie, B.R., Logan, B., Lawlor, P. and Dragunow, M. (1994). Immediate early gene expression associated with the persistence of heterosynaptic long-term depression in the hippocampus. *Proc Natl Acad Sci USA* **91**, 10049–10053.

Abraham, W.C. (2003). How long will long-term potentiation last? *Philos Trans R Soc Lond B* **358**, 735–744.

Abraham, W.C., Logan, B., Wolff, A., Benuskova, L. (2007). "Heterosynaptic" LTD in the dentate gyrus of anesthetized rat requires homosynaptic activity. *J Neurophysiol* **98**, 1048–1051.

Aniksztejn, L. and Ben-Ari, Y. (1991). Novel form of long-term potentiation produced by a K^+ channel blocker in the hippocampus. *Nature* **349**, 67–69.

Applegate, M.D., Kerr D.S. and Landfield, P.W. (1987). Redistribution of synaptic vesicles during long-term potentiation in the hippocampus. *Brain Res* **401**, 401–406.

Applegate, M.D. and Landfield, P.W. (1988). Synaptic vesicle redistribution during hippocampal frequency potentiation and depression in young and aged rats. *J Neurosci* **8**, 1096–1111.

Bastrikova, N., Gardner, G.A., Reece, J.M., Jeromin, A. and Dudek, S.M. (2008). Synapse elimination accompanies functional plasticity in hippocampal neurons. *Proc Natl Acad Sci USA* **105**, 3123–3127.

Becker, N., Wierenga, C.J., Fonseca, R., Bonhoeffer, T. and Nägerl UV. (2008). LTD induction causes morphological changes of presynaptic boutons and reduces their contacts with spines. *Neuron* **60**, 590–597.

Bhatt, D.H., Zhang, S. and Gan, W.B. (2009). Dendritic spine dynamics. *Annu Rev Physiol* **71**, 261–282.

Blanpied, T.A., Scott, D.B. and Ehlers, M. (2002). Dynamics and regulation of clathrin coats at specialized endocytic zones of dendrites and spines. *Neuron* **36**, 435–449.

Bliss, T.V.P. and Lømo, T. (1973). Long-lasting potentiation of synaptic transmission in the dentate area of the anaesthetized rabbit following stimulation of the perforant path. *J Physiol (Lond)* **232**, 331–356.

Bliss, T.V.P. and Collingridge, G.L. (1993). A synaptic model of memory: long term potentiation in the hippocampus. *Nature* **361**, 31–39.

Bourne, J. and Harris, K.M. (2007). Do thin spines learn to be mushroom spines that remember? *Curr Opin Neurobiol* **17**, 381–386.

Bourne, J. and Harris K.M. (2008). Balancing structure and function at hippocampal dendritic spines. *Annu Rev Neurosci* **31**, 47–67.

Bourne, J. and Harris, K.M. (2011). Coordination of size and number of excitatory and inhibitory synapses results in a balanced structural plasticity along mature hippocampal CA1 dendrites during LTP. *Hippocampus* **21**, 354–373.

Bradshaw, K.D., Emptage, N.J. and Bliss, T.V.P. (2003). A role for local protein synthesis is required for hippocampal late LTP. *Eur J Neurosci* **11**, 3150–3152.

Bramham, C.R., Alme, M.N., Bittins, M., Kuipers,S.D., Nair RR, Pai, B., Panja, D., Schubert M, Soule, J., Tiron, A. and Wibrand, K. (2010). The arc of synaptic memory. *Exp Brain Res* **220**, 125–140.

Cai, F., Frey, J.U., Sanna, P.P. and Behnisch, T. (2010). Protein degradation by the proteasome is required for synaptic tagging and the heterosynaptic stabilization of hippocampal late-phase long-term potentiation. *Neuroscience* **169**, 1520–1526.

Chen, J.R., Yan, Y.T., Wang, T.J., Chen, L.J., Wang, Y.J. and Tseng, G.F. (2009). Gonadal hormones modulate the dendritic spine densities of primary cortical pyramidal neurons in adult female rat. *Cerebr Cortex* **19**, 2719–2712.

Chen, Y., Rex, C.S., Rice, C.J., Dubé C.M., Gall, C.M., Lynch, G. and Baram, T.Z. (2010). Correlated memory defects and hippocampal dendritic spine loss after acute stress involve corticotropin-releasing hormone signaling. *Proc Natl Acad Sci USA* **107**, 13123–13128.

Christie, B.R. and Abraham, W.C. (1992). Priming of associative long-term depression in the dentate gyrus by theta frequency synaptic activity. *Neuron* **9**, 79–84.

Chung, C. and Kavalali, E.T. (2009). Synaptic vesicle endocytosis: Get two for the price of one? *Neuron* **61**, 333–334.

Cooke, B.M. and Woolley, C.S. (2005). Gonadal hormone modulation of dendrites in the mammalian CNS. *J Neurobiol* **64**, 34–46.

DeFelipe, J. (2006). Brain plasticity and mental processes: Cajal again. *Nat Rev Neurosci* **10**, 811–817.

De Roo, M., Klauster, P. and Muller, D. (2008). LTP promotes a selective long-term stabilization and clustering of dendritic spines. *PLoS Biol* **6**, 1850–1860.

De Simoni, A., Fernandes, F. and Edwards, F.A. (2004). Spines and dendrites cannot be assumed to distribute dye evenly. *Trends Neurosci* **27**, 15–16.

Desmond, N. and Levy, W.B. (1986a). Changes in the numerical density of synaptic contacts with long-term potentiation in the hippocampal dentate gyrus. *J Comp Neurol* **253**, 466–475.

Desmond, N. and Levy, W.B. (1986b). Changes in the postsynaptic density with long-term potentiation in the dentate gyrus. *J Comp Neurol* **253**, 476–482.

Desmond, N. and Levy, W.B. (1988). Synaptic interface surface area increases with long-term potentiation in the hippocampal dentate gyrus. *Brain Res* **453**, 308–314.

Dhanrajan, T.M., Lynch, M.A., Kelly, A., Popov, V.I., Rusakov, D.A. and Stewart, M.G. (2003). Expression of long term potentiation in aged rats involves perforated synapses but dendritic spine branching results from high frequency stimulation alone. *Hippocampus* **14**, 255–264.

Dillon, C., Goda, Y. (2005). The actin cytoskeleton: Integrating form and function at the synapse. *Annu Rev Neurosci* **28**, 25–55.

Dityatev, A., Dityateva, G., Sytnyk, V., Delling, M., Toni, N., Nikonenk, I., Muller, D. and Schachner, M., (2004). Polysialylated neural cell adhesion molecule promotes remodeling and formation of hippocampal synapses. *J Neurosci* **24**, 9372–9382.

Doyère, V., Burette, F., Negro, C.R. and Laroche, S. (1993). Long-term potentiation of hippocampal afferents and efferents to prefrontal cortex: Implications for associative learning. *Neuropsychologia* **31**, 1031–1053.

Doyère, V., Srebro, B. and Laroche, S. (1997). Heterosynaptic LTD and depotentiation in the medial perforant path of the dentate gyrus in the freely moving rat. *J Neurophysiol* **77**, 571–578.

Edwards, F.A. (1998). Dancing dendrites. *Nature* **394**, 129–130.

Ehlers, M.D. (2000). Reinsertion or degradation of AMPA receptors determined by activity-dependent endocytic sorting. *Neuron* **28**, 511–525.

Engert, F. and Bonhoeffer, T. (1999). Dendritic spine changes associated with hippocampal long-term synaptic plasticity. *Nature* **399**, 66–70.

Fiala, J.C., Feinberg, M., Popov, V. and Harris, K.M. (1998). Synaptogenesis via dendritic filopodia in developing hippocampal area CA1. *J Neurosci* **18**, 8900–8911.

Fiala, J.C. and Harris, K.M. (2001). Extending unbiased stereology of brain ultrastructure to three-dimensional volumes. *J Am Med Inform Assoc* **8**, 1–16.

Fischer, M., Kaech, S., Knutti, D. and Matus, A. (1998). Rapid actin-based plasticity in dendritic spines. *Neuron* **20**, 847–854.

Frey, U., Krug, M., Reymann, K.G. and Matthies, H. (1988). Anisomycin, an inhibitor of protein synthesis, blocks late phases of LTP phenomena in the hippocampal CA1 region *in vitro*. *Brain Res* **452**, 57–65.

Frey, U. and Morris, R.G. (1997). Synaptic tagging and long-term potentiation. *Nature* **385**, 533–536.

Frost, N.A., Kerr, J. M., Lu, H.E. and Blanpied, T.A. (2010). A network of networks: cytoskeletal control of compartmentalized function within dendritic spines. *Curr Opin Neurobiol*. in press.

Fukazawa, Y., Saitoh, Y., Ozawa, F., Ohta, Y., Mizuno, K. and Inokuchi, K. (2003). Hippocampal LTP is accompanied by enhanced F-actin content within the dendritic spine that is essential for late LTP maintenance *in vivo*. *Neuron* **38**, 447–456.

Ganeshina, O., Berry, R.W., Petralia R.S., Nicholson, D.A. and Geinisman, Y. (2004a). Synapses with a segmented, completely partitioned postsynaptic density express more AMPA receptors than other axospinous synaptic junctions. *Neuroscience* **125**, 615–623.

Ganeshina, O., Berry, R.W., Petralia R.S., Nicholson, D.A. and Geinisman, Y. (2004b). Differences in the expression of AMPA and NMDA receptors between axospinous perforated and nonperforated synapses are related to the configuration and size of postsynaptic densities. *J Comp Neurol* **468**, 86–95.

Garcıa-Lopez, P., Garcıa-Marın, V. and Freire, M. (2006). Three-dimensional reconstruction and quantitative study of a pyramidal cell of a Cajal histological preparation. *J Neurosci* **26**, 11249–11252.

Geinisman Y., de Toledo-Morrell, L., Morrell, F., Heller, R.E., Rossi, M. and Parshall, R.F. (1993). Structural synaptic correlate of long-term potentiation: Formation of axospinous synapses with multiple, completely partitioned transmission zones. *Hippocampus* **3**, 435–445.

Geinisman, Y. (2000). Structural synaptic modifications associated with hippocampal LTP and behavioural learning. *Cereb Cortex* **10**, 952–962.

Geinisman, Y., Berry, R.W., John F. Disterhoft, J., Power, J.M. and. Van der Zee, E.A., (2001). Associative learning elicits the formation of multiple-synapse boutons. *J Neurosci* **21**, 5568–5573.

Genoud, C., Knott, G.W., Sakata, K., Lu, B. and Welker, E. (2004). Altered synapse formation in the adult somatosensory cortex of brain-derived neurotrophic factor heterozygote mice. *J Neurosci* **24**, 2394–2400.

Goldin, M., Segal, M. and Avignone, E. (2001). Functional plasticity triggers formation and pruning of dendritic spines in cultured hippocampal networks. *J Neurosci* **21**, 186–193.

Golshani, P. and Portera-Cailliau, C. (2008). *In vivo* 2-photon calcium imaging in layer 2/3 of mice. *J Vis Exp* **13**. 681.

Gu, J. and Aheng, J.Q. (2009). Microtubules in dendritic spine development and plasticity. *The Open Neuroscience Journal* **3**, 128–133.

Harris, K.M. and Stevens, J.K. (1989). Dendritic spines of CA 1 pyramidal cells in the rat hippocampus: Serial electron microscopy with reference to their biophysical characteristics. *J Neurosci* **9**, 2982–2997.

Harris, K.M., Jensen, F.E. and Tsao, B., (1992). Three-dimensional structure of dendritic spines and synapses in rat hippocampus (CA1) at postnatal day 15 and adult ages: Implications for the maturation of synaptic physiology and long term potentiation. *J Neurosci* **12**, 2685–2705.

Harris, K.M. (1994). Serial electron microscopy as an alternative or compliment to confocal microscopy for the study of synapses and dendritic spines in the central nervous system. In: *Three-Dimensional Confocal Microscopy, Volume Investigation of Biological Specimens* (Stevens, J.K, Mills, L.R., Trogadis, J.E., eds), New York: Academic Press. pp. 421–445

Harris, K.M. (1999). Calcium from internal stores modifies dendritic spine shape. *Proc Natl Acad Sci USA* **96**, 12213–12215.

Harshad Bhatt, D., Zhang, S. and Gan, W. (2009). Dendritic spine dynamics. *Annu Rev Physiol* **71**, 261–282.

Hering, H. and Sheng, M. (2001). Dendritic spines: Structure, dynamics and regulation. *Nat Rev Neurosci* **2**, 880–888.

Holtmat, A.J., Trachtenberg, J.T., Wilbrecht, L., Shepherd, G.M., Zhang, X., Knott, G. and Svodoba, K. (2005). Transient and persistent dendritic spines in the neocortex *in vivo*. *Neuron* **45**, 279–291.

Honkura, N., Matsuzaki, M., Noguchi, J., Ellis-Davies, G.C. and Kasai, H. (2008). The subspine organization of actin fibers regulates the structure and plasticity of dendritic spines. *Neuron* **57**, 719–729.

Hosokawa, T., Rusakov, D.A., Bliss, T.V. and Fine, A. (1995). Repeated confocal imaging of individual dendritic spines in the living hippocampal slice: Evidence for changes in length and orientation associated with chemically induced LTP. *J Neurosci* **15**, 5560–5573.

Hotulainen, P. and Hoogenraad, C.C. (2010). Actin in dendritic spines: Connecting dynamics to function. *J Cell Biol* **189**, 619–629.

Jedlicka, P., Vlachos, A., Schwarzacher, S.W. and Deller, T. (2008). A role for the spine apparatus in LTP and spatial learning. *Behav Brain Res* **192**, 12–19.

Jedlicka, P., Schwarzacher, S.W., Winkels, R., Kienzler, F., Frotscher, M., Bramham, C.R., Schultz, C., Bas Orth, C. and Deller, T. (2009). Impairment of *in vivo* theta-burst long-term potentiation and network excitability in the dentate gyrus of synaptopodin-deficient mice lacking the spine apparatus and the cisternal organelle. *Hippocampus* **19**, 130–140.

Karpova, A., Mikhaylova M., Thomas, U., Knöpfel, T. and Behnisch, T. (2006). Involvement of protein synthesis and degradation in long-term potentiation of Schaffer collateral CA1 synapses. *J Neurosci* **26**, 4949–4955.

Kasai, H., Hayama, T., Ishikawa, M., Watanabe, S., Yagishita, S., Noguchi, J. (2010a). Learning rules and persistence of dendritic spines. *Eur J Neurosci* **32**, 241–249.

Kasai, H., Fukuda, M., Watanabe, S., Hayashi-Takagi, A., Noguchi, J., Structural dynamics of dendritic spines in memory and cognition. (2010b). *Trends Neurosci* **33**, 121–129.

Kavalali, E.T., (2007). Multiple vesicle recycling pathways in central synapses and their impact on neurotransmission. *J Physiol* **585**, 669–679.

Kemp, A., Manahan-Vaughan D. (2007). Hippocampal long-term depression: Master or minion in declarative memory processes? *Trends Neurosci* **30**, 111–118.

Kim, C.H. and Lisman J.E. (1999). A role of actin filament in synaptic transmission and long-term potentiation. *J Neurosci* **19**, 4314–4324.

Kirov, S.A. and Harris, K.M. (1999). Dendrites are more spiny on mature hippocampal neurons when synapses are inactivated. *Nat Neurosci* **2**, 878–883.

Korkotian, E. and Segal, M. (2001). Regulation of dendritic spine motility in cultured hippocampal neurons. *J Neurosci* **21**, 6115–6124.

Kosub, K.A., D.O, V.H. and Derrick, B.E. (2005). NMDA receptor antagonists block heterosynaptic long-term depression (LTD) but not long-term potentiation (LTP) in the CA3 region following lateral perforant path stimulation. *Neurosci Lett* **374**, 29–34.

Knott, G.W., Holtmaat, A., Trachtenberg, J.T., Svoboda, K. and Welker, E. (2009). A protocol for preparing GFP-labeled neurons previously imaged *in vivo* and in slice preparations for light and electron microscopic analysis. *Nat Protoc* **4**,1145–1156.

Kraev, I.V., Godukhin, O.V., Patrushev, I.V., Davies, H.A., Popov, V.I and Stewart.M.G. (2009). Partial kindling induces neurogenesis, activates astrocytes and alters synaptic morphology in the dentate gyrus of freely moving adult rats. *Neuroscience* **162**, 254–267.

Lushnikova, I.V., Nikonenko, I.R., Nikonenko, O.H. and Skibo, H.H. (2008). Spatial distribution of synaptic vesicles in CA1 hippocampal synapses under conditions of induced long-term. *Fiziol Zh* **54**, 35–44.

Malenka, R.C. and Nicoll, R.A. (1999). Long-term potentiation: A decade of progress? *Science* **285**, 1870–1874.

Malenka, R.C. and Bear, M.F. (2004). LTP and LTD: An embarrassment of riches. *Neuron* **44**, 5–21.

Manahan-Vaughan, D., von Haebler, D., Winter, C., Juckel, G., Heinemann, U. (2008). A single application of MK801 causes symptoms of acute psychosis, deficits in spatial memory and impairment of synaptic plasticity in rats. *Hippocampus* **18**, 125–134.

Markus, E.J. and Petit, T.L. (1989). Synaptic structural plasticity: Role of synaptic shape. *Synapse* **3**, 1–11.

Marrone, D.F. (2007). Ultrastructural plasticity associated with hippocampal-dependent learning: A meta-analysis. *Neurobiol Learn Mem* **87**, 361–371.

Marrone, D.F. and Petit, T.L. (2002). The role of synaptic morphology in neural plasticity: Structural interactions underlying synaptic power. *Brain Res Rev* **38**, 291–308.

Marrone, D.F., LeBoutillier, J.C. and Petit, T.L. (2004). Changes in synaptic ultrastructure during reactive synaptogenesis in the rat dentate gyrus. *Brain Res* **1005**, 124–136.

Martin, K.C. and Kosik, K.S. (2002). Synaptic tagging — who's it? *Nat Rev Neurosci* **3**, 813–820.

Matsuzaki, M., Ellis-Davies, G.C., Nemoto T., Miyashita, Y., Iino, M. and Kasai, H. (2001). Dendritic spine geometry is critical for AMPA receptor expression in hippocampal CA1 pyramidal neurons. *Nat Neurosci* **4**, 1086–1092.

Matsuzaki, M., Honkura, N., Ellis-Davies, G.C. and Kasai, H. (2004). Structural basis of long-term potentiation in single dendritic spines. *Nature* **429**, 761–766.

Matus, A., Brinkhaus, H. and Wagner, U. (2000). Actin dynamics in dendritic spines: A form of regulated plasticity at excitatory synapses. *Hippocampus* **10**, 555–560.

Matus, A. (2000). Actin-based plasticity in dendritic spines. *Science* **290**, 754–758.

Matus, A. (2005). Growth of dendritic spines: A continuing story. *Curr Opin Neurobiol* 15, 67–72.

Matz, J., Gilyan, A., Kolar, A., McCarvill, T. and Krueger, S.R. (2010). Rapid structural alterations of the active zone lead to sustained changes in neurotransmitter release. *Proc Natl Acad Sci USA* **107**, 8836–8841.

Medvedev, N.I., Popov V.I., Rodriguez Arellano, J.J., Dallérac, G, Davies, H.A., Gabbott, P.L, Laroche, S., Kraev, I.V, Doyère, V. and Stewart, M.G. (2010a). The NMDA receptor antagonist CPP alters synapse and spine structure and impairs LTP and LTD induced morphological plasticity in dentate gyrus of the awake rat. *Neuroscience* **165**, 1170–1181.

Medvedev N.I., Popov, V.I., Rodriguez Arellano, J.J., Dallérac, G., Davies H.A., Laroche S., Kraev, I.V., Doyère V. and Stewart M.G. (2010b). Alterations in synaptic curvature in the dentate gyrus following induction of long-term potentiation, long-term depression and treatment with the N-methyl-d-aspartate receptor antagonist CPP. *Neuroscience* **171,** 390–397.

Messaoudi, E., Kanhema, T., Soulé, J., Tiron, A., Dagyte, G., da, Silva, B. and Bramham, C.R. (2007). Sustained Arc/Arg3.1 synthesis controls long-term potentiation consolidation through regulation of local actin polymerization in the dentate gyrus *in vivo. J Neurosci* **27**, 10445–10455.

Mezey, S., Doyère, V., De Souza, I., Harrison, E., Cambon, C., Kendal, C.E., Davies, H.A., Laroche, S. and Stewart, M.G. (2004). Long-term synaptic morphometry changes after induction of LTP and LTD in the dentate gyrus of awake rats are not simply mirror phenomena. *Eur J Neurosci* **19**, 2310–2318.

Mizuno, K., Ris, L., Sánchez-Capelo, A., Godaux, E. and Giese, K.P. (2006). Ca^{2+}/calmodulin kinase kinase alpha is dispensable for brain development but is required for distinct memories in male, though not in female, mice. *Mol Cell Biol* **26**, 9094–9104.

Mizuno, K. and Giese, K.P. (2010). Towards a molecular understanding of sex differences in memory formation. *Trends Neurosci* **33**, 285–291.

Mochida, H., Sato, K., Sasaki, S., Yazawa, I., Kamino, K., Momose-Sato, Y. (2001). Effects of anisomycin on LTP in the hippocampal CA1: Long-term analysis using optical recording. *Neuroreport* **12**, 987–991.

Morris, R.G. (2006). Elements of a neurobiological theory of hippocampal function: The role of synaptic plasticity, synaptic tagging and schemas. *Eur J Neurosci* **23**, 2829–2846.

Moser, M.B., Trommald, M., Egeland, T. and Andersen, P. (1997). Spatial training in a complex environment and isolation alter the spine distribution differently in rat CA1 pyramidal cells. *J Comp Neurol* **380**, 373–381.

Muller, D., Wang, C., Skibo, G., Toni, N., Cremer, H., Calaora, V., Rougon, G. and Kiss, J.Z. (1996). PSA-NCAM is required for activity-induced synaptic plasticity. *Neuron* **17**, 413–422.

Muller, D., Toni, N. and Buchs, P-A. (2000). Spine changes associated with long-term potentiation. *Hippocampus* **10**, 596–604.

Muller, D., Mendez, P., Deroo, M., Klauser, P., Steen, S. and Poglia, L. (2010). Role of NCAM in spine dynamics and synaptogenesis. *Adv Exp Med Biol* **663**, 245–256 (Stevens, JK, Mills, LR, Trogadis, JE, eds), New York: Academic Press, pp. 421–445.

Murk, J.L., Humbel, B.M., Ziese, U., Griffith, J.M., Posthuma, G., Slot, J.W., Koster, A.J., Verkleij, A.J., Geuze, H.J. and Kleijmeer, M.J. (2003). Endosomal compartmentalization in three dimensions: Implications for membrane fusion. *Proc Natl Acad Sci USA* **100**, 13332–13337.

Nägerl, U.V., Willig, K.I., Hein, B., Hell, S.W. and Bonhoeffer, T. (2008). Live-cell imaging of dendritic spines by STED microscopy. *Proc Natl Acad Sci USA* **105**, 18982–18987.

Nägerl, U.V. and Bonhoeffer, T. (2010). Imaging living synapses at the nanoscale by STED microscopy. *J Neurosci* **30**, 9341–9346.

Nicholson, D.A., Yoshida, R., Berry, R.W., Gallagher, M. and Geinisman, Y. (2004). Reduction in size of perforated postsynaptic densities in hippocampal axospinous synapses and age-related spatial learning impairments. *J. Neurosci* **24**, 7648–7653.

Nicholson, D.A., Trana, R., Katz, Y., Kath, W.L., Spruston, N. and Geinisman, Y. (2006). Distance-dependent differences in synapse number and AMPA receptor expression in hippocampal CA1 pyramidal neurons. *Neuron* **50**, 431–442.

Nicholson, .DA. and Geinisman, Y. (2009). Axospinous synaptic subtype specific differences in structure, size, ionotropic receptor expression and connectivity in apical dendritic regions of rat hippocampal CA1 pyramidal neurons. *J Comp Neurol* **512**, 399–418.

Nikonenko, I., Jourdain. P. and Muller, D. (2003). Presynaptic remodeling contributes to activity-dependent synaptogenesis. *J Neurosci* **23**, 8498–8505.

Ostroff, L.E., Fiala, J.C., Allwardt, B. and Harris, K.M. (2002) Polyribosomes redistribute from dendritic shafts into spines with enlarged synapses during LTP in developing rat hippocampal slices. *Neuron* **35**, 535–545.

Otmakhov, N., Tao-Cheng, J.H., Carpenter, S., Asrican, B., Dosemeci, A., Reese, T.S. and Lisman, J. (2004) Persistent accumulation of calcium/calmodulin- dependent protein kinase II in dendritic spines after induction of NMDA receptor-dependent chemical long-term potentiation. *J Neurosci* **24**, 9324–9331.

Peters, A. and Kaiserman-Abramof, I.R. (1970). The small pyramidal neuron of the rat cerebral cortex: the perikaryon, dendrites and spines. *Am J Anat* **127**, 321–356.

Popov, V.I., Bocharova, L.S. and Bragin, A.G. (1992). Repeated changes of dendritic morphology in the hippocampus of ground squirrels in the course of hibernation. *Neuroscience* **48**, 45–51.

Popov, V.I., Davies, H.A., Rogachevskii, V.V., Errington, M.L., Gabbott, P.L.A, Bliss, TVP and Stewart M.G. (2004). Remodelling of synaptic morphology but unchanged synaptic density during late phase LTP: A serial section EM study of the dentate gyrus in the anaesthetised rat. *Neuroscience* **28**, 251–262.

Popov, V.I., Medvedev, N.,I., Davies, H.A. and Stewart, M.G. (2005). Mitochondria form a filamentous reticular network in hippocampal dendrites but are present as discrete bodies in axons: A three-dimensional ultrastructural study. *J Comp Neurol* **492**, 50–65.

Popov,V.I., Medvedev, N.I., Patrushev, I., IIgnat'ev, D.A., Morenkov, E.D.. and Stewart, M.G. (2007). Reversible reduction in dendritic spines in CA1 of rat and ground squirrel subjected to hypothermia-normothermia *in vivo*: A 3-dimensional electron microscope study. *Neuroscience* **149**, 549–560.

Popov, V.I., Medvedev, N.I., Kraev I.V., Gabbott, P.L., Davies, H.A., Lynch, M., Cowley, T.R., Berezin, V., Bock, E. and Stewart, M.G. (2008). A cell adhesion molecule mimetic, FGL peptide, induces alterations in synapse and dendritic spine structure in the dentate gyrus of aged rats: A three-dimensional ultrastructural study. *Eur J Neurosci* **2**, 301–314.

Popov, V.I. and Stewart, M.G. (2009). Complexity of contacts between synaptic boutons and dendritic spines in adult mammalian hippocampus: Three-dimensional Reconstructions from serial Ultrathin sections *in vivo*. *Synapse* **63**, 369–377.

Popov, V.I., Medvedev, N.I., Patrushev, I., IIgnat'ev, D.A., Morenkov, E.D. and Stewart, M.G. (2008). Reversible reduction in dendritic spines in CA1 of rat and ground squirrel subjected to hypothermia-normothermia *in vivo*: A 3-dimensional electron microscope study. *Neuroscience* **149**, 549–560.

Redondo, R.L., Okuno, H., Spooner, P.A., Frenguelli, B.G., Bito, H. and Morris R.G. (2010). Synaptic tagging and capture: Differential role of distinct calcium/calmodulin kinases in protein synthesis-dependent long-term potentiation. *J Neurosci* 30, 4981–4989.

Rodríguez, J.J., Davies, H.A., Silva, A.T., De Souza, I.E., Peddie, C.J., Colyer, F.M., Lancashire, C.L., Fine, A., Errington, M.L., Bliss, T.V. and Stewart, M.G. (2005). Long-term potentiation in the rat dentate gyrus is associated with enhanced Arc/Arg3.1 protein expression in spines, dendrites and glia. *Eur J Neurosci* **21**, 2384–2396.

Rusakov, D.A., Richter-Levin, G., Stewart, M.G. and Bliss, T.V.P. (1997). Reduction in spine density associated with long-term potentiation in the dentate gyrus suggests a spine fusion model of potentiation. *Hippocampus* **7**, 1–12.

Segal, M., (2005). Dendritic spines and long-term plasticity. *Nat Rev Neurosci* **6**, 277–284.

Segal, M., Vlachos, A. and Korkotian, E. (2010). The spine apparatus, synaptopodin and dendritic spine plasticity. *Neuroscientist* **16**, 125–131.

Sharma, N, Deppmann, C.D., Harrington, A.W, St Hillaire, C., Chen, Z.Y., Lee, F.S. and Ginty, D.D. (2010). Long-distance control of synapse assembly by target-derived NGF. *Neuron* **67**, 422–434.

Senkov, O., Sun, M., Weinhold, B., Gerardy-Schahn, R, Schachner, M. and Dityatev, A. (2006). Polysialylated neural cell adhesion molecule is involved in induction of long-term potentiation and memory acquisition and consolidation in a fear-conditioning paradigm. *J Neurosci* **26**, 10888–109898.

Spacek, J. and Harris, K.M. (1997). Three-dimensional organization of smooth endoplasmic reticulum in hippocampal CA1 dendrites and dendritic spines of the immature and mature rat. *J Neurosci* **17**, 190–203.

Schuster, T., Krug, M. and Wenzel, J. (1990). Spinules in axospinous synapses of the rat dentate gyrus: Changes in density following long-term potentiation. *Brain Res* **523**, 171–174.

Sorra, K.E and Harris, K.M. (1998). Stability in synapse number and size at 2 hr after long-term potentiation in hippocampal area CA1. *J Neurosci* **18**, 658–671.

Spacek, J. and Harris, K.M. (1997). Three-dimensional organization of smooth endoplasmic reticulum in hippocampal CA1 dendrites and dendritic spines of the immature and mature rat. *J Neurosci* **17**,190–203.

Spacek, J. and Harris, K.M. (2004). Trans-endocytosis via spinules in adult rat hippocampus. *J Neurosci* **24**, 4233–4241.

Staubli, U. and Lynch, G. (1987). Stable hippocampal long-term potentiation elicited by "theta" pattern stimulation. *Brain Res* **435**, 227–234.

Sterio, D.C. (1984). The unbiased estimation of number and sizes of arbitrary particles using the disector. *J Microsc* **134**, 127–136.

Steward, O. and Worley, P.F. (2001). Selective targeting of newly synthesized Arc mRNA to active synapses requires NMDA receptor activation. *Neuron* **30**, 227–240.

Steward, O. and Schuman, E.M. (2001). Protein synthesis at synaptic sites on dendrites. *Ann Rev Neurosci* **24**, 299–325.

Stewart, M.G., Medvedev, N.I., Popov, V.I., Schoepfer, R., Davies, H.A., Murphy, K, Dallérac, G.M., Kraev, I.V. and Rodríguez, J.J. (2005a). Chemically induced long-term potentiation increases the number of perforated and complex postsynaptic densities but does not alter dendritic spine volume in CA1 of adult mouse hippocampal slices. Eur *J Neurosci* **21**, 3368–3378.

Stewart, M.G., Davies, H.A., Sandi, C., Kraev, I.V., Rogachevsky, V.V., Peddie, C.J., Rodriguez, J.J., Cordero, M.I., Donohue, H.S., Gabbott, P.L. and Popov, V.I. (2005b). Stress suppresses and learning induces plasticity in CA3 of rat hippocampus: A three-dimensional ultrastructural study of thorny excrescences and their postsynaptic densities. *Neuroscience* **131**, 43–54.

Svoboda, K., Tank, D.W. and Denk, W. (1996). Direct measurement of coupling between dendritic spines and shafts. *Science* **272**, 716–719.

Svoboda, K., Tao-Cheng, J.H., Dosemeci, A., Gallant, P.E., Miller, S., Galbraith, J.A., Winters, C.A., Azzam, R. and Reese, T.S. (2009). Rapid turnover of spinules at synaptic terminals. *Neuroscience* **160**, 42–50.

Toni, N., Buchs, P.A., Nikonenko, I., Bron, C.R., Muller, D. (1999). LTP promotes formation of multiple spine synapses between a single axon terminal and a dendrite. *Nature* **402**, 421–425.

Toni, N., Buchs, P-A., Nikonenko, I., Povilaitite, P., Parisi, P. and Muller, D. (2001). Remodeling of synaptic membranes after induction of long-term potentiation. *J Neurosci* **21**, 6245–6251.

Trommald, M. and Hulleberg, G. (1997). Dimensions and density of dendritic spines from rat dentate granule cells based on reconstructions from serial electron micrographs. *J Comp Neurol* **353**, 260–274.

Van Rossum, D., Hanisch, U.K. (1999). Cytoskeletal dynamics in dendritic spines: Direct modulation by glutamate receptors? *Trends Neurosci.* **22**, 290–295.

Wang, X.B., Yang, Y., Zhou, Q. (2007). Independent expression of synaptic and morphological plasticity associated with long-term depression. *J Neurosci* **27**, 12419–12429.

Weeks, A.C., Ivanco, T.L., Leboutillier, J.C., Racine, R.J. and Petit, T.L. (1999). Sequential changes in the synaptic structural profile following long term potentiation in the rat dentate gyrus: I. The intermediate maintenance phase. *Synapse* **31**, 97–107.

Weeks, A., Ivanco, T., LeBoutillier, J., Racine, R. and Petit, T. (2000). Sequential changes in the synaptic structural profile following long-term potentiation in the rat dentate gyrus. II. Induction/early maintenance phase. *Synapse* **36**, 286–296.

Weeks, A., Ivanco, T., LeBoutillier, J., Racine, R. and Petit, T. (2001). Sequential changes in the synaptic structural profile following long-term potentiation in the rat dentate gyrus. III. Long-term maintenance phase. *Synapse* **40**, 74–84.

Wilt, B.A., Burns, L.D., Wei Ho, E.T., Ghosh, K.K., Mukamel, E.A. and Schnitzer, M.J. (2009). Advances in light microscopy for neuroscience. *Annu Rev Neurosci* **32**, 435–506.

Yasumatsu, N., Matsuzaki. M., Miyazaki, T., Noguchi, J. and Kasai, H. (2008). Principles of long-term dynamics of dendritic spines. *J Neurosci* **28**, 13592–13608.

Yuste, R. and Bonhoeffer, T. (2001) Morphological changes in dendritic spines associated with long-term synaptic plasticity. *Annu Rev Neurosci* **24**, 1071–1089.

Zakharenko, S.S., Zablow L. and Siegelbaum, S.A. (2001). Visualization of changes in presynaptic function during long-term synaptic plasticity. *Nat Neurosci* **4**, 711–717.

Zhou, Q., Homma, K.J. and Poo, M.M. (2004). Shrinkage of dendritic spines associated with long-term depression of hippocampal synapses. *Neuron* **44**, 749–757.

Zhu, Y., Xu, J. and Heinemann, S.F. (2009). Two pathways of synaptic vesicle retrieval revealed by single-vesicle imaging. *Neuron* **61**, 397–411.

Zucker, R.S. and Regehr, W.G. (2002). Short-term synaptic plasticity. *Annu Rev Physiol* **64**, 355–405.

Memory Beyond Synaptic Plasticity: The Role of Intrinsic Neuronal Excitability

3

Athanasia Papoutsi*,†, Kyriaki Sidiropoulou*,†
and Panayiota Poirazi*

1. Introduction

Memory is the process by which our experiences modify neurons and neuronal networks in such a way that influence future behaviour, and can be classified in two very broad categories: (a) short-term (or working) memory, referring to the temporary storage of information used to guide future actions, and (b) long-term memory, referring to more permanent information storage.

Identifying the mechanisms by which memories are allocated, stored, consolidated, and retrieved within individual neurons and neuronal networks is a very exciting field of contemporary neuroscience and has provided valuable insights regarding the ways in which neurons maximize information processing and storage. Decades of research have identified different mechanisms for memory formation: (a) physiological synaptic plasticity, (b) structural plasticity, and (c) plasticity of intrinsic excitability. In this chapter, we will focus on the behavioural and activity-induced plasticity of intrinsic neuronal excitability and we will discuss how this type of excitability modulation is involved in different aspects of short- and long-term memory processes.

* Institute of Molecular Biology and Biotechnology (IMBB), Foundation for Research and Technology (FORTH) Heraklion, Crete, Greece.
† Department of Biology, University of Crete, Heraklion, Crete, Greece.

2. General aspects of plasticity of intrinsic neuronal excitability

2.1. *Biophysical and morphological properties shape intrinsic excitability*

Neuronal excitability, in general, refers to the number of action potentials (APs) a neuron fires in response to a given stimulus although the shape and temporal occurrence of the action potentials is frequently considered as well. Initially, neurons were thought to act as simple summation devices that generated an AP when the additive combination of synaptic inputs reached a certain threshold (McCulloch and Pitts, 1943). However, a multitude of electrophysiological, morphological, and theoretical studies unraveled great complexity in neuronal integrative properties, far outreaching the digital output signal. First, the passive properties and complex geometry of dendrites and axonal arborizations allow neurons to integrate their inputs in a non-linear manner (Rall *et al.*, 1967) and alter the shape and propagation properties of APs (Debanne, 2004). Thus, the neuron can no longer be considered a linear summing device since the detailed morphology significantly contributes to neuronal output. Second, the discovery of voltage-dependent ion channels with different densities and biophysical properties along the axo-somato-dendritic axis (Spruston, 2008) shows that not only the soma, but the dendrites as well, have an active role in shaping neuronal excitability. Finally, the combination of a complex dendritic morphology with a location-specific distribution of ionic mechanisms results in the formation of distinct dendritic subunits which can integrate inputs in a quasi-independent manner (Losonczy and Magee, 2006; Poirazi *et al.*, 2003a, b; Polsky *et al.*, 2004). This compartmentalization of information processing endows neurons with additional computational power which exceeds by at least one order of magnitude that of a "point neuron" model (London and Hausser, 2005; Poirazi and Mel, 2001).

Thus, neuronal excitability is not only affected by the axosomatic biophysical properties but also by the geometry and ion channel repertoire of dendritic arborizations, which greatly influence neural integration.

2.2. *Plasticity of intrinsic excitability depends on ionic mechanisms*

Plasticity of neuronal excitability refers to changes in the number of APs, the AP shape or the temporal pattern of AP trains, which occurs in response to a specific input and results from plastic changes in ion channel function. Such changes include alterations in the number of ion channels, their subunit composition and/or conductance, as well as their activation or inactivation kinetics.

Regulation of ion channel properties could differentially occur along the axo-somato-dendritic axis, resulting in location-dependent changes in neuronal excitability. Modulation of ion channels located at the soma or the spike initiation zone will globally alter the throughput of the neuron as it will influence the potency of all synaptic inputs, irrespective of their location, to generate APs (Zhang and Linden, 2003). On the other hand, ion channel plasticity within specific dendritic branches will result in excitability changes that are seen only when the affected branches are activated. Both synaptic integration and the propensity of an input to generate a somatic AP will be altered but in a branch-specific (local) manner. It should be pointed out that although long-lasting local modifications enhance information-storage capacity, global modifications of a neuron's firing properties or changes that affect an entire neuronal population generally decrease storage capacity (Frick and Johnston, 2005).

Regulation of ion channels could also occur on different time scales: *in vitro*-induced changes in excitability could last a few seconds, a few hours/days or persist for longer periods (Magee and Johnston, 2005). A similar time-dependent regulation of intrinsic excitability has been documented *in vivo* in response to learning (Moyer *et al.*, 1996; Saar *et al.*, 1998; Schreurs *et al.*, 1998) or under pathological conditions (Chen *et al.*, 2002).

Regulation of ion channels on the shorter timescale (that of seconds) seems to be more relevant for processes associated with the temporary storage of information, as for example sustained activity during working memory tasks (Major and Tank, 2004) (see Section 6.2). Transient channel modifications are usually a result of altered activation and inactivation properties of membrane ion channels due to spiking activity or neuromodulatory inputs. Prolonged (short-term) channel modifications that last for a few hours up to a few days following the inducing stimulus could have a permissive role for other more permanent changes, such as long-lasting modifications in synaptic strength or structure. This type of changes are usually a result of post-translational phosphorylation-dependent modifications of ion channels and ion channel trafficking (Shah *et al.*, 2010) or due to differences in the distribution of ion channels (diffused versus clustered) within the dendritic arbors (Misonou *et al.*, 2004). Finally, ion channel modifications that persist for more than a few days may comprise part of the memory trace *per se*: just as the remodeling of synaptic connections underlies memory consolidation, altered intrinsic properties that persist beyond the learning process could also act as encoding mechanisms and/or storage reservoirs for a given memory. Activity-dependent regulation of ion channel protein synthesis is very likely to mediate these long-lasting changes in intrinsic excitability, since memory consolidation depends on protein-synthesis processes. Indeed, recent data from the lateral amygdala and the hippocampus suggest that upregulation of the cAMP response element-binding (CREB) protein, a well-known molecule implicated in the regulation of gene transcription, results in the enhancement of neuronal excitability (Lopez de Armentia *et al.*, 2007; Zhou *et al.*, 2009). Moreover, activity-induced local suppression of protein synthesis of voltage-gated potassium channels in the dendrites has been documented in the hippocampus (Raab-Graham *et al.*, 2006), while a recent report using protein synthesis inhibitors showed that persistent long-term plasticity of intrinsic excitability requires protein synthesis (Cohen-Matsliah *et al.*, 2010).

Although synaptic plasticity has long been thought as the primary input-specific information storage mechanism, plasticity of intrinsic neuronal excitability could allow for a new form of memory storage that does not necessarily depend on the specific incoming signal but more on the overall history of the cell. Currently, studies of plasticity of excitability involve investigations of two basic issues: (a) what exactly is altered when a neuron undergoes plasticity of its intrinsic mechanisms, and (b) what are the additional computational capabilities that this type of plasticity endows the cell. In the following sections, we will discuss *in vivo* and *in vitro* experiments associated with these issues.

3. Learning and experience-dependent plasticity of neuronal excitability

In vivo studies of invertebrates were the first to reveal the link between learning processes and changes in intrinsic excitability. The pioneering work of Alkon and co-workers demonstrated that following associative learning in the mollusk *Hermissenda crassicornis*, neurons display enhanced excitability that persists for several weeks. This increase in excitability was due to reduced A-type potassium current (I_A) and calcium-dependent potassium (K_{Ca}) currents and was mediated by increased intracellular calcium concentration and protein phosphorylation (Alkon, 1984).

Subsequent studies showed that training to several conditional learning tasks affects neuronal excitability specifically in circuits involved in each task (Brons and Woody, 1980; Woody and Engel, 1972). Acquisition of hippocampal-dependent associative tasks, such as trace eye-blink conditioning and contextual or trace cued fear conditioning, result in prolonged increased excitability of CA1 and CA3 hippocampal pyramidal neurons (McKay *et al.*, 2009; Moyer *et al.*, 1996; Thompson *et al.*, 1996) accompanied by a decrease in the post-burst afterhyperpolarisation (AHP). This type of plasticity probably provides the necessary time window for the consolidation of the memory in other brain areas (Disterhoft and Oh, 2006; Moyer *et al.*, 1996). Interestingly, this change in excitability could be reversed with extinction of the contextual or trace cued fear memory (McKay *et al.*, 2009). Pyramidal neurons of the infralimbic cortex, which mediate the extinction of fear memory (Quirk *et al.*, 2006), were initially rendered less excitable following fear conditioning, but then, extinction training returned their average firing levels to baseline due to a smaller fast AHP (fAHP) and a subsequent increase in bursting (Santini *et al.*, 2008). Taken together, these findings suggest that conditioning tasks can differentially alter intrinsic neuronal excitability in brain areas associated with different stages of memory formation (i.e. acquisition or extinction).

Changes in excitability following learning are not limited to the classical conditioning protocols. Training in rule-learning tasks, such as odor discrimination, resulted in increased excitability and decreased slow AHP (sAHP) in piriform cortical neurons (Saar *et al.*, 1998), that lasted for several days, during which the performance of learning for subsequent odor pairs was also improved (Saar and Barkai, 2003). Once excitability returned to baseline levels (after about 5 days), however, learning performance declined. The enhanced ability to acquire new memories rapidly following learning of similar tasks indicates that plasticity of excitability may have a permissive role for establishing novel but similar memories, possibly through facilitation of other types of plasticity. A similar phenomenon is the facilitation of re-learning, termed "savings"; that is, following the extinction of a memory, retraining within a specific time window results in faster acquisition of the memory than in the initial training. Savings have been associated with plasticity of excitability in cases where, although modulation of excitability is transient, its time course correlates with the time course of enhanced learning (Zhang and Linden, 2003). In many cases, neuronal plasticity in response to learning affects the overall neuronal output of the neuronal circuits involved in the given task (Frick and Johnston, 2005). This type of plasticity globally increases or decreases the gain in specific neuronal circuits, allowing these neurons to enhance their learning capacity for the specific tasks over a short period of time following the initial training process.

In addition to the well-documented changes in axosomatic mechanisms that globally influence neuronal excitability, recent data from behavioural studies have also identified localized dendritic changes in the expression level of small conductance calcium-dependent potassium channels (SK2 channels). Interestingly, no change in the expression levels of SK3 channels, which are localized at the axon terminals was observed (Brosh *et al.*, 2007), suggesting that plasticity of intrinsic mechanisms in response to learning may occur not only in a time-restricted but also in a spatially-restricted manner at the single-neuron level. Recent data from *in vitro* preparations discussed later provide further evidence in support of this claim.

Besides the short-lasting changes in neuronal excitability, a few behavioural studies have reported learning-induced persistent changes in excitability. Apart from the pioneering study of Alkon (1984), increased neuronal excitability accompanied by a decreased sAHP that persists for over a month, has been observed in Purkinje cerebellar neurons of rabbits following training to a cerebellum-dependent conditioning task (delay eyelid conditioning) (Schreurs *et al.*, 1998). In this case, plasticity was restricted to microzones of the cerebellar lobule (Schreurs *et al.*, 1997), suggesting that these localized persistent changes in intrinsic excitability may encode for at least a part of the memory.

Experience has also been found to alter neuronal excitability in different brain areas. For example, light presentation to Xenopus tadpoles increases tectal neuronal excitability due to an increase in sodium currents. This increase is a result of a polyamine-mediated decrease of synaptic drive whose goal is to maintain optimal neuronal responsiveness in the new environment (Aizenman *et al.*, 2003). Moreover, forepaw denervation in rats that results in reorganization of the primary somatosensory area (SI) is also accompanied by a decrease in intrinsic excitability through increased AHP (Hickmott, 2005). Similar results (i.e. decrease of intrinsic excitability) were obtained from neurons of the sensorimotor cortex of rats that were subject to hindpaw sensorimotor deprivation (Canu *et al.*, 2010), substantiating the role of plastic excitability in the sensory mapping (Feldman, 2009). It should be noted that neurons of the visual cortex in brain slices may also undergo activity-dependent modifications of their intrinsic excitability (Cudmore and Turrigiano, 2004). Finally, modulation of single cell ionic currents can trigger plasticity of motor programs. *In vivo* intracellular recordings revealed that the neuronal excitability of the motor cortex is modulated in an activity-dependent bidirectional manner. This plasticity of motor neurons is global, long-lasting, and is a result of the altered properties of voltage-gated ion channels (Paz *et al.*, 2009). Overall, experience-driven plasticity of intrinsic excitability seems to be another mechanism utilized by the brain to facilitate adjustment in novel environments.

These learning and experience dependent paradigms have substantiated the role of plasticity of excitability to either the learning process or memory storage. More analytic investigations of what exactly is altered when a neuron undergoes plasticity of excitability are made by *in vitro* experiments under various conditions.

4. Activity-dependent plasticity of intrinsic excitability

Synaptic plasticity in the form of long-term potentiation (LTP) and long-term depression (LTD) remains the most widely accepted cellular correlate of learning and memory (Siegelbaum and Kandel, 1991). LTP is induced following strong, correlated input to the postsynaptic neuron and results in synapse-specific increases in synaptic strength (Frey and Morris, 1997). Although most studies of LTP have focused on the mechanisms underlying the changes induced in synaptic strength (Kandel, 2001; Kelleher *et al.*, 2004), it has become evident that the same stimuli that result in synaptic LTP have additional effects on intrinsic properties, which in turn influence neuronal excitability (Campanac and Debanne, 2007). Furthermore, ion channel function is also altered under non-Hebbian homeostatic protocols, such as long-lasting stimulation or stimulus deprivation (Turrigiano and Nelson, 2000). In this section, we will focus on the modulation of intrinsic excitability as this is observed *in vitro* and in particular when (a) it serves as a long-term memory reservoir, (b) has a role in inducing long-term synaptic plasticity, and (c) participates in short-term memory processes.

4.1. *The E-S coupling phenomenon*

Early long-term potentiation studies were the first to suggest that LTP influences intrinsic excitability. This is achieved by increasing the propensity of a postsynaptic neuron to fire APs in response to a given excitatory postsynaptic potential (EPSP) (Bliss and Lomo, 1973), a phenomenon termed EPSP-spike (E-S) potentiation. E-S potentiation has been primarily attributed to modifications of the inhibitory drive (Chavez-Noriega *et al.*, 1990; Lu *et al.*, 2000), although it can still be induced in the absence of inhibition. For example, in lesioned rats with decreased hippocampal IPSPs, both E-S potentiation and depression can be induced (Bernard and Wheal, 1995). Moreover, LTP induction in the medial amygdala results in E-S potentiation under conditions of blocked GABAergic inhibition (Noguchi *et al.*, 1998).

Stimuli known to induce LTP, such as activation of NMDA receptors and elevated intracellular calcium concentration (Aizenman and Linden, 2000) as well as an overall increase in intrinsic excitability (Pugliese *et al.*, 1994), also trigger E-S potentiation. Furthermore, E-S depression following induction of LTD has been observed postsynaptically in CA1 pyramidal neurons (Daoudal *et al.*, 2002), suggesting that, just like synaptic plasticity, E-S coupling can be bidirectional. Since E-S coupling refers primarily to the plasticity of intrinsic excitability, changes in E-S coupling are attributed to the spatial and temporal modulation of various ionic mechanisms along the somato-dendritic axis. In the following paragraphs, we will discuss cases in which E-S coupling results from:

(1) changes in ion channel expression/function at the soma and the axon hillock, thus affecting the probability of the neuron to spike, and
(2) changes in ion channel expression/function locally at the stimulated dendrite or across multiple dendrites, thus affecting the magnitude of local EPSPs, their integration and propagation to the soma.

4.1.1. *Activity-dependent changes in intrinsic mechanisms located at the soma*

Similar to changes following learning, high frequency or correlated pre- and postsynaptic stimulation in acute or organotypic brain slices induce E-S potentiation that is accompanied by a significantly decreased sAHP (Jung and Hoffman, 2009; Sourdet *et al.*, 2003; Xu *et al.*, 2005), possibly due to a decrease in small-conductance calcium-activated potassium (SK) channels (Sourdet *et al.*, 2003) (Figure 3.1D). Often, the decrease in sAHP is accompanied by a lowered AP threshold. The latter is due to either a shift in the voltage-dependent activation of sodium channels (Xu *et al.*, 2005), or an initial shift of the voltage-dependent activation of A-type K^+ current (I_A) towards more hyperpolarised potentials, followed by a decrease in the peak I_A due to increased endocytosis of the Kv4.2 channel subunit (Jung and Hoffman, 2009). Thus, activity-induced enhancement of excitability can result from changes in different ionic mechanisms located in close proximity to the cell body.

In addition to postsynaptic neurons, presynaptic neurons may also undergo bidirectional changes in excitability following a plasticity induction protocol (Li *et al.*, 2004). Specifically, enhancement of synaptic plasticity leads to a prolonged increase in excitability manifested as a reduction in the AP threshold due to less inactivation of the presynaptic sodium channels (Ganguly *et al.*, 2000), while depression of synaptic function decreases excitability through increased activation of slow-inactivating potassium currents (Li *et al.*, 2004). Taken together, these findings suggest that plasticity of the presynaptic neuron excitability acts as a feedback mechanism that greatly influences network dynamics.

However, activity does not only result in enhanced neuronal excitability, but also in a change in the firing pattern of APs. Theta-burst stimulation, for example, has been shown to enhance bursting behaviour of subicular neurons (Moore *et al.*, 2009). On the other hand, long-lasting depolarizations decrease the spike after-depolarization (ADP) due to altered properties of the M-type potassium current, resulting in decreased bursting in CA3 pyramidal neurons (Brown and Randall,

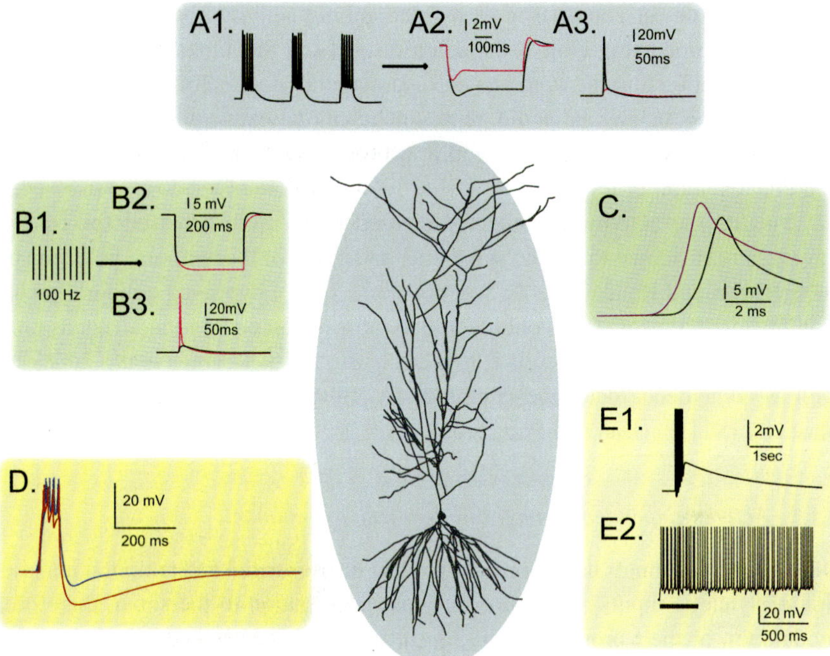

Figure 3.1. Spatial characteristics of different types of intrinsic plasticity.

(A) Homeostatic regulation of I_h in response to maximal LTP induction. Traces were generated using the model of (Poirazi *et al.*, 2003a). (A1) Neuronal response following theta-burst pairing of bAPs and synaptic stimulation. (A2) The protocol in A1 induces global reduction of excitability, mediated by upregulation of I_h, evident by a decrease in the voltage response at the soma to the same current injection (black trace: control, red trace: after protocol in A1). (A3) The same stimulation protocol that induces a spike under control conditions (black) results in a subthreshold EPSP after application of the protocol used in A1 (red trace), i.e. decreased E-S coupling.

(B) Spatially-restricted enhancement of excitability in response to induction of moderate LTP. Traces were generated using the model of (Poirazi *et al.*, 2003a). (B1) High-frequency synaptic stimulation used to induce moderate LTP. (B2) The protocol in B1 results in a decrease of dendritic I_h, evident by an increase in the voltage response at the stimulated dendrite in response to the same current injection (black trace: control, purple trace: after protocol in B1). Voltage response at the soma remains unaffected. (B3) Stimulation of the same dendrite that induces a subthreshold EPSP under control conditions (black) results in a spike after application of the protocol in A1 (purple trace), i.e. increased E-S coupling.

2009). Bursting, in comparison to regular-spiking activity, results in enhanced information processing both at the dendritic-end and the axonal-end of the neuron, without affecting the average neuronal output (Cooper, 2002). Thus, plasticity in bursting behaviour could represent a more flexible way of modifying information processing capacity within a neuron without changing the overall neuronal excitability level.

Finally, the activity-dependent decreased sAHP and increased spike-ADP have also been shown to reduce spike jitter and enhance the temporal precision of APs (Brown and Randall, 2009; Sourdet *et al.*, 2003). Improved temporal precision of signal transmission could facilitate information coding that depends on a temporal rather than a rate code (Pissadaki *et al.*, 2010; Tiesinga *et al.*, 2008), as well as coincidence detection (Konig *et al.*, 1996) and spike-timing-dependent plasticity (STDP) (Dan and Poo, 2006).

4.1.2. *Activity-dependent changes in dendritic ion channels*

Dendritic ion channels have a profound effect on the spatial and temporal integration of synaptic inputs, as well as on input propagation to the soma; therefore, modulation of their properties could result in plasticity of dendritic excitability that modifies the rules of information processing in an input specific manner. Dendritic channels that affect signal integration include the hyperpolarisation-activated potassium current (I_h), the A-type potassium current, (I_A) and the persistent sodium current (I_{NaP}) (Sidiropoulou *et al.*, 2006). LTP-inducing stimuli also modify dendritic intrinsic properties, triggering plasticity of dendritic integration, that may follow the same timing rules as STDP (Campanac and Debanne, 2008). Positively correlated pre- and postsynaptic activity in CA1 neurons

---

Figure 3.1. (*Continued*) (C) Activity-dependent alterations of dendritic I_A kinetics modulate bAPs. A hyperpolarised shift in the inactivation half potential of I_A from −56 mV (black trace) to −61mV (purple trace) decreases the latency and increases the duration of bAPs in the apical dendrite. Traces were generated using the model of (Poirazi *et al.*, 2003a).

(D) Conditional learning results in decreased sAHP. The sAHP was measured after a 50-ms current injection to the soma in a control (red trace) and conditioned (blue trace) rat. Traces were generated using the model of (Markaki *et al.*, 2005).

(E) Somatic dADP following a burst of APs and activation of mGluRs mediates an enhancement of post-burst excitability (E1) as well as sustained firing (E2) after excitatory stimulation (indicated by the black bar). Traces were generated using the model of (Papoutsi *et al.*, 2009).

induces facilitation of dendritic integration, whereas negatively correlated pre- and postsynaptic activity induces dendritic signalling depression, within a stringent time-window.

Among the ion channels that are modulated during the induction of plasticity of excitability, the hyperpolarisation-activated cation channel (HCN) that underlies the *h* current is of special importance. I_h is active at the resting membrane potential and depolarizes the cell after a hyperpolarisation. Its presence along the dendritic tree has been found to favor sublinear summation of spatially-distributed inputs and to regulate the temporal window of such summation (Magee, 2000). High-frequency stimulation that results in modest LTP has been found to decrease I_h in CA1 pyramidal neurons in an input-specific and spatially restricted manner (Campanac *et al.*, 2008; Wang *et al.*, 2003) (Figure 3.1B), thus promoting the linear summation of synaptic inputs (Wang *et al.*, 2003). This LTP-induced facilitation of linear summation differentially affects temporal summation along the dendritic tree, allowing only a narrow time window for integration at distal dendrites and a broader time window at proximal dendrites. According to these findings, LTP seems to enhance coincidence detection at distal dendrites and temporal integration at proximal dendrites (Xu *et al.*, 2006). It should be noted, however, that all of the aforementioned studies were performed under conditions of suppressed GABAergic inhibition. Given that activation of the perforant pathway leads to strong inhibitory input to the distal apical dendrites of CA1 pyramidal neurons that greatly abridges the contribution of excitatory postsynaptic potentials to the output of the neurons (Buzsaki *et al.*, 1995; Pissadaki *et al.*, 2010), the exact role of plasticity of I_h under physiological conditions remains to be elucidated.

Various ionic conductances in the dendrites, such as sodium (Stuart and Sakmann, 1994), calcium (Yuste *et al.*, 1994), NMDA (Larkum *et al.*, 2009; Schiller *et al.*, 2000), or a combination of the above (Ariav *et al.*, 2003; Wei *et al.*, 2001) have been shown to underlie back-propagated or forward-propagated dendritic spikes, which allow for supra-linear summation of synaptic inputs (Losonczy and Magee, 2006; Poirazi *et al.*, 2003a, b; Polsky *et al.*, 2009), coincidence detection (Larkum *et al.*, 1999), and LTP induction (Magee and Johnston, 1997; Remy and Spruston, 2007). Potassium currents, and in particular I_A, modulate the amplitude of dendritic spikes (Gasparini *et al.*, 2007; Hoffman *et al.*, 1997; Migliore *et al.*, 2005) in an activity-dependent way, thus influencing their propagation and facilitating dendritic compartmentalization (Frick *et al.*, 2004; Hoffman and Johnston, 1998) (Figure 3.1C). For example, associative pairing of dendritic and somatic depolarization in CA1 pyramidal neurons results in a prolonged I_A-dependent enhancement of dendrosomatic coupling of regenerative events that is restricted to the activated branch, a phenomenon called Branch-Strength potentiation (Losonczy *et al.*, 2008). In addition, interneurons of the stratum oriens/alveus border may also undergo activity-dependent potentiation of the L-type voltage-sensitive calcium channels (VSCCs) (Topolnik *et al.*, 2009). This potentiation is restricted to specific dendritic branches that also express mGluR5 receptors. This specificity in regulating the integration properties of individual dendrites provides an additional reservoir for memory storage within a single neuron, whereby different memories are represented not only by a set of synaptic weights but also by combinations of synaptic and dendritic strengths. In favour of this view, animals that are exposed to an enriched environment display enhanced compartmentalization in the propagation of dendritic spikes within individual branches, indicating that modulation of dendritic properties is an efficient memory-storage mechanism (Makara *et al.*, 2009).

An additional level of complexity stems from the short-term plasticity (up to seconds) of dendritic spikes: activity-dependent and branch-specific short-term attenuation of sodium currents in basal dendrites has been shown to modulate dendritic spike generation (Remy *et al.*, 2009). Specifically, prior activity in the range of ~1 Hz does not affect subsequent dendritic spike generation, whereas prior activity in the range of 5–10 Hz attenuates the fast component, and subsequently the amplitude of dendritic spikes only in the same dendritic branch. If the dendritic spike succeeds in generating a somatic AP, attenuation of dendritic spikes occurs globally throughout the dendritic tree due to the backpropagating spikes that invade the dendritic tree. This study indicates a short-term mechanism for switching between local and global regulation of excitability in response to different frequencies of the inputs. As the authors point out, long-term plasticity of intrinsic excitability that depends on dendritic spikes, as in the case of (Losonczy *et al.*, 2008), is effective only when the frequency of the input pattern is below 1 Hz, thus constraining the number of input patterns that can be stored by means of dendritic spikes. In accordance, the speed by which information is retrieved cannot surpass 1 Hz, (one pattern per second) since larger frequencies lead to the attenuation of dendritic spikes and, presumably, loss of branch-specific information (Branco and Hausser, 2009). Although lacking experimental verification, the aforementioned scenarios, are a good example of how short-term memory participates in long-term storage processes.

In light of these evidence, the old view that a neuron with an elaborate dendritic morphology acts as a linear summing device (point neuron-Hebb, 1949) has been replaced by a new theory according to which information processing and storage can take place independently at the ionic, the synaptic, the dendritic, and/or the somatic levels within a cell, thus rendering a neuron a multi-layer processing device (London and Hausser, 2005; Poirazi *et al.*, 2003b; Poirazi and Mel, 2001; Sidiropoulou *et al.*, 2006).

4.2. *Homeostatic plasticity*

While learning or activity-induced changes in neuronal excitability and/or synaptic function are necessary for information processing and memory storage, at the same time, neuronal and network activity should be constrained within certain limits so as not to become overly excited or completely silenced. Homeostatic regulation of excitatory and inhibitory synaptic currents is the primary mechanism used to normalize neuronal activity (Turrigiano and Nelson, 2004). Recent evidence, however, indicates that plasticity of intrinsic excitability may also contribute to stabilizing network properties while homeostatic regulation of intrinsic excitability helps exploit the optimal firing range of neurons in response to stimuli (Stemmler and Koch, 1999). This is achieved via the activation of homeostatic mechanisms which provide negative feedback in response to changes in neuronal excitability. For example, a long-lasting reduction in the overall neuronal activity and/or the lack of synaptic drive lead to either a change in the firing pattern of pyramidal neurons from tonic to bursting (Turrigiano *et al.*, 1994) or an increase in the firing frequency and a decrease in the AP threshold of both pyramidal neurons and interneurons. The latter is due to the upregulation of sodium currents and/or the downregulation of potassium currents and I_h (Desai *et al.*, 1999; Gibson *et al.*, 2006). In agreement, long-lasting depolarization of hippocampal neurons leads to L-type calcium channel-induced downregulation of intrinsic excitability through the enhancement of potassium channel permeability (O'Leary *et al.*, 2010).

In addition, short-term activity modulation has also been shown to rapidly affect intrinsic neuronal excitability. Brief periods of synaptic inhibition or hyperpolarisation in vestibular neurons induce firing rate potentiation due to a reduction of the BK-type calcium-activated potassium current which results from a decline in the intracellular calcium accumulation (Nelson *et al.*, 2003). Enhanced glutamatergic activity on the other hand, decreases the persistent sodium current (Carlier *et al.*, 2006) and upregulates I_h (van Welie *et al.*, 2004), leading to reduction of neuronal excitability. Thus, neurons respond to brief changes in synaptic activity by altering the level of intrinsic excitability in the opposite direction.

Furthermore, stimuli that induce input-specific synaptic modifications also result in homeostatic alterations of ion channels. In CA1 pyramidal neurons, LTP induction causes homeostatic upregulation of I_h and subsequent decrease of excitability (Fan *et al.*, 2005), while the opposite occurs with LTD induction

(Brager and Johnston, 2007). As noted previously, however, I_h is also involved in plasticity of dendritic integration following LTP induction (Campanac *et al.*, 2008). So, how does the neuron decide to modulate its intrinsic mechanisms in a reinforcing or homeostatic manner? The answer to this question lies in the strength of LTP. High-frequency synaptic stimulation induces moderate LTP and downregulates I_h, thus further enhancing synaptic potentiation (Figure 3.1B), while theta-burst pairing of backpropagating APs (bAPs) and synaptic stimulation of the Schaffer collaterals induces maximal LTP and turns down the gain by homeostatically regulating neuronal excitability (Figure 3.1A). Overall, by maintaining a stable activity level in neuronal networks, homeostatic plasticity could have a critical role in memory formation.

In summary, activity-dependent manipulations of the various axo-soma-dendritic conductances form a part of the memory engram. Via enhancing the output of the neuron (E-S potentiation), promoting network stability (homeostatic plasticity), or altering the output characteristics, information is stored in the differential membrane excitability and enables the neuron to perform the most appropriate computation for a given behavioural task. In the next section, we will investigate the role of plastic excitability in cases where it does not constitute part of a memory storage system, but is part of the learning process itself.

5. Interaction between intrinsic and synaptic plasticity

As mentioned above, learning-induced increases in neuronal excitability enhance subsequent learning capabilities, suggesting that learning-induced plasticity of excitability may have a gating or permissive role for the induction of synaptic plasticity in an experience-dependent manner. Accordingly, treatments that enhance neuronal excitability through downregulation of sAHP in brain slices also facilitate LTP induction (Cohen *et al.*, 1999). Thus, it is possible that plasticity of intrinsic neuronal excitability serves as a mechanism for inducing metaplasticity, a term used to describe the dynamic nature of the rules to which synaptic plasticity abides to (Abraham, 2008). Metaplasticity may in turn allow the "savings" or "rule-learning" phenomena observed. Since plasticity of ion channels underlies the plasticity of excitability, we will discuss several studies supporting the role of ion channel modulation in facilitating or inhibiting synaptic plasticity.

I_A, one of the intrinsic currents involved in shaping the AP and influencing dendritic spikes, has been shown to be modulated by neuronal activity (Frick *et al.*, 2004; Jung and Hoffman, 2009). Both the elimination of I_A expression in mice (by knocking-out the Kv4.2 subunit) and the downregulation of the

Kv4.2 subunit facilitate LTP induction (Chen *et al.*, 2006; Jung *et al.*, 2008). Studies using knockout mice for the Kvβ1.1 auxiliary subunit of A-type channels demonstrate how the AHP correlates with learning and memory. Removing the Kvβ1.1 subunit slows down the inactivation kinetics of I_A and reduces the frequency-dependent spike broadening. This reduction leads to decreased calcium influx during spikes and a subsequent decrease of the sAHP (Giese *et al.*, 2001). Interestingly, the effect of decreased sAHP in learning performance is age-dependent. In young mice, this decrease is associated with impaired flexibility of learning as observed in water maze and social transmission of food preference (STFP) tasks (Giese *et al.*, 1998). On the other hand, wild-type aged mice display deficits in learning and memory which are associated with an increased density of voltage-gated calcium channels and therefore, increased sAHP (Thibault *et al.*, 2001). In this case, removing the Kvβ1.1 auxiliary subunit ameliorates age-related learning deficiencies by readjusting the sAHP to its physiological level (Murphy *et al.*, 2004). In agreement, the impaired performance in the STFP task observed in aged mice is rescued in Kvβ1.1 mutants and this rescue is hippocampus dependent (Need *et al.*, 2003). Since a decreased I_A has a permissive role for LTP induction, and its activation helps maintain sAHP in optimal levels for learning and memory, we conclude that I_A modulation affects learning through synaptic plasticity in an age-dependent manner.

Interactions between ion channel and synaptic plasticity have also been documented within single spines. Dendritic spines bear voltage-gated channels (Nimchinsky *et al.*, 2002; Segev and Rall, 1998) that can boost or dampen incoming synaptic inputs (Bloodgood and Sabatini, 2007). Induction of LTP has been shown to rely on activity-dependent and spine-specific internalization of the potassium subunit Kv4.2 (Kim *et al.*, 2007) or the small conductance, calcium-activated potassium channels (Lin *et al.*, 2008). In agreement, brief trains of high-frequency APs result in a local, long-lasting (tens of minutes) reduction in the opening of the R-type calcium currents (Yasuda *et al.*, 2003) and decrease the efficacy of LTP induction in these spines. Finally, calcium pumps or sodium/calcium exchangers in dendritic spines also display activity-dependent downregulation (Scheuss *et al.*, 2006). Taken together, these findings indicate that the ionic properties of single spines can be regulated independently, thus enabling a differential role in the induction of synaptic plasticity.

One of the key molecules that seem to be implicated both in synaptic plasticity and excitability is the CREB protein. CREB is a transcription factor that has a long-standing role in synaptic plasticity and memory but also a recently discovered role in regulating neuronal excitability (Benito and Barco, 2010). Neuronal excitability is upregulated by CREB (Dong *et al.*, 2006; Zhou *et al.*, 2009), while neurons that artificially express more CREB are selected to code for a specific fear memory (Han *et al.*, 2007). In agreement, neurons of the lateral amygdala with increased CREB and increased excitability also exhibit enhanced E-S coupling (Zhou *et al.*, 2009) and hippocampal neurons overexpressing CREB display reduced medium and slow AHP (Lopez de Armentia *et al.*, 2007). Thus, CREB could be an intracellular link between increased excitability and induction of synaptic plasticity. In addition, both intrinsic excitability and CREB seem to have a role in determining whether a neuron will undergo plasticity and whether it will be selected for encoding a memory. Their specific contribution to each of these functions remains to be elucidated.

6. Contribution of plasticity of excitability in short-term memory

Over the last couple of decades, changes in intrinsic excitability have been shown to play a key role in learning and memory processes (Zhang and Linden, 2003). Here, we chose to focus our attention on the role of intrinsic excitability in short-term memory formation. In the following sections, we will discuss the contribution of transient or prolonged plasticity of excitability in short-term synaptic plasticity and working memory.

6.1. *Interplay of plasticity of excitability and short-term synaptic plasticity*

In short-term synaptic plasticity processes, such as short-term facilitation, plasticity of excitability is often observed in the presynaptic neuron where the shape of the presynaptic spike is affected. Modulation of AP shape in a frequency-dependent manner was first reported in invertebrates (Aldrich *et al.*, 1979) and has thereafter also been observed in vertebrates. For example, high-frequency stimulation of the mossy fibers in hippocampal slices induced a threefold broadening of the presynaptic spike. This spike broadening enhanced the depolarization and subsequent calcium influx to a given action potential at the mossy fiber boutons (Geiger and Jonas, 2000). Activity-induced broadening of presynaptic APs has been attributed to the inactivation of potassium channels, such as the delayed-rectifier potassium currents (Geiger and Jonas, 2000; Nick and Ribera, 2000), the BK channels (Faber and Sah, 2003; Shao *et al.*, 1999), the A-type potassium channels (Sonner *et al.*, 2008) or the *h* current (Beaumont and Zucker, 2000). Spike broadening and the successive rise of the intracellular calcium concentration at the presynaptic nerve terminals enhances the duration of neurotransmitter release (Augustine, 1990; Beaumont and Zucker, 2000; Geiger and Jonas, 2000), inducing short-term facilitation of the synaptic strength. Therefore, transient increases of excitability at the presynaptic terminal enhance synaptic plasticity and its contribution to short-term memory processes.

6.2. *Plasticity of excitability and working memory*

Working memory is a special type of short-term memory that involves the "online" monitoring and manipulation of information or actions in real time. A classical experimental procedure used to evaluate working memory is the delayed match to sample task, during which a transient cue is presented and after a delay period (during which the individual holds the cue features in mind), the individual is asked to execute a specific action (for example, locate the cue) (Goldman-Rakic, 1992). *In vivo* experiments conducted in the 1970s showed that neurons in the prefrontal cortex (PFC) maintain their activity for several seconds, outlasting the presence of the transient environmental stimulus (Fuster and Alexander, 1971; Kubota and Niki, 1971), thus providing the necessary neuronal processing to establish the behavioural continuity over time. This maintenance of excitability is basically a short-term enhancement of excitability that allows neurons to fire in the absence of incoming external stimuli.

The cellular substrate that supports sustained firing entails both synaptic and intrinsic mechanisms (Wang, 2001). Recurrent synaptic activity mediated by NMDA receptors (Shu *et al.*, 2003; Wang, 1999) has been shown to be important for proper generation and maintenance of sustained firing. Specifically, NMDA-dependent plateau depolarizations that initiate at the basal dendrites and propagate to the soma (Milojkovic *et al.*, 2005) could be involved in sustained firing. Furthermore, local ionic conductances, such as the SK channels, could regulate these NMDA-dependent regenerative events (Faber, 2010) and modulate sustained firing.

The intrinsic mechanism underlying sustained firing involves the generation of a plateau potential, namely the delayed after-depolarization (dADP) (Figure 3.1E). The dADP supports regenerative events in single pyramidal neurons in the absence of AMPA and NMDA-mediated synaptic activity (Egorov *et al.*, 2002; Sidiropoulou *et al.*, 2009), but following pharmacological or synaptic activation of G_q-coupled receptors, such as mGluRs, mAchR and 5-HT receptors (Constanti and Libri, 1992; Greene *et al.*, 1994; Haj-Dahmane and Andrade, 1998). The dADP is activated following an increase of intracellular calcium, through IP_3-dependent calcium release from the endoplasmic reticulum as well as activation of voltage-gated calcium channels (Fowler *et al.*, 2007; Hagenston *et al.*, 2008). This calcium influx then activates a calcium-activated nonselective cation current (I_{CAN}) mediated by the transient receptor potential channels (TRPCs) (Fowler *et al.*, 2007).

The dADP provides a short temporal window (2–3 s) during which incoming inputs may reach threshold easier due to the underlying depolarizing envelope, converting subthreshold inputs to suprathreshold responses (Sidiropoulou *et al.*, 2009). Thus, the dADP temporally increases neuronal excitability due to the depolarized membrane potential. Furthermore, it provides a biophysical substrate for sustained firing, allowing the neuron to increase its firing rate and sustain its activity in the presence of the dADP (Figure 3.1E) (Sidiropoulou *et al.*, 2009). In addition, sequential dADP-inducing stimuli result in an increased firing rate during the sustained activity with each successive stimulus, a phenomenon known as graded persistent activity, and provide a mechanism by which a neuron can remember its previous activation history (Egorov *et al.*, 2002). Graded changes in the firing rate of tonic activity have also been shown to occur in response to successive hyperpolarising inputs due to a calcium-dependent modulation of the I_h current (Winograd *et al.*, 2008).

To sum up, both intrinsic somatic mechanisms and dendritic ionic currents shape the activity-dependent sustained spiking of the neurons that support working memory processes. In working memory, the excitability modulation of prefrontal neurons through the regulation of ion channels seems to provide the fundamental neuronal substrate for the online processing of information associated with the behavioural output.

7. Conclusion

> *"If an extraterrestrial neuroscientist managed to obtain a badge and abstract book and attend the Society for Neuroscience annual meeting, she could be forgiven for concluding that humans believe that memory storage is solely accomplished through LTP/LTD of fast (ionotropic) neurotransmission at excitatory, glutamatergic synapses."*
>
> *Text quoted by Kim and Linden (2007).*

In continuation to Kim and Linden's quote, our knowledge of how our brain adapts to a given environment is mainly determined by the experiments we choose to conduct. Over the last 50 years, intensive investigations of learning and memory processes focused almost exclusively on synaptic modifications. As a result, the most widely accepted notion remains the one poised by Hebb: that memory is synonymous to modifications in synaptic strength between interconnected neurons.

The now well-established fact that neurons are not just a "node" inside a neuronal network that sum synaptic input, but consist of submodules with quasi-independent integrative properties (dendrites, parts of dendrites, spines) has extended the Hebbian conjecture to include these submodules as an additional memory storage reservoir. Under this view, neuronal excitability no longer refers to the cell's ability to generate a spike, but also the transformations that the inputs leading to this spike undergo during their propagation to the cell body. In accordance, memory formation and storage does not only involve the synaptic component of neurons but also the mechanisms and phenomena whose properties change in response to learning protocols. Among these, neuronal excitability is a very successful candidate for the transient or permanent storage of the history of the neuron.

Long-lasting changes in intrinsic neuronal excitability can result in enhancement (or suppression) of the neuronal throughput (plasticity of E-S coupling), homeostatic plasticity or plasticity of other firing characteristics, such as bursting, time precision and coincidence detection. In these cases, modulation of excitability may serve as a memory coding mechanism. In cases where plasticity of intrinsic excitability is transient, it may contribute to the induction of other forms of plasticity, such as metaplasticity, where priming activity regulates both the efficiency and direction of synaptic plasticity and possibly serves as the neuronal substrate for cognitive phenomena, such as "rule-learning" and "savings".

In light of the biophysical complexity and augmented information processing capabilities of individual neurons, the field of learning and memory storage is expanding beyond the traditional synaptic plasticity theories in search of additional memory substrates. Modulation of intrinsic neuronal excitability, like synaptic plasticity, is a learning-induced phenomenon associated with both short- and long-term memory processes and offers an alternative and possibly complementary means for information storage.

Acknowledgments

The authors would like to thank members of their lab (in particular George Kastellakis) for their help and input. This work was supported by the Marie Curie Outgoing Fellowship of the European Commission [PIOF-GA-2008–219622].

References

Abraham, W.C. (2008). Metaplasticity: Tuning synapses and networks for plasticity. *Nat Rev Neurosci* **9**, 387.

Aizenman, C.D., Akerman, C.J., Jensen, K.R., and Cline, H.T. (2003). Visually driven regulation of intrinsic neuronal excitability improves stimulus detection *in vivo*. *Neuron* **39**, 831–842.

Aizenman C.D. and Linden, D.J. (2000). Rapid, synaptically driven increases in the intrinsic excitability of cerebellar deep nuclear neurons. *Nat Neurosci* **3**, 109–111.

Aldrich, R.W., Jr., Getting, P.A. and Thompson, S.H. (1979). Mechanism of frequency-dependent broadening of molluscan neurone soma spikes. *J Physiol* **291**, 531–544.

Alkon, D.L. (1984). Calcium-mediated reduction of ionic currents: A biophysical memory trace. *Science* **226**, 1037–1045.

Ariav, G., Polsky, A. and Schiller, J. (2003). Submillisecond precision of the input-output transformation function mediated by fast sodium dendritic spikes in basal dendrites of CA1 pyramidal neurons. *J Neurosci* **23**, 7750–7758.

Augustine, G.J. (1990). Regulation of transmitter release at the squid giant synapse by presynaptic delayed rectifier potassium current. *J Physiol* **431**, 343–364.

Beaumont, V. and Zucker, R.S. (2000). Enhancement of synaptic transmission by cyclic AMP modulation of presynaptic Ih channels. *Nat Neurosci* **3**, 133–141.

Benito, E. and Barco, A. (2010). CREB's control of intrinsic and synaptic plasticity: Implications for CREB-dependent memory models. *Trends Neurosci* **33**, 230–240.

Bernard, C. and Wheal, H.V. (1995). Simultaneous expression of excitatory postsynaptic potential/spike potentiation and excitatory postsynaptic potential/spike depression in the hippocampus. *Neuroscience* **67**, 73–82.

Bliss, T.V. and Lomo, T. (1973). Long-lasting potentiation of synaptic transmission in the dentate area of the anaesthetized rabbit following stimulation of the perforant path. *J Physiol* **232**, 331–356.

Bloodgood, B.L. and Sabatini, B.L. (2007). Nonlinear regulation of unitary synaptic signals by CaV(2.3) voltage-sensitive calcium channels located in dendritic spines. *Neuron* **53**, 249–260.

Brager, D.H. and Johnston, D. (2007). Plasticity of intrinsic excitability during long-term depression is mediated through mGluR-dependent changes in I(h) in hippocampal CA1 pyramidal neurons. *J Neurosci* **27**, 13926–13937.

Branco, T. and Hausser, M. (2009). The selfish spike: Local and global resets of dendritic excitability. *Neuron* **61**, 815–817.

Brons, J.F. and Woody, C.D. (1980). Long-term changes in excitability of cortical neurons after Pavlovian conditioning and extinction. *J Neurophysiol* **44**, 605–615.

Brosh, I., Rosenblum, K. and Barkai, E. (2007). Learning-induced modulation of SK channels-mediated effect on synaptic transmission. *Eur J Neurosci* **26**, 3253–3260.

Brown, J.T. and Randall, A.D. (2009). Activity-dependent depression of the spike after-depolarization generates long-lasting intrinsic plasticity in hippocampal CA3 pyramidal neurons. *J Physiol* **587**, 1265–1281.

Buzsaki, G., Penttonen, M., Bragin, A., Nadasdy, Z. and Chrobak, J.J. (1995). Possible physiological role of the perforant path-CA1 projection. *Hippocampus* **5**, 141–146.

Campanac, E., Daoudal, G., Ankri, N. and Debanne, D. (2008). Downregulation of dendritic I(h) in CA1 pyramidal neurons after LTP. *J Neurosci* **28**, 8635–8643.

Campanac, E. and Debanne, D. (2007). Plasticity of neuronal excitability: Hebbian rules beyond the synapse. *Arch Ital Biol* **145**, 277–287.

Campanac, E. and Debanne, D. (2008). Spike timing-dependent plasticity: A learning rule for dendritic integration in rat CA1 pyramidal neurons. *J Physiol* **586**, 779–793.

Canu, M.H., Picquet, F., Bastide, B. and Falempin, M. (2010). Activity-dependent changes in the electrophysiological properties of regular spiking neurons in the sensorimotor cortex of the rat *in vitro*. *Behav Brain Res* **209**, 289–294.

Carlier, E., Sourdet, V., Boudkkazi, S., Deglise, P., Ankri, N., Fronzaroli-Molinieres, L. and Debanne, D. (2006). Metabotropic glutamate receptor subtype 1 regulates sodium currents in rat neocortical pyramidal neurons. *J Physiol* **577**, 141–154.

Chavez-Noriega, L.E., Halliwell, J.V. and Bliss, T.V. (1990). A decrease in firing threshold observed after induction of the EPSP-spike (E-S) component of long-term potentiation in rat hippocampal slices. *Exp Brain Res* **79**, 633–641.

Chen, K., Aradi, I., Santhakumar, V. and Soltesz, I. (2002). H-channels in epilepsy: New targets for seizure control? *Trends Pharmacol Sci* **23**, 552–557.

Chen, X., Yuan, L.L., Zhao, C., Birnbaum, S.G., Frick, A., Jung, W.E., Schwarz, T.L., Sweatt, J.D. and Johnston, D. (2006). Deletion of Kv4.2 gene eliminates dendritic A-type K+ current and enhances induction of long-term potentiation in hippocampal CA1 pyramidal neurons. *J Neurosci* **26**, 12143–12151.

Cohen-Matsliah, S.I., Motanis, H., Rosenblum, K. and Barkai, E. (2010). A novel role for protein synthesis in long-term neuronal plasticity: Maintaining reduced postburst afterhyperpolarization. *J Neurosci* **30**, 4338–4342.

Cohen, A.S., Coussens, C.M., Raymond, C.R. and Abraham, W.C. (1999). Long-lasting increase in cellular excitability associated with the priming of LTP induction in rat hippocampus. *J Neurophysiol* **82**, 3139–3148.

Constanti, A. and Libri, V. (1992). Trans-ACPD induces a slow post-stimulus inward tail current (IADP) in guinea-pig olfactory cortex neurones in vitro. *Eur J Pharmacol* **216**, 463–464.

Cooper, D.C. (2002). The significance of action potential bursting in the brain reward circuit. *Neurochem Int* **41**, 333–340.

Cudmore, R.H. and Turrigiano, G.G. (2004). Long-term potentiation of intrinsic excitability in LV visual cortical neurons. *J Neurophysiol* **92**, 341–348.

Dan, Y. and Poo, M.M. (2006). Spike timing-dependent plasticity: Fom synapse to perception. *Physiol Rev* **86**, 1033–1048.

Daoudal, G., Hanada, Y. and Debanne, D. (2002). Bidirectional plasticity of excitatory postsynaptic potential (EPSP)-spike coupling in CA1 hippocampal pyramidal neurons. *Proc Natl Acad Sci USA* **99**, 14512–14517.

Debanne, D. (2004). Information processing in the axon. *Nat Rev Neurosci* **5**, 304–316.

Desai, N.S., Rutherford, L.C. and Turrigiano, G.G. (1999). Plasticity in the intrinsic excitability of cortical pyramidal neurons. *Nat Neurosci* **2**, 515–520.

Disterhoft, J.F. and Oh, M.M. (2006). Learning, aging and intrinsic neuronal plasticity. *Trends Neurosci* **29**, 587–599.

Dong, Y., Green, T., Saal, D., Marie, H., Neve, R., Nestler, E.J. and Malenka, R.C. (2006). CREB modulates excitability of nucleus accumbens neurons. *Nat Neurosci* **9**, 475–477.

Egorov, A.V., Hamam, B.N., Fransen, E., Hasselmo, M.E. and Alonso, A.A. (2002). Graded persistent activity in entorhinal cortex neurons. *Nature* **420**, 173–178.

Faber, E.S. (2010). Functional interplay between NMDA receptors, SK channels and voltage-gated Ca2+ channels regulates synaptic excitability in the medial prefrontal cortex. *J Physiol* **588**, 1281–1292.

Faber, E.S. and Sah, P. (2003). Ca2+-activated K+ (BK) channel inactivation contributes to spike broadening during repetitive firing in the rat lateral amygdala. *J Physiol* **552**, 483–497.

Fan, Y., Fricker, D., Brager, D.H., Chen, X., Lu, H.C., Chitwood, R.A. and Johnston, D. (2005). Activity-dependent decrease of excitability in rat hippocampal neurons through increases in I(h). *Nat Neurosci* **8**, 1542–1551.

Feldman, D.E. (2009). Synaptic mechanisms for plasticity in neocortex. *Annu Rev Neurosci* **32**, 33–55.

Fowler, M.A., Sidiropoulou, K., Ozkan, E.D., Phillips, C.W. and Cooper, D.C. (2007). Corticolimbic expression of TRPC4 and TRPC5 channels in the rodent brain. *PLoS One* **2**, e573.

Frey, U. and Morris, R.G. (1997). Synaptic tagging and long-term potentiation. *Nature* **385**, 533–536.

Frick, A. and Johnston, D. (2005). Plasticity of dendritic excitability. *J Neurobiol* **64**, 100–115.

Frick, A., Magee, J. and Johnston, D. (2004). LTP is accompanied by an enhanced local excitability of pyramidal neuron dendrites. *Nat Neurosci* **7**, 126–135.

Fuster, J.M. and Alexander, G.E. (1971). Neuron activity related to short-term memory. *Science* **173**, 652–654.

Ganguly, K., Kiss, L. and Poo, M. (2000). Enhancement of presynaptic neuronal excitability by correlated presynaptic and postsynaptic spiking. *Nat Neurosci* **3**, 1018–1026.

Gasparini, S., Losonczy, A., Chen, X., Johnston, D. and Magee, J.C. (2007). Associative pairing enhances action potential back-propagation in radial oblique branches of CA1 pyramidal neurons. *J Physiol* **580**, 787–800.

Geiger, J.R. and Jonas, P. (2000). Dynamic control of presynaptic Ca(2+) inflow by fast-inactivating K(+) channels in hippocampal mossy fiber boutons. *Neuron* **28**, 927–939.

Gibson, J.R., Bartley, A.F. and Huber, K.M. (2006). Role for the subthreshold currents ILeak and IH in the homeostatic control of excitability in neocortical somatostatin-positive inhibitory neurons. *J Neurophysiol* **96**, 420–432.

Giese, K.P., Peters, M. and Vernon, J. (2001). Modulation of excitability as a learning and memory mechanism: A molecular genetic perspective. *Physiol Behav* **73**, 803–810.

Giese, K.P., Storm, J.F., Reuter, D., Fedorov, N.B., Shao, L.R., Leicher, T., Pongs, O. and Silva, A.J. (1998). Reduced K+ channel inactivation, spike broadening and after-hyperpolarization in Kvbeta1.1-deficient mice with impaired learning. *Learn Mem* **5**, 257–273.

Goldman-Rakic, P.S. (1992). Working memory and the mind. *Sci Am* **267**, 110–117.

Greene, C.C., Schwindt, P.C. and Crill, W.E. (1994). Properties and ionic mechanisms of a metabotropic glutamate receptor-mediated slow after depolarization in neocortical neurons. *J Neurophysiol* **72**, 693–704.

Hagenston, A.M., Fitzpatrick, J.S. and Yeckel, M.F. (2008). MGluR-mediated calcium waves that invade the soma regulate firing in layer V medial prefrontal cortical pyramidal neurons. *Cereb Cortex* **18**, 407–423.

Haj-Dahmane, S. and Andrade, R. (1998). Ionic mechanism of the slow after depolarization induced by muscarinic receptor activation in rat prefrontal cortex. *J Neurophysiol* **80**, 1197–1210.

Han, J.H., Kushner, S.A., Yiu, A.P., Cole, C.J., Matynia, A., Brown, R.A., Neve, R.L., Guzowski, J.F., Silva, A.J. and Josselyn, S.A. (2007). Neuronal competition and selection during memory formation. *Science* **316**, 457–460.

Hickmott, P.W. (2005). Changes in intrinsic properties of pyramidal neurons in adult rat S1 during cortical reorganization. *J Neurophysiol 94*, 501–511.

Hoffman, D.A. and Johnston, D. (1998). Downregulation of transient K+ channels in dendrites of hippocampal CA1 pyramidal neurons by activation of PKA and PKC. *J Neurosci* **18**, 3521–3528.

Hoffman, D.A., Magee, J.C., Colbert, C.M. and Johnston, D. (1997). K^+ channel regulation of signal propagation in dendrites of hippocampal pyramidal neurons. *Nature* **387**, 869–875.

Jung, S.C. and Hoffman, D.A. (2009). Biphasic somatic A-type K channel downregulation mediates intrinsic plasticity in hippocampal CA1 pyramidal neurons. *PLoS One* **4**, e6549.

Jung, S.C., Kim, J. and Hoffman, D.A. (2008). Rapid, bidirectional remodeling of synaptic NMDA receptor subunit composition by A-type K^+ channel activity in hippocampal CA1 pyramidal neurons. *Neuron* **60**, 657–671.

Kandel, E.R. (2001). The molecular biology of memory storage: A dialogue between genes and synapses. *Science* **294**, 1030–1038.

Kelleher, R.J., 3rd, Govindarajan, A. and Tonegawa, S. (2004). Translational regulatory mechanisms in persistent forms of synaptic plasticity. *Neuron* **44**, 59–73.

Kim, J., Jung, S.C., Clemens, A.M., Petralia, R.S. and Hoffman, D.A. (2007). Regulation of dendritic excitability by activity-dependent trafficking of the A-type K+ channel subunit Kv4.2 in hippocampal neurons. *Neuron* **54**, 933–947.

Kim, S.J. and Linden, D.J. (2007). Ubiquitous plasticity and memory storage. *Neuron* **56**, 582–592.

Konig, P., Engel, A.K. and Singer, W. (1996). Integrator or coincidence detector? The role of the cortical neuron revisited. *Trends Neurosci* **19**, 130–137.

Kubota, K. and Niki, H. (1971). Prefrontal cortical unit activity and delayed alternation performance in monkeys. *J Neurophysiol* **34**, 337–347.

Larkum, M.E., Nevian, T., Sandler, M., Polsky, A. and Schiller, J. (2009). Synaptic integration in tuft dendrites of layer 5 pyramidal neurons: A new unifying principle. *Science* **325**, 756–760.

Larkum, M.E., Zhu, J.J. and Sakmann, B. (1999). A new cellular mechanism for coupling inputs arriving at different cortical layers. *Nature* **398**, 338–341.

Li, C.Y., Lu, J.T., Wu, C.P., Duan, S.M. and Poo, M.M. (2004). Bidirectional modification of presynaptic neuronal excitability accompanying spike timing-dependent synaptic plasticity. *Neuron* **41**, 257–268.

Lin, M.T., Lujan, R., Watanabe, M., Adelman, J.P. and Maylie, J. (2008). SK2 channel plasticity contributes to LTP at Schaffer collateral-CA1 synapses. *Nat Neurosci* **11**, 170–177.

London, M. and Hausser, M. (2005). Dendritic computation. *Annu Rev Neurosci* **28**, 503–532.

Lopez de Armentia, M., Jancic, D., Olivares, R., Alarcon, J.M., Kandel, E.R. and Barco, A. (2007). cAMP response element-binding protein-mediated gene expression increases the intrinsic excitability of CA1 pyramidal neurons. *J Neurosci* **27**, 13909–13918.

Losonczy, A. and Magee, J.C. (2006). Integrative properties of radial oblique dendrites in hippocampal CA1 pyramidal neurons. *Neuron* **50**, 291–307.

Losonczy, A., Makara, J.K. and Magee, J.C. (2008). Compartmentalized dendritic plasticity and input feature storage in neurons. *Nature* **452**, 436–441.

Lu, Y.M., Mansuy, I.M., Kandel, E.R. and Roder, J. (2000). Calcineurin-mediated LTD of GABAergic inhibition underlies the increased excitability of CA1 neurons associated with LTP. *Neuron* **26**, 197–205.

Magee, J.C. (2000). Dendritic integration of excitatory synaptic input. *Nat Rev* **1**, 181–190.

Magee, J.C. and Johnston, D. (1997). A synaptically controlled, associative signal for Hebbian plasticity in hippocampal neurons. *Science* **275**, 209–213.

Magee, J.C. and Johnston, D. (2005). Plasticity of dendritic function. *Current Opin Neurobiol* **15**, 334–342.

Major, G. and Tank, D. (2004). Persistent neural activity: Prevalence and mechanisms. *Current Opin Neurobiol* **14**, 675–684.

Makara, J.K., Losonczy, A., Wen, Q. and Magee, J.C. (2009). Experience-dependent compartmentalized dendritic plasticity in rat hippocampal CA1 pyramidal neurons. *Nat Neurosci* **12**, 1485–1487.

Markaki, M., Orphanoudakis, S. and Poirazi, P. (2005). Modelling reduced excitability in aged CA1 neurons as a calcium-dependent process. *Neurocomputing* **65–66**, 305–314.

McCulloch, W.S. and Pitts, W. (1943). A logical calculus of the ideas immanent in nervous activity. *Bull Math Biophys* **5**, 115:133.

McKay, B.M., Matthews, E.A., Oliveira, F.A. and Disterhoft, J.F. (2009). Intrinsic neuronal excitability is reversibly altered by a single experience in fear conditioning. *J Neurophysiol* **102**, 2763–2770.

Migliore, M., Ferrante, M. and Ascoli, G.A. (2005). Signal propagation in oblique dendrites of CA1 pyramidal cells. *J Neurophysiol* **94**, 4145–4155.

Milojkovic, B.A., Radojicic, M.S. and Antic, S.D. (2005). A strict correlation between dendritic and somatic plateau depolarizations in the rat prefrontal cortex pyramidal neurons. *J Neurosci* **25**, 3940–3951.

Misonou, H., Mohapatra, D.P., Park, E.W., Leung, V., Zhen, D., Misonou, K., Anderson, A.E. and Trimmer, J.S. (2004). Regulation of ion channel localization and phosphorylation by neuronal activity. *Nat Neurosci* **7**, 711–718.

Moore, S.J., Cooper, D.C. and Spruston, N. (2009). Plasticity of burst firing induced by synergistic activation of metabotropic glutamate and acetylcholine receptors. *Neuron* **61**, 287–300.

Moyer, J.R., Jr., Thompson, L.T. and Disterhoft, J.F. (1996). Trace eyeblink conditioning increases CA1 excitability in a transient and learning-specific manner. *J Neurosci* **16**, 5536–5546.

Murphy, G.G., Fedorov, N.B., Giese, K.P., Ohno, M., Friedman, E., Chen, R. and Silva, A.J. (2004). Increased neuronal excitability, synaptic plasticity, and learning in aged Kvbeta1.1 knockout mice. *Curr Biol* **14**, 1907–1915.

Need, A.C., Irvine, E.E. and Giese, K.P. (2003). Learning and memory impairments in Kv beta 1.1-null mutants are rescued by environmental enrichment or ageing. *Eur J Neurosci* **18**, 1640–1644.

Nelson, A.B., Krispel, C.M., Sekirnjak, C. and du Lac, S. (2003). Long-lasting increases in intrinsic excitability triggered by inhibition. *Neuron* **40**, 609–620.

Nick, T.A. and Ribera, A.B. (2000). Synaptic activity modulates presynaptic excitability. *Nat Neurosci* **3**, 142–149.

Nimchinsky, E.A., Sabatini, B.L. and Svoboda, K. (2002). Structure and function of dendritic spines. *Annu Rev Physiol* **64**, 313–353.

Noguchi, K., Saito, H. and Abe, K. (1998). Medial amygdala stimulation produces a long-lasting excitatory postsynaptic potential/spike dissociation in the dentate gyrus in vivo. *Brain Res* **794**, 151–154.

O'Leary, T., van Rossum, M.C. and Wyllie, D.J. (2010). Homeostasis of intrinsic excitability in hippocampal neurones: Dynamics and mechanism of the response to chronic depolarization. *J Physiol* **588**, 157–170.

Papoutsi, A., Sidiropoulou, K. and Poirazi, P. (2009). Mechanisms underlying persistent activity in a model PFC microcircuit. *BMC Neuroscience* **10**, p. 42.

Paz, J.T., Mahon, S., Tiret, P., Genet, S., Delord, B. and Charpier, S. (2009). Multiple forms of activity-dependent intrinsic plasticity in layer V cortical neurones *in vivo*. *J Physiol* **587**, 3189–3205.

Pissadaki, E., Sidiropoulou, K., Reczko, M. and Poirazi, P. (2010). Encoding of spatiotemporal input characteristics by a CA1 pyramidal neuron model. *PloS Comput Biol* **6**, e1001038.

Poirazi, P., Brannon, T. and Mel, B.W. (2003a). Arithmetic of subthreshold synaptic summation in a model CA1 pyramidal cell. *Neuron* **37**, 977–987.

Poirazi, P., Brannon, T. and Mel, B.W. (2003b). Pyramidal neuron as two-layer neural network. *Neuron* **37**, 989–999.

Poirazi, P. and Mel, B.W. (2001). Impact of active dendrites and structural plasticity on the memory capacity of neural tissue. *Neuron* **29**, 779–796.

Polsky, A., Mel, B. and Schiller, J. (2009). Encoding and decoding bursts by NMDA spikes in basal dendrites of layer 5 pyramidal neurons. *J Neurosci* **29**, 11891–11903.

Polsky, A., Mel, B.W. and Schiller, J. (2004). Computational subunits in thin dendrites of pyramidal cells. *Nat Neurosci* **7**, 621–627.

Pugliese, A.M., Ballerini, L., Passani, M.B. and Corradetti, R. (1994). EPSP-spike potentiation during primed burst-induced long-term potentiation in the CA1 region of rat hippocampal slices. *Neuroscience* **62**, 1021–1032.

Quirk, G.J., Garcia, R. and Gonzalez-Lima, F. (2006). Prefrontal mechanisms in extinction of conditioned fear. *Biol Psychiatry* **60**, 337–343.

Raab-Graham, K.F., Haddick, P.C., Jan, Y.N. and Jan, L.Y. (2006). Activity- and mTOR-dependent suppression of Kv1.1 channel mRNA translation in dendrites. *Science* **314**, 144–148.

Rall, W., Burke, R.E., Smith, T.G., Nelson, P.G. and Frank, K. (1967). Dendritic location of synapses and possible mechanisms for the monosynaptic EPSP in motoneurons. *J Neurophysiol* **30**, 1169–1193.

Remy, S., Csicsvari, J. and Beck, H. (2009). Activity-dependent control of neuronal output by local and global dendritic spike attenuation. *Neuron* **61**, 906–916.

Remy, S. and Spruston, N. (2007). Dendritic spikes induce single-burst long-term potentiation. *Proc Natl Acad Sci USA* **104**, 17192–17197.

Saar, D. and Barkai, E. (2003). Long-term modifications in intrinsic neuronal properties and rule learning in rats. *Eur J Neurosci* **17**, 2727–2734.

Saar, D., Grossman, Y. and Barkai, E. (1998). Reduced after-hyperpolarization in rat piriform cortex pyramidal neurons is associated with increased learning capability during operant conditioning. *Eur J Neurosci* **10**, 1518–1523.

Santini, E., Quirk, G.J. and Porter, J.T. (2008). Fear conditioning and extinction differentially modify the intrinsic excitability of infralimbic neurons. *J Neurosci* **28**, 4028–4036.

Scheuss, V., Yasuda, R., Sobczyk, A. and Svoboda, K. (2006). Nonlinear [Ca2+] signaling in dendrites and spines caused by activity-dependent depression of Ca^{2+} extrusion. *J Neurosci* **26**, 8183–8194.

Schiller, J., Major, G., Koester, H.J. and Schiller, Y. (2000). NMDA spikes in basal dendrites of cortical pyramidal neurons. *Nature* **404**, 285–289.

Schreurs, B.G., Gusev, P.A., Tomsic, D., Alkon, D.L. and Shi, T. (1998). Intracellular correlates of acquisition and long-term memory of classical conditioning in Purkinje cell dendrites in slices of rabbit cerebellar lobule HVI. *J Neurosci* **18**, 5498–5507.

Schreurs, B.G., Tomsic, D., Gusev, P.A. and Alkon, D.L. (1997). Dendritic excitability microzones and occluded long-term depression after classical conditioning of the rabbit's nictitating membrane response. *J Neurophysiol* **77**, 86–92.

Segev, I. and Rall, W. (1998). Excitable dendrites and spines: Earlier theoretical insights elucidate recent direct observations. *Trends Neurosci* **21**, 453–460.

Shah, M.M., Hammond, R.S. and Hoffman, D.A. (2010). Dendritic ion channel trafficking and plasticity. *Trends Neurosci* **33**, 307–316.

Shao, L.R., Halvorsrud, R., Borg-Graham, L. and Storm, J.F. (1999). The role of BK-type Ca2+-dependent K+ channels in spike broadening during repetitive firing in rat hippocampal pyramidal cells. *J Physiol* **521** Pt 1, 135–146.

Shu, Y., Hasenstaub, A. and McCormick, D.A. (2003). Turning on and off recurrent balanced cortical activity. *Nature* **423**, 288–293.

Sidiropoulou, K., Lu, F.M., Fowler, M.A., Xiao, R., Phillips, C., Ozkan, E.D., Zhu, M.X., White, F.J. and Cooper, D.C. (2009). Dopamine modulates an mGluR5-mediated depolarization underlying prefrontal persistent activity. *Nat Neurosci* **12**, 190–199.

Sidiropoulou, K., Pissadaki, E.K. and Poirazi, P. (2006). Inside the brain of a neuron. *EMBO Rep* **7**, 886–892.

Siegelbaum, S.A. and Kandel, E.R. (1991). Learning-related synaptic plasticity: LTP and LTD. *Curr Opin Neurobiol* **1**, 113–120.

Sonner, P.M., Filosa, J.A. and Stern, J.E. (2008). Diminished A-type potassium current and altered firing properties in presympathetic PVN neurones in renovascular hypertensive rats. *J Physiol* **586**, 1605–1622.

Sourdet, V., Russier, M., Daoudal, G., Ankri, N. and Debanne, D. (2003). Long-term enhancement of neuronal excitability and temporal fidelity mediated by metabotropic glutamate receptor subtype 5. *J Neurosci* **23**, 10238–10248.

Spruston, N. (2008). Pyramidal neurons: Dendritic structure and synaptic integration. *Nat Rev* **9**, 206–221.

Stemmler, M. and Koch, C. (1999). How voltage-dependent conductances can adapt to maximize the information encoded by neuronal firing rate. *Nat Neurosci* **2**, 521–527.

Stuart, G.J. and Sakmann, B. (1994). Active propagation of somatic action potentials into neocortical pyramidal cell dendrites. *Nature* **367**, 69–72.

Thibault, O., Hadley, R. and Landfield, P.W. (2001). Elevated postsynaptic [Ca2+]i and L-type calcium channel activity in aged hippocampal neurons: Relationship to impaired synaptic plasticity. *J Neurosci* **21**, 9744–9756.

Thompson, L.T., Moyer, J.R., Jr. and Disterhoft, J.F. (1996). Transient changes in excitability of rabbit CA3 neurons with a time course appropriate to support memory consolidation. *J Neurophysiol* **76**, 1836–1849.

Tiesinga, P., Fellous, J.M. and Sejnowski, T.J. (2008). Regulation of spike timing in visual cortical circuits. *Nat Rev* **9**, 97–107.

Topolnik, L., Chamberland, S., Pelletier, J.G., Ran, I. and Lacaille, J.C. (2009). Activity-dependent compartmentalized regulation of dendritic Ca^{2+} signaling in hippocampal interneurons. *J Neurosci* **29**, 4658–4663.

Turrigiano, G., Abbott, L.F. and Marder, E. (1994). Activity-dependent changes in the intrinsic properties of cultured neurons. *Science* **264**, 974–977.

Turrigiano, G.G. and Nelson, S.B. (2000). Hebb and homeostasis in neuronal plasticity. *Current Opin Neurobiol* **10**, 358–364.

Turrigiano, G.G. and Nelson, S.B. (2004). Homeostatic plasticity in the developing nervous system. *Nat Rev* **5**, 97–107.

van Welie, I., van Hooft, J.A. and Wadman, W.J. (2004). Homeostatic scaling of neuronal excitability by synaptic modulation of somatic hyperpolarization-activated Ih channels. *Proc Natl Acad Sci USA* **101**, 5123–5128.

Wang, X.J. (1999). Synaptic basis of cortical persistent activity: The importance of NMDA receptors to working memory. *J Neurosci* **19**, 9587–9603.

Wang, X.J. (2001). Synaptic reverberation underlying mnemonic persistent activity. *Trends Neurosci.* **24**, 455–463.

Wang, Z., Xu, N.L., Wu, C.P., Duan, S. and Poo, M.M. (2003). Bidirectional changes in spatial dendritic integration accompanying long-term synaptic modifications. *Neuron* **37**, 463–472.

Wei, D.S., Mei, Y.A., Bagal, A., Kao, J.P., Thompson, S.M. and Tang, C.M. (2001). Compartmentalized and binary behavior of terminal dendrites in hippocampal pyramidal neurons. *Science* **293**, 2272–2275.

Winograd, M., Destexhe, A. and Sanchez-Vives, M.V. (2008). Hyperpolarization-activated graded persistent activity in the prefrontal cortex. *Proc Nat Acad Sci USA* **105**, 7298–7303.

Woody, C.D. and Engel, J., Jr. (1972). Changes in unit activity and thresholds to electrical microstimulation at coronal-pericruciate cortex of cat with classical conditioning of different facial movements. *J Neurophysiol* **35**, 230–241.

Xu, J., Kang, N., Jiang, L., Nedergaard, M. and Kang, J. (2005). Activity-dependent long-term potentiation of intrinsic excitability in hippocampal CA1 pyramidal neurons. *J Neurosci* **25**, 1750–1760.

Xu, N.L., Ye, C.Q., Poo, M.M. and Zhang, X.H. (2006). Coincidence detection of synaptic inputs is facilitated at the distal dendrites after long-term potentiation induction. *J Neurosci* **26**, 3002–3009.

Yasuda, R., Sabatini, B.L. and Svoboda, K. (2003). Plasticity of calcium channels in dendritic spines. *Nat Neurosci* **6**, 948–955.

Yuste, R., Gutnick, M.J., Saar, D., Delaney, K.R. and Tank, D.W. (1994). Ca^{2+} accumulations in dendrites of neocortical pyramidal neurons: An apical band and evidence for two functional compartments. *Neuron* **13**, 23–43.

Zhang, W. and Linden, D.J. (2003). The other side of the engram: Experience-driven changes in neuronal intrinsic excitability. *Nat Rev* **4**, 885–900.

Zhou, Y., Won, J., Karlsson, M.G., Zhou, M., Rogerson, T., Balaji, J., Neve, R., Poirazi, P. and Silva, A.J. (2009). CREB regulates excitability and the allocation of memory to subsets of neurons in the amygdala. *Nat Neurosci* **12**, 1438–1443.

Adult Hippocampal Neurogenesis and Memory

4

Scellig S.D. Stone* and Paul W. Frankland*

1. Introduction

The continued generation of neurons within the adult central nervous system was identified over 40 years ago in rodents (Altman and Das, 1965) and subsequently confirmed in other mammals including humans (Eriksson *et al.*, 1998). Adult neurogenesis is consistently reported in the subventricular zone (SVZ) of the lateral ventricles and the subgranular zone (SGZ) of the hippocampal dentate gyrus (DG), where adult-generated dentate granule cells (DGCs) eventually integrate into the local neural network (Deng *et al.*, 2010). Given the widely accepted importance of the hippocampus in forming new memories (Moscovitch *et al.*, 2005; Scoville and Milner, 1957; Squire and Zola-Morgan, 1991), the possibility that hippocampal neurogenesis represents a form of structural plasticity in memory processes has garnered considerable attention (Shors, 2008). However, precisely how adult-generated DGCs contribute to these cognitive processes is less clear. While alterations in hippocampal neurogenesis have been associated with changes in behavioural measures of learning and memory, inconsistent and conflicting data has frequently emerged.

In this chapter, we review the literature concerning the putative role(s) for adult hippocampal neurogenesis in learning and memory. Also discussed are unresolved issues that impede our progress towards a clearer understanding, and potential technical and interpretative reasons underlying conflicting results. We begin by first summarizing key biological aspects: where neurogenesis occurs within hippocampal memory circuits, what characterizes adult-generated DGCs, and how this process is regulated. Second, we discuss evidence for functional integration of adult-generated DGCs into learning and memory processes at the cellular and behavioural level, including evidence for learning- and memory-specific activation

* Neurosciences & Mental Health, The Hospital for Sick Children, 555 University Ave., 6018 McMaster Building, Toronto, Ontario, M5G 1X8, CANADA.

of newborn neurons and the impact of manipulating the rate of adult hippocampal neurogenesis on behaviour. Third, we consider theorized roles for new DGCs at the network or cognitive level with special attention to computational modelling research. Our concluding comments touch upon the potential for our evolving understanding of this phenomenon to ultimately provide the foundation for therapeutic advancements in the clinical realm of memory impairment.

2. Biology of adult hippocampal neurogenesis

2.1. *Anatomy of the dentate gyrus*

The hippocampal formation includes several distinct cytoarchitectural regions (Figure 4.1A): the entorhinal cortex (EC), DG, hippocampus proper (including the CA1-3 and sometimes -4 subfields), and subicular complex (Amaral and

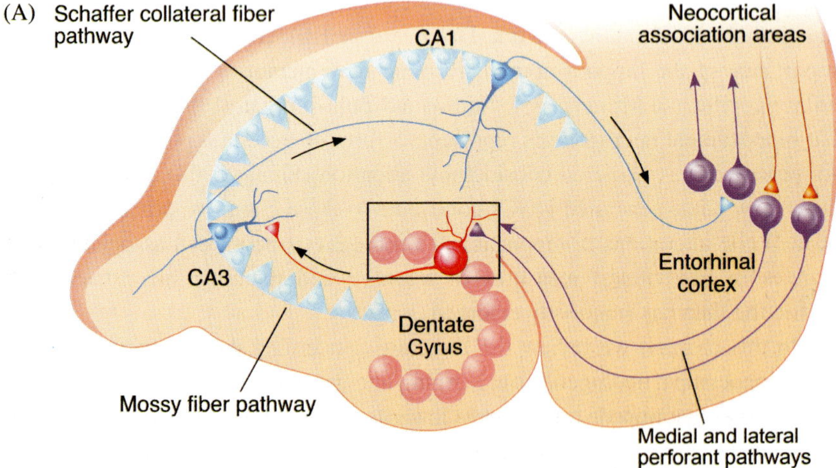

Figure 4.1. Continuous generation of dentate granule cells in the adult mammalian hippocampal formation.

(A) Adult-generated dentate granule cells (red) integrate into the hippocampal formation. Modified from Lie *et al.* (2004).

(B) A close-up of the dentate gyrus illustrating how adult-generated dentate granule cells arise from a neural stem/progenitor cell type lineage within the subgranular zone. Maturation proceeds through a series of phenotypic, morphologic, and electrophysiologic stages. Modified from Bischofberger (2007).

(B)

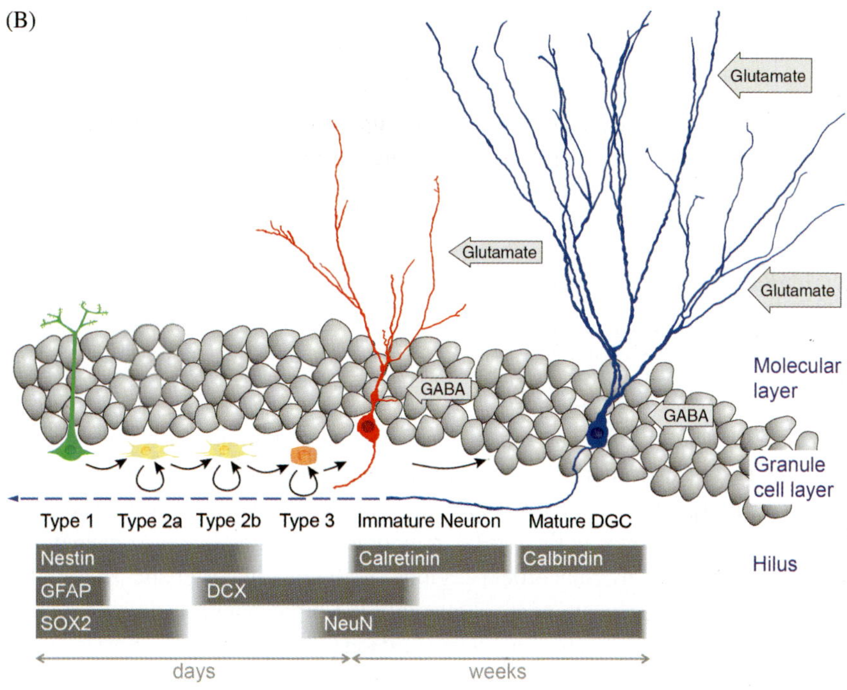

Figure 4.1. (*Continued*)

Lavenex, 2007; Amaral and Witter, 1989). Table 4.1, listing approximate numbers of neurons in major regions of the rat and human hippocampus, indicates that the DG contains a relatively large pool of hippocampal neurons, of which the majority are DGCs (Mulders *et al.*, 1997; West and Gundersen, 1990; West and Slomianka, 1998; West *et al.*, 1991). Although these regional totals are less precisely defined in mice, strains commonly used in neurogenesis studies harbor roughly 500,000–1,000,000 DGCs (Abusaad *et al.*, 1999; Kempermann *et al.*, 1997a). DGCs are the principal excitatory neurons of the DG, receiving inputs from layer II of the medial and lateral EC via the medial and lateral perforant pathways respectively, and sending mossy fiber terminals to CA3 pyramidal cells (Amaral and Lavenex, 2007). The fact that a small number of EC neurons project to a relatively larger number of DGCs (Table 4.1) has been argued to allow the DG to separate multiple EC inputs (O'Reilly and McClelland, 1994). This coding property will be discussed in greater detail later. These DG connections, along with Schaffer collateral fibers from CA3 pyramidal neurons to CA1 neurons, participate within the classically described trisynaptic circuit of the hippocampus

Table 4.1. Approximate number of neurons in major regions of each adult rat and human hippocampal formation.

Major hippocampal regions	Approximate number of neurons in rat	Approximate number of neurons in human
Entorhinal cortices layer 2	200,000[a]	2,200,000[b]
Dentate gyri	2,400,000[c]	30,000,000[d]
CA2/3 subfields	500,000[c]	5,400,000[d]
CA1 subfields	800,000[c]	32,000,000[d]
Subiculi	600,000[c]	9,000,000[d]

[a]Mulders *et al.*, 1997; [b]West and Slomianka, 1998; [c]West *et al.*, 1991; [d]West and Gundersen, 1990.

(Andersen *et al.*, 1969). The hippocampal formation communicates extensively with the rest of the brain. The most numerous hippocampal-cortical connections comprise a loop of EC-mediated neocortical interactions with unimodal and polymodal association areas in the frontal, temporal, and parietal lobes (Lavenex and Amaral, 2000). Many subcortical regions, including diencephalic and limbic structures, also reciprocate connections with the hippocampi. Damage to either the hippocampi or any of these associated regions can produce anterograde amnesia, indicating their importance in memory (Aggleton and Saunders, 1997).

The DG contains three main layers (Figure 4.1B): an inner hilus (also referred to as either the polymorphic layer or CA4 if considered a hippocampal subfield rather than a portion of the DG (Amaral, 1978)), middle granule cell layer, and outer molecular layer (Amaral and Lavenex, 2007). The hilus primarily contains inhibitory interneurons, glutamatergic mossy cells, and DGC axons. Perforant path fibers synapse upon apical dendrites of DGCs in the outer region of the molecular layer, with lateral EC originating fibers arriving outermost and medial EC fibers incoming just below (van Groen *et al.*, 2002). Some commissural fibres from the contralateral DG have also been demonstrated medially in the molecular layer (Blackstad, 1956). DGC bodies populate the densely packed granule cell layer, and it is at its interface with the hilus, termed the SGZ, where adult hippocampal neurogenesis occurs.

2.2. *Identification and quantification of neurogenesis*

Cellular proliferation within the SGZ of the DG has been characterized using a variety of methods, including the detection of exogenously delivered and endogenously expressed markers of cell division. Altman and Das first demonstrated adult hippocampal neurogenesis in rodents by detecting exogenously administered tritiated thymidine (Altman and Das, 1965). Nucleotide derivatives such as these can be given *in vivo*, for example to rats, are available (for up to 2 h in DG cells) for permanent incorporation within the 7-h S phase (DNA replication) of the 25-h cell cycle, and can be subsequently detected at later time-points (Cameron and McKay, 2001; Kempermann *et al.*, 2003). Using a short interval between analogue delivery and animal sacrifice generates a labelled cell count that estimates proliferation. A long delay, during which some labelled cells may die, provides an opportunity to estimate cell survival.

Electron microscopy of autoradiographically labelled new DGCs in adult rodents demonstrated neuronal morphology, confirming neurogenesis (Kaplan and Bell, 1984; Kaplan and Hinds, 1977). Combining this technique with immunohistochemical detection of the relatively selective neuron marker neuron-specific enolase (Deloulme *et al.*, 1996; Schmechel *et al.*, 1980) and tracer injection further revealed that the majority of adult-generated cells were likely DGCs extending mossy fibres to CA3 (Cameron *et al.*, 1993; Stanfield and Trice, 1988). While studying adult neurogenesis in the higher vocal center of adult songbirds, Barnea and Nottenbohm separately confirmed new neuron production in the avian equivalent of the hippocampal formation and noted an increase in neurogenesis with high environmental novelty and spatial memory demand (Barnea and Nottebohm, 1994; 1996).

Despite its successes, tritiated thymidine labelling use was inherently limited by its radioactivity, requirement for very thin tissue sections and difficulty with multi-labelling, and soon gave way to the thymidine analogue BrdU (Figure 4.2A) (Miller and Nowakowski, 1988). BrdU is detectable immunohistochemically in thick sections required for stereology, readily amenable to multi-labelling, and capable of signal amplification (West *et al.*, 1991). Accordingly, BrdU became the primary agent for identifying adult-generated cells in the 1990s (Kempermann *et al.*, 1997b; Kuhn *et al.*, 1996). Compelling evidence for neurogenesis was obtained by co-staining tissue for BrdU and one of many endogenous neuron-specific markers. These included NeuN, a nuclear-specific protein (Figure 4.2A) (Mullen *et al.*, 1992), and calbindin, a vitamin D-dependent calcium-binding protein (Jande *et al.*, 1981). Using BrdU incorporation to evaluate cell cycle kinetics in the DG, investigators determined that young adult mice and rats gain roughly 3000 and 9000 new DGCs per day respectively, or roughly 0.4% of their total granule cell populations (Cameron and McKay, 2001; Hayes and Nowakowski, 2002). Rodents then lose half or more of these new cells by 1 month, after which most survive for at least several months (Dayer *et al.*, 2003; Kempermann *et al.*, 2003). This rate of production declines precipitously with age, but does not cease (Drapeau *et al.*, 2003; Kempermann *et al.*, 1998b; Kuhn

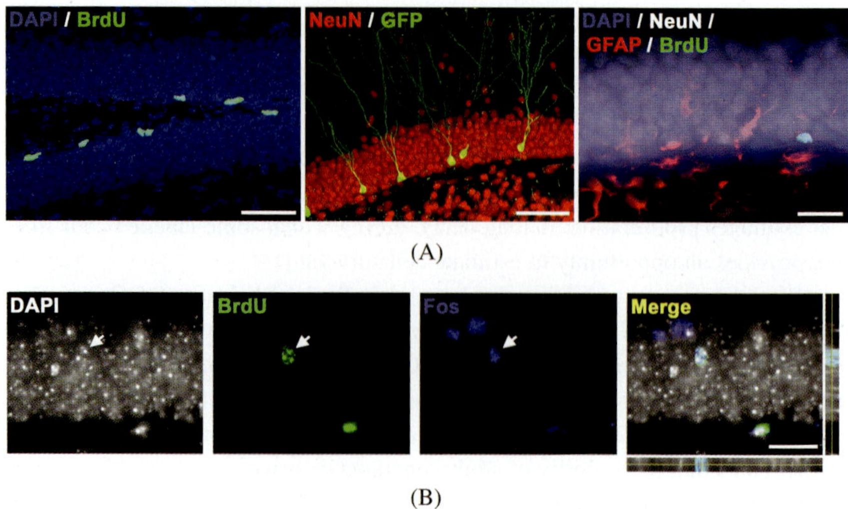

Figure 4.2. Sample immunohistochemical techniques for labelling adult-generated dentate granule cells.

(A) Adult-generated cells in the dentate gyrus can be visualised through immunohistochemical staining for exogenously administered and incorporated BrdU (left panel, scale bar = 50 mm) and for GFP expression following retroviral labelling (middle panel, scale bar = 50 mm). These can be combined with stains for endogenous markers, such as the neuronal stain NeuN (middle panel) and glial stain GFAP (right panel, scale bar = 20 mm).

(B) Activation of adult-generated dentate granule cells, such as following behavioural testing, can be visualized using immunohistochemical detection of immediate-early gene expression, such as Fos, co-labelled with cell markers, such as BrdU (scale bar = 20 mm).

et al., 1996). BrdU labelling also confirmed adult hippocampal neurogenesis in numerous other mammals including primates and humans (Eriksson *et al.*, 1998; Gould *et al.*, 1997; Gould *et al.*, 1999b; Gould *et al.*, 1998; Kornack and Rakic, 1999). In Eriksson *et al.*'s seminal study, newly generated DGCs were present in five cancer patients that received BrdU for diagnostic purposes; however, their rate of production in humans remains unknown. The striking degree of evolutionary conservation suggests, at least in general terms, that this process carries some degree of biological importance.

A collective disadvantage of exogenously delivered cellular proliferation markers is the inherent requirement to deliver a foreign compound, potentially influencing the proliferation, maturation, and survival of new neurons. For instance, stress that animals may experience while receiving injections can alter normal cellular processes including neurogenesis (Gould *et al.*, 1991; Heine *et al.*, 2004). Thymidine analogues also, in principle, can incorporate into any cell undergoing DNA repair and thus falsely identify non-mitotic cells (Kuan *et al.*, 2004). However, mitosis involves the synthesis of the entire genome and typically utilizes several orders of magnitude more nucleotides than DNA repair (Kee *et al.*, 2007a). Consistent with this, the risk of false-positive labelling associated with DNA repair is negligible even at high analogue doses and under conditions that stimulate DNA repair such as inflammation and brain irradiation (Cameron and McKay, 2001; Palmer *et al.*, 2000).

Certain methodological considerations remain important caveats to thymidine analogue use. These include the effect of administration timing and dosage on the number of labelled cells (Cameron and McKay, 2001; Mandyam *et al.*, 2007) and impact of relatively harsh epitope retrieval steps on multi-labelling success (Wojtowicz and Kee, 2006). Although temporally limited by their natural expression patterns, endogenous markers can obviate these potential confounds and some provide useful cell-cycle kinetic information (Eisch and Mandyam, 2007). Ki-67 expression is a common example used to verify thymidine analogue-detected proliferation rates (Dayer *et al.*, 2003; Eisch and Mandyam, 2007; Kee *et al.*, 2002). This nuclear protein has a short half-life and is expressed in dividing cells primarily during the S, G2 and mitosis phases (Brown and Gatter, 2002). Doublecortin (DCX) is a microtubule-associated protein expressed in the cytoplasm for roughly 2 weeks and thus can be used as an endogenous marker of new immature neurons (Brown *et al.*, 2003; Nacher *et al.*, 2001; Sisti *et al.*, 2007).

2.3. *Adult-generated DGC development*

2.3.1. *Technical considerations*

Three types of approaches have been used to investigate the origins and progressive development of adult-generated DGCs. Initial studies characterized overlapping phases of proliferation, differentiation, and maturation by mapping the temporal expression of cell markers in BrdU-labelled cells (Kempermann *et al.*, 2004; Steiner *et al.*, 2004). More recent work has applied genetic lineage tracking techniques to facilitate the focused study of newly generated cells across their entire lifespan. Specifically, *in vivo* delivery of reporter transgenes using retroviruses to selectively infect dividing cells has since become a popular means of labelling and tracking newly generated cells. The signal(s) can be targeted to either the nucleus or cytoplasm, facilitating multi-labelling and morphological analyses. Despite its continued utility, concerns have arisen over the possibility for reporter protein transfer to post-mitotic neurons (Ackman *et al.*, 2006), requirement for invasive surgical delivery, potential for inefficient cell labelling at a distance from the infection site(s), and lack of temporal specificity due to potentially continued slow and fast proliferation of some infected stem and progenitor cells (Breunig *et al.*, 2008). Some of these issues can be overcome by using transgenic reporter animals to genetically label the adult-generated progeny of neural precursors. For example, mice that express markers under nestin gene regulatory control elements have been used to quantify neural progenitor and stem cells with extremely high labelling efficiency (Imayoshi *et al.*, 2008; Lagace *et al.*, 2007; Ninkovic *et al.*, 2007). Reporter gene expression can also be temporally activated in animals expressing Cre recombinase bound to a receptor, such as the ligand-binding domain of the estrogen receptor (Hayashi and McMahon, 2002). In this case, Cre-mediated recombination and hence marker expression only occurs when in the presence of tamoxifen. Accordingly, new neuron labelling can be initiated at a desired adult time-point.

2.3.2. *Stem and progenitor origins*

Adult neural stem cells should be capable of self-renewal and differentiation into all types of neural cells, including neurons, astrocytes, and oligodendrocytes (Gage, 2000). While at present no definitive marker for neural stem cells has been identified, candidate multipotent neurogenic radial glia-like cells within the SGZ (termed type 1 cells) have been uniquely identified as glial fibrillary acidic protein (GFAP), nestin, and sex-determining region Y-box 2 transcription factor (SOX2) positive (Figure 4.1B) (Fukuda *et al.*, 2003; Garcia *et al.*, 2004; Suh *et al.*, 2007). They extend a radial process through the granule cell layer that arborizes in the molecular layer along with a smaller horizontal process within the SGZ, and possess some astrocytic features (Filippov *et al.*, 2003; Fukuda *et al.*, 2003; Seri *et al.*, 2001). It is possible that these type 1 cells are relatively quiescent neural stem cells and are able to generate self-renewing progenitor cells (Mathews *et al.*, 2010; Mu *et al.*, 2010). Highly proliferative type 2a and 2b cells are distinguished by initially remaining nestin positive but lacking GFAP staining and having only short processes (type 2a), then becoming DCX positive (type 2b). This early detection of DCX importantly indicates lineage commitment prior to cell cycle exit. Whether or not type 2 cells arise from type 1 cells is uncertain, but type 2 nestin/SOX2 positive cells can self-renew and generate both neurons and astrocytes, suggesting stem cell properties and possibly a reciprocal relationship with type 1 cells (Suh *et al.*, 2007). Type 2 cells give rise to DCX positive/

nestin negative type 3 cells or neuroblasts, which continue to cycle. Cells become post-mitotic immature neurons in as little as 3 days, transiently express calretinin, and can already express NeuN. Ultimately DCX expression ceases and calbindin replaces calretinin as a mature DGC phenotype is assumed.

2.3.3. *Morphological development*

Post-mitotic adult-generated DGCs progressively attain mature granule cell morphology, implying functional capacity (Figure 4.1B). During this process, primarily elucidated in rodents using green fluorescent protein (GFP) retroviral labelling (Figure 4.2A), most structural modification occurs over roughly 2 months (Esposito *et al.*, 2005; Zhao *et al.*, 2006). Within days, some cells begin migrating into the GCL, although most still remain near the hilar border even months later (Esposito *et al.*, 2005; Zhao *et al.*, 2006). Axonal projections reach CA3 in 4 to 11 days (Hastings and Gould, 1999; Zhao *et al.*, 2006) and continue to extend distally thereafter. Apical dendrites grow and arborize into the molecular layer over the first 3 to 4 weeks (Esposito *et al.*, 2005; Zhao *et al.*, 2006). Dendritic spines are apparent in 2 weeks and display synaptic terminals, with subsequent modification extending beyond this period (van Praag *et al.*, 2002; Zhao *et al.*, 2006). Structurally mature axosomatic, axodentric, and axospinous synaptic inputs have been visualized by 30 days, with the suggestion that some arise from the EC, along with similar output synapses on appropriate CA3 and hilar target neurons (Toni *et al.*, 2008; Toni *et al.*, 2007; van Praag *et al.*, 2002). These findings suggest that adult-generated DGCs possess the requisite morphology to assume some degree of circuit function within a few weeks of birth.

2.3.4. *Electrophysiologic development*

Developing adult-generated DGCs begin to exhibit functional amino butyric acid (GABA) and glutamate afferents by roughly 1 and 2 weeks of age respectively (Figure 4.1B) (Ambrogini *et al.*, 2010; Esposito *et al.*, 2005; Ge *et al.*, 2006). Relative to mature granule cells, immature cells have distinct electrophysiological properties at around 2–6 weeks of age, prompting some to propose this represents a period of enhanced plasticity and distinct function (Ge *et al.*, 2007). Features identified include: lower activation thresholds, higher resting potentials, and a greater propensity for long-term potentiation (LTP) (Ambrogini *et al.*, 2004a; Ge *et al.*, 2007; Schmidt-Hieber *et al.*, 2004; Wang *et al.*, 2000). Indeed, eliminating adult-DGC production leads to a significant reduction in perforant pathway-induced DG LTP *in vitro*, underscoring the important contribution of adult neurogenesis to DG synaptic plasticity (Snyder *et al.*, 2001; Zhao *et al.*, 2007). These electrophysiologic parameters approach those of mature granule cells by approximately 6 to 8 weeks in rodents (Ge *et al.*, 2007; Laplagne *et al.*, 2006; Laplagne *et al.*, 2007; van Praag *et al.*, 2002). Collectively, the developmental sequence of electrophysiologic maturation in adult-generated cells follows closely that of cells in the developing brain (Ambrogini *et al.*, 2004a; Esposito *et al.*, 2005).

Mature new adult hippocampal neurons have similar electrophysiological properties to developmentally generated DGCs (Esposito *et al.*, 2005; Laplagne *et al.*, 2006; van Praag *et al.*, 2002). Hippocampal slice recordings find mature cell-like membrane potentials, action potentials, and spontaneous postsynaptic currents. Moreover, these cells show postsynaptic responses to perforant path stimulation with similar properties to cells generated during development (Laplagne *et al.*, 2006; van Praag *et al.*, 2002), and essentially identical responses to numerous GABAergic and glutamatergic input parameters (Laplagne *et al.*, 2006; Laplagne *et al.*, 2007). This compelling evidence implies that new granule cells become functional members of the hippocampal circuit akin to their developmentally generated counterparts.

2.3.5. *Contribution to the adult DGC population*

Long-term fate mapping studies using transgenic reporter animals demonstrate that adult-generated DGCs primarily reside in the innermost granule cell layers and constitute 10–15% of the entire DGC population (Imayoshi *et al.*, 2008; Ninkovic *et al.*, 2007). This proportion initially increases over several months, as does the density of DGCs (Imayoshi *et al.*, 2008), reflecting a net addition of granule cells over time. The proportion of DGCs that are generated during adulthood eventually reaches a plateau, likely reflecting declining neurogenesis with age; however, whether a certain degree of dynamic equilibrium between new neuron addition and death is partly responsible for the plateau remains a matter of debate (Ninkovic *et al.*, 2007).

2.4. *Regulation of neurogenesis*

The microenvironment of the SGZ, including its neural, glial and vascular components, grants a wide variety of influences access to neural stem/progenitor cells (Ma *et al.*, 2008; Morrison and Spradling, 2008). These elements influence the proliferation, differentiation, survival, and integration of adult-generated DGCs both positively and negatively through a large repertoire of mechanisms, including cell surface receptor-mediated signalling pathways, transcription factor modulation, and epigenetic regulation. Known factors and their putative mechanisms of action are extensively reviewed elsewhere (Zhao *et al.*, 2008). In summary, neurotransmitters and neural peptides, growth factors and neurotrophins, morphogens, cytokines, hormones (including steroids), and drugs (including antidepressants) have all been shown to modulate aspects of adult hippocampal neurogenesis. Collectively, these numerous influences and their mechanisms of action permit the system to respond to complex internal and external stimuli and experiences, such as environmental enrichment, ageing, stress and physical activity (Drapeau *et al.*, 2003; Kempermann *et al.*, 1997b; Kuhn *et al.*, 1996; Mineur *et al.*, 2007; van Praag *et al.*, 1999b; van Praag *et al.*, 2005).

Since adult-generated DGCs join hippocampal networks, as suggested by their phenotypic, morphologic, and electrophysiologic development, it is fitting that their generation and integration responds to activity within those networks. Seizures and electroconvulsive shock increase DGC generation, although resultant cells can exhibit altered morphology and may arise ectopically in the hilus and inner molecular layer (Jessberger *et al.*, 2007; Ma *et al.*, 2009; Madsen *et al.*, 2000; Overstreet-Wadiche *et al.*, 2006; Parent *et al.*, 1997). Electrical stimulation of the perforant pathway promotes both the proliferation and survival of adult-generated granule cells *in vivo* (Bruel-Jungerman *et al.*, 2006; Derrick *et al.*, 2000; Kitamura *et al.*, 2010; Toda *et al.*, 2008). Neurotransmitters are likely mediators of activity-dependent influences on hippocampal neurogenesis, including GABA and glutamate. Although type 2 progenitor cells respond to GABA receptor antagonism and agonism with increased and decreased proliferation respectively, GABAergic activation promotes their neuronal differentiation possibly by inducing NeuroD transcription factor expression (Tozuka *et al.*, 2005). GABA's influence also extends to facilitating eventual glutamatergic innervation of newborn DGCs through its ambient release from interneurons, potentially directing DGC integration into the excitatory hippocampal circuit in an activity-dependent fashion (Ge *et al.*, 2006). Conflicting reports of glutamate either increasing or decreasing progenitor cell proliferation suggest a complex or indirect influence (Jang *et al.*, 2008). However, direct excitatory stimulation does encourage neuronal differentiation (Deisseroth *et al.*, 2004). Moreover, retrovirus-mediated gene knockout of N-methyl-D-aspartic acid receptor (NMDAR) expression in proliferating DG cells impairs their survival during the third week of development, possibly denoting input-specific selection of developing new neurons (Tashiro *et al.*, 2006). These linkages with activity suggest mechanisms through which hippocampal neurogenesis could respond to network activity arising from hippocampal-dependent cognitive processes.

3. Relationships between adult hippocampal neurogenesis and learning and memory

The preceding sections suggest that DGCs are produced in adulthood and join the hippocampal circuit, and that this process is regulated by various intrinsic and extrinsic variables. Considerable evidence also suggests that learning impacts neurogenesis, and adult-generated cells functionally integrate into memory networks and contribute to learning and memory.

3.1. *Learning impacts neurogenesis*

Learning promotes the survival of adult-generated DGCs. Studies suggest this is characteristic of hippocampal-dependent tasks, correlates with performance, biases precursors towards neuronal differentiation, and is temporally specific. For example, survival of roughly week-old DGCs is enhanced by Morris water maze training and trace eyeblink conditioning in young adult rodents (Anderson *et al.*, 2011; Epp *et al.*, 2010, 2011a; Epp *et al.*, 2007; Gould *et al.*, 1999a; Leuner *et al.*, 2004; Sisti *et al.*, 2007; Trouche *et al.*, 2009; Waddell and Shors, 2008). This effect is most prominent in the outer blade of the DG (Ambrogini *et al.*, 2000) and does not extend to the SVZ (Gould *et al.*, 1999a). Interestingly, DG LTP induced by *in vivo* stimulation of the perforant pathway has comparable effects (Bruel-Jungerman *et al.*, 2006; Kitamura *et al.*, 2010). The extent of survival also correlates with performance in some reports (Ambrogini *et al.*, 2000; Drapeau *et al.*, 2003; Epp *et al.*, 2011a; Leuner *et al.*, 2004; Sisti *et al.*, 2007; Waddell and Shors, 2008). Since the rate of hippocampal neurogenesis is activity dependent, it seems logical that multiple forms of learning that engage — but are not necessarily dependent on — the hippocampus could promote survival. Consistent with this reasoning, some propose that increased survival occurs with more robust forms of learning regardless of whether they are hippocampal dependent (Dalla *et al.*, 2007; Epp *et al.*, 2010).

Adult DGC proliferation may also be stimulated by learning and correlate with performance. For instance, BrdU incorporation studies report increases in proliferation associated with water maze training in rats of varying ages (Drapeau *et al.*, 2003; Lemaire *et al.*, 2000). Greater increases correlated with faster maze learning and better recall. Since BrdU incorporation is permanent, increased cell numbers after even short delays between labelling and detection can partially represent enhanced survival rather than solely proliferation (Prickaerts *et al.*, 2004). Dupret *et al.* mitigated this potential confound by detecting endogenous expression of Ki67 and confirmed elevated DG proliferation after 5 days of water maze training (Dupret *et al.*, 2007).

Despite evidence for a pro-survival and pro-proliferative influence of learning on adult hippocampal neurogenesis, many studies have failed to detect these effects. In some cases, discrepant results arise when similar procedures are tested in different animal strains (Olariu *et al.*, 2005; Snyder *et al.*, 2005) or ages (Drapeau *et al.*, 2003). Inconsistent findings could also be explained by the complex interplay of cognitive load and stress levels associated with training protocols. For example, one study demonstrated that water maze task difficulty was inversely proportional to survival, and suggested this might reflect animals adopting nonsurvival-promoting learning strategies when given less spatial cues (Epp *et al.*, 2010). However, it is difficult to separate the potential anti-survival influence of greater task difficulty from that of higher levels of psychological stress that may be associated with more challenging tasks (Thomas *et al.*, 2007). Stress is a pivotal factor in determining whether or not learning modulates neurogenesis, as evidenced by a lack of benefit of water maze training on the survival of week-old DGCs with protocols that significantly elevate circulating corticosterone levels (Mohapel *et al.*, 2006). Moreover, some experimental procedures require pre-exposing mice to the water maze days before training in order to demonstrate learning-induced increased cell survival (Ehninger and Kempermann, 2006), possibly because pre-exposure to the maze environment lowers stress levels during training (Beiko *et al.*, 2004).

Studies examining several time points in and around behavioural tasks propose a learning-induced selective process of adult-generated neuron retention and removal. For instance, "snapshots" of neurogenesis taken during different periods of learning demonstrate opposing phases of enhanced survival and cell death in adult born DGC populations of different ages (Ambrogini *et al.*, 2004b; Anderson *et al.*, 2011; Epp *et al.*, 2007; Olariu *et al.*, 2005). A homeostatic regulation of adult-generated DGC production and survival that is modulated by learning has been proposed by studies labelling newborn DGCs at multiple time points during extended water maze training regimens. During the early learning phase, corresponding to rapid performance improvement, neurogenesis was reduced and DG apoptosis rates were elevated relative to controls (Dobrossy *et al.*, 2003; Drapeau *et al.*, 2007; Dupret *et al.*, 2007). Interestingly, this reduced neurogenesis also correlated with better performance (Dobrossy *et al.*, 2003; Drapeau *et al.*, 2007). Cells born during or soon after the second phase, when asymptotic performance levels were reached, were more numerous than controls (Dobrossy *et al.*, 2003; Drapeau *et al.*, 2007). Blocking apoptosis during this second phase of learning, by intraventricular injection of a pan-caspase inhibitor, prevented this spike in neurogenesis and impaired memory during learning and testing (Dupret *et al.*, 2007). Collectively, these results suggest that certain forms of learning promote the survival of selected new granule cells, while removing others; however, the precise utility of this process, and why it would occur in such immature cell populations, remain to be determined.

Beyond influencing their number, learning can also potentiate the differentiation and morphological development of new neurons. Water maze training was associated with increases in neuronal differentiation (Epp *et al.*, 2007), as well as dendrite length and branching, and spine density and maturation (Tronel *et al.*, 2010), of DGCs generated either 1 week prior to or during the early phase of water maze training. These morphological effects were not seen in primarily developmentally generated DGCs, identified using a reporter viral vector that infects all neurons, suggesting adult-born DGCs were specifically influenced. Structural changes were also abolished by blocking either apoptosis, pointing to a relationship between the learning-induced homeostatic regulation of cell survival and that of morphology, or NMDARs, implicating activity-dependent regulation. Alterations were enhanced by a more challenging delayed matching-to-place task (Tronel *et al.*, 2010), and were not seen in a separate study when the water maze training regimen resulted in rapid acquisition (Ambrogini *et al.*, 2010), thus suggesting greater task difficulty promotes learning-induced changes in morphology.

3.2. *Adult-generated neurons functionally integrate into hippocampal circuits and contribute to learning and memory*

3.2.1. *Functional activation*

Recruitment of adult-generated DGCs into hippocampal memory circuits in intact animals has been demonstrated using immunohistochemical approaches. In order to identify learning- and memory-induced activation of adult born DGCs, immediate early gene (IEG) products are co-detected with new neuron markers, such as BrdU, in a temporally specific fashion (Figure 4.2B) (Guzowski *et al.*, 2005; Kee *et al.*, 2007a). Rapid and temporary upregulation of IEGs (such as *Fos, Arc, zif268*) is induced by DGC activation (Farivar *et al.*, 2004), for e.g. following seizures (Dragunow and Robertson, 1987; Jessberger and Kempermann, 2003) and LTP induction (Cole *et al.*, 1989). Accordingly, animals can be given BrdU to label adult-generated DGCs, subsequently learn a hippocampal-dependent task (such as the hidden platform version of the water maze), and be sacrificed when learning- or testing-induced IEG expression reaches its peak (Kee *et al.*, 2007a). Quantifying the likelihood of IEG expression within labelled new neurons, and comparing these values to appropriate controls in order to subtract activity arising from nonspecific aspects of the testing procedure, provides a measure of adult-generated granule cell involvement in memory processes.

Spatial exploration (Alme *et al.*, 2010; Chawla *et al.*, 2005; Ramirez-Amaya *et al.*, 2006), water maze learning and recall (Jessberger and Kempermann, 2003; Kee *et al.*, 2007b; Snyder *et al.*, 2009a; Snyder *et al.*, 2009b; Tashiro *et al.*, 2007; Trouche *et al.*, 2009), and contextual fear conditioning (Stone *et al.*, 2010) increase IEG expression rates in DGCs, suggesting functional activation. This expression occurs in a relatively small proportion of DGCs, consistent with the concept of sparse encoding by the DG (Jung and McNaughton, 1993). More importantly, hippocampal-dependent learning and recall increases activation rates in adult-generated DGCs only once they are at least several weeks old at the time of learning, suggesting age-dependent functional integration of adult-generated granule cells into hippocampal-dependent memory circuits in the DG (Kee *et al.*, 2007b; Snyder *et al.*, 2009a; Stone *et al.*, 2010). The timeline of this cell age-dependent recruitment tracks the morphological and physiological maturation of adult-generated neurons.

A key question for understanding how functionally integrated adult-generated DGCs contribute to memory is whether they are functionally distinct from DGCs generated during development. As previously discussed, adult born DGCs experience a transitory developmental period of distinct electrophysiologic properties prior to eventually assuming similar cellular phenotypes to developmentally generated DGCs (Ge *et al.*, 2007; Schmidt-Hieber *et al.*, 2004). This transient state of increased plasticity may promote integration of newborn neurons into hippocampal memory networks by allowing them to either temporarily or permanently out-compete their developmentally generated neighbors. Along these lines, water maze training or environmental exposure when adult-born cells are 1–2 weeks old, followed by re-exposure to that same specific training or environment once those cells had reached maturity, resulted in proportional increases in surviving and activated cells and suggested that experiences during a critical developmental period could promote experience-specific functional integration (Tashiro *et al.*, 2007; Trouche *et al.*, 2009). Alternatively, developmentally and adult-generated DGCs may be functionally equivalent and integrate into hippocampal memory circuits at equivalent rates.

Stone *et al.* demonstrated functional convergence at the behavioural level by comparing integration rates of directly labelled developmentally and adult-generated DGCs into hippocampal memory networks supporting hippocampal-dependent memory: the water maze and contextual fear conditioning (Stone *et al.*, 2010). Individual neuron cohorts were directly labelled at separate developmental stages with different thymidine analogues (BrdU, IdU, CldU), permitting comparisons between cell populations marked by equivalent techniques. The rate of IEG expression within each cohort did not differ, indicating equivalent integration rates. Indeed, such phenotypic convergence typifies many other organs and tissues with regenerative capacity, including skin, blood, muscle, liver etc., and would be advantageous for potential cell replacement therapies: That is, stimulation of adult neurogenesis would lead to the repopulation of the DG with functionally equivalent cell types.

3.2.2. *Changes in neurogenic rate influence learning and memory*

Associations between neurogenesis levels and mnemonic performance are frequently observed. Many correlative examples have been demonstrated by behavioural studies of animals with differing baseline levels of neurogenesis and following manipulations that increase or decrease the adult-generated DGC population.

Animals with a relatively high rate of endogenous neurogenesis generally outperform similar animals with lower neurogenic rates on hippocampal-dependent learning tasks. For instance, neurogenesis levels in several recombinant strains of mice positively correlate with acquisition in the Morris water maze (Kempermann and Gage, 2002a). Furthermore, declining hippocampal neurogenesis with advancing age has been linked to the emergence of a variety of memory deficits (Aizawa *et al.*, 2009; Kim *et al.*, 2010; Montaron *et al.*, 2006; van Praag *et al.*, 2005). However, the age-related decline in neurogenesis significantly precedes the onset of cognitive impairment, arguing against a straightforward relationship between ageing, memory impairment, and lowered neurogenesis.

Experimentally augmenting adult hippocampal neurogenesis correlates with improved hippocampal-dependent learning in many instances. For example, subjecting unaltered young and aged animals to neurogenesis enhancing manipulations including environmental enrichment, physical activity, or proneurogenic drug treatment, and subsequently allowing sufficient time for these adult-generated DGCs to mature, enhances spatial learning in the Morris water maze (Clark *et al.*, 2008; Kempermann *et al.*, 1998a; Kempermann *et al.*, 2002; Kempermann *et al.*, 1997b, 1998b; Kobilo *et al.*, 2011; Koo *et al.*, 2003; Li *et al.*, 2011; Nilsson *et al.*, 1999; Pieper *et al.*, 2010; van Praag *et al.*, 1999a), Y-maze (Van der Borght *et al.*, 2007), and radial arm maze (Diederich *et al.*, 2009). One report also demonstrates increased contextual fear memory after running-enhanced neurogenesis (Clark *et al.*, 2008). A recent study using a genetic knockout means of increasing neurogenesis revealed similar findings (Li *et al.*, 2011). However, these studies cannot exclude the possibility that other consequences of experimental manipulations lead to behavioural improvements rather than increased neurogenesis. For instance, blocking the environmental enrichment-induced increase in neurogenesis did not prevent improved water maze performance following enrichment (Meshi *et al.*, 2006). A stronger link between neurogenesis and learning was demonstrated for a non-spatial novel object recognition task (Bruel-Jungerman *et al.*, 2005). Here, blocking the environmental enrichment-induced increase in neurogenesis, while maintaining baseline levels, prevented the improvement in novel object recognition.

Many studies have sought a causal link by examining the effects of knocking down adult hippocampal neurogenesis (Table 4.2A). A variety of nongenetic means of neurogenesis suppression lead to hippocampal-dependent memory impairments, which in some cases can be rescued by neurogenesis-promoting manipulations. Treatments that downregulate neurogenesis, including the use of stress, social isolation, perinatal inflammation, constant light exposure, brain irradiation, forebrain cholinergic lesioning, vitamin A deficiency, and toxins (including lead, isoflurane, NMDAR antagonists, temozolamide, methotrexate, and methylazoxymethanol (MAM)), have impaired spatial learning in behavioural paradigms that include the Morris water maze, Barnes maze, radial arm maze, Y-maze, and place recognition tasks (Bonnet *et al.*, 2008; Fan *et al.*, 2007; Fujioka *et al.*, 2011; Garthe *et al.*, 2009; Goodman *et al.*, 2010; Hu *et al.*, 2009; Ibi *et al.*, 2008; Lemaire *et al.*, 2000; Lyons *et al.*, 2010; Madsen *et al.*, 2003; Mohapel *et al.*, 2005; Raber *et al.*, 2004; Rola *et al.*, 2004b; Snyder *et al.*, 2005; Veena *et al.*, 2009; Wojtowicz *et al.*, 2008). Similar results are reported for other forms of hippocampal-dependent learning, including contextual fear conditioning (Hernandez-Rabaza *et al.*, 2009; Jaako-Movits and Zharkovsky, 2005; Jaako-Movits *et al.*, 2005; Ko *et al.*, 2009; Saxe *et al.*, 2006; Snyder *et al.*, 2009a; Warner-Schmidt *et al.*, 2008; Winocur *et al.*, 2006; Wojtowicz *et al.*, 2008; Yun *et al.*, 2010), novel object recognition (Graciarena *et al.*, 2010), delayed nonmatching-to-sample and place (Clelland *et al.*, 2009; Winocur *et al.*, 2006), touchscreen 2-point discrimination (Clelland *et al.*, 2009), and trace conditioning (Achanta *et al.*, 2009; Shors *et al.*, 2001; Shors *et al.*, 2002). In some cases, these behavioural deficits have been subsequently ameliorated by restoring neurogenesis rates through exercise, environmental enrichment, electroconvulsive seizure, adrenalectomy, selective serotonin reuptake inhibitors, or reversal of the neurogenesis-suppressing treatment (Bonnet *et al.*, 2008; Fan *et al.*, 2007; Graciarena *et al.*, 2010; Ibi *et al.*, 2008; Kim *et al.*, 2010; Lyons *et al.*, 2010; Montaron *et al.*, 2006; van Praag *et al.*, 2005; Veena *et al.*, 2009; Warner-Schmidt *et al.*, 2008).

Genetic methods of disrupting adult hippocampal neurogenesis generate similar behavioural impairments to those induced by nongenetic techniques

Table 4.2. Summary of studies directly investigating the effect of depletion of adult-generated dentate granule cells on hippocampal-dependent learning and memory. (A) Studies showing effect. (B) Studies showing no effect.

(A) Behavioural tasks	Species	Strain	Sex	Ablation method	Animal age at ablation	Timing of behavioural test	Phenotype	References
Morris water maze								
Hidden pf 4 trials/d × 7 d	Rat	Wistar	M	Vitamin A-free diet × 13 w	3 w	During last w of diet	Deficits in acquisition & retention	Bonnet *et al.*, 2008
Hidden pf 4 trials/d × 8 d or 4 trials/d × 6 d before extinction	Mouse	Nestin-tk	B	Genetic, GCV induced	8 w	1, 3.5, or 9 w(s) post-GCV	No effects on acquisition & short-term retention, deficits in long-term retention & extinction for 1 w only	Deng *et al.*, 2009
Visible then hidden pf 3 trials/d × 9 d	Mouse	Nestin-rtTA; TRE-Bax	M	Genetic, Dox induced	2 m	6 or > w post-Dox	Deficits in acquisition & memory retention	Dupret *et al.*, 2008
Visible then hidden pf 6 trials/d × 5 d	Gerbil	Mongolian	M	Irradiation	2 m	13 w post-irradiation	Deficits in acquisition & retention	Fan *et al.*, 2007
Hidden pf 6 trials/d × 3 d, then 2 reversals 6 trials/d × 2 d each	Mouse	Nestin-rtTA; TRE-PC3	B	Genetic, Dox induced	30 or 138 d	65 or 22 d post-Dox	Deficits in acquisition, reversal learning & retention	Farioli-Vecchioli *et al.*, 2008
Hidden pf 3 trials/d × 5 d	Mouse	C57BL/6J	M	Constant light exposure × 3 w	8–10 w	1 d post-light	Deficits in acquisition	Fujioka *et al.*, 2011

(Continued)

Table 4.2. (*Continued*)

(A) Behavioural tasks	Species	Strain	Sex	Ablation method	Animal age at ablation	Timing of behavioural test	Phenotype	References
Hidden pf 6 trials/d × 3 d, then reversal 6 trials/d × 2 d	Mouse	C57BL/6	F	TMZ treatment × 4 w	6–8 w	4 w post-TMZ	Deficits in acquisition, learning strategy selection during acquisition & reversal, & retention of reversal	Garthe *et al.*, 2009
Visible pf 3 trials × 1 d, then hidden pf 12 trials × 1 d	Mouse	C57BL/6J	M	MAM treatment × 14 d	9 w	1 d post-MAM	No effect on acquisition, deficits in short- & long-term retention	Goodman *et al.*, 2010
Visible & hidden pf, details n/a	Mouse	C57BL/6J	M	NVP-AAM077 treatment × 2 d	6–7 w	1 & 29 d post-NVP AAM077	No effects on acquisition initially, deficits in delayed re-acquisition & retention	Hu *et al.*, 2009
Hidden pf trials n/a × 6 d	Mouse	ICR	M	Social isolation × 5 w	3 w	During last w of social isolation	Deficits in acquisition & memory retention	ibi *et al.*, 2008
Hidden pf 4 trials/d × 7 d, then reversal	Rat	Sprague-Dawley	B	Lentivirus-dnWnt	7–8 w	8-9 w post-virus injection	No effects on acquisition, short-term retention or reversal learning, deficit in long-term retention	Jessberger *et al.*, 2009
Hidden pf 4 trials/d × 5 d	Rat	Sprague-Dawley	B	Prenatal stress in last w of gestation	sometime post-E15	4 m post-stress	Deficits in acquisition	Lemaire *et al.*, 2000

(*Continued*)

Table 4.2. *(Continued)*

(A) Behavioural tasks	Species	Strain	Sex	Ablation method	Animal age at ablation	Timing of behavioural test	Phenotype	References
Hidden pf 4 trials/d × 7 d	Rat	Sprague-Dawley	F	Saporin lesion of basal forebrain	12–14 w	3–4 w post-lesion	Deficits in acquisition & memory retention	mohapel *et al.*, 2005
Hidden pf 6 trials/d × 3 d	Mouse	C57BL/6J	M	Irradiation	21 d	3 m post-irradiation	No effects on acquisition, deficit in short-term retention	Rola *et al.*, 2004
Hidden pf 2 trials/d × 7 d	Mouse	Nestin-Cre NT3	M	Genetic knockout	Constitutive	2–4 m of age	Deficits in acquisition & memory retention	Shimazu *et al.*, 2006
Hidden pf 8 trials/d × 6 d	Rat	Long Evans	M	Irradiation	40 d	2 d before, 3–4 d after, or 4 w post-irradiation	No effects on acquisition or short-term retention, deficit in retention 4 w later only	Synder *et al.*, 2005
Hidden pf 8 trials/d × 6 d	Rat	Long Evans	M	Irradiation	40 d	2 d before, 3–4 d after, or 4 w post-irradiation	No effects on acquisition or short-term retention, deficit in retention 4 w later only	Synder *et al.*, 2005
Hidden pf 4 trials/d × 6 d	Mouse	Liver IGF-1	M	Genetic knockout	Constitutive	3–4 m of age	Deficits in acquisition & memory retention	Trejo *et al.*, 2008
Hidden pf 4-5 trials/d × 6 d	Rat	Long Evans	M	Irradiation	4.5 m	5 w post-irradiation	No effects on acquisition or short-term retention, deficits in reversal learning	Wojtowicz *et al.*, 2008

(Continued)

Table 4.2. (*Continued*)

(A) Behavioural tasks	Species	Strain	Sex	Ablation method	Animal age at ablation	Timing of behavioural test	Phenotype	References
Hidden pf 4 trials/d × 7 d, then reversal	Mouse	Tlxfl/fl; CMV-CreER	B	Genetic, TM induced	8 w	4 w post-TM	Deficits in acquisition, reversal learning & retention	Zhang *et al.*, 2008
Hidden pf 4 trials/d × 9 d	Mouse	MBD1	M	Genetic knockout	Constitutive	2–5 m of age	Deficits in acquisition & memory retention	Zhao *et al.*, 2003
Hidden pf 8 trials/d × 4 d	Mouse	Nestin-Cre FGFR1	B	Genetic knockout	Constitutive	6–8 w of age	Deficits in acquisition with moving pf, not stationary	Zhao *et al.*, 2007
Barnes maze								
Spatial version	Mouse	Nes-CreER, NSE-DTA	M	Genetic, TM induced	8 w	41 ds post-TM	Deficits in acquisition & long-term retention, no effect on short-term retention	Imayoshi *et al.*, 2008
Spatial then visual version	Mouse	C57BL/6J	M	Irradiation	2 m	3 m post-irradiation	Deficits on spatial version & searching strategies, but not in visual version	Raber *et al.*, 2004
Contextual fear conditioning								
1 shock	Mouse	Nestin-tk	B	Genetic, GCV induced	8 w	1 or 5 w post-GCV	No effect on context conditioning, deficits in extinction × 5 w only	Deng *et al.*, 2009

(*Continued*)

Table 4.2. (*Continued*)

(A) Behavioural tasks	Species	Strain	Sex	Ablation method	Animal age at ablation	Timing of behavioural test	Phenotype	References
1 tone-shock pair	Mouse	Nestin-rtTA; TRE-PC3	B	Genetic, Dox induced	30 or 138 d	65 or 22 d post Dox	Deficits in freezing	Farioli-Vecchioli *et al.*, 2008
1 unsignalled shock	Mouse	PC3/Tis21	M	Genetic knockout	Constitutive	2 m of age	Deficits in freezing	Farioli-Vecchioli *et al.*, 2009
1 unsignalled shock	Rat	Long Evans	M	Irradiation	9–11 w	9 w post irradiation	Deficits in freezing	Hernandez-Rabaza *et al.*, 2009
3 tone-shock pairs	Mouse	Nes-CreER, NSE-DTA	M	Genetic, TM induced	8 w	59 d post TM	Deficits in freezing	Imayoshi *et al.*, 2008
3 tone-shock pairs	Rat	Wistar	M	Olfactory bulbectomy	2 m	6 w post surgery	Deficits in freezing	Jaako-Movitz & Zharovsky, 2005
3 unsignaled shocks	Rat	Wistar	M	Lead treatment × 30 d	1 d	50 d post lead treatment	Deficits in freezing	Jaako-Movitz *et al.*, 2005
3 tone-shock pairs	Mouse	C57BL/6	M	Irradiation	5 or 8 w	5 w or 11 d post irradiation	No effect on context conditioning, but prolonged hippocampal-dependency of memory	Kitamura *et al.*, 2009
1 unsignalled shock	Mouse	C57BL/6	M	Irradiation	7–10 w	3 m post irradiation	Deficits in contextual conditioning, no effects on extinction	Ko *et al.*, 2009

(*Continued*)

Table 4.2. (*Continued*)

(A) Behavioural tasks	Species	Strain	Sex	Ablation method	Animal age at ablation	Timing of behavioural test	Phenotype	References
3 tone-shock pairs	Mouse	129/SvEv	M	Irradiation	12-25 w	3 m post-irradiation	Deficits in freezing	Saxe *et al.*, 2006
5 tone-shock pairs	Mouse	GFAP-tk	M	Genetic, GCV induced	12-20 w	After 6-w GCV treatment	Deficits in freezing	Saxe *et al.*, 2006
3 tone-shock pairs	Mouse	FoxG1	B	Genetic knockout	constitutive	12–14 w of age	Deficits in freezing	Shen *et al.*, 2006
7 tone-shock pairs	Rat	Sprague-Dawley	M	Irradiation	8-9 w	3, 4, or 8 w post-irradiation	Deficits in freezing × 4 & 8 w only	Snyder *et al.*, 2009
3 tone-shock pairs	Rat	Sprague-Dawley	M	Irradiation	n/a, adult	6 w post-irradiation	Deficits in freezing	Warner-Schmidt *et al.*, 2008
10 tone-shock pairs	Rat	Long Evans	M	Irradiation	4 m	4 w post-irradiation	Deficits in freezing	Winocur *et al.*, 2006
7 tone-shock pairs	Rat	Long Evans	M	Irradiation	4.5 m	5 w post-irradiation	Deficits in freezing	Wojtowicz *et al.*, 2008
1 tone-shock pair	Mouse	ICR	M	Restraint stress × 4 w	3 w	1 d post stress	Deficits in freezing	Yun *et al.*, 2010

(*Continued*)

Table 4.2. (*Continued*)

(A) Behavioural tasks	Species	Strain	Sex	Ablation method	Animal age at ablation	Timing of behavioural test	Phenotype	References
Trace fear conditioning								
10 tone-shock pairs following pre-exposure trace paradigm	Rat	Sprague-Dawley	M	Irradiation	P21, P50, & P70	90 d post-irradiation	Deficits in freezing	Achanta *et al.*, 2009
10 tone-shock pairs following pre-exposure trace paradigm	Rat	Sprague-Dawley	M	MAM treatment × 14 d	n/a, adult	1 d post-MAM	Deficits in mobility	Shors *et al.*, 2002
Eye blinking conditioning								
Trace paradigm	Rat	Sprague-Dawley	M	MAM treatment × 6 or 14 d	n/a, adult	2 d post, 6 d, or 2 or 21 d post 14 d of MAM	Deficits in acquisition of conditioned response 2 d post 14 d of MAM only	Shors *et al.*, 2001
Working memory								
Radial maze: working memory	Mouse	Nestin-rtTA; TRE-PC3	B	Genetic, Dox induced	30 or 138 d	65 or 22 d post-Dox	Deficits in working memory	Farioli-Vecchioli *et al.*, 2008
Radial maze: high memory load/no interference	Mouse	129/SvEv	M	Irradiation	10 w	3 m post-irradiation	Improvement at long but not short delays	Saxe *et al.*, 2007

(*Continued*)

Table 4.2. *(Continued)*

(A) Behavioural tasks	Species	Strain	Sex	Ablation method	Animal age at ablation	Timing of behavioural test	Phenotype	References
Radial maze: low memory load/limited interference	Mouse	129/SvEv	M	Irradiation	10 w	3 m post-irradiation	Improvement at relatively long delay	Saxe *et al.*, 2007
Radial maze: low memory load/limited interference	Mouse	GFAP-tk	M	Genetic, GCV induced	12–20 w	After 10 w GCV treatment	Improvement at long but not short delays	Saxe *et al.*, 2007
Radial maze	Rat	Wistar	M	Restraint stress × 21 d	2–2.5 m	1 or 10 d post-stress	Deficits in acquisition & reference errors	Veena *et al.*, 2009
Cued water maze: DNMS	Rat	Long Evans	M	Irradiation	4 m	4 w post-irradiation	Deficits at long but not short delays	Winocur *et al.*, 2006
Object recognition memory								
Object exploration in an arena	Rat	Wistar	B	Prenatal LPS treatment × E14-E20	n/a adult	60 d post-birth	Deficits in retention with 3 h delay interval, not 1 h	Graciarena *et al.*, 2010
Object exploration in an arena	Rat	Sprague-Dawley	B	Lentiviral dnWnt	7–8 w	8–9 w post-virus injection	Deficits in long-term retention	Jessberger *et al.*, 2009

(Continued)

Table 4.2. (*Continued*)

(A) Behavioural tasks	Species	Strain	Sex	Ablation method	Animal age at ablation	Timing of behavioural test	Phenotype	References
Place recognition memory								
Y maze	Mouse	NF-kB	M	Genetic knockout	Constitutive	3–4 m of age	Deficits in exploring novel arm	Denis-Donini *et al.*, 2008
Novel location in an arena	Mouse	C57BL/6J	M	MAM treatment × 14 d	9 w	1 d post-MAM	Deficits in exploring novel location	Goodman *et al.*, 2010
Novel location in an arena	Rat	Lister hooded	M	MTX treatment × 1 w	N/A, adult	1 m post-MTX	Deficits in exploring novel location	Lyons *et al.*, 2010
T-maze	Rat	Wistar	M	Irradiation	2 m	8 d post-irradiation	Deficits in exploring novel arm	Madsen *et al.*, 2003
T-maze	Rat	Wistar	M	Irradiation	2 m	21 d post-irradiation	Deficits in exploring novel arm	Madsen *et al.*, 2003
Spatial discrimination								
Radial arm maze: DNMP	Mouse	C56BL/6	F	Irradiation	8 w	8 w post-irradiation	Deficits in proximal but not distal spatial discrimination	Clelland *et al.*, 2009
Radial arm maze: DNMP	Mouse	C56BL/6	F	Lentivirus-dnWnt	8 w	8 w post-virus injection	Deficits in proximal but not distal spatial discrimination	Clelland *et al.*, 2009
Touch screen: 2 choice discrimination	Mouse	C56BL/6	F	Irradiation	8 w	16 w post-irradiation	Deficits in proximal but not distal spatial discrimination	Clelland *et al.*, 2009

(*Continued*)

Table 4.2. (*Continued*)

(B) Behavioural tasks	Species	Strain	Sex	Ablation method	Animal age at ablation	Timing of behavioural test	Phenotype	References
Morris water maze								
Hidden pf 2 trials/d × 5 d	Mouse	C57BL/6J	B	Irradiation	50-66 d	3–4 m post-irradiation	Acquisition & retention	Clark *et al.*, 2008
Hidden pf 1 trial/d × 10 d	Mouse	NF-kB	M	Genetic knockout	Constitutive	3–4 m of age	Acquisition & short-term retention	Denis-Donini *et al.*, 2008
Hidden pf 4 trials/d × 6 d	Mouse	PC3/Tis21	M	Genetic knockout	Constitutive	2 m of age	Acquisition & short-term retention	Farioli-Vecchioli *et al.*, 2009
Hidden pf 4 trials/d × 11 d	Mouse	D2	B	Genetic knockout	Constitutive	6–8 m of age	Acquisition & retention	Jaholkowski *et al.*, 2009
Hidden pf 5 trials/d × 5 d	Rat	Wistar	M	Irradiation	2 m	10 d post-irradiation	Acquisition	Madsen *et al.*, 2003
Visible then hidden pf 3 trials/d × 7d	Mouse	129/SvEv	F	Irradiation	10 w	104 d post-irradiation	Acquisition, short-term retention, & no interaction with enrichment	Meshi *et al.*, 2006
Visible then hidden pf 6 trials/d × 5d	Mouse	C57BL/6J	M	Irradiation	2 m	3 m post-irradiation	Acquisition & short term retention	Raber *et al.*, 2004
Visible then hidden pf 4 trials/d × 12 d	Mouse	129/SvEv	M	Irradiation	12-25 w	3 m post-irradiation	Acquisition, short- & long-term retention	Saxe *et al.*, 2006
Hidden pf 4 trials/d × 4 d	Rat	Sprague-Dawley	M	MAM treatment × 14 d	N/A, adult	2 d post-MAM	Acquisition & short term retention	Shors *et al.*, 2002

(*Continued*)

Table 4.2. (*Continued*)

(B) Behavioural tasks	Species	Strain	Sex	Ablation method	Animal age at ablation	Timing of behavioural test	Phenotype	References
Barnes maze								
Spatial then visual version	Mouse	C57BL/6J	M	Irradiation	21 d of age	3 m post-irradiation	Spatial version, visual version, & searching strategies	Rola *et al.*, 2004
Contextual fear conditioning								
10 tone-shock pairs following pre-exposure trace paradigm	Rat	Sprague-Dawley	M	Irradiation	P21, P50, & P70	90 d post-irradiation	Freezing	Achanta *et al.*, 2009
2 unsignalled shocks	Mouse	C57BL/6J	B	Irradiation	50–66 d	3–4 m post-irradiation	Freezing	Clark *et al.*, 2008
2 shocks (unpaired)	Mouse	Nestin-rtTA, TRE-Bax	M	Genetic, Dox induced	2 m	> 6 w post-Dox	Freezing	Dupret *et al.*, 2008
3 tone-shock pairs	Mouse	D2	B	Genetic knockout	Constitutive	4–6 m of age	Freezing 2 d later	Jaholkowski *et al.*, 2009
5 unsignalled shocks	Mouse	D2	B	Genetic knockout	Constitutive	3–4 m of age	Freezing 36 d later	Jaholkowski *et al.*, 2009
1 unsignalled shock	Mouse	C57BL/6	M	MAM treatment × 14 d	7–10 w	3 d post-irradiation	Contextual conditioning & extinction	Ko *et al.*, 2009

(*Continued*)

Table 4.2. *(Continued)*

(B) Behavioural tasks	Species	Strain	Sex	Ablation method	Animal age at ablation	Timing of behavioural test	Phenotype	References
10 tone-shock pairs following pre-exposure trace paradigm	Rat	Sprague-Dawley	M	MAM treatment × 14 d	N/A, adult	1 d post-MAM	Mobility	Shors *et al.*, 2002
7 tone-shock pairs	Mouse	C57BL/6	M	Irradiation	8–9 w	4 or 8 w post-irradiation	Freezing	Snyder *et al.*, 2009
3 tone-shock pairs	Mouse	Tlxfl/fl; CMV-CreER	B	Genetic, TM induced	8 w	4 w post-TM	Freezing	Zhang *et al.*, 2008
Cued fear conditioning								
1 tone-shock pair	Mouse	Nestin-rtTA; TRE-PC3	B	Genetic, Dox induced	30 or 138 d	65 or 22 d post-Dox	Freezing during tone	Farioli-Vecchioli *et al.*, 2008
3 tone-shock pairs	Mouse	Nes-CreER, NSE-DTA	M	Genetic, TM induced	8 w	59 d post-TM	Freezing during tone	Imayoshi *et al.*, 2008
3 tone-shock pairs	Rat	Wistar	M	Olfactory bulbectomy	2 m	6 w post-surgery	Freezing during tone	Jaako-Movitz & Zharovsky 2005

(Continued)

Table 4.2. *(Continued)*

(B) Behavioural tasks	Species	Strain	Sex	Ablation method	Animal age at ablation	Timing of behavioural test	Phenotype	References
3 tone-shock pairs	Mouse	D2	B	Genetic knockout	Constitutive	4-6 m of age	Freezing	Jaholkowski *et al.*, 2009
3 tone-shock pairs	Mouse	129/SvEv	M	Irradiation	12–25 w	3 m post-irradiation	Freezing during tone	Saxe *et al.*, 2006
5 tone-shock pairs	Mouse	GFAP-tk	M	Genetic, GCV induced	12–20 w	after 6 w GCV treatment	Freezing during tone	Saxe *et al.*, 2006
3 tone-shock pairs	Mouse	FoxG1	B	Genetic knockout	Constitutive	12–14 w of age	Freezing during tone	Shen *et al.*, 2006
3 tone-shock pairs	Rat	Sprague-Dawley	M	Irradiation	N/A, adult	6 w post-irradiation	Freezing during tone	Warner-Schmidt *et al.*, 2008
10 tone-shock pairs	Rat	Long Evans	M	Irradiation	4 m	4 w post-irradiation	Freezing during tone	Winocur *et al.*, 2006
7 tone-shock pairs	Rat	Long Evans	M	Irradiation	4.5 m	5 w post-irradiation	Freezing during tone	Wojtowicz *et al.*, 2008
1 tone-shock pair	Mouse	ICR	M	Restraint stress × 4 w	3 w	1 d post-stress	Freezing during tone	Yun *et al.*, 2010
3 tone-shock pairs	Mouse	Tlxfl/fl; CMV-CreER	B	Genetic, TM induced	8 w	4 w post-TM	Freezing during tone	Zhang *et al.*, 2008

(Continued)

Table 4.2. (*Continued*)

(B) Behavioural tasks	Species	Strain	Sex	Ablation method	Animal age at ablation	Timing of behavioural test	Phenotype	References
Trace fear conditioning								
10 tone-shock pairs following pre-exposure trace paradigm	Mouse	D2	B	Genetic knockout	Constitutive	6–7 m of age	Increased freezing	Jaholkowski *et al.*, 2009
1 tone-shock pair in trace paradigm	Mouse	D2	B	Genetic knockout	Constitutive	6–8 m of age	Freezing	Jaholkowski *et al.*, 2009
Working memory								
T-maze: MTP & NMTP tasks	Rat	Long Evans	M	Irradiation	9-11 w	2–8 w post-irradiation	No effect	Hernandez-Rabaza *et al.*, 2009
MWM: DMP	Mouse	129/SvEv	M	Irradiation	12-25 w	3 m post-irradiation	No effect	Saxe *et al.*, 2006
Radial maze:low memory load/high interference	Mouse	129/SvEv	M	Irradiation	10 w	3 m post-irradiation	No effect	Saxe *et al.*, 2007
Object recognition memory								
Object exploration in an arena	Rat	Sprague-Dawley	M	MAM treatment × 14 d	N/A adult	2 d post-MAM	No effect on retention, deficits in enrichment induced long-term retention	Bruel-Jungerman *et al.*, 2005

(*Continued*)

Table 4.2. (*Continued*)

(B) **Behavioural tasks**	**Species**	**Strain**	**Sex**	**Ablation method**	**Animal age at ablation**	**Timing of behavioural test**	**Phenotype**	**References**
Touch screen: object-place association	Mouse	C56BL/6	F	Irradiation	8 w	8 w post-irradiation	No effect	Clelland *et al.*, 2009
Object exploration in an arena	Mouse	C57BL/6J	M	MAM treatment × 14 d	9 w	1 d post-MAM	No effect	Goodman *et al.*, 2010
Object exploration in an arena	Mouse	D2	B	Genetic knockout	Constitutive	4–5 m of age	No effect	Jaholkowski *et al.*, 2009
Object exploration in an arena	Rat	Wistar	M	Irradiation	2 m	5 or 18 d post	No effect	Madsen *et al.*, 2003
Object exploration in an arena	Mouse	C57BL/6J	M	Irradiation	2 m	3 m post-irradiation	No effect	Raber *et al.*, 2004
Object exploration in an arena	Mouse	C57BL/6J	M	Irradiation	21 d	3 m post-irradiation	No effect	Rola *et al.*, 2004
Place recognition memory								
T-maze	Rat	Wistar	M	Irradiation	2 m	42 d post-irradiation	No effect	Madsen *et al.*, 2003
Novel location in an arena	Mouse	C57BL/6J	M	Irradiation	2 m	3 m post-irradiation	No effect	Raber *et al.*, 2004

(*Continued*)

Table 4.2. (*Continued*)

(B) Behavioural tasks	Species	Strain	Sex	Ablation method	Animal age at ablation	Timing of behavioural test	Phenotype	References
Novel location in an arena	Mouse	C57BL/6J	M	Irradiation	21 d	3 m post-irradiation	No effect	Rola *et al.*, 2004
Y maze	Mouse	129/SvEv	M	Irradiation	12–25 w	3 m post-irradiation	No effect	Saxe *et al.*, 2006

Abreviations:

B: Both male and female
D2: Cyclin D2
DMP: Delayed matching to place
DNMP: Delayed non-matching to place
DNMS: Delayed non-match to sample
Dox: Doxycycline
DTA: Diphtheria toxin fragment A
E: Embryonic day
F: Female
FGFR1: Fibroblast growth factor receptor protein 1
GCV: Ganciclovir
GFAP: Glial fibrillary acidic protein
LPS: Lipopolysaccharide
M: Male
MAM: Methylazoxymethanol acetate
MBD1: methyl-CpG binding protein 1
MTP: Matching to place
MTX: Methotrexate
MWM: Morris water maze
n/a: Not available
NMTP: Non-matching to place
NSE: Neuronal specific enolase 2
NVP-AAM077: NR2A antagonist
P: Postnatal day
pf: Platform
TM: Tamoxifen
TMZ: Temozolomide
TRE: Tetracycline responsive elements

(A) Studies showing an effect. Modified and updated from Deng *et al.* (2010).
(B) Studies showing no effect. Modified and updated from Deng *et al.* (2010).

(Table 4.2A). In theory, genetic manipulations can lead to a more selective effect on neurogenesis, thereby lessening the possibility for confounding links between neurogenic rates and memory function. For example, stereotactic DG infections with dominant-negative WNT-expressing lentivirus disrupts the WNT signalling pathway specifically in the DG, reduces neurogenesis, and impairs water maze, delayed nonmatching-to-place, and object recognition learning (Clelland *et al.*, 2009; Jessberger *et al.*, 2009). Several transgenic knockout mouse lines with constitutive genomic alterations in transcription factors, epigenetic regulators, and growth factors exhibit diminished adult hippocampal neurogenesis and deficits in spatial (Denis-Donini *et al.*, 2008; Trejo *et al.*, 2008; Zhao *et al.*, 2003) and contextual fear (Farioli-Vecchioli *et al.*, 2009; Shen *et al.*, 2006) memory. Using genetic recombination under the control of the nestin promoter to specifically target genes involved in brain development, fibroblast growth factor receptor 1 or neurotrophin-3 knockout impedes adult hippocampal neurogenesis, partially alters DG LTP and impairs spatial memory (Shimazu *et al.*, 2006; Zhao *et al.*, 2007).

Inducible, and in some cases reversible, genetic techniques of specific neurogenesis suppression are the latest tools available to tease out functional contributions for adult-generated DGCs in learning and memory (Table 4.2A). In principle, these approaches retain many of the benefits of other techniques, but additionally offer three advantages: (1) Animals develop undisturbed until a desired age, attaining normal adult neural anatomy and function without potentially generating compensatory mechanisms during development. (2) DG neurogenesis can be suppressed with cell-type specificity at a desired age and/or time point relative to memory acquisition, retention, or recall. (3) In some systems, this effect is reversible such that the potential for rescue of function can be examined.

A variety of transgenic mice that enable inducible neurogenesis ablation have been examined, including those that exploit the expression of herpes simplex virus-thymidine kinase (HSV-TK), Bax protein, PC3 transgene, diphtheria toxin, and which knockout a key mediator of neural stem cell maintenance. Mice that express HSV-TK under the control of cell-type–specific promoters can be treated with the antiviral ganciclovir, eliminating dividing TK positive cells (Garcia *et al.*, 2004). Saxe *et al.* utilized HSV-TK under GFAP promoter control such that GFAP positive progenitor cells were ablated in adulthood upon drug administration (Saxe *et al.*, 2006). Following treatment, freezing was significantly reduced in contextual fear conditioning. In another transgenic mouse line, HSV-TK is expressed under nestin promoter control in the brain only (Deng *et al.*, 2009). Here, ablation of dividing TK-expressing cells ceases upon drug withdrawal, restoring adult hippocampal neurogenesis. Using this method, a selective partial removal of adult-generated DGCs that would have been 1–4 weeks old during water maze training impaired memory retention during subsequent testing. This memory deficit could not be explained by enhanced extinction. Furthermore, the fact that the memory impairment withstood reversal of the neurogenesis suppression specifically points to the importance of newly generated DGCs during spatial learning.

Ablating neurogenesis through induced expression of transgenes, under nestin promoter control, enables relatively specific knockdown of hippocampal neurogenesis at desired adult time-points. Transcription factors that activate the transgene expression without such systems utilize transcription factors to control transgene expression knockdown with drug exposure.

"Bax" and "TgPC3" mice overexpress the pro-apoptotic Bax protein (Dupret *et al.*, 2008) or pro-neuronal differentiation PC3 gene (Farioli-Vecchioli *et al.*, 2008) respectively, using a transcription factor that is activated by exogenous tetracycline analogues. The transcription factor itself is expressed under nestin promoter control, thus enabling relatively specific adult hippocampal neurogenesis ablation (Bax expression) or altered maturation (PC3 expression) with drug exposure. Upon induction in adulthood, these animals exhibit deficits in hippocampal-dependent water maze learning (Dupret *et al.*, 2008; Farioli-Vecchioli *et al.*, 2008), as well as radial arm maze and contextual fear learning (Farioli-Vecchioli *et al.*, 2008).

Transgenic mice that express diphtheria toxin in newly generated neurons following tamoxifen exposure can be generated by crossing nestin promoter-controlled inducible Cre recombinase mice with those carrying a reversibly silenced diphtheria toxin gene coupled to a neuron-specific promoter (Imayoshi *et al.*, 2008). In these animals, neurogenesis ablation resulted in impairments of both Barnes maze and contextual fear conditioning. The tamoxifen inducible Cre recombinase system was also utilized in another study to knockout expression of the orphan nuclear receptor TLX from the adult neural stem cell pool (Zhang *et al.*, 2008). TLX is known to maintain the proliferative capacity of neural stem cells and is critical for normal neurologic development (Shi *et al.*, 2004). Defective adult neurogenesis in these animals impaired spatial learning in the Morris water maze.

3.3.3. *Inconsistencies in the literature*

Despite the considerable mass of studies supporting an important role for adult-generated DGCs in hippocampal-dependent learning and memory, there are many conflicting data (Table 4.2B). While these discrepancies signal the need for cautious interpretation of data and remind us to avoid the trappings of dogma, careful examination of these studies underscores several important technical considerations.

The false attribution of behavioural phenotypes to neurogenesis suppression can arise due to several confounding factors. The specificity of neurogenesis knockdown as well as possible developmental consequences of treatment in early life are potentially important alternative factors in determining behavioural phenotypes. For instance, irradiation interferes with cell cycle proteins and instigates DNA damage, preferentially targeting proliferating cells, thus depleting new neurons in a dose-dependent fashion (Wojtowicz, 2006). The effects of radiation, however, are not neurogenesis specific (Leuner *et al.*, 2006a), and include an inflammatory reaction which may impact memory performance (Monje *et al.*, 2002; Rola *et al.*, 2004a; Rola *et al.*, 2004b). Some evidence also suggests that contextual fear memory deficits are only seen when a sufficiently high dose of radiation is used (Ko *et al.*, 2009). This could indicate the necessity for a profound decline in neurogenesis to generate a behavioural effect; however, it could also reflect increased nonspecific toxicity associated with higher radiation exposure. Analogous side effect and dose-response concerns have been raised for other anti-neurogenic treatments, such as MAM (Dupret *et al.*, 2005), as well as for viral vector-mediated genetic manipulations (Jessberger *et al.*, 2009; Thomas *et al.*, 2003). Lastly, transgenic animals inherently carry a risk for developmental abnormalities, including subtle differences in brain maturation, leading to unique susceptibilities to neurologic impairments (Breunig *et al.*, 2008).

While each individual method of suppressing neurogenesis has its own set of drawbacks, the fact that suppressing neurogenesis using such a wide array of methods produces similar phenotypic impairments provides considerable support for an important role for adult-generated DGCs in hippocampal-dependent memories. In spite of this, many studies utilizing similar techniques do not demonstrate a critical role for adult hippocampal neurogenesis in some hippocampal-dependent learning paradigms (Achanta *et al.*, 2009; Bruel-Jungerman *et al.*, 2005; Clark *et al.*, 2008; Clelland *et al.*, 2009; Denis-Donini *et al.*, 2008; Dupret *et al.*, 2008; Farioli-Vecchioli *et al.*, 2009; Goodman *et al.*, 2010; Hernandez-Rabaza *et al.*, 2009; Jaholkowski *et al.*, 2009; Ko *et al.*, 2009; Madsen *et al.*, 2003; Meshi *et al.*, 2006; Raber *et al.*, 2004; Rola *et al.*, 2004b; Saxe *et al.*, 2006; Saxe *et al.*, 2007; Shors *et al.*, 2002; Snyder *et al.*, 2009a; Zhang *et al.*, 2008). These studies could, of course, show that neurogenesis is not critical to these tasks; however, the volume of work indicating its importance suggests otherwise. Several factors could diminish the likelihood of detecting a phenotypic effect of suppressing neurogenesis, including differences between species and strains, compensatory mechanisms following neurogenesis suppression, variations in behavioural testing protocols, and the age of new neurons at training/testing.

First, it is possible that in the minority of cases, discordant results could be obtained between studies using separate species and strains. These different animals can exhibit distinct baseline rates of adult-generated DGC production and maturation, behavioural phenotypes in hippocampal-dependent learning and memory paradigms, and fundamental differences in the patterns and extent of adult born DGC integration into memory networks (Cameron *et al.*, 1993; Epp *et al.*, 2011b; Hayes and Nowakowski, 2002; Kempermann and Gage, 2002b; Schauwecker, 2006; Snyder *et al.*, 2009a).

Second, a more significant reason that some studies suppressing neurogenesis do not reveal behavioural effects relates to compensatory mechanisms. Virtually all studies knocking down neurogenesis do so prior to training (Table 4.2), creating a potential opportunity for remaining adult- or developmentally-generated DGCs to compensate for this loss. Residual adult-generated DGCs following incomplete knockdown techniques might adequately fulfill functional requirements (Dupret *et al.*, 2005; Jessberger *et al.*, 2009; Ko *et al.*, 2009). This could explain why the profound age-related decline in neurogenesis precedes the appearance of cognitive deficits. In addition, similar incorporation rates of adult- and developmentally-generated DGCs into networks supporting hippocampal-dependent memories suggest their functional equivalence (Stone *et al.*, 2010); accordingly, developmentally-generated DGCs might assume the learning function of their missing adult-generated counterparts. One strategy to test whether compensation occurs when adult-generated DGCs are removed is to lower their functional, rather than physical, presence. For instance, Farioli-Vecchioli *et al.* accelerated the maturational process of immature adult-generated DGCs by conditional expression of a pro-neuronal differentiation gene in progenitor cells (Farioli-Vecchioli *et al.*, 2008). These animals were selectively deficient in immature neurons, and harbored normal numbers but altered morphologies of mature adult-generated neurons. These changes were accompanied by spatial and contextual fear memory deficits for both new and previously learned information, reduced DG LTP, and less testing-induced activation of adult-generated DGCs. One interpretation of these results is that the continued presence of functionally impaired adult-generated DGCs prevented the intiation of compensatory processes by other DGCs and thus led to the observed phenotype. A more definitive test of whether adult-generated DGCs contribute to learning and memory would be to have animals learn during a period of normal neurogenesis, subsequently eliminate those newly born cells that should have been recruited into hippocampal memory networks, and then look for impaired recall.

Third, protocol variations for hippocampal-dependent memory tasks between studies could significantly influence results by potentially altering underlying neural mechanisms and the performance demands placed on animals. For instance, task difficulty can influence hippocampal dependency, as demonstrated by classical conditioning paradigms becoming hippocampal dependent only once rendered sufficiently difficult by an extended insterstimulus interval (Beylin *et al.*, 2001). It is also possible that hippocampal-dependent tasks must be sufficiently difficult in order to detect subtle differences in performance. Given the relatively small proportional contribution of adult-generated DGCs to hippocampal-dependent memory networks (Kee *et al.*, 2007b; Snyder *et al.*, 2009a; Stone *et al.*, 2010; Tashiro *et al.*, 2007; Trouche *et al.*, 2009) and functional equivalence between adult- and developmentally-generated DGCs (Stone *et al.*, 2010), it is reasonable to expect that suppressing neurogenesis might only translate into detectable phenotypes when animals are subjected to tasks with sufficiently high hippocampal performance demands. For instance, suppressing neurogenesis may reduce contextual fear conditioning-induced freezing only when brief training is provided (Drew *et al.*, 2010). This is also supported by the fact that Morris water maze acquisition deteriorated with reduced neurogenesis when a moving, rather than fixed, hidden platform location was used in otherwise identical protocols (Zhao *et al.*, 2007), and improved following exercise-induced enhanced neurogenesis only when animals received a small number of daily trials (van Praag *et al.*, 1999a), possibly relating to greater task difficulty. Clearly, relationships and interactions between neurogenesis, task difficulty, and hippocampal dependency require further clarification (Shors *et al.*, 2002).

Fourth, the age of removed adult-generated DGCs during behavioural testing could impact the likelihood for detecting, and the nature of, phenotypic consequences. Given the aforementioned developmental timeline of adult-generated DGCs, it is unlikely that cells merely days old could functionally participate in local networks supporting memory. In contrast, as many week-old adult-generated DGCs approach full maturity, participation resembling that of mature developmentally generated DGCs is relatively likely. Inconsistent results between these time-points (Table 4.2) could, in part, reflect variability in the age of deleted new neurons at the time of behavioural training/testing (Deng *et al.*, 2010). Indeed, it is likely that painstaking highly-resolved and carefully controlled experiments to delineate an evolving effect of knocking out differentially aged adult-generated cells are needed to better define this issue.

4. Contributions of neurogenesis to memory at the network level

Several theories propose to explain how adult-generated DGCs contribute to memory formation at the network or cognitive level. Prior to discussing these theories, it is worth noting several general complicating factors. First, our evolving biological understanding of adult neurogenesis creates a moving target for theorists. For instance, recent evidence indicates that adult hippocampal neurogenesis primarily leads to a net increase in the DGC population over time (Imayoshi *et al.*, 2008); however, older studies pointed to a greater role in DGC replacement and thus promoted theories emphasizing neural turnover (Dayer *et al.*, 2003; Lledo *et al.*, 2006). Second, different theories are not necessarily mutually exclusive, and it is possible that adult-generated neurons, or subsets of them, contribute to multiple, possibly even opposing, cognitive processes during different periods of their development (Leuner *et al.*, 2006a). Finally, our incomplete understanding of hippocampal function also hampers attempts to propose functional roles for new neurons (Kempermann, 2002). General consensus posits that the hippocampal formation initially encodes episodic memories regarding context-specific information, and spatial memories involving environmental relations and orientations. However, the roles of mesial temporal lobe structures in memory, and their interaction with other brain regions, continue to be debated. Accordingly, this review attempts to relate neurogenesis to memory processes without explicitly entering this debate. The interested reader is invited to compare and contrast theories of mesial temporal lobe and neocortical interactions in learning and memory (Frankland and Bontempi, 2005; Moscovitch *et al.*, 2006; Squire and Zola-Morgan, 1991; Winocur *et al.*, 2010).

The following sections discuss potential cognitive roles for adult-generated DGCs in increasing the capacity for information processing, providing memory storage, clearing memory traces, associating memories, and discriminating between similar patterns.

4.1. *Increasing information processing capability*

Regardless of their details, most published theories propose that generating more DGCs in adulthood is beneficial to learning and memory. One such theory asserts that the primary purpose of adult neurogenesis is to allow the brain to increase mnemonic capacity by assisting the hippocampus to accommodate greater levels of novelty and complexity (Kempermann, 2002; 2008). In doing so, new neurons could act as new "gatekeepers" (Kempermann, 2002), increasing the processing ability of the DG to prepare information for memory storage. A "neurogenic reserve", or greater potential to remain flexible and plastic, is created when neurogenesis increases in response to functional stimulation (such as enrichment) that signifies probable future increased cognitive demand (Kempermann, 2008). Accordingly, increased neurogenesis associated with certain experiences, such as enrichment and exercise, might non-specifically contribute to learning through a positive feedback loop (Kempermann, 2002). This would predict that the functional benefit of increasing adult hippocampal neurogenesis develops in a delayed fashion, because it takes time for adult-generated DGCs to mature and functionally integrate into hippocampal circuits (Kee *et al.*, 2007b; van Praag *et al.*, 2002). While nicely accounting for the activity-dependent regulation of neurogenesis rates and delayed maturation of adult-generated DGCs, this theory has yet to be extensively tested directly. For instance, it is not known if manipulations that increase neurogenesis only improve performance in memory tasks after a sufficient cell maturational delay. In addition, one would need to demonstrate that increasing or decreasing neurogenesis expands or contracts memory processing capacity in the long term through an experimental design that reliably stresses information input capacity.

4.2. *Providing memory storage*

A changing population of adult-generated neurons might provide transient storage of information (Barnea and Nottebohm, 1996; Gould *et al.*, 1999c; Gross, 2000). This possibility is analogous to the annual renewal of song repertoire in adult canaries, linked to the temporary incorporation of new neurons into their high vocal center (Kirn *et al.*, 1991). The plasticity features of newborn neurons might support flexible roles in temporary memory storage that older cells are less able to perform, such as the detection and encoding of novel stimuli (Leuner *et al.*, 2006a). Computational simulations have elaborated upon this role for adult hippocampal neurogenesis in short-term information coding, namely by providing distinct codes for highly similar events that ultimately permit an increase in memory capacity (Becker, 2005). Using another network model, Weisz and Argibay demonstrated that a lump increase in the DGC population of 30% significantly improved the model's ability to accurately retrieve a specific recently learned input pattern, and this benefit was enhanced when the system was exposed to greater numbers of competing input patterns that simulate increasing storage requirements (Weisz and Argibay, 2009). In keeping with concepts of memory reorganization and eventual transfer to extra-hippocampal regions (Squire and Zola, 1997), the turnover of some new DGCs could also be linked to early transient processes in memory formation, after which those neurons may no longer be required to support consolidated memories (Barnea and Nottebohm, 1996; Gould *et al.*, 1999c; Gross, 2000). However, the long-term survival of some adult-generated DGCs suggests they could then still serve as yet undetermined longer-lasting functions.

Another potential role for adult-generated DGCs is in providing additional plastic substrates for long-term memory storage. Continued modification of existing neural elements, as they become increasingly shared amongst different representations, could eventually create interference between those representations and render them indistinguishable from one another (Carpenter and Grossberg, 1987; Nottebohm, 2002). The necessity to avoid such "catastrophic interference" (McCloskey and Cohen, 1989) is particularly relevant to the DG in most computational simulations of hippocampal function (Bakker *et al.*, 2008; Leutgeb *et al.*, 2007; O'Reilly and McClelland, 1994; Treves and Rolls, 1994). Overlapping EC inputs must be highly separated in order to encode different memories in CA3, and it is believed that this separation task occurs primarily in the DG. Continually adding new neurons to the DG could provide additional structural plasticity, thus increasing the potential for generating non-interfering memory traces when creating new long-term memory stores. Nottebohm suggests that the newly born neurons may be preferentially selected for storing new memories, thereby protecting old memories from interference (Deng *et al.*, 2010; Nottebohm, 2002). Consistent with this hypothesis, both simplified neural network and hippocampal memory models demonstrate that adding highly plastic new neurons to encode novel information can prevent new learning from interfering catastrophically with older memories, in-turn preserved by stable older neurons (Appleby and Wiskott, 2009; Wiskott *et al.*, 2006). This would predict that animals who learnt a given task in the remote past, underwent neurogenesis suppression in the present, and subsequently learnt a new highly similar task would experience a significant disruption of the older memory because the new memory would have no alternative but to use and modify some neurons critical to maintaining the older memory.

Despite the potential utility of new neurons in reducing interference between long-term memories, some data suggest that adult neurogenesis can actually increase interference under certain short-term memory conditions. In a relevant study, focal hippocampal irradiation or an inducible transgenic knockout of hippocampal neurogenesis actually improved performance in a radial maze task (Saxe *et al.*, 2007). This task was designed to promote interference, requiring the animal to disregard conflicting non-relevant information from previous trials. Unlike controls, neurogenesis-deficient animals were relatively insensitive to this interference. The authors speculated that blocking neurogenesis may reduce short-term memory capacity sufficiently to prevent the formation of interfering memories.

4.3. *Clearing memory traces*

In contrast to increasing processing or storage capacity, other models propose that DG neuronal turnover clears older memories and in so doing potentially increases capacity for new learning (Chambers *et al.*, 2004). Some animal data support this concept, including a first study using presenilin-1 conditional knockout mice (Feng *et al.*, 2001). These transgenic AD model animals exhibit normal baseline hippocampal neurogenesis, deficient enrichment-induced increased neurogenesis, and no identified alterations in basal synaptic function or plasticity. Knockout mice suffered no deficits during hippocampal-dependent contextual fear learning relative to controls; however, when subjected to a sequential paradigm of learning, enrichment for 2 weeks, and retrieval, knockout mice outperformed controls in the retention test. Feng *et al.* postulated that enrichment-induced increased neurogenesis in the DG of controls facilitated the clearance of hippocampal memory traces (Feng *et al.*, 2001). In the absence of this clearance, memory retention was superior in knockout mice. They also propose that DGC turnover could purge hippocampal memory traces after they are consolidated into long-term memory, enabling the system to process new memories and avoid interference. Indeed, the time it takes for hippocampal-cortical transfer of contextual fear memories in rodents matches the baseline survival period for the majority of their adult-generated DGCs (Cameron *et al.*, 1993; Kim and Fanselow, 1992). Increasing the rate of adult neurogenesis, and presumably therefore DGC turnover, could accelerate this clearance process. In a more direct test of this theory, *in vivo* radiation-induced depression of hippocampal neurogenesis promoted the maintenance of DG LTP and extended the period of hippocampal dependency for contextual fear memories in rodents (Kitamura *et al.*, 2009). Pharmacological hippocampal inactivation during retrieval impaired remote memory expression specifically in animals with previously suppressed neurogenesis (either by radiation or transgenic means) (Kitamura *et al.*, 2009). Moreover, exercise-induced enhanced neurogenesis accelerated the decay of hippocampal dependency without altering retrieval. Collectively, these studies not only point to a role for adult-generated DGCs in degrading existing memories, but also suggest that such degradation may promote the transformation of memories from hippocampal-dependent to hippocampal-independent forms in the cortex.

Ascribing a role for memory clearance to adult-generated DGCs warrants caution for at least two reasons. First, broad assertions regarding hippocampal memory processing from these well-designed studies are inherently hampered by many potential confounds. Effects of presenilin-1 knockout on memory could be mediated through several known non-neurogenesis–related mechanisms, including modulation of neuronal calcium entry (Yoo *et al.*, 2000), alterations in amyloid precursor protein processing or accumulation (McGuire and Davis, 2001), and secondary effects of presenilins on other proteins via their secretase activity (Brown *et al.*, 2000). Likewise, as previously discussed, radiation-induced brain inflammation can impact memory performance (Monje *et al.*, 2002; Rola *et al.*, 2004a; Rola *et al.*, 2004b). Second, balancing the afore mentioned memory-promoting influence of neurogenesis with this idea of memory disruption likely necessitates some form(s) of homeostatic control mechanism(s) that adaptively regulate(s) the system (Chambers and Conroy, 2007; Chambers *et al.*, 2004; Meltzer *et al.*, 2005). Further studies will be required in order to establish the generalizeability and biological control of this hippocampal memory clearance phenomenon.

4.4. *Encoding associations*

A suggested critical learning and memory function of the hippocampus is to construct new associations between items separated in time or space (Eichenbaum *et al.*, 1990; Rawlins, 1985; Wallenstein *et al.*, 1998). Shors *et al.* proposed a specific role for adult-generated DGCs in forming a subset of associations, namely trace memories (Shors *et al.*, 2001; Shors *et al.*, 2002). Trace memories are maintained associations between discontinuous stimuli that were experienced at different times (Pavlov, 1927). These associations are studied using trace conditioning (TC), a type of classical conditioning where an interval of time or a "trace" separates the unconditioned and conditioned stimuli. In contrast, delayed conditioning (DC) occurs when the unconditioned and conditioned stimuli overlap. Unlike DC, evidence suggests that TC is dependent on the hippocampus (McEchron *et al.*, 1998; Moyer *et al.*, 1990; Shors, 2004; Solomon *et al.*, 1986). While traditional mechanisms of hippocampal neural plasticity have been implicated in TC, including postsynaptic density remodeling and increased dendritic spine density of CA1 pyramidal neurons (Geinisman *et al.*, 2000; Leuner *et al.*, 2003), so has adult hippocampal neurogenesis (Shors *et al.*, 2001). Shors *et al.* partially reduced neurogenesis and demonstrated a selective impairment for TC, not DC, which was subsequently rescued as neurogenesis recovered. Related studies demonstrate that associative learning in general increases the survival of immature new neurons in the adult hippocampus, regardless of stimulus timing, as long as a hippocampal-dependent behavioural paradigm is employed (Gould *et al.*, 1999a; Leuner *et al.*, 2006b).

Another theoretical associative function for new hippocampal neurons is in encoding time in new episodic memories (Aimone *et al.*, 2006, 2009). Although there is general consensus that time is associated with episodic memories, how this occurs remains undetermined (Friedman, 1993). It is plausible that the hippocampus encodes temporal associations between new episodic memories given its central role in their formation (Moscovitch *et al.*, 2005; Scoville and Milner, 1957; Squire and Zola-Morgan, 1991). Aimone *et al.*'s theory presupposes that immature adult-generated DGCs preferentially integrate into memory networks (Aimone *et al.*, 2006). Assuming that mature granule neurons encode events by converting EC inputs into sparse and distinct downstream representations (McNaughton and Morris, 1987; Treves and Rolls, 1994), Aimone *et al.* hypothesized that the constantly renewing population of immature DGCs encode inputs in a relatively less distinct manner (Aimone *et al.*, 2006). Temporally associated representations could harbour partially overlapped immature granule cell-derived code, computationally modelled as pattern integration. This fits with the suggestion that experiences during a critical developmental period can promote experience-specific functional integration (Tashiro *et al.*, 2007; Trouche *et al.*, 2009). This could also account for the temporal properties of recency judgments in humans that are consistent with the time course for the maturation of newborn granule cells (Friedman, 1993, 2001). There is a rapid increase in the subjectively determined ages of events in the first weeks following their experience, when immature DGCs are undergoing considerable developmental changes, but small changes in perceived age beyond about 1–2 months when cells become less distinct from each other and developmentally generated DGCs.

Preferential incorporation of adult-generated DGCs into hippocampal memory networks is critical to the temporal association theory for adult hippocampal neurogenesis. As previously discussed, preferential selection of these neurons during encoding is suggested by their transiently enhanced plasticity relative to their developmentally generated counterparts (Ge *et al.*, 2007; Schmidt-Hieber *et al.*, 2004). While previous behavioural studies have also supported preferential integration by indirectly comparing functional incorporation of BrdU-labelled adult-generated DGCs to that of NeuN-labelled cells in the same animals (Kee *et al.*, 2007b; Ramirez-Amaya *et al.*, 2006; Tashiro *et al.*, 2007), a direct comparison of equivalently labelled adult- and developmentally generated DGCs found similar integration rates in hippocampal memory networks, suggesting a functional equivalence between DGCs generated at different developmental stages (Stone *et al.*, 2010). Moreover, mature adult-generated DGCs exhibit similar neuronal connectivity, morphology, and electrophysiologic properties as developmentally generated DGCs (Ge *et al.*, 2007; Laplagne *et al.*, 2006; Laplagne *et al.*, 2007; Toni *et al.*, 2008; Toni *et al.*, 2007; Zhao *et al.*, 2006). Although this evidence for functional convergence does not support a critical contribution of adult neurogenesis through outcompeting neighbours, it is not incompatible with a critical role via some other mechanism. Important contributions are still possible, especially in light of evidence that a normal neurogenesis level is necessary for hippocampal memory function in some conditions (Deng *et al.*, 2009; Shors, 2008). Incorporation studies solely indicate the number of adult-generated DGCs involved in memory processes, thus only permitting assessment of theories that hinge upon alterations in cellular incorporation rates. The absence of unique recruitment rates for adult-generated DGCs into larger memory networks still does not tell us what qualitative or content aspects of memory might be specifically mediated by these neurons.

The specific idea that cohorts of immature postnatally generated DGCs are permanently tuned to temporally linked events is also challenged by a series of experiments exposing rats to several environments at different times (Alme *et al.*, 2010). If adult-generated DGCs preferentially integrate into memories during a highly plastic period of development, experiencing multiple environments at separate points in time should generate network representations for each experience that include distinct cohorts of adult-generated DGCs. These representations should overlap less than those created by the same environments if experienced around the same time, because these temporally linked experiences should share temporally tuned adult-generated DGCs. Accordingly, the total population of DGCs activated upon re-exposure to those experiences should be larger if they were initially separated by time. Both expression of the IEG *Arc* in, and electrophysiologic recordings from, putative DGCs indicate no such cumulative recruitment (Alme *et al.*, 2010). In addition, the authors suggest that instead of permanently tagging memories with a temporal context through integrating into networks, recently born DGCs might preferentially respond to new stimuli in a temporary fashion in order to sculpt CA3 activity and generate time-encoded downstream representations. New DGCs would then become less excitable with age and respond minimally either to reinstatement of remote experiences or to new, even highly similar, events. This functional "retirement" of DGCs would lessen network overlap between temporally dispersed events because their respective CA3 representations would be created by different cohorts of adult-generated DGCs. Collectively, these findings suggest that if adult neurogenesis specifically provides a mechanism for encoding time in new memories, this likely does not occur at the cellular integration level.

Despite the aforementioned links between immature adult-generated-neurons and creating new memory associations, their role once mature remains poorly understood. For instance, the function of long-term survivors beyond the

1-week period of apparent hippocampal dependency for TC is unclear (Shors, 2004; Takehara *et al.*, 2003). Whether this indicates a transient purpose for these new neurons in trace associations, followed by other functions, remains unknown. Similarly unanswered is if immature granule neurons encode time, would they do so permanently or would downstream representations produced by these cells eventually do so independently. Aimone *et al.* theorizes that early pattern integration functions could eventually be complemented by the development of pattern separation capabilities (Aimone *et al.*, 2009). They propose that adult-generated DGCs form specialised dimensions based on environmental experiences during maturation and eventually encode new information using those predefined dimensions, allowing the network to processes new information in the context of past experiences.

4.5. *Discrimination between similar patterns*

Pattern separation, the process of transforming similar representations or memories into highly dissimilar, non-overlapping representations, is considered a critical step in information processing within the hippocampal formation (McNaughton and Morris, 1987). Computational and behavioural studies assign this role to the DG given its relatively large neuronal population, sparse encoding, and ability to activate specific CA3 pyramidal neurons using single DGCs (Bakker *et al.*, 2008; Gilbert *et al.*, 2001; Leutgeb *et al.*, 2007; O'Reilly and McClelland, 1994; Treves and Rolls, 1994). Not surprisingly, adult hippocampal neurogenesis has recently been implicated in this process.

In the first direct experimental evaluation of adult hippocampal neurogenesis in pattern separation, mice receiving focal hippocampal brain irradiation suffered a specific depression of DG neurogenesis and impaired performance in a delayed nonmatching-to-place radial arm maze task in which locations were closely spaced (Clelland *et al.*, 2009). The radial arm maze findings were also verified substituting radiation with DG lentiviral vector delivery of dominant negative WNT protein to partially knock down neurogenesis. Furthermore, this behavioural phenotype, with low but not high degrees of spatial separation, generalised to a touch screen spatial task only when a pattern separation paradigm was employed. More recently, enhanced neurogenesis following voluntary running improved spatial pattern discrimination performance (Creer *et al.*, 2010). An important confound, however, is the difference in task difficulty between low and high spatial separation tasks. It is possible that the low spatial separation tasks were sufficiently challenging to reveal a small memory deficit, but the less difficult high separation trials were insensitive to any difference.

6. Conclusions and clinical implications

Decades of research have demonstrated that dentate granule cells are generated throughout mammalian adulthood. Moreover, their assumption of neural function within the hippocampal circuit and involvement to some degree in hippocampal memory processing is reasonably established. In spite of these monumental achievements, the specific purpose and overall significance of this phenomenon remains somewhat elusive. Although answers to these questions are unlikely to come easily, they are the targets of intensive research focus across the globe.

Motivations for understanding the relationships between adult hippocampal neurogenesis and memory function partly stem from the growing unmet clinical need for cognitive-restoring therapies for neurodegenerative diseases. Roughly 1 in 11 North Americans over age 65 have Alzheimer's disease (AD) or a related dementia, costing over 150 billion dollars annually along with an immeasurable caregiver burden (2010; Dudgeon, 2010). Current treatments have limited efficacy and no reparative capability, prompting a need for new therapies including regenerative strategies. Some evidence suggests deficient neurogenesis underlies cognitive deficits in AD model mice (Feng *et al.*, 2001; Fiorentini *et al.*, 2010; Wang *et al.*, 2010), and pharmacologically restoring neurogenesis in these mice confers functional benefit (Fiorentini *et al.*, 2010; Wang *et al.*, 2010). Limited human studies suggest AD patients have increased hippocampal proliferation and immature neuron generation, but overall reduced neuronal maturation culminating in less neurogenesis (Jin *et al.*, 2004; Li *et al.*, 2008). Interestingly, neurogenesis may remain intact at presenile stages (Boekhoorn *et al.*, 2006). Deficient neurogenesis might also contribute to dementia associated with late stage Parkinson's disease (PD). For instance, a study depleting dopaminergic input to the SGZ found reduced neurogenesis in rodents (Hoglinger *et al.*, 2004). Alpha synuclein, a protein whose accumulation in brain is a key pathologic feature of PD (Lang and Lozano, 1998), also impairs neurogenesis *in vitro* and in animal models (Crews *et al.*, 2008). One human study demonstrated reduced numbers of proliferating neural progenitor cells isolated from the SGZ of PD patients post-mortem, though the reduction did not correlate with the presence or absence of dementia (Hoglinger *et al.*, 2004). Several Huntington's Disease models also exhibit reduced neurogenesis (Gil *et al.*, 2005; Lazic *et al.*, 2004; Simpson *et al.*, 2011), although verification in human patients is still lacking. Should emerging non-invasive techniques for visualizing neurogenesis *in vivo* (Manganas *et al.*, 2007; Pereira *et al.*, 2007) prove reproducible, they should greatly facilitate additional human studies examining perturbations of hippocampal neurogenesis in diseases of cognitive impairment. It is perhaps equally relevant at this point in time to consider how understanding this intrinsic regenerative capability could lead to new concepts in restoring brain function. The fact that the adult brain can generate and integrate new neurons autonomously at least raises the possibility of exploiting this biology for hippocampal, or even brain, regenerative therapies.

References

(2010). 2010 Alzheimer's disease facts and figures. *Alzheimers Dement* **6**, 158–194.

Abusaad, I., MacKay, D., Zhao, J., Stanford, P., Collier, D.A. and Everall, I.P. (1999). Stereological estimation of the total number of neurons in the murine hippocampus using the optical disector. *J Comp Neurol* **408**, 560–566.

Achanta, P., Fuss, M. and Martinez, J.L., Jr. (2009). Ionizing radiation impairs the formation of trace fear memories and reduces hippocampal neurogenesis. *Behav Neurosci* **123**, 1036–1045.

Ackman, J.B., Siddiqi, F., Walikonis, R.S. and LoTurco, J.J. (2006). Fusion of microglia with pyramidal neurons after retroviral infection. *J Neurosci* **26**, 11413–11422.

Aggleton, J.P. and Saunders, R.C. (1997). The relationships between temporal lobe and diencephalic structures implicated in anterograde amnesia. *Memory* **5**, 49–71.

Aimone, J.B., Wiles, J. and Gage, F.H. (2006). Potential role for adult neurogenesis in the encoding of time in new memories. *Nat Neurosci* **9**, 723–727.

Aimone, J.B., Wiles, J. and Gage, F.H. (2009). Computational influence of adult neurogenesis on memory encoding. *Neuron* **61**, 187–202.

Aizawa, K., Ageyama, N., Yokoyama, C. and Hisatsune, T. (2009). Age-dependent alteration in hippocampal neurogenesis correlates with learning performance of macaque monkeys. *Exp Anim* **58**, 403–407.

Alme, C.B., Buzzetti, R.A., Marrone, D.F., Leutgeb, J.K., Chawla, M.K., Schaner, M.J., Bohanick, J.D., Khoboko, T., Leutgeb, S., Moser, E.I., *et al.* (2010). Hippocampal granule cells opt for early retirement. *Hippocampus* **20**, 1109–1123.

Altman, J. and Das, G.D. (1965). Post-natal origin of microneurones in the rat brain. *Nature 207*, 953–956.

Amaral, D. and Lavenex, P. (2007). Hippocampal neuroanatomy. In Anderson, P. Morris, R.G., Amaral, D.G., Bliss, T and O'Keefe, J. eds. *The Hippocampus Book.* (New York: Oxford University Press).

Amaral, D.G. (1978). A Golgi study of cell types in the hilar region of the hippocampus in the rat. *J Comp Neurol* **182**, 851–914.

Amaral, D.G. and Witter, M.P. (1989). The three-dimensional organization of the hippocampal formation: A review of anatomical data. *Neuroscience* **31**, 571–591.

Ambrogini, P., Cuppini, R., Cuppini, C., Ciaroni, S., Cecchini, T., Ferri, P., Sartini, S. and Del Grande, P. (2000). Spatial learning affects immature granule cell survival in adult rat dentate gyrus. *Neurosci Lett* **286**, 21–24.

Ambrogini, P., Cuppini, R., Lattanzi, D., Ciuffoli, S., Frontini, A. and Fanelli, M. (2010). Synaptogenesis in adult-generated hippocampal granule cells is affected by behavioural experiences. *Hippocampus* **20**, 799–810.

Ambrogini, P., Lattanzi, D., Ciuffoli, S., Agostini, D., Bertini, L., Stocchi, V., Santi, S. and Cuppini, R. (2004a). Morpho-functional characterization of neuronal cells at different stages of maturation in granule cell layer of adult rat dentate gyrus. *Brain Res* **1017**, 21–31.

Ambrogini, P., Orsini, L., Mancini, C., Ferri, P., Ciaroni, S. and Cuppini, R. (2004b). Learning may reduce neurogenesis in adult rat dentate gyrus. *Neurosci Lett* **359**, 13–16.

Andersen, P., Bliss, T.V., Lomo, T., Olsen, L.I. and Skrede, K.K. (1969). Lamellar organization of hippocampal excitatory pathways. *Acta Physiol Scand* **76**, 4A–5A.

Anderson, M.L., Sisti, H.M., Curlik, D.M., 2nd, and Shors, T.J. (2011). Associative learning increases adult neurogenesis during a critical period. *Eur J Neurosci* **33**, 175–181.

Appleby, P.A. and Wiskott, L. (2009). Additive neurogenesis as a strategy for avoiding interference in a sparsely-coding dentate gyrus. *Network* **20**, 137–161.

Bakker, A., Kirwan, C.B., Miller, M. and Stark, C.E. (2008). Pattern separation in the human hippocampal CA3 and dentate gyrus. *Science* **319**, 1640–1642.

Barnea, A. and Nottebohm, F. (1994). Seasonal recruitment of hippocampal neurons in adult free-ranging black-capped chickadees. *Proc Natl Acad Sci USA* **91**, 11217–11221.

Barnea, A. and Nottebohm, F. (1996). Recruitment and replacement of hippocampal neurons in young and adult chickadees: An addition to the theory of hippocampal learning. *Proc Natl Acad Sci USA* **93**, 714–718.

Becker, S. (2005). A computational principle for hippocampal learning and neurogenesis. *Hippocampus* **15**, 722–738.

Beiko, J., Lander, R., Hampson, E., Boon, F. and Cain, D.P. (2004). Contribution of sex differences in the acute stress response to sex differences in water maze performance in the rat. *Behav Brain Res* **151**, 239–253.

Beylin, A.V., Gandhi, C.C., Wood, G.E., Talk, A.C., Matzel, L.D. and Shors, T.J. (2001). The role of the hippocampus in trace conditioning: Temporal discontinuity or task difficulty? *Neurobiol Learn Mem* **76**, 447–461.

Bischofberger, J. (2007). Young and excitable: New neurons in memory networks. *Nat Neurosci* **10**, 273–275.

Blackstad, T.W. (1956). Commissural connections of the hippocampal region in the rat, with special reference to their mode of termination. *J Comp Neurol* **105**, 417–537.

Boekhoorn, K., Joels, M. and Lucassen, P.J. (2006). Increased proliferation reflects glial and vascular-associated changes, but not neurogenesis in the presenile Alzheimer hippocampus. *Neurobiol Dis* **24**, 1–14.

Bonnet, E., Touyarot, K., Alfos, S., Pallet, V., Higueret, P. and Abrous, D.N. (2008). Retinoic acid restores adult hippocampal neurogenesis and reverses spatial memory deficit in vitamin A deprived rats. *PLoS One* **3**, e3487.

Breunig, J.J., Rakic, P. and Macklis, J.D. (2008). Evolving methods for the labelling and mutation of postnatal neuronal precursor cells: A critical review. In Gage, F.H., Kempermann, G and Song, H. eds. *Adult Neurogenesis*. (Cold Spring Harbor, New York: Cold Spring Harbor Laboratory Press), pp. 49–80.

Brown, D.C. and Gatter, K.C. (2002). Ki67 protein: The immaculate deception? *Histopathology* **40**, 2–11.

Brown, J.P., Couillard-Despres, S., Cooper-Kuhn, C.M., Winkler, J., Aigner, L. and Kuhn, H.G. (2003). Transient expression of doublecortin during adult neurogenesis. *J Comp Neurol* **467**, 1–10.

Brown, M.S., Ye, J., Rawson, R.B. and Goldstein, J.L. (2000). Regulated intramembrane proteolysis: A control mechanism conserved from bacteria to humans. *Cell* **100**, 391–398.

Bruel-Jungerman, E., Davis, S., Rampon, C. and Laroche, S. (2006). Long-term potentiation enhances neurogenesis in the adult dentate gyrus. *J Neurosci* **26**, 5888–5893.

Bruel-Jungerman, E., Laroche, S. and Rampon, C. (2005). New neurons in the dentate gyrus are involved in the expression of enhanced long-term memory following environmental enrichment. *Eur J Neurosci* **21**, 513–521.

Cameron, H.A. and McKay, R.D. (2001). Adult neurogenesis produces a large pool of new granule cells in the dentate gyrus. *J Comp Neurol* **435**, 406–417.

Cameron, H.A., Woolley, C.S., McEwen, B.S. and Gould, E. (1993). Differentiation of newly born neurons and glia in the dentate gyrus of the adult rat. *Neuroscience* **56**, 337–344.

Carpenter, G.A. and Grossberg, S. (1987). Discovering order in chaos: Stable self-organization of neural recognition codes. *Ann NY Acad Sci* **504**, 33–51.

Chambers, R.A. and Conroy, S.K. (2007). Network modeling of adult neurogenesis: Shifting rates of neuronal turnover optimally gears network learning according to novelty gradient. *J Cogn Neurosci* **19**, 1–12.

Chambers, R.A., Potenza, M.N., Hoffman, R.E. and Miranker, W. (2004). Simulated apoptosis/neurogenesis regulates learning and memory capabilities of adaptive neural networks. *Neuropsychopharmacology* **29**, 747–758.

Chawla, M.K., Guzowski, J.F., Ramirez-Amaya, V., Lipa, P., Hoffman, K.L., Marriott, L.K., Worley, P.F., McNaughton, B.L. and Barnes, C.A. (2005). Sparse, environmentally selective expression of Arc RNA in the upper blade of the rodent fascia dentata by brief spatial experience. *Hippocampus* **15**, 579–586.

Clark, P.J., Brzezinska, W.J., Thomas, M.W., Ryzhenko, N.A., Toshkov, S.A. and Rhodes, J.S. (2008). Intact neurogenesis is required for benefits of exercise on spatial memory but not motor performance or contextual fear conditioning in C57BL/6J mice. *Neuroscience* **155**, 1048–1058.

Clelland, C.D., Choi, M., Romberg, C., Clemenson, G.D., Jr., Fragniere, A., Tyers, P., Jessberger, S., Saksida, L.M., Barker, R.A., Gage, F.H. and Bussey, T.J. (2009). A functional role for adult hippocampal neurogenesis in spatial pattern separation. *Science* **325**, 210–213.

Cole, A.J., Saffen, D.W., Baraban, J.M. and Worley, P.F. (1989). Rapid increase of an immediate early gene messenger RNA in hippocampal neurons by synaptic NMDA receptor activation. *Nature* **340**, 474–476.

Creer, D.J., Romberg, C., Saksida, L.M., van Praag, H. and Bussey, T.J. (2010). Running enhances spatial pattern separation in mice. *Proc Natl Acad Sci USA* **107**, 2367–2372.

Crews, L., Mizuno, H., Desplats, P., Rockenstein, E., Adame, A., Patrick, C., Winner, B., Winkler, J. and Masliah, E. (2008). Alpha-synuclein alters Notch-1 expression and neurogenesis in mouse embryonic stem cells and in the hippocampus of transgenic mice. *J Neurosci* **28**, 4250–4260.

Dalla, C., Bangasser, D.A., Edgecomb, C. and Shors, T.J. (2007). Neurogenesis and learning: Acquisition and asymptotic performance predict how many new cells survive in the hippocampus. *Neurobiol Learn Mem* **88**, 143–148.

Dayer, A.G., Ford, A.A., Cleaver, K.M., Yassaee, M. and Cameron, H.A. (2003). Short-term and long-term survival of new neurons in the rat dentate gyrus. *J Comp Neurol* **460**, 563–572.

Deisseroth, K., Singla, S., Toda, H., Monje, M., Palmer, T.D. and Malenka, R.C. (2004). Excitation-neurogenesis coupling in adult neural stem/progenitor cells. *Neuron* **42**, 535–552.

Deloulme, J.C., Lucas, M., Gaber, C., Bouillon, P., Keller, A., Eclancher, F. and Sensenbrenner, M. (1996). Expression of the neuron-specific enolase gene by rat oligodendroglial cells during their differentiation. *J Neurochem* **66**, 936–945.

Deng, W., Aimone, J.B. and Gage, F.H. (2010). New neurons and new memories: How does adult hippocampal neurogenesis affect learning and memory? *Nat Rev Neurosci* **11**, 339–350.

Deng, W., Saxe, M.D., Gallina, I.S. and Gage, F.H. (2009). Adult-born hippocampal dentate granule cells undergoing maturation modulate learning and memory in the brain. *J Neurosci* **29**, 13532–13542.

Denis-Donini, S., Dellarole, A., Crociara, P., Francese, M.T., Bortolotto, V., Quadrato, G., Canonico, P.L., Orsetti, M., Ghi, P., Memo, M., *et al.* (2008). Impaired adult neurogenesis associated with short-term memory defects in NF-kappaB p50-deficient mice. *J Neurosci* **28**, 3911–3919.

Derrick, B.E., York, A.D. and Martinez, J.L., Jr. (2000). Increased granule cell neurogenesis in the adult dentate gyrus following mossy fiber stimulation sufficient to induce long-term potentiation. *Brain Res* **857**, 300–307.

Diederich, K., Schabitz, W.R., Kuhnert, K., Hellstrom, N., Sachser, N., Schneider, A., Kuhn, H.G. and Knecht, S. (2009). Synergetic effects of granulocyte-colony stimulating factor and cognitive training on spatial learning and survival of newborn hippocampal neurons. *PLoS One* **4**, e5303.

Dobrossy, M.D., Drapeau, E., Aurousseau, C., Le Moal, M., Piazza, P.V. and Abrous, D.N. (2003). Differential effects of learning on neurogenesis: Learning increases or decreases the number of newly born cells depending on their birth date. *Mol Psychiatry* **8**, 974–982.

Dragunow, M. and Robertson, H.A. (1987). Kindling stimulation induces c-fos protein(s) in granule cells of the rat dentate gyrus. *Nature* **329**, 441–442.

Drapeau, E., Mayo, W., Aurousseau, C., Le Moal, M., Piazza, P.V. and Abrous, D.N. (2003). Spatial memory performances of aged rats in the water maze predict levels of hippocampal neurogenesis. *Proc Natl Acad Sci USA* **100**, 14385–14390.

Drapeau, E., Montaron, M.F., Aguerre, S. and Abrous, D.N. (2007). Learning-induced survival of new neurons depends on the cognitive status of aged rats. *J Neurosci* **27**, 6037–6044.

Drew, M.R., Denny, C.A. and Hen, R. (2010). Arrest of adult hippocampal neurogenesis in mice impairs single- but not multiple-trial contextual fear conditioning. *Behav Neurosci* **124**, 446–454.

Dudgeon, S. (2010). *Rising Tide: The Impact of Dementia on Canadian Society*. p. 65.

Dupret, D., Fabre, A., Dobrossy, M.D., Panatier, A., Rodriguez, J.J., Lamarque, S., Lemaire, V., Oliet, S.H., Piazza, P.V. and Abrous, D.N. (2007). Spatial learning depends on both the addition and removal of new hippocampal neurons. *PLoS Biol* **5**, e214.

Dupret, D., Montaron, M.F., Drapeau, E., Aurousseau, C., Le Moal, M., Piazza, P.V. and Abrous, D.N. (2005). Methylazoxymethanol acetate does not fully block cell genesis in the young and aged dentate gyrus. *Eur J Neurosci* **22**, 778–783.

Dupret, D., Revest, J.M., Koehl, M., Ichas, F., De Giorgi, F., Costet, P., Abrous, D.N. and Piazza, P.V. (2008). Spatial relational memory requires hippocampal adult neurogenesis. *PLoS One* **3**, e1959.

Ehninger, D. and Kempermann, G. (2006). Paradoxical effects of learning the Morris water maze on adult hippocampal neurogenesis in mice may be explained by a combination of stress and physical activity. *Genes Brain Behav* **5**, 29–39.

Eichenbaum, H., Stewart, C. and Morris, R.G. (1990). Hippocampal representation in place learning. *J Neurosci* **10**, 3531–3542.

Eisch, A.J. and Mandyam, C.D. (2007). Adult neurogenesis: Can analysis of cell cycle proteins move us "Beyond BrdU"? *Curr Pharm Biotechnol* **8**, 147–165.

Epp, J.R., Haack, A.K. and Galea, L.A. (2010). Task difficulty in the Morris water task influences the survival of new neurons in the dentate gyrus. *Hippocampus* **20**, 866–876.

Epp, J.R., Haack, A.K. and Galea, L.A. (2011a). Activation and survival of immature neurons in the dentate gyrus with spatial memory is dependent on time of exposure to spatial learning and age of cells at examination. *Neurobiol Learn Mem*.

Epp, J.R., Scott, N.A. and Galea, L.A. (2011b). Strain differences in neurogenesis and activation of new neurons in the dentate gyrus in response to spatial learning. *Neuroscience* **172**, 342–354.

Epp, J.R., Spritzer, M.D. and Galea, L.A. (2007). Hippocampus-dependent learning promotes survival of new neurons in the dentate gyrus at a specific time during cell maturation. *Neuroscience* **149**, 273–285.

Eriksson, P.S., Perfilieva, E., Bjork-Eriksson, T., Alborn, A.M., Nordborg, C., Peterson, D.A. and Gage, F.H. (1998). Neurogenesis in the adult human hippocampus. *Nat Med* **4**, 1313–1317.

Esposito, M.S., Piatti, V.C., Laplagne, D.A., Morgenstern, N.A., Ferrari, C.C., Pitossi, F.J. and Schinder, A.F. (2005). Neuronal differentiation in the adult hippocampus recapitulates embryonic development. *J Neurosci* **25**, 10074–10086.

Fan, Y., Liu, Z., Weinstein, P.R., Fike, J.R. and Liu, J. (2007). Environmental enrichment enhances neurogenesis and improves functional outcome after cranial irradiation. *Eur J Neurosci* **25**, 38–46.

Farioli-Vecchioli, S., Saraulli, D., Costanzi, M., Leonardi, L., Cina, I., Micheli, L., Nutini, M., Longone, P., Oh, S.P., Cestari, V. and Tirone, F. (2009). Impaired terminal differentiation of hippocampal granule neurons and defective contextual memory in PC3/Tis21 knockout mice. *PLoS One* **4**, e8339.

Farioli-Vecchioli, S., Saraulli, D., Costanzi, M., Pacioni, S., Cina, I., Aceti, M., Micheli, L., Bacci, A., Cestari, V. and Tirone, F. (2008). The timing of differentiation of adult hippocampal neurons is crucial for spatial memory. *PLoS Biol* **6**, e246.

Farivar, R., Zangenehpour, S. and Chaudhuri, A. (2004). Cellular-resolution activity mapping of the brain using immediate-early gene expression. *Front Biosci* **9**, 104–109.

Feng, R., Rampon, C., Tang, Y.P., Shrom, D., Jin, J., Kyin, M., Sopher, B., Miller, M.W., Ware, C.B., Martin, G.M., *et al.* (2001). Deficient neurogenesis in forebrain-specific presenilin-1 knockout mice is associated with reduced clearance of hippocampal memory traces. *Neuron* **32**, 911–926.

Filippov, V., Kronenberg, G., Pivneva, T., Reuter, K., Steiner, B., Wang, L.P., Yamaguchi, M., Kettenmann, H. and Kempermann, G. (2003). Subpopulation of nestin-expressing progenitor cells in the adult murine hippocampus shows electrophysiological and morphological characteristics of astrocytes. *Mol Cell Neurosci* **23**, 373–382.

Fiorentini, A., Rosi, M.C., Grossi, C., Luccarini, I. and Casamenti, F. (2010). Lithium improves hippocampal neurogenesis, neuropathology and cognitive functions in APP mutant mice. *PLoS One* **5**, e14382.

Frankland, P.W. and Bontempi, B. (2005). The organization of recent and remote memories. *Nat Rev Neurosci* **6**, 119–130.

Friedman, W.J. (1993). Memory for the time of past events. *Psychological Bulletin* **113**, 44–66.

Friedman, W.J. (2001). Memory processes underlying humans' chronological sense of the past. In Hoerl, C and McCormack, T. eds. *Time and Memory: Issues in Philosophy and Psychology*. (Oxford: Oxford University Press), pp. 139–167.

Fujioka, A., Fujioka, T., Tsuruta, R., Izumi, T., Kasaoka, S. and Maekawa, T. (2011). Effects of a constant light environment on hippocampal neurogenesis and memory in mice. *Neurosci Lett* **488**, 41–44.

Fukuda, S., Kato, F., Tozuka, Y., Yamaguchi, M., Miyamoto, Y. and Hisatsune, T. (2003). Two distinct subpopulations of nestin-positive cells in adult mouse dentate gyrus. *J Neurosci* **23**, 9357–9366.

Gage, F.H. (2000). Mammalian neural stem cells. *Science* **287**, 1433–1438.

Garcia, A.D., Doan, N.B., Imura, T., Bush, T.G. and Sofroniew, M.V. (2004). GFAP-expressing progenitors are the principal source of constitutive neurogenesis in adult mouse forebrain. *Nat Neurosci* **7**, 1233–1241.

Garthe, A., Behr, J. and Kempermann, G. (2009). Adult-generated hippocampal neurons allow the flexible use of spatially precise learning strategies. *PLoS One* **4**, e5464.

Ge, S., Goh, E.L., Sailor, K.A., Kitabatake, Y., Ming, G.L. and Song, H. (2006). GABA regulates synaptic integration of newly generated neurons in the adult brain. *Nature* **439**, 589–593.

Ge, S., Yang, C.H., Hsu, K.S., Ming, G.L. and Song, H. (2007). A critical period for enhanced synaptic plasticity in newly generated neurons of the adult brain. *Neuron* **54**, 559–566.

Geinisman, Y., Disterhoft, J.F., Gundersen, H.J., McEchron, M.D., Persina, I.S., Power, J.M., van der Zee, E.A. and West, M.J. (2000). Remodeling of hippocampal synapses after hippocampus-dependent associative learning. *J Comp Neurol* **417**, 49–59.

Gil, J.M., Mohapel, P., Araujo, I.M., Popovic, N., Li, J.Y., Brundin, P. and Petersen, A. (2005). Reduced hippocampal neurogenesis in R6/2 transgenic Huntington's disease mice. *Neurobiol Dis* **20**, 744–751.

Gilbert, P.E., Kesner, R.P. and Lee, I. (2001). Dissociating hippocampal subregions: Double dissociation between dentate gyrus and CA1. *Hippocampus* **11**, 626–636.

Goodman, T., Trouche, S., Massou, I., Verret, L., Zerwas, M., Roullet, P. and Rampon, C. (2010). Young hippocampal neurons are critical for recent and remote spatial memory in adult mice. *Neuroscience* **171**, 769–778.

Gould, E., Beylin, A., Tanapat, P., Reeves, A. and Shors, T.J. (1999a). Learning enhances adult neurogenesis in the hippocampal formation. *Nat Neurosci* **2**, 260–265.

Gould, E., McEwen, B.S., Tanapat, P., Galea, L.A. and Fuchs, E. (1997). Neurogenesis in the dentate gyrus of the adult tree shrew is regulated by psychosocial stress and NMDA receptor activation. *J Neurosci* **17**, 2492–2498.

Gould, E., Reeves, A.J., Fallah, M., Tanapat, P., Gross, C.G. and Fuchs, E. (1999b). Hippocampal neurogenesis in adult Old World primates. *Proc Natl Acad Sci USA* **96**, 5263–5267.

Gould, E., Tanapat, P., Hastings, N.B. and Shors, T.J. (1999c). Neurogenesis in adulthood: A possible role in learning. *Trends Cogn Sci* **3**, 186–192.

Gould, E., Tanapat, P., McEwen, B.S., Flugge, G. and Fuchs, E. (1998). Proliferation of granule cell precursors in the dentate gyrus of adult monkeys is diminished by stress. *Proc Natl Acad Sci USA* **95**, 3168–3171.

Gould, E., Woolley, C.S., Cameron, H.A., Daniels, D.C. and McEwen, B.S. (1991). Adrenal steroids regulate postnatal development of the rat dentate gyrus: II. Effects of glucocorticoids and mineralocorticoids on cell birth. *J Comp Neurol* **313**, 486–493.

Graciarena, M., Depino, A.M. and Pitossi, F.J. (2010). Prenatal inflammation impairs adult neurogenesis and memory related behaviour through persistent hippocampal TGFbeta1 downregulation. *Brain Behav Immun* **24**, 1301–1309.

Gross, C.G. (2000). Neurogenesis in the adult brain: Death of a dogma. *Nat Rev Neurosci* **1**, 67–73.

Guzowski, J.F., Timlin, J.A., Roysam, B., McNaughton, B.L., Worley, P.F. and Barnes, C.A. (2005). Mapping behaviourally relevant neural circuits with immediate-early gene expression. *Curr Opin Neurobiol* **15**, 599–606.

Hastings, N.B. and Gould, E. (1999). Rapid extension of axons into the CA3 region by adult-generated granule cells. *J Comp Neurol* **413**, 146–154.

Hayashi, S. and McMahon, A.P. (2002). Efficient recombination in diverse tissues by a tamoxifen-inducible form of Cre: A tool for temporally regulated gene activation/inactivation in the mouse. *Dev Biol* **244**, 305–318.

Hayes, N.L. and Nowakowski, R.S. (2002). Dynamics of cell proliferation in the adult dentate gyrus of two inbred strains of mice. *Brain Res Dev Brain Res* **134**, 77–85.

Heine, V.M., Maslam, S., Zareno, J., Joels, M. and Lucassen, P.J. (2004). Suppressed proliferation and apoptotic changes in the rat dentate gyrus after acute and chronic stress are reversible. *Eur J Neurosci* **19**, 131–144.

Hernandez-Rabaza, V., Llorens-Martin, M., Velazquez-Sanchez, C., Ferragud, A., Arcusa, A., Gumus, H.G., Gomez-Pinedo, U., Perez-Villalba, A., Rosello, J., Trejo, J.L., *et al.* (2009). Inhibition of adult hippocampal neurogenesis disrupts contextual learning but spares spatial working memory, long-term conditional rule retention and spatial reversal. *Neuroscience* **159**, 59–68.

Hoglinger, G.U., Rizk, P., Muriel, M.P., Duyckaerts, C., Oertel, W.H., Caille, I. and Hirsch, E.C. (2004). Dopamine depletion impairs precursor cell proliferation in Parkinson disease. *Nat Neurosci* **7**, 726–735.

Hu, M., Sun, Y.J., Zhou, Q.G., Auberson, Y.P., Chen, L., Hu, Y., Luo, C.X., Wu, J.Y., Zhu, D.Y. and Li, L.X. (2009). Reduced spatial learning in mice treated with NVP-AAM077 through down-regulating neurogenesis. Eur *J Pharmacol* **622**, 37–44.

Ibi, D., Takuma, K., Koike, H., Mizoguchi, H., Tsuritani, K., Kuwahara, Y., Kamei, H., Nagai, T., Yoneda, Y., Nabeshima, T. and Yamada, K. (2008). Social isolation rearing-induced impairment of the hippocampal neurogenesis is associated with deficits in spatial memory and emotion-related behaviours in juvenile mice. *J Neurochem* **105**, 921–932.

Imayoshi, I., Sakamoto, M., Ohtsuka, T., Takao, K., Miyakawa, T., Yamaguchi, M., Mori, K., Ikeda, T., Itohara, S. and Kageyama, R. (2008). Roles of continuous neurogenesis in the structural and functional integrity of the adult forebrain. *Nat Neurosci* **11**, 1153–1161.

Jaako-Movits, K. and Zharkovsky, A. (2005). Impaired fear memory and decreased hippocampal neurogenesis following olfactory bulbectomy in rats. *Eur J Neurosci* **22**, 2871–2878.

Jaako-Movits, K., Zharkovsky, T., Romantchik, O., Jurgenson, M., Merisalu, E., Heidmets, L.T. and Zharkovsky, A. (2005). Developmental lead exposure impairs contextual fear conditioning and reduces adult hippocampal neurogenesis in the rat brain. *Int J Dev Neurosci* **23**, 627–635.

Jaholkowski, P., Kiryk, A., Jedynak, P., Ben Abdallah, N.M., Knapska, E., Kowalczyk, A., Piechal, A., Blecharz-Klin, K., Figiel, I., Lioudyno, V., *et al.* (2009).

New hippocampal neurons are not obligatory for memory formation; cyclin D2 knockout mice with no adult brain neurogenesis show learning. *Learn Mem* **16**, 439–451.

Jande, S.S., Maler, L. and Lawson, D.E. (1981). Immunohistochemical mapping of vitamin D-dependent calcium-binding protein in brain. *Nature* **294**, 765–767.

Jang, M.H., Song, H. and Ming, G.L. (2008). Regulation of adult neurogenesis by neurotransmitters. In Gage, F.H., Kempermann, G and Song, H. eds. *Adult Neurogenesis*. (Cold Spring Harbor, New York: Cold Spring Harbor Laboratory Press), pp. 397–423.

Jessberger, S., Clark, R.E., Broadbent, N.J., Clemenson, G.D., Jr., Consiglio, A., Lie, D.C., Squire, L.R. and Gage, F.H. (2009). Dentate gyrus-specific knockdown of adult neurogenesis impairs spatial and object recognition memory in adult rats. *Learn Mem* **16**, 147–154.

Jessberger, S. and Kempermann, G. (2003). Adult-born hippocampal neurons mature into activity-dependent responsiveness. *Eur J Neurosci* **18**, 2707–2712.

Jessberger, S., Zhao, C., Toni, N., Clemenson, G.D., Jr., Li, Y. and Gage, F.H. (2007). Seizure-associated, aberrant neurogenesis in adult rats characterized with retrovirus-mediated cell labelling. *J Neurosci* **27**, 9400–9407.

Jin, K., Peel, A.L., Mao, X.O., Xie, L., Cottrell, B.A., Henshall, D.C. and Greenberg, D.A. (2004). Increased hippocampal neurogenesis in Alzheimer's disease. *Proc Natl Acad Sci USA* **101**, 343–347.

Jung, M.W. and McNaughton, B.L. (1993). Spatial selectivity of unit activity in the hippocampal granular layer. *Hippocampus* **3**, 165–182.

Kaplan, M.S. and Bell, D.H. (1984). Mitotic neuroblasts in the 9-day-old and 11-month-old rodent hippocampus. *J Neurosci* **4**, 1429–1441.

Kaplan, M.S. and Hinds, J.W. (1977). Neurogenesis in the adult rat: Electron microscopic analysis of light radioautographs. *Science* **197**, 1092–1094.

Kee, N., Sivalingam, S., Boonstra, R. and Wojtowicz, J.M. (2002). The utility of Ki-67 and BrdU as proliferative markers of adult neurogenesis. *J Neurosci Methods* **115**, 97–105.

Kee, N., Teixeira, C.M., Wang, A.H. and Frankland, P.W. (2007a). Imaging activation of adult-generated granule cells in spatial memory. *Nat Protoc* **2**, 3033–3044.

Kee, N., Teixeira, C.M., Wang, A.H. and Frankland, P.W. (2007b). Preferential incorporation of adult-generated granule cells into spatial memory networks in the dentate gyrus. *Nat Neurosci* **10**, 355–362.

Kempermann, G. (2002). Why new neurons? Possible functions for adult hippocampal neurogenesis. *J Neurosci* **22**, 635–638.

Kempermann, G. (2008). The neurogenic reserve hypothesis: What is adult hippocampal neurogenesis good for? *Trends Neurosci* **31**, 163–169.

Kempermann, G., Brandon, E.P. and Gage, F.H. (1998a). Environmental stimulation of 129/SvJ mice causes increased cell proliferation and neurogenesis in the adult dentate gyrus. *Curr Biol* **8**, 939–942.

Kempermann, G. and Gage, F.H. (2002a). Genetic determinants of adult hippocampal neurogenesis correlate with acquisition, but not probe trial performance, in the water maze task. *Eur J Neurosci* **16**, 129–136.

Kempermann, G. and Gage, F.H. (2002b). Genetic influence on phenotypic differentiation in adult hippocampal neurogenesis. *Brain Res Dev Brain Res* **134**, 1–12.

Kempermann, G., Gast, D. and Gage, F.H. (2002). Neuroplasticity in old age: Sustained fivefold induction of hippocampal neurogenesis by long-term environmental enrichment. *Ann Neurol* **52**, 135–143.

Kempermann, G., Gast, D., Kronenberg, G., Yamaguchi, M. and Gage, F.H. (2003). Early determination and long-term persistence of adult-generated new neurons in the hippocampus of mice. *Development* **130**, 391–399.

Kempermann, G., Jessberger, S., Steiner, B. and Kronenberg, G. (2004). Milestones of neuronal development in the adult hippocampus. *Trends Neurosci* **27**, 447–452.

Kempermann, G., Kuhn, H.G. and Gage, F.H. (1997a). Genetic influence on neurogenesis in the dentate gyrus of adult mice. *Proc Natl Acad Sci USA* **94**, 10409–10414.

Kempermann, G., Kuhn, H.G. and Gage, F.H. (1997b). More hippocampal neurons in adult mice living in an enriched environment. *Nature* **386**, 493–495.

Kempermann, G., Kuhn, H.G. and Gage, F.H. (1998b). Experience-induced neurogenesis in the senescent dentate gyrus. *J Neurosci* **18**, 3206–3212.

Kim, J.J. and Fanselow, M.S. (1992). Modality-specific retrograde amnesia of fear. *Science* **256**, 675–677.

Kim, S.E., Ko, I.G., Kim, B.K., Shin, M.S., Cho, S., Kim, C.J., Kim, S.H., Baek, S.S., Lee, E.K. and Jee, Y.S. (2010). Treadmill exercise prevents aging-induced failure of memory through an increase in neurogenesis and suppression of apoptosis in rat hippocampus. *Exp Gerontol* **45**, 357–365.

Kirn, J.R., Alvarez-Buylla, A. and Nottebohm, F. (1991). Production and survival of projection neurons in a forebrain vocal center of adult male canaries. *J Neurosci* **11**, 1756–1762.

Kitamura, T., Saitoh, Y., Murayama, A., Sugiyama, H. and Inokuchi, K. (2010). LTP induction within a narrow critical period of immature stages enhances the survival of newly generated neurons in the adult rat dentate gyrus. *Mol Brain* **3**, 13.

Kitamura, T., Saitoh, Y., Takashima, N., Murayama, A., Niibori, Y., Ageta, H., Sekiguchi, M., Sugiyama, H. and Inokuchi, K. (2009). Adult neurogenesis modulates the hippocampus-dependent period of associative fear memory. *Cell* **139**, 814–827.

Ko, H.G., Jang, D.J., Son, J., Kwak, C., Choi, J.H., Ji, Y.H., Lee, Y.S., Son, H. and Kaang, B.K. (2009). Effect of ablated hippocampal neurogenesis on the formation and extinction of contextual fear memory. *Mol Brain* **2**, 1.

Kobilo, T., Yuan, C. and van Praag, H. (2011). Endurance factors improve hippocampal neurogenesis and spatial memory in mice. *Learn Mem* **18**, 103–107.

Koo, J.W., Park, C.H., Choi, S.H., Kim, N.J., Kim, H.S., Choe, J.C. and Suh, Y.H. (2003). The postnatal environment can counteract prenatal effects on cognitive ability, cell proliferation and synaptic protein expression. *FASEB J* **17**, 1556–1558.

Kornack, D.R. and Rakic, P. (1999). Continuation of neurogenesis in the hippocampus of the adult macaque monkey. *Proc Natl Acad Sci USA* **96**, 5768–5773.

Kuan, C.Y., Schloemer, A.J., Lu, A., Burns, K.A., Weng, W.L., Williams, M.T., Strauss, K.I., Vorhees, C.V., Flavell, R.A., Davis, R.J., *et al.* (2004). Hypoxia-ischemia induces DNA synthesis without cell proliferation in dying neurons in adult rodent brain. *J Neurosci* **24**, 10763–10772.

Kuhn, H.G., Dickinson-Anson, H. and Gage, F.H. (1996). Neurogenesis in the dentate gyrus of the adult rat: Age-related decrease of neuronal progenitor proliferation. *J Neurosci* **16**, 2027–2033.

Lagace, D.C., Whitman, M.C., Noonan, M.A., Ables, J.L., DeCarolis, N.A., Arguello, A.A., Donovan, M.H., Fischer, S.J., Farnbauch, L.A., Beech, R.D., *et al.* (2007). Dynamic contribution of nestin-expressing stem cells to adult neurogenesis. *J Neurosci* **27**, 12623–12629.

Lang, A.E. and Lozano, A.M. (1998). Parkinson's disease. First of two parts. *N Engl J Med* **339**, 1044–1053.

Laplagne, D.A., Esposito, M.S., Piatti, V.C., Morgenstern, N.A., Zhao, C., van Praag, H., Gage, F.H. and Schinder, A.F. (2006). Functional convergence of neurons generated in the developing and adult hippocampus. *PLoS Biol* **4**, e409.

Laplagne, D.A., Kamienkowski, J.E., Esposito, M.S., Piatti, V.C., Zhao, C., Gage, F.H. and Schinder, A.F. (2007). Similar GABAergic inputs in dentate granule cells born during embryonic and adult neurogenesis. *Eur J Neurosci* **25**, 2973–2981.

Lavenex, P. and Amaral, D.G. (2000). Hippocampal-neocortical interaction: a hierarchy of associativity. *Hippocampus* **10**, 420–430.

Lazic, S.E., Grote, H., Armstrong, R.J., Blakemore, C., Hannan, A.J., van Dellen, A. and Barker, R.A. (2004). Decreased hippocampal cell proliferation in R6/1 Huntington's mice. *Neuroreport* **15**, 811–813.

Lemaire, V., Koehl, M., Le Moal, M. and Abrous, D.N. (2000). Prenatal stress produces learning deficits associated with an inhibition of neurogenesis in the hippocampus. *Proc Natl Acad Sci USA* **97**, 11032–11037.

Leuner, B., Falduto, J. and Shors, T.J. (2003). Associative memory formation increases the observation of dendritic spines in the hippocampus. *J Neurosci* **23**, 659–665.

Leuner, B., Gould, E. and Shors, T.J. (2006a). Is there a link between adult neurogenesis and learning? *Hippocampus* **16**, 216–224.

Leuner, B., Mendolia-Loffredo, S., Kozorovitskiy, Y., Samburg, D., Gould, E. and Shors, T.J. (2004). Learning enhances the survival of new neurons beyond the time when the hippocampus is required for memory. *J Neurosci* **24**, 7477–7481.

Leuner, B., Waddell, J., Gould, E. and Shors, T.J. (2006b). Temporal discontiguity is neither necessary nor sufficient for learning-induced effects on adult neurogenesis. *J Neurosci* **26**, 13437–13442.

Leutgeb, J.K., Leutgeb, S., Moser, M.B. and Moser, E.I. (2007). Pattern separation in the dentate gyrus and CA3 of the hippocampus. *Science* **315**, 961–966.

Li, B., Yamamori, H., Tatebayashi, Y., Shafit-Zagardo, B., Tanimukai, H., Chen, S., Iqbal, K. and Grundke-Iqbal, I. (2008). Failure of neuronal maturation in Alzheimer disease dentate gyrus. *J Neuropathol Exp Neurol* **67**, 78–84.

Li, Y.F., Cheng, Y.F., Huang, Y., Conti, M., Wilson, S.P., O'Donnell, J.M. and Zhang, H.T. (2011). Phosphodiesterase-4D knock-out and RNA interference-mediated knock-down enhance memory and increase hippocampal neurogenesis via increased cAMP signaling. *J Neurosci* **31**, 172–183.

Lie, D.C., Song, H., Colamarino, S.A., Ming, G.L. and Gage, F.H. (2004). Neurogenesis in the adult brain: new strategies for central nervous system diseases. *Annu Rev Pharmacol Toxicol* **44**, 399–421.

Lledo, P.M., Alonso, M. and Grubb, M.S. (2006). Adult neurogenesis and functional plasticity in neuronal circuits. *Nat Rev Neurosci* **7**, 179–193.

Lyons, L., Elbeltagy, M., Umka, J., Markwick, R., Startin, C., Bennett, G. and Wigmore, P. (2010). Fluoxetine reverses the memory impairment and reduction in proliferation and survival of hippocampal cells caused by methotrexate chemotherapy. *Psychopharmacology (Berl)*.

Ma, D.K., Jang, M.H., Guo, J.U., Kitabatake, Y., Chang, M.L., Pow-Anpongkul, N., Flavell, R.A., Lu, B., Ming, G.L. and Song, H. (2009). Neuronal activity-induced Gadd45b promotes epigenetic DNA demethylation and adult neurogenesis. *Science* **323**, 1074–1077.

Ma, D.K., Ming, G.L., Gage, F.H. and Song, H. (2008). Neurogenic niches in the adult mammalian brain. In Gage, F.H., Kempermann, G. and Song, H. eds. Adult Neurogenesis. (Cold Spring Harbor, New York: Cold Spring Harbor Laboratory Press), pp. 207–225.

Madsen, T.M., Kristjansen, P.E., Bolwig, T.G. and Wortwein, G. (2003). Arrested neuronal proliferation and impaired hippocampal function following fractionated brain irradiation in the adult rat. *Neuroscience* **119**, 635–642.

Madsen, T.M., Treschow, A., Bengzon, J., Bolwig, T.G., Lindvall, O. and Tingstrom, A. (2000). Increased neurogenesis in a model of electroconvulsive therapy. *Biol Psychiatry* **47**, 1043–1049.

Mandyam, C.D., Harburg, G.C. and Eisch, A.J. (2007). Determination of key aspects of precursor cell proliferation, cell cycle length and kinetics in the adult mouse subgranular zone. *Neuroscience* **146**, 108–122.

Manganas, L.N., Zhang, X., Li, Y., Hazel, R.D., Smith, S.D., Wagshul, M.E., Henn, F., Benveniste, H., Djuric, P.M., Enikolopov, G. and Maletic-Savatic, M. (2007). Magnetic resonance spectroscopy identifies neural progenitor cells in the live human brain. *Science* **318**, 980–985.

Mathews, E.A., Morgenstern, N.A., Piatti, V.C., Zhao, C., Jessberger, S., Schinder, A.F. and Gage, F.H. (2010). A distinctive layering pattern of mouse dentate granule cells is generated by developmental and adult neurogenesis. *J Comp Neurol* **518**, 4479–4490.

McCloskey, M. and Cohen, N.J. (1989). Catastrophic interference in connectionist networks: The sequential learning problem. In Bower, G.H. ed. *The psychology of learning and motivation* (New York: Academic Press), pp. 109–165.

McEchron, M.D., Bouwmeester, H., Tseng, W., Weiss, C. and Disterhoft, J.F. (1998). Hippocampectomy disrupts auditory trace fear conditioning and contextual fear conditioning in the rat. *Hippocampus* **8**, 638–646.

McGuire, S.E. and Davis, R.L. (2001). Presenilin-1 and memories of the forebrain. *Neuron* **32**, 763–765.

McNaughton, B.L. and Morris, R.G.M. (1987). Hippocampal synaptic enhancement and information storage within a distributed memory system. *Trends in Neurosciences* **10**, 408–415.

Meltzer, L.A., Yabaluri, R. and Deisseroth, K. (2005). A role for circuit homeostasis in adult neurogenesis. *Trends Neurosci* **28**, 653–660.

Meshi, D., Drew, M.R., Saxe, M., Ansorge, M.S., David, D., Santarelli, L., Malapani, C., Moore, H. and Hen, R. (2006). Hippocampal neurogenesis is not required for behavioural effects of environmental enrichment. *Nat Neurosci* **9**, 729–731.

Miller, M.W. and Nowakowski, R.S. (1988). Use of bromodeoxyuridine-immunohistochemistry to examine the proliferation, migration and time of origin of cells in the central nervous system. *Brain Res* **457**, 44–52.

Mineur, Y.S., Belzung, C. and Crusio, W.E. (2007). Functional implications of decreases in neurogenesis following chronic mild stress in mice. *Neuroscience* **150**, 251–259.

Mohapel, P., Leanza, G., Kokaia, M. and Lindvall, O. (2005). Forebrain acetylcholine regulates adult hippocampal neurogenesis and learning. *Neurobiol Aging* **26**, 939–946.

Mohapel, P., Mundt-Petersen, K., Brundin, P. and Frielingsdorf, H. (2006). Working memory training decreases hippocampal neurogenesis. *Neuroscience* **142**, 609–613.

Monje, M.L., Mizumatsu, S., Fike, J.R. and Palmer, T.D. (2002). Irradiation induces neural precursor-cell dysfunction. *Nat Med* **8**, 955–962.

Montaron, M.F., Drapeau, E., Dupret, D., Kitchener, P., Aurousseau, C., Le Moal, M., Piazza, P.V. and Abrous, D.N. (2006). Lifelong corticosterone level determines age-related decline in neurogenesis and memory. *Neurobiol Aging* **27**, 645–654.

Morrison, S.J. and Spradling, A.C. (2008). Stem cells and niches: mechanisms that promote stem cell maintenance throughout life. *Cell* **132**, 598–611.

Moscovitch, M., Nadel, L., Winocur, G., Gilboa, A. and Rosenbaum, R.S. (2006). The cognitive neuroscience of remote episodic, semantic and spatial memory. *Curr Opin Neurobiol* **16**, 179–190.

Moscovitch, M., Rosenbaum, R.S., Gilboa, A., Addis, D.R., Westmacott, R., Grady, C., McAndrews, M.P., Levine, B., Black, S., Winocur, G. and Nadel, L. (2005). Functional neuroanatomy of remote episodic, semantic and spatial memory: A unified account based on multiple trace theory. *J Anat* **207**, 35–66.

Moyer, J.R., Jr., Deyo, R.A. and Disterhoft, J.F. (1990). Hippocampectomy disrupts trace eye-blink conditioning in rabbits. *Behav Neurosci* **104**, 243–252.

Mu, Y., Lee, S.W. and Gage, F.H. (2010). Signaling in adult neurogenesis. *Curr Opin Neurobiol* **20**, 416–423.

Mulders, W.H., West, M.J. and Slomianka, L. (1997). Neuron numbers in the presubiculum, parasubiculum and entorhinal area of the rat. *J Comp Neurol* **385**, 83–94.

Mullen, R.J., Buck, C.R. and Smith, A.M. (1992). NeuN, a neuronal specific nuclear protein in vertebrates. *Development* **116**, 201–211.

Nacher, J., Crespo, C. and McEwen, B.S. (2001). Doublecortin expression in the adult rat telencephalon. *Eur J Neurosci* **14**, 629–644.

Nilsson, M., Perfilieva, E., Johansson, U., Orwar, O. and Eriksson, P.S. (1999). Enriched environment increases neurogenesis in the adult rat dentate gyrus and improves spatial memory. *J Neurobiol* **39**, 569–578.

Ninkovic, J., Mori, T. and Gotz, M. (2007). Distinct modes of neuron addition in adult mouse neurogenesis. *J Neurosci* **27**, 10906–10911.

Nottebohm, F. (2002). Why are some neurons replaced in adult brain? *J Neurosci* **22**, 624–628.

O'Reilly, R.C. and McClelland, J.L. (1994). Hippocampal conjunctive encoding, storage and recall: avoiding a trade-off. *Hippocampus* **4**, 661–682.

Olariu, A., Cleaver, K.M., Shore, L.E., Brewer, M.D. and Cameron, H.A. (2005). A natural form of learning can increase and decrease the survival of new neurons in the dentate gyrus. *Hippocampus* **15**, 750–762.

Overstreet-Wadiche, L.S., Bromberg, D.A., Bensen, A.L. and Westbrook, G.L. (2006). Seizures accelerate functional integration of adult-generated granule cells. *J Neurosci* **26**, 4095–4103.

Palmer, T.D., Willhoite, A.R. and Gage, F.H. (2000). Vascular niche for adult hippocampal neurogenesis. *J Comp Neurol* **425**, 479–494.

Parent, J.M., Yu, T.W., Leibowitz, R.T., Geschwind, D.H., Sloviter, R.S. and Lowenstein, D.H. (1997). Dentate granule cell neurogenesis is increased by seizures and contributes to aberrant network reorganization in the adult rat hippocampus. *J Neurosci* **17**, 3727–3738.

Pavlov, I.P. (1927). Conditioned reflexes; an investigation of the physiological activity of the cerebral cortex. Translated and edited by G.V. Anrep (London: Oxford University Press).

Pereira, A.C., Huddleston, D.E., Brickman, A.M., Sosunov, A.A., Hen, R., McKhann, G.M., Sloan, R., Gage, F.H., Brown, T.R. and Small, S.A. (2007). An *in vivo* correlate of exercise-induced neurogenesis in the adult dentate gyrus. *Proc Natl Acad Sci USA* **104**, 5638–5643.

Pieper, A.A., Xie, S., Capota, E., Estill, S.J., Zhong, J., Long, J.M., Becker, G.L., Huntington, P., Goldman, S.E., Shen, C.H., *et al.* (2010). Discovery of a proneurogenic, neuroprotective chemical. *Cell* **142**, 39–51.

Prickaerts, J., Koopmans, G., Blokland, A. and Scheepens, A. (2004). Learning and adult neurogenesis: survival with or without proliferation? *Neurobiol Learn Mem* **81**, 1–11.

Raber, J., Rola, R., LeFevour, A., Morhardt, D., Curley, J., Mizumatsu, S., VandenBerg, S.R. and Fike, J.R. (2004). Radiation-induced cognitive impairments are associated with changes in indicators of hippocampal neurogenesis. *Radiat Res* **162**, 39–47.

Ramirez-Amaya, V., Marrone, D.F., Gage, F.H., Worley, P.F. and Barnes, C.A. (2006). Integration of new neurons into functional neural networks. *J Neurosci* **26**, 12237–12241.

Rawlins, J.N.P. (1985). Associations across time: The hippocampus as a temporary memory store. *Behavioural and Brain Sciences* **8**, 479–497.

Rola, R., Otsuka, S., Obenaus, A., Nelson, G.A., Limoli, C.L., VandenBerg, S.R. and Fike, J.R. (2004a). Indicators of hippocampal neurogenesis are altered by 56Fe-particle irradiation in a dose-dependent manner. *Radiat Res* **162**, 442–446.

Rola, R., Raber, J., Rizk, A., Otsuka, S., VandenBerg, S.R., Morhardt, D.R. and Fike, J.R. (2004b). Radiation-induced impairment of hippocampal neurogenesis is associated with cognitive deficits in young mice. *Exp Neurol* **188**, 316–330.

Saxe, M.D., Battaglia, F., Wang, J.W., Malleret, G., David, D.J., Monckton, J.E., Garcia, A.D., Sofroniew, M.V., Kandel, E.R., Santarelli, L., *et al.* (2006). Ablation of hippocampal neurogenesis impairs contextual fear conditioning and synaptic plasticity in the dentate gyrus. *Proc Natl Acad Sci USA* **103**, 17501–17506.

Saxe, M.D., Malleret, G., Vronskaya, S., Mendez, I., Garcia, A.D., Sofroniew, M.V., Kandel, E.R. and Hen, R. (2007). Paradoxical influence of hippocampal neurogenesis on working memory. *Proc Natl Acad Sci USA* **104**, 4642–4646.

Schauwecker, P.E. (2006). Genetic influence on neurogenesis in the dentate gyrus of two strains of adult mice. *Brain Res* **1120**, 83–92.

Schmechel, D.E., Brightman, M.W. and Marangos, P.J. (1980). Neurons switch from non-neuronal enolase to neuron-specific enolase during differentiation. *Brain Res* **190**, 195–214.

Schmidt-Hieber, C., Jonas, P. and Bischofberger, J. (2004). Enhanced synaptic plasticity in newly generated granule cells of the adult hippocampus. *Nature* **429**, 184–187.

Scoville, W.B. and Milner, B. (1957). Loss of recent memory after bilateral hippocampal lesions. *J Neurol Neurosurg Psychiatry* **20**, 11–21.

Seri, B., Garcia-Verdugo, J.M., McEwen, B.S. and Alvarez-Buylla, A. (2001). Astrocytes give rise to new neurons in the adult mammalian hippocampus. *J Neurosci* **21**, 7153–7160.

Shen, L., Nam, H.S., Song, P., Moore, H. and Anderson, S.A. (2006). FoxG1 haploinsufficiency results in impaired neurogenesis in the postnatal hippocampus and contextual memory deficits. *Hippocampus* **16**, 875–890.

Shi, Y., Chichung Lie, D., Taupin, P., Nakashima, K., Ray, J., Yu, R.T., Gage, F.H. and Evans, R.M. (2004). Expression and function of orphan nuclear receptor TLX in adult neural stem cells. *Nature* **427**, 78–83.

Shimazu, K., Zhao, M., Sakata, K., Akbarian, S., Bates, B., Jaenisch, R. and Lu, B. (2006). NT-3 facilitates hippocampal plasticity and learning and memory by regulating neurogenesis. *Learn Mem* **13**, 307–315.

Shors, T.J. (2004). Memory traces of trace memories: Neurogenesis, synaptogenesis and awareness. *Trends Neurosci* **27**, 250–256.

Shors, T.J. (2008). From stem cells to grandmother cells: How neurogenesis relates to learning and memory. *Cell Stem Cell* **3**, 253–258.

Shors, T.J., Miesegaes, G., Beylin, A., Zhao, M., Rydel, T. and Gould, E. (2001). Neurogenesis in the adult is involved in the formation of trace memories. *Nature* **410**, 372–376.

Shors, T.J., Townsend, D.A., Zhao, M., Kozorovitskiy, Y. and Gould, E. (2002). Neurogenesis may relate to some but not all types of hippocampal-dependent learning. *Hippocampus* **12**, 578–584.

Simpson, J.M., Gil-Mohapel, J., Pouladi, M.A., Ghilan, M., Xie, Y., Hayden, M.R. and Christie, B.R. (2011). Altered adult hippocampal neurogenesis in the YAC128 transgenic mouse model of Huntington disease. *Neurobiol Dis* **41**, 249–260.

Sisti, H.M., Glass, A.L. and Shors, T.J. (2007). Neurogenesis and the spacing effect: Learning over time enhances memory and the survival of new neurons. *Learn Mem* **14**, 368–375.

Snyder, J.S., Choe, J.S., Clifford, M.A., Jeurling, S.I., Hurley, P., Brown, A., Kamhi, J.F. and Cameron, H.A. (2009a). Adult-born hippocampal neurons are more numerous, faster maturing and more involved in behaviour in rats than in mice. *J Neurosci* **29**, 14484–14495.

Snyder, J.S., Hong, N.S., McDonald, R.J. and Wojtowicz, J.M. (2005). A role for adult neurogenesis in spatial long-term memory. *Neuroscience* **130**, 843–852.

Snyder, J.S., Kee, N. and Wojtowicz, J.M. (2001). Effects of adult neurogenesis on synaptic plasticity in the rat dentate gyrus. *J Neurophysiol* **85**, 2423–2431.

Snyder, J.S., Radik, R., Wojtowicz, J.M. and Cameron, H.A. (2009b). Anatomical gradients of adult neurogenesis and activity: Young neurons in the ventral dentate gyrus are activated by water maze training. *Hippocampus* **19**, 360–370.

Solomon, P.R., Vander Schaaf, E.R., Thompson, R.F. and Weisz, D.J. (1986). Hippocampus and trace conditioning of the rabbit's classically conditioned nictitating membrane response. *Behav Neurosci* **100**, 729–744.

Squire, L.R. and Zola, S.M. (1997). Amnesia, memory and brain systems. *Philos Trans R Soc Lond B Biol Sci* **352**, 1663–1673.

Squire, L.R. and Zola-Morgan, S. (1991). The medial temporal lobe memory system. *Science* **253**, 1380–1386.

Stanfield, B.B. and Trice, J.E. (1988). Evidence that granule cells generated in the dentate gyrus of adult rats extend axonal projections. *Exp Brain Res* **72**, 399–406.

Steiner, B., Kronenberg, G., Jessberger, S., Brandt, M.D., Reuter, K. and Kempermann, G. (2004). Differential regulation of gliogenesis in the context of adult hippocampal neurogenesis in mice. *Glia* **46**, 41–52.

Stone, S.S., Teixeira, C.M., Zaslavsky, K., Wheeler, A.L., Martinez-Canabal, A., Wang, A.H., Sakaguchi, M., Lozano, A.M. and Frankland, P.W. (2010). Functional convergence of developmentally and adult-generated granule cells in dentate gyrus circuits supporting hippocampus-dependent memory. *Hippocampus*.

Suh, H., Consiglio, A., Ray, J., Sawai, T., D'Amour, K.A. and Gage, F.H. (2007). In vivo fate analysis reveals the multipotent and self-renewal capacities of Sox2+ neural stem cells in the adult hippocampus. *Cell Stem Cell* **1**, 515–528.

Takehara, K., Kawahara, S. and Kirino, Y. (2003). Time-dependent reorganization of the brain components underlying memory retention in trace eyeblink conditioning. *J Neurosci* **23**, 9897–9905.

Tashiro, A., Makino, H. and Gage, F.H. (2007). Experience-specific functional modification of the dentate gyrus through adult neurogenesis: A critical period during an immature stage. *J Neurosci* **27**, 3252–3259.

Tashiro, A., Sandler, V.M., Toni, N., Zhao, C. and Gage, F.H. (2006). NMDA-receptor-mediated, cell-specific integration of new neurons in adult dentate gyrus. *Nature* **442**, 929–933.

Thomas, C.E., Ehrhardt, A. and Kay, M.A. (2003). Progress and problems with the use of viral vectors for gene therapy. *Nat Rev Genet* **4**, 346–358.

Thomas, R.M., Hotsenpiller, G. and Peterson, D.A. (2007). Acute psychosocial stress reduces cell survival in adult hippocampal neurogenesis without altering proliferation. *J Neurosci* **27**, 2734–2743.

Toda, H., Hamani, C., Fawcett, A.P., Hutchison, W.D. and Lozano, A.M. (2008). The regulation of adult rodent hippocampal neurogenesis by deep brain stimulation. *J Neurosurg* **108**, 132–138.

Toni, N., Laplagne, D.A., Zhao, C., Lombardi, G., Ribak, C.E., Gage, F.H. and Schinder, A.F. (2008). Neurons born in the adult dentate gyrus form functional synapses with target cells. *Nat Neurosci* **11**, 901–907.

Toni, N., Teng, E.M., Bushong, E.A., Aimone, J.B., Zhao, C., Consiglio, A., van Praag, H., Martone, M.E., Ellisman, M.H. and Gage, F.H. (2007). Synapse formation on neurons born in the adult hippocampus. *Nat Neurosci* **10**, 727–734.

Tozuka, Y., Fukuda, S., Namba, T., Seki, T. and Hisatsune, T. (2005). GABAergic excitation promotes neuronal differentiation in adult hippocampal progenitor cells. *Neuron* **47**, 803–815.

Trejo, J.L., Llorens-Martin, M.V. and Torres-Aleman, I. (2008). The effects of exercise on spatial learning and anxiety-like behaviour are mediated by an IGF-I-dependent mechanism related to hippocampal neurogenesis. *Mol Cell Neurosci* **37**, 402–411.

Treves, A. and Rolls, E.T. (1994). Computational analysis of the role of the hippocampus in memory. *Hippocampus* **4**, 374–391.

Tronel, S., Fabre, A., Charrier, V., Oliet, S.H., Gage, F.H. and Abrous, D.N. (2010). Spatial learning sculpts the dendritic arbor of adult-born hippocampal neurons. *Proc Natl Acad Sci USA* **107**, 7963–7968.

Trouche, S., Bontempi, B., Roullet, P. and Rampon, C. (2009). Recruitment of adult-generated neurons into functional hippocampal networks contributes to updating and strengthening of spatial memory. *Proc Natl Acad Sci USA* **106**, 5919–5924.

Van der Borght, K., Havekes, R., Bos, T., Eggen, B.J. and Van der Zee, E.A. (2007). Exercise improves memory acquisition and retrieval in the Y-maze task: Relationship with hippocampal neurogenesis. *Behav Neurosci* **121**, 324–334.

van Groen, T., Kadish, I. and Wyss, J.M. (2002). Species differences in the projections from the entorhinal cortex to the hippocampus. *Brain Res Bull* **57**, 553–556.

van Praag, H., Christie, B.R., Sejnowski, T.J. and Gage, F.H. (1999a). Running enhances neurogenesis, learning and long-term potentiation in mice. *Proc Natl Acad Sci USA* **96**, 13427–13431.

van Praag, H., Kempermann, G. and Gage, F.H. (1999b). Running increases cell proliferation and neurogenesis in the adult mouse dentate gyrus. *Nat Neurosci* **2**, 266–270.

van Praag, H., Schinder, A.F., Christie, B.R., Toni, N., Palmer, T.D. and Gage, F.H. (2002). Functional neurogenesis in the adult hippocampus. *Nature* **415**, 1030–1034.

van Praag, H., Shubert, T., Zhao, C. and Gage, F.H. (2005). Exercise enhances learning and hippocampal neurogenesis in aged mice. *J Neurosci* **25**, 8680–8685.

Veena, J., Srikumar, B.N., Mahati, K., Bhagya, V., Raju, T.R. and Shankaranarayana Rao, B.S. (2009). Enriched environment restores hippocampal cell proliferation and ameliorates cognitive deficits in chronically stressed rats. *J Neurosci Res* **87**, 831–843.

Waddell, J. and Shors, T.J. (2008). Neurogenesis, learning and associative strength. *Eur J Neurosci* **27**, 3020–3028.

Wallenstein, G.V., Eichenbaum, H. and Hasselmo, M.E. (1998). The hippocampus as an associator of discontiguous events. *Trends Neurosci* **21**, 317–323.

Wang, J.M., Singh, C., Liu, L., Irwin, R.W., Chen, S., Chung, E.J., Thompson, R.F. and Brinton, R.D. (2010). Allopregnanolone reverses neurogenic and cognitive deficits in mouse model of Alzheimer's disease. *Proc Natl Acad Sci USA* **107**, 6498–6503.

Wang, S., Scott, B.W. and Wojtowicz, J.M. (2000). Heterogenous properties of dentate granule neurons in the adult rat. *J Neurobiol* **42**, 248–257.

Warner-Schmidt, J.L., Madsen, T.M. and Duman, R.S. (2008). Electroconvulsive seizure restores neurogenesis and hippocampus-dependent fear memory after disruption by irradiation. *Eur J Neurosci* **27**, 1485–1493.

Weisz, V.I. and Argibay, P.F. (2009). A putative role for neurogenesis in neurocomputational terms: inferences from a hippocampal model. *Cognition* **112**, 229–240.

West, M.J. and Gundersen, H.J. (1990). Unbiased stereological estimation of the number of neurons in the human hippocampus. *J Comp Neurol* **296**, 1–22.

West, M.J. and Slomianka, L. (1998). Total number of neurons in the layers of the human entorhinal cortex. *Hippocampus* **8**, 69–82.

West, M.J., Slomianka, L. and Gundersen, H.J. (1991). Unbiased stereological estimation of the total number of neurons in thesubdivisions of the rat hippocampus using the optical fractionator. *Anat Rec* **231**, 482–497.

Winocur, G., Moscovitch, M. and Bontempi, B. (2010). Memory formation and long-term retention in humans and animals: Convergence towards a transformation account of hippocampal-neocortical interactions. *Neuropsychologia* **48**, 2339–2356.

Winocur, G., Wojtowicz, J.M., Sekeres, M., Snyder, J.S. and Wang, S. (2006). Inhibition of neurogenesis interferes with hippocampus-dependent memory function. *Hippocampus* **16**, 296–304.

Wiskott, L., Rasch, M.J. and Kempermann, G. (2006). A functional hypothesis for adult hippocampal neurogenesis: Avoidance of catastrophic interference in the dentate gyrus. *Hippocampus* **16**, 329–343.

Wojtowicz, J.M. (2006). Irradiation as an experimental tool in studies of adult neurogenesis. *Hippocampus* **16**, 261–266.

Wojtowicz, J.M., Askew, M.L. and Winocur, G. (2008). The effects of running and of inhibiting adult neurogenesis on learning and memory in rats. *Eur J Neurosci* **27**, 1494–1502.

Wojtowicz, J.M. and Kee, N. (2006). BrdU assay for neurogenesis in rodents. *Nat Protoc* **1**, 1399–1405.

Yoo, A.S., Cheng, I., Chung, S., Grenfell, T.Z., Lee, H., Pack-Chung, E., Handler, M., Shen, J., Xia, W., Tesco, G., *et al.* (2000). Presenilin-mediated modulation of capacitative calcium entry. *Neuron* **27**, 561–572.

Yun, J., Koike, H., Ibi, D., Toth, E., Mizoguchi, H., Nitta, A., Yoneyama, M., Ogita, K., Yoneda, Y., Nabeshima, T., *et al.* (2010). Chronic restraint stress impairs neurogenesis and hippocampus-dependent fear memory in mice: Possible involvement of a brain-specific transcription factor Npas4. *J Neurochem* **114**, 1840–1851.

Zhang, C.L., Zou, Y., He, W., Gage, F.H. and Evans, R.M. (2008). A role for adult TLX-positive neural stem cells in learning and behaviour. *Nature* **451**, 1004–1007.

Zhao, C., Deng, W. and Gage, F.H. (2008). Mechanisms and functional implications of adult neurogenesis. *Cell* **132**, 645–660.

Zhao, C., Teng, E.M., Summers, R.G., Jr., Ming, G.L. and Gage, F.H. (2006). Distinct morphological stages of dentate granule neuron maturation in the adult mouse hippocampus. *J Neurosci* **26**, 3–11.

Zhao, M., Li, D., Shimazu, K., Zhou, Y.X., Lu, B. and Deng, C.X. (2007). Fibroblast growth factor receptor-1 is required for long-term potentiation, memory consolidation, and neurogenesis. *Biol Psychiatry* **62**, 381–390.

Zhao, X., Ueba, T., Christie, B.R., Barkho, B., McConnell, M.J., Nakashima, K., Lein, E.S., Eadie, B.D., Willhoite, A.R., Muotri, A.R., *et al.* (2003). Mice lacking methyl-CpG binding protein 1 have deficits in adult neurogenesis and hippocampal function. *Proc Natl Acad Sci USA* **100**, 6777–6782.

Memory Consolidation and Its Underlying Mechanisms

5

Cristina M. Alberini*,†, Dhananjay Bambah-Mukku* and Dillon Y. Chen*

1. Defining memory consolidation

1.1. *The historical beginning*

A fascination with the understanding of memory formation and persistence has accompanied much of human history. The idea that memories take time in order to become long term was noted by early philosophers such as Quintillian in the 1st century AD. However, the first concrete steps toward an understanding of how long-term memories form came in the nineteenth century when the studies left the realm of the metaphysical and philosophical and entered the realm of the biological.

Until the 19th century, most knowledge about memory involved either anecdotal evidence or descriptions of striking cases of amnesia which only yielded sketchy maxims such as "*He who learns quickly also forgets quickly*", "*Relatively long series of ideas are retained better than relatively short ones*", "*Old people forget most quickly the things they learned last*" (from Ebbinghaus in the 19th century), and so forth. Thus, the psychological study of memory remained limited and yielded little information about the mechanisms of memory formation. Around the 1880s, the French psychologist Theodule Ribot studied the clinical phenomenon of "organic" amnesia, which is manifested as a gradient of memory loss of past events in patients with brain injury, and proposed what is now known as Ribot's law of regression: "*Progressive destruction advances progressively from the unstable to the stable*" which refers to progressive amnesia as a temporal gradient going from the most recent to the oldest memories. For Ribot, this law implied that memory depends upon permanent modifications and organization of

* Department of Neuroscience, Mount Sinai School of Medicine, New York, NY 10029, USA
† Psychiatry, Mount Sinai School of Medicine, New York, NY 10029, USA

neurons, and it is their disorganization that leads to amnesia. Ribot's other contribution was to describe memory in evolutionary terms, which had a profound impact on the idea that simple organisms could be used as models to study psychological phenomena in more complex animals (Nicolas and Murray 1999).

1.2. *The experimental revolution*

In 1885, Hermann Ebbinghaus published his seminal work entitled "*Memory: A Contribution to Experimental Psychology*". In this work, he detailed some of the very first experimental studies on the nature of memory and forgetting. Ebbinghaus realised that the only way forward towards a systematic understanding of memory was to find a way to quantify it. He outlined a sophisticated statistical toolkit to analyze memory numerically and chose an experimental system that cleverly controlled for many variables. His experimental system consisted of subjects remembering lists of words that were composed of "*nonsense syllables*" — words that had no meaning. This negated any interference that prior knowledge would have on the task. Using this system, Ebbinghaus made several important findings. He famously established the "*forgetting curve*", which was a graphical representation of the exponential decline in memory retention as a function of time. He also discovered the "*spacing effect*" that refers to the phenomenon that spaced learning — learning that includes intervals of relaxation — is better for memory retention than massed learning, learning in a crammed sequence over a short period of time.

All these contributions laid the foundations for the studies of Georg Elias Müller and his student Alfons Pilzecker who confirmed the existence of a post-acquisition phase during which newly acquired information becomes a long-term memory through a process they termed "*consolidation*" (Müller and Pilzecker, 1900), which, as we will see below, has been studied for the last century and is still the subject of many investigations and discussions (Lechner *et al.*, 1999; McGaugh, 2000). In a landmark paper, Müller and Pilzecker described a series of 40 experiments carried out between 1892 and 1900. They used Ebbinghaus' system of nonsense syllables where human subjects were asked to remember pairs of syllables. Memory was tested by presenting one member of each pair shown earlier and asking for the other member. They quantified memory by determining the percentages of correct and incorrect responses as well as recall failures in a given session. Using this method, Müller and Pilzecker identified a phenomenon they termed "*perseveration*" which was characterized as their subjects' tendency to recall unrelated syllable pairs. This effect occurred more during the first few hours of training and tapered off after the first day. This simple observation led to their discovery of *retroactive inhibition*, wherein they hypothesized that the memorization of paired associations took some time after the pairs were first presented, and that the presentation of new pairs before the "consolidation" of the first pair led to the inhibition of the formation of the memory of the first pair. Thus, they defined, for the first time, consolidation as the process by which the memory strengthens as a function of time and becomes resilient to interferences. This observation became central to the many decades of research that followed. It also provided an attractive explanation for the temporally graded nature of retrograde amnesia, or loss of memories of previous facts, that was noted in clinical studies following traumatic brain injuries. These conclusions, from clinical and human studies, were then confirmed in animal models by Carl Duncan in the 1940s and 1950s. Duncan studied several learning paradigms in rats and reported that electroconvulsive shock (ECS) given minutes after training disrupts memory retention, but if given hours after training had no effect, and the memory formed normally.

1.3. *Modern times: Memory consolidation and the temporal lobe*

In addition to time, one important question that arose from the 19th century consolidation studies was: where is memory stored? In the early part of the 20th century, the prevailing view of brain function was that the various mental faculties were localised to specific regions of the brain. This view extended to the idea of localising memory to a single brain region. Karl Lashley performed a series of experiments in the 1920s to identify the site of memory storage. Rats trained to run a maze were tested again after different areas of the cortex were removed. Lashley found that the resulting amnesia for the training was proportional to the amount of cortical tissue removed and not its location, and he summarised his findings as follows — "*The degree of retardation seems proportionate to the amount of tissue destroyed, irrespective of the locus of injury*", known as the *law of mass action.*

A few decades later, an early theory by Donald Hebb attempted to explain memory consolidation in terms of network activity. In 1949, Hebb hypothesised that the representation of the memory is stored in distributed networks of interconnected cells that widen with the passage of time. This hypothesis has proven prescient in the wake of several recent studies that also posit a more widely distributed memory trace. These theories influenced a number of Hebb's colleagues, most notably, Brenda Milner and William Scoville, a student of the neurosurgeon Wilder Penfield. Penfield studied patients with various brain lesions, and together with Scoville, performed frontal and temporal lobectomies to relieve seizures in epileptic patients. Penfield had noticed early on that a few of these patients with unilateral temporal lobe lesions displayed striking anterograde amnesia — a lack of ability to form new memories. These findings were followed then by a landmark study in the history of memory consolidation. In 1957, Brenda Milner and William Scoville described the memory deficit of the patient H.M., now revealed to be Henry Gustav Molaison (February 26, 1926–December 2, 2008), who had undergone a bilateral transection of the entire hippocampus and amygdala as a treatment for severe temporal lobe epilepsy in 1953. The patient suffered a serious form of anterograde amnesia. He also suffered moderate retrograde amnesia, and could not remember most events in the 1–2-year period before the surgery, nor some events up to 11 years before, suggesting that his amnesia was temporally graded. However, his ability to form long-term *procedural* memories was intact; thus he could (for example) learn new motor skills, despite not being able to remember having learned them. Thus, studies on H.M.'s memory deficits also uncovered evidence for the existence of multiple memory systems. They, in fact, clearly showed that sensorimotor skill learning was independent of the function of the medial temporal lobe which was instead critical for processing memories of facts, people and events, or also known as *explicit* or *declarative memories*. Procedural memory involves a gradual increase in performance with repeated practice and is recalled unconsciously whereas declarative memory refers to the conscious recollection of factual or autobiographical information. From their studies on H.M., Milner and Scoville concluded that the hippocampal region in particular is critical for the formation of declarative memories. The temporal limits of the retrograde amnesia for declarative memories also suggested that the storage and/or recall of the older memories that were spared do not require the hippocampal region any longer. Later studies on other patients found that cortical damage could result in flat, extensive retrograde amnesia lacking a temporal gradient.

Based on several human retrograde and anterograde amnesia studies, including that of H.M., as well as animal lesion studies, three main theories were put forward to explain consolidation at the brain-wide or system level — standard consolidation theory, cognitive map theory and the multiple trace theory. It must be noted before defining these theories that all three have been based on data concerned mostly with the role of the hippocampus and on spatial declarative memory. Therefore, the forms of memory that are less hippocampus dependent are given a less rigorous treatment in these theories.

Larry Squire and others distinguished between the formation of declarative and non-declarative memories, and their studies revealed that, while declarative memories require the operation of storage mechanisms involving the cortex, procedural memories involve more sub-cortical structures and online processing of perceived stimuli rather than active recollection (Squire *et al.*, 2004). Based on a number of investigations in both humans and animal models, Squire *et al.* proposed that long-term memory consolidation involves a progressive reorganization of the memory trace, which initially requires the hippocampus and other subcortical structures, but later becomes stabilised through the involvement of cortical processes. This hypothesis, which is the dominant model of memory consolidation, is known as standard consolidation theory, and has gained momentum in the last decade with the advent of brain imaging studies as well as more sophisticated techniques that measure brain activity in animal models. Functional magnetic resonance imaging (fMRI) studies reported that declarative memories in humans recruit the hippocampus for only a few years while the recall of remote memories (up to 20 years old) correlated with activity in the entorhinal cortex. Similarly, using markers of metabolic activity in animal models, Bontempi *et al.* found that recalling older memories in rats recruited the frontal, temporal, and anterior cingulate cortices but not the hippocampus which, however, showed activation when young memories were recalled (Frankland and Bontempi, 2005).

Although this theory is quite comprehensive and succeeds in explaining many different data from human and animal models, it fails to entirely explain the existence of cases where flat amnesia is produced as a result of hippocampal damage. An explanation for this outcome is that the lesion involved may also have included cortical damage that affected the storage of the memory. Other proponents argue that the hippocampus may have been involved in retrieval of those tasks and was not involved in storage *per se*, thus causing flat amnesia.

Another model, known as cognitive map theory has been proposed by Lynn Nadel and John O'Keefe (1978) and was put forward soon after the discovery of place cells in the hippocampus, which are groups of cells that fire action potentials only when the animal is in a specific location in space. This theory posits that the hippocampus is involved in creating allocentric representations of an animal's environment which provide the contextual basis for the formation of episodic memories. One important prediction made by this theory is that a hippocampal lesion would lead to loss of episodic memories but not semantic memories (memories concerning facts as well as conceptual knowledge of objects and grammar), as these are context independent. Thus, this theory predicts that the hippocampus would always be required for episodic memory representation and that hippocampal lesions should cause flat retrograde amnesia for these memories. This, however, does not account for many studies that do not observe flat amnesia despite complete lesions of the hippocampus.

Based on the discrepancies in both the cognitive map and standard consolidation theories, Nadel and Moscovitch proposed the multiple trace theory of memory consolidation (Nadel *et al.*, 2000). This theory, while agreeing with cognitive map theory with respect to the permanent requirement of the hippocampus, also agreed with some aspects of the standard model that dealt with the requirement of reactivation for memory strengthening. According to this model, whenever a memory is reactivated, it forms a new distinct trace, leading to the formation of multiple traces for any given memory. Thus, older memories would be represented more widely in cortico-hippocampal networks than younger ones. Additionally, a prediction made by the theory is that vivid episodic memories would always require the hippocampus whereas other types of memory would become hippocampus independent and become largely semantic. This prediction would explain some of the cases in which graded amnesia is seen.

2. Mechanisms of memory consolidation

By the mid-20th century fundamental insights into the mechanisms of memory consolidation were significantly advanced by pharmacological research. In the last 2 decades, an explosion of studies, based on molecular and cellular investigations, begun elucidating the biological mechanisms of memory formation. As we will see, these studies mainly focused on the identification and functional characterization of molecular pathways that are critically involved in long-term memory and long-term plasticity.

2.1. *Memory modulation*

The knowledge on the temporal evolution of memory consolidation led to attempts to identify the mechanisms by which post-training treatments could impair or enhance memory retention in a time-dependent manner. The question led to investigations of hormonal responses to experience, such as release of epinephrine and corticosterone into the circulation and the release of neurotransmitters and modulators such as norepinephrine, dopamine and acetylcholine. Based on the common observation that stressful situations could enhance or impair memory performance, work by McGaugh, Gold, and others (McGaugh and Roozendaal, 2002) demonstrated the involvement of epinephrine in these effects of stress. They showed that administration of epinephrine could enhance an inhibitory avoidance memory in rats in a dose-dependent fashion and that β-adrenergic receptors in the amygdala play an essential role in the memory-modulating effects of epinephrine. β-adrenergic receptors belong to a class of receptors called G-protein–coupled receptors (GPCRs) that are present on the membranes of neurons and have seven transmembrane domains. β-adrenergic receptors transduce the signal generated by the binding of epinephrine by activating intracellular G-proteins that, in turn, activate an enzyme called adenylyl cyclase, which catalyzes the formation of cyclic adenosine monophosphate (cAMP). The increase in cAMP activates a protein kinase called cAMP-dependent protein kinase A (PKA), which then activates a number of downstream molecules through phosphorylation. The activation of the cAMP-dependent pathway, as will be discussed in detail later, has proven to be critically recruited in memory consolidation.

Similarly, it was shown that the hypothalamic-pituitary-adrenal (HPA) axis is activated during emotional arousal. This led to the investigation of the role adrenal hormones such as glucocorticoids in memory consolidation. Glucocorticoids such as corticosterone, similar to epinephrine, display an inverted U-shaped dose-response relationship with memory consolidation. Low to moderate doses of glucorticoids, like low to moderate levels of stress, enhance memory, whereas high doses/levels impair memory. Moreover, blocking glucocorticoid receptors (GRs) in brain regions such as the hippocampus and amygdala using the GR antagonist RU38486 disrupts memory consolidation. The amygdala emerged as an important player in the modulatory role of glucocorticoids on memory, as lesions of the amygdala block the memory enhancing effects of glucocorticoids. GRs belong to a class of receptors called nuclear receptors that bind membrane permeable steroid hormones. These receptors are special in that, once bound by their ligand, they can translocate from the cytoplasm to the nucleus to direct gene transcription. In addition to its classical role as a transcription factor, GR may also have non-genomic effects that have been implicated in modulating memory consolidation but these effects have not yet been conclusively demonstrated. The molecular mechanisms of these effects are not yet well understood and are the subject of much current research.

The cholinergic system has also been shown to be an important modulator of memory consolidation. The hippocampus receives extensive cholinergic innervation from a brain region called the medial septum. These neuronal projections release the neurotransmitter acetylcholine which can bind to two kinds of receptors, nicotinic receptors, which are ligand-gated ion channels, and muscarinic receptors, which are GPCRs. Depletion of the cholinergic projections to the hippocampus impair memory consolidation while stimulating these projections enhances consolidation. Given its critical role in memory and cognition, the cholinergic system is an important target for the development of novel drugs to treat dementia and Alzheimer's disease (AD). The release of another catecholamine neurotransmitter, dopamine, largely from neurons in a brain region called the ventral tegmental area (VTA) has also been shown to modulate memory consolidation. Dopamine can act by binding to a variety of dopamine receptors, named D1–5, with its actions on memory primarily occurring via D1 and D2 receptors. However, the effects of dopamine on memory consolidation depend upon several factors such as the brain region and temporal window of action. Its effects have been shown to be mediated by a region called the nucleus accumbens (NAc) which receives extensive projections from the VTA. Intra-NAc infusions of the D2 antagonist sulpiride impair consolidation of a spatial water maze task while injections of D1 and D2 receptor agonists enhance memory consolidation.

These studies laid the foundation toward the identification of molecular pathways involved in memory consolidation.

2.2. *De novo protein and RNA syntheses are required for memory consolidation*

Around the same time as the studies on H.M. emerged, the first clues toward the biochemical understanding of long-term memory formation were offered by studies from Louis Flexner and others in mice, who reported that long-term memory retention was impaired by the administration of antibiotics that block new protein synthesis immediately after training (Davis and Squire, 1984). Soon thereafter, Agranoff *et al.* observed that, in goldfish, intracranial injections of the RNA synthesis inhibitor actinomycin D or of the protein synthesis inhibitor puromycin produce an impairment of an avoidance task but do not affect performance in naïve goldfish. Importantly, these treatments did not affect short-term memory and did not affect long-term memory if administered days after training, demonstrating that long-term, not short-term, memory formation requires an initial and temporally limited phase of gene expression. This initial discovery was followed by numerous other studies throughout the last 50 years that showed the requirement of *de novo* protein and RNA synthesis in a multitude of species and memory tasks. Despite the disagreement of some authors (Gold, 2008), who argued against the requirement of *de novo* protein synthesis due to concerns over the specificity of pharmacological reagents used for these studies, these findings, based on the multitude of different inhibitors used producing similar outcomes led to the conclusion that memory consolidation is a process that initially depends on gene expression. Moreover, recent genetic studies, which will be discussed later, complemented the pharmacological results and provided additional support for the important role of *de novo* protein synthesis in memory consolidation.

What does *initially* refer to? How long does the new RNA and protein synthesis last to allow for memory consolidation? The temporal evolution of the protein synthesis requirement for memory consolidation is a very important and complex question that needs to be addressed. Different types of memories seem to have different temporal profiles of gene expression requirement. Furthermore, the existence of multiple waves of RNA or protein synthesis following learning underlies the complexity of the new macromolecular synthetic process that accompanies memory consolidation.

Additionally, an important discovery that was first reported in the 1960s, and has received much attention in the last decade, is that the gene expression requirement is not restricted to the learning phase. In fact, many studies have reported that new RNA and protein synthesis are required again to stabilize the memory, even days or weeks after the memory has been formed, if the memory itself is retrieved. Thus, retrieval reactivates the memory trace, which remains in a fragile state for a limited time, and, like a new memory after learning, is sensitive to interference, but becomes resistant to disruption over time. The gene expression-dependent phase induced by memory reactivation and required for the memory to persist has been termed *reconsolidation* (Nader and Einarsson, 2010). This rediscovery of memory reconsolidation has challenged the classical consolidation theory, which believed that the strengthening or consolidation of memory occurs only once (Alberini, 2005). Understanding reconsolidation is the subject of a number of current studies and will not be further discussed in this chapter. In Chapter 10, more details on reconsolidation are provided.

More recent studies that targeted translation mechanisms or interfered with the functional role of components of the cellular translational machinery confirmed and extended the conclusion that translation is critical for memory consolidation. Signalling pathways controlled by mammalian target of rapamycin (mTOR), which regulates translation by controlling the phosphorylation of eukaryotic initiation factor 4E (eIF4E)-binding protein and p70S6 kinase (p70S6k), two proteins that are critical for the initiation of protein synthesis, have been shown to play an essential role in rodent memory formation in a variety of tasks including inhibitory avoidance, novel object recognition, and associative fear conditioning (Gkogkas *et al.*, 2010). Furthermore, the requirement of protein synthesis and the molecular mechanisms required for memory formation have also been investigated in cellular models believed to underlie long-term memory, including long-term facilitation (LTF), long-term potentiation (LTP), and long-term depression (LTD). LTP and LTD are strengthening or weakening of synapses induced by patterns of electrical stimulation of nerve fibers.

In summary, since the initial discovery in the 1960s, decades of animal experimental evidence ranging from invertebrates to mammals, in a multitude of learning paradigms and brain areas, have provided strong support that the expression of new RNAs and proteins during an early and temporally limited phase after learning, and in certain cases after memory reactivation, are necessary for the formation of long-term memory. This raises the important question about the nature and identity of the RNA and proteins that are synthesized and required during memory consolidation. Although some pathways have been identified, which will be described below, a comprehensive list of proteins required for memory consolidation still remains to be identified. The first elucidation of the molecular mechanisms underlying long-term memory formation came from studies in invertebrate model systems.

2.3. *The nature of molecular mechanisms underlying memory consolidation: Invertebrates provide the first understanding*

The complexity of mammalian systems presented an experimental obstacle to the identification of the molecular mechanisms that accompany memory consolidation. This obstacle was overcome by the identification of simple model systems with rudimentary neural circuits that lent themselves to easy manipulation. The two model systems that led the field, and continue to provide fundamental insights into the molecular mechanisms of memory formation, are the invertebrate species *Aplysia californica* and *Drosophila melanogaster*.

2.3.1. *Aplysia californica*

In the 1960s, Eric Kandel *et al.* began to characterize simple forms of memory in the marine snail *Aplysia californica*. In contrast to the brain of a mammal with 100 billion nerve cells, the central nervous system of *Aplysia* has only approximately 20,000 neural cells. The relatively large size and the identifiable nature of their neurons make them particularly suitable for electrophysiological and cell biological investigations. Several simple forms of learning and memory were characterized in *Aplysia*, including habituation, sensitisation and classical conditioning of the gill and siphon withdrawal reflex. The form of learning that has then been studied more in detail at the behavioural, cellular and molecular levels is the sensitisation of the gill and siphon withdrawal reflex (Kandel *et al.*, 2000). In this type of learning, an animal's encounter with an aversive stimulus increases its subsequent defensive response to a neutral stimulus that acts as a reminder of the aversive experience. In a series of studies, it was shown that long-term sensitisation of the gill and siphon withdrawal reflex is a function of the number of training trials. Furthermore, it was found that, short-term sensitisation depends on post-translational modifications, whereas long-term sensitisation requires *de novo* RNA and protein synthesis. In a subsequent series of fundamental investigations, the *Aplysia* memory system was translated into an electrophysiological and cellular model system. Kandel *et al.* showed that it was possible to reconstitute *in vitro* the main components of the neuronal circuit underlying sensitization *in vivo,* and to reproduce in a dish the synaptic responses induced *in vivo* by learning. This was accomplished by simply co-culturing a sensory neuron, which *in vivo* responds to the aversive stimulus, with a motor neuron, which *in vivo* mediates the siphon and gill withdrawal. Using this *in vitro* system, Kandel *et al.* showed that, similar to behavioural sensitisation, one application of serotonin (5-hydroxytryptamine, 5-HT) leads to a short-term enhancement in the excitatory postsynaptic potential (EPSP) or short-term facilitation (STF) that lasts for minutes, whereas four to five applications of serotonin results in a long-term enhancement in the EPSP, or LTF, which lasts for more than one day. Importantly, like the memories they reflect, whereas STF depends on post-translational modifications, LTF requires *de novo* RNA and protein synthesis. Extensive biochemical investigations then revealed that sensory neurons respond to 5-HT with an increase in cAMP level and activation of PKA, both of which were found to be critical for both short-term facilitation and LTF.

In the 1990s, the *Aplysia* system was ready to be investigated at the molecular level. Numerous studies were carried out to identify the gene expression cascade in long-term facilitation: it emerged that the activation of the cAMP pathway results in the activation of the transcription regulators belonging to the cAMP response element-binding protein (CREB) and CCAAT enhancer-binding protein (C/EBP) families. In brief, it was found that, in response to repeated pulses of 5-HT, the PKA catalytic subunit and mitogen-activated protein kinase (MAPK) translocate to the nucleus and mediate the derepression of the transcription repressor CREB-2 while leading to the activation of transcription activator CREB-1. Thus, CREB1 and CREB2 appear to work in a concerted manner to regulate gene expression necessary for LTF. This concerted action results in the expression regulation of immediate early genes, which also were found to play a critical role in long-term facilitation: one example is the regulatory immediate early gene *Aplysia* C/EBP (ApC/EBP). Like CREB, this transcription factor is required for LTF but not STF and its requirement is limited to the first 9 hours after 5-HT application. These results revealed that the long-term facilitation underlying long-term sensitisation of the gill and siphon withdrawal reflex uses a cascade of gene expression, with transcription factors regulating the expression of downstream genes including other transcription factors, which in turn regulate the expression of effector target genes. These types of gene expression cascades generally accompany changes in the gene expression pattern of cells, which are processes that, for example, occur during development or cell differentiation. Another immediate early gene required for LTF was found to be the ubiquitin C-terminal hydrolase, an enzyme associated with proteasome-dependent proteolysis, which promotes LTF by degrading the regulatory subunit of PKA, thus prolonging the activity of the PKA catalytic subunit. Following these initial findings, numerous genes were identified and studied in the *Aplysia* system and greatly contributed to our current understanding of the molecular bases of long-term changes underlying memory formation. Notably, because of the complexity of the pattern of gene expression regulation and their temporal changes, a comprehensive understanding of target genes regulated by both CREBs and ApC/CEBP still remains to be identified.

The characterization of the *Aplysia* system also provided an opportunity to examine the synaptic structural changes that accompany long-term memory, something that was hypothesized many years previously by Santiago Ramon y Cajal. Cajal reasoned that in order for long-term changes to last a lifetime, neurons must form new connections via the growth of new synaptic boutons. Following sensitisation training or LTF in culture, it was found that there is a significant increase in the number of synaptic terminals per neuron. These anatomical changes occur at both pre- and postsynaptic terminals, which suggests that long-term synaptic plasticity requires coordinated structural changes at both pre- and postsynaptic cells (Kandel, 2001).

Another important question that the *Aplysia in vitro* system is uniquely suited to address is whether and how transcription and translation regulations induced during LTF may or may not be compartmentalised. As we have seen, long-term synaptic plasticity and long-term memory require new transcription, which occurs in the nucleus, yet each neuron forms numerous synaptic connections with other neurons, some of which are distant from the nucleus. Thus, an important question that arose from the gene expression studies was whether synaptic plasticity occurs in a cell-wide manner or whether there are mechanisms that selectively target activated synapses. If this was the case, how does the cell determine the correct synapses to which the newly synthesised proteins necessary for long-term plasticity be delivered? To address these questions, Kelsey Martin in Kandel's laboratory developed a modified *in vitro Aplysia* culture system: she cultured a single sensory neuron with bifurcated neurites targeting two separated motor neurons. Using this system, it was found that branch/synapse-specific application of 5-HT elicits a branch-specific LTF, which is CREB dependent is and accompanied by the branch-specific growth of new synapses. Thus, these studies showed that synaptic plasticity begins in a compartmentalised fashion and indicated that there is a retrograde signal from the stimulated synapse that travels to the nucleus following stimulation, which then leads to the activation of processes that cross-talk with the activated synapses.

To begin understanding the mechanism of retrograde signaling from distal synapses to nucleus, Kelsey Martin *et al.* then asked the question: how does the stimulation signal occurring distally at synapses convey information to the nucleus to initiate transcriptional changes necessary for long-term plasticity? They found that a component of the classical active nuclear import pathway, importin $\alpha 3$, is localised in *Aplysia* distal neurites; repeated application of 5-HT trigger its nuclear translocation, and the blockade of the active nuclear import pathway prevents the formation of LTF. They further demonstrated that mRNAs are translated at synapses at which repeated applications of serotonin induce LTF, but not at unstimulated synapses and not at synapses where LTD was induced (Wang *et al.*, 2010).

Because the induction of a gene expression cascade is required for LTF, parallel investigations explored the relevance of chromatin changes in the nervous system of *Aplysia* in response to long-term synaptic plasticity and memory formation. Investigating chromatin changes that accompany memory consolidation is at the center of important, current studies. Taking advantage of the *Aplysia* bifurcated sensory-motor co-culture system, it was found that 5-HT stimulation at one branch results in histone acetylation at the ApC/EBP promoter, whereas LTD-inducing stimulation at the other branch leads to histone deacetylation at the ApC/EBP promoter. A sensory neuron receiving both LTF-inducing excitatory inputs and LTD-inducing inhibitory inputs simultaneously expresses depressed EPSP and an inhibition of histone acetylation. These studies revealed that neurons integrate different signals at the level of chromatin modification mechanisms in a coordinated fashion to regulate long-term synaptic plasticity (Guan *et al.*, 2002).

In summary, the *Aplysia* cell culture model system has fundamentally contributed to our molecular and cellular understanding of learning-induced changes underlying memory formation, and particularly those mediating memory consolidation. Since several signaling cascades underlying long-term memory formation seem well conserved throughout evolution, insights from *Aplysia* studies continue to provide critical insights for understanding memory consolidation in the mammalian brain. These studies also underscore the value of simple systems like *Aplysia* and, as we will see in the following section, *Drosophila melanogaster,* as groundbreaking platforms for uncovering fundamental, evolutionarily conserved mechanisms underlying memory formation.

2.3.2. *Drosophila melanogaster*

The neurobiological and molecular techniques that were applied to *Aplysia* were contemporarily complemented by genetic approaches in *Drosophila melanogaster*. By the mid 1960s, the fruit fly had already emerged as a leading model system for studying the molecular genetics of a variety of physiological systems, because it enables unbiased screens for gene mutations that affect behavior, including learning and memory. In fact, *Drosophila* has been shown to be capable of a multitude of learning tasks. One of the most robust tasks, developed by Seymour Benzer *et al.* is a Pavlovian olfactory learning task, in which flies learn to avoid an odor (conditioned stimulus or CS) that is paired with an electrical shock (unconditioned stimulus or US). Olfactory learning in flies follows many of the behavioural properties described for Pavlovian learning in other animals, for example acquisition, extinction, CS/US pre-exposure effect, conditioned inhibition and temporal specificity (CS/US need to paired within a specific time window in order for learning to occur). Using these learning paradigms, it was found that *Drosophila* has at least four types of memories: (1) short-term memory (STM) that lasts less than 1 h, (2) middle-term memory (MTM) that lasts less than 4–5 h, (3) anaesthesia-resistant memory (ARM) that lasts for more than 24 h after one single CS/US pairing and is deficient in the Drosophila mutant *radish*, and (4) a protein synthesis-dependent long-term memory (LTM) that is induced by multiple spaced training trials.

Using the odor discrimination task and chemically mutated flies, a number of learning mutants were isolated and studied. The first two characterized were *rutabaga (rut)* and *dunce*. Both were revealed to have defective acquisition and STM and were found to have mutations in the genes that regulate the synthesis and degradation of the cAMP respectively (Dubnau and Tully, 1998). Subsequent genetic studies generated a long list of memory mutants, which provided mechanistic separations between the different memory types and shed light on their underlying molecular mechanisms. For example, *amnesiac* (*amn*; homologous to pituitary adenylyl cyclase activating peptide) mutants have impaired MTM: they show near-normal retention immediately after training and 7 h later, but have appreciably lower retention in between. Notably, mutations in the transcription factors CREB2 (dCREB2) were also found to selectively block LTM without affecting ARM. Most of the memory mutants initially isolated using the odor discrimination task also performed poorly in other memory tasks, suggesting that the underlying biochemical mechanisms and neural circuitry are conserved for other learned behaviors. In conclusion, similarly to the cellular and molecular mechanisms uncovered in *Aplysia*, many of the genes found to be mutated in learning and memory mutants were part of the cAMP-PKA-CREB–dependent gene cascade, which is highly conserved in higher vertebrates.

The *Drosophila* model system has then offered the opportunity to anatomically map the cells that contribute to memory formation. As in vertebrate animals, the central nervous system of *Drosophila* is organized into highly specialized sets of neuronal structures. One brain area that has emerged as critical for Pavlovian learning and memory is the mushroom bodies. The first evidence for a role of mushroom bodies in olfactory memory came from memory mutants with neuroanatomical defects. Mutants with defects in mushroom bodies were unable to form an associative memory even though odor and shock-sensations were intact. Histological analyses revealed that many genes found to be required for learning and memory were expressed at high levels in the mushroom bodies, supporting the conclusion that mushroom bodies are principle centers for learned information (Davis, 2005). Powerful molecular genetic methods such as temperature-sensitive temporal and regional gene expression targeting (TARGET) in Drosophila offer the unique advantage of selectively control gene expression in region-, cell- and time-specific manner, a technical tool that is ahead of any other behavioural system (Keene and Waddell, 2007). For example, TARGET allowed the selective expression of *rut* in mushroom bodies in adult flies and revealed that the transient expression of *rut* restores the memory deficit in *rut* mutants, suggesting that *rut*-dependent plasticity in the mushroom bodies is necessary and sufficient for olfactory memory formation in adulthood. This powerful method avoids the serious problems of using mutant animals that carry mutations from birth, which generally result in compensatory regulations during development. Interestingly, the use of TARGET recently showed that selective inhibition of rac GTPase in a specific set of mushroom body neurons in adulthood leads to active forgetting, which is a process that appeared to be dissociable from impaired memory consolidation.

A second technique that has been developed recently along the same lines of thought is the mosaic analysis with a repressible cell marker (MARCM) technique. This technique allows for the study of homozygous mutations in specific subsets of neurons in an otherwise wild type fly. A series of these types of studies revealed that within the mushroom bodies, there are subtypes of neurons that are necessary for different phases of memory including acquisition, consolidation, early memory retrieval and older memory retrieval.

Furthermore, powerful optogenetic tools have been developed in the *Drosophila* system to monitor neural physiological processes *in vivo*. With these tools, it is now possible to genetically express one memory gene coupled to one optical reporter in one set of neurons and a different gene in another set of neurons.

Thus, great advances have been made in illuminating the mechanism underlying CS-US association and memory formation and the powerful genetic approaches available in *Drosophila* continue to shed light and revolutionize the field of learning and memory. The *Drosophila* system is no longer just a tool to identify genes and molecules necessary for memory formation. Recent advances in genetic and optical recording technologies transformed *Drosophila* from a useful organism for gene discovery to a model of studying where and how neuronal signalling pathways interact to encode memory.

2.4. *Mechanisms of memory consolidation in mammals*

The landmark studies in *Drosophila melanogaster* and *Aplysia Californica* laid the foundation for studying mechanisms of memory consolidation in mammals.

2.4.1. *Molecular genetic approaches identify molecules and pathways essential for long-term memory formation*

By the late 1980s, mouse transgenic technology had been developed, which allowed for molecular genetic studies of memory formation in mammals. This technology overcame many of the specificity issues of pharmacological manipulations and provided targeted manipulations of specific genes. In 1992, the first two studies on transgenic mice reported that mutations of the genes α-calcium calmodulin kinase II (α-CaMKII) or the tyrosine kinase Fyn impair LTP and long-term memory formation. The α-CaMKII mutant was studied by Alcino Silva in Susumu Tonegawa's laboratory. These authors showed that LTP and several types of long-term memories, including contextual and spatial, were impaired in the knockout (KO) mice. Similar results were reported by Seth Grant in Kandel's laboratory.

The findings on the cAMP-dependent gene cascade in both *Aplysia* and *Drosophila* also led to the question of whether the CREB-dependent pathway plays an evolutionarily conserved function in long-term memory formation. Thus, KO mice of the CREB isoforms most abundantly expressed in the brain, the α and Δ isoforms (CREBα/Δ) were analyzed behaviourally and electrophysiologically by Bourtchuladze *et al.* (1994). It was found that these mutants have significant deficits in several types of long-term memories, including spatial, contextual, and cued as well as impaired hippocampal LTP, with intact short-term memory, indicating that indeed the CREB pathway plays a conserved role in memory formation from invertebrates to mammals (Matynia *et al.*, 2002).

These results marked the beginning of the era of genetic studies in learning and memory. However, the field was aware that using traditional KO, in which the targeted gene is deleted throughout the life of the animal in all the tissues, poses a number of problems for understanding the mechanisms of a temporally limited behavioral function such as learning and memory. One major drawback of the classical transgenic and knockout type of approach is that the gene deletions or mutations may affect developmental processes and may also lead to compensatory changes that may confound the interpretation of the results. A second issue is regional control: as classical KO/transgenics affect the entire system, it is impossible to draw conclusions about where in the brain the expression of a specific gene contributes to memory formation. To overcome these limitations, Mark Mayford in Kandel's laboratory developed brain region-specific conditional knockouts that were also temporally regulated. These authors used a tetracycline-controlled transactivator (tTA) to regulate the expression of a constitutively active form of CaMKII (CaMKIIT286D) (Mayford *et al.*, 1997). When present in the system, a tetracycline analogue (doxycycline) represses the expression of a transgene under the regulation of the tet operon promoter, which results in its repression. By applying this repression during development and allowing the expression at desired times, it was possible to obtain temporally regulated gene expression. Subsequently, a modified tetracycline-inducible system was developed, in which the tTA has been mutated: thus, instead of repressing transcription, the mutant transactivator activates it and doxycycline induces the transgene, an approach that avoids the toxicity of maintaining an animal on doxycycline for extended periods of time. Similar, although more precise systems were further developed.

To address the anatomical issue, in the mid 1990s, Joe Tsien in Tonegawa's laboratory successfully developed a region-specific gene knockout technique by employing the phage P1-derived Cre/loxP recombination system. Using a brain-specific CaMKII promoter to drive the Cre transgene, these authors were able to delete the NMDA receptor 1 (NR1) gene specifically in the CA1 region of the hippocampus. These anatomically restricted KO mice were viable and developed normally. However, they showed impaired synaptic plasticity in the CA1 region and were impaired in many memory tests (Kuang *et al.*, 2009).

All these approaches, however, require weeks-long regulation of the transgene, which is a temporal window incompatible for the study of the phases underlying memory consolidation. Hence, to achieve flexible temporal control of genetic modification, an inducible, reversible, and region-specific knockout technique, also known as the third-generation gene knockout technique, was developed, which combines the Cre/loxP-mediated recombination system with the tTA system. The use of these third-generation transgenic mice, has led to important conclusions: for example, it has long been assumed that the NMDA receptor was required only for memory acquisition; however, the third generation transgenic mice revealed that switching off the NMDA receptor during the consolidation or storage stage impairs long-term memory.

Additionally, many transgenic mice with overexpression of genes or dominant negative constructs, as well as point mutations complemented the studies of the KO mice in identifying molecular pathways and targets important for memory formation.

Molecular genetics has also been useful in confirming the role of protein synthesis in memory formation and has circumvented many concerns about the specificity of pharmacological reagents. Major components of translational control including eukaryotic initiation factor 2α (eIF2α) and mammalian target of rapamycin (mTOR) have been investigated in long-term memory formation using transgenic mouse models. eIF2α is a translation initiation factor that controls the rate of protein synthesis through its effect on translation initiation. Phosphorylation of eIF2α at ser-51 decreases general translation initiation. Using a point mutation in ser-51 in a transgenic system and combining with pharmacological approaches, Costa-Mattioli *et al.* (2009) demonstrated that associative fear conditioning, spatial learning and memory, and novel taste memory, require the phosphorylation of eIF2α.

Despite the importance of mouse molecular genetic studies, one issue that these systems cannot address is that of the identification and precise temporal evolution of the endogenous molecular changes that accompany memory consolidation. In the next section, will see how some alternative approaches have provided information in this regard.

2.4.2. *The temporal evolution of the endogenous molecular changes accompanying memory consolidation*

One approach that has been utilized for decades in biology to selectively block the synthesis of specific transcripts is that based on the use of antisense sequences. Antisense sequences and small interfering RNA (siRNA) technologies inhibit the translation or the transcription of targeted sequences. Because of the small sequences used as antisense or siRNA, and their rapid turnover, these technologies when combined with anatomically restricted delivery, offer the advantage of targeting specific gene regulation in an anatomically and temporally controlled fashion. Using these types of approaches, a wide array of proteins have been identified as regulated following learning in specific brain regions, including the hippocampus and amygdala, and found to be necessary for long-term memory consolidation. They include: transcription factors such as CREB, C/EBP, Zif 268, c-Fos, Nur 77, neurotransmitter receptor subunits like NR1 and acetylcholine receptors, growth factors like brain-derived neutrophic factor (BDNF) and insulin-like growth factor II (IGF-II) and many other molecules, some of which are explored in more detail below (Alberini, 2009).

2.4.2.1. The CREB-C/EBP pathway and memory consolidation

CREB was first discovered as a key transcription factor mediating the cellular responses to various hormonal stimuli that lead to the upregulation of cAMP in various cell types and tissues. This in turn leads to the activation of PKA. PKA has been shown to be critical for memory consolidation in a variety of species and across levels of analysis. Mutant mice expressing a dominant negative version of PKA have deficits in memory retention. The inhibition of PKA in the hippocampus after training impairs memory consolidation in a variety of tasks. A major target of PKA is the transcription factor CREB. The phosphorylation of CREB at ser-133 leads to the recruitment of co-activators of CREB including the histone acetyltransferases CREB-binding protein (CBP) and p300 (Mayr and Montminy, 2001) leading to the activation of transcription. CREB can be phosphorylated by a number of protein kinases including PKA, CaMKIV (calcium/calmodulin-dependent protein kinase IV) and MAPK. Each of these kinases has been shown to be required for long-term memory consolidation.

As seen earlier, molecular deletion studies in *Aplysia*, *Drosophila* and mice showed that the members of the CREB family of transcription factors are necessary for memory formation, because their elimination result in impaired memories. However, an important question that needs to be addressed is the following: Is endogenously expressed CREB activated by learning? In which brain regions? For how long? The temporal evolution of CREB activation, measured as ser-133 phosphorylation (pCREB), and of CREB requirement in the hippocampus have been determined, respectively, by detecting pCREB changes in the brain of trained animals and by stereotactically injecting antisense oligodeoxynucleotides (ODNs) into the regions of interest at different times before or after training. Using an inhibitory avoidance task, it was found that pCREB increases in the hippocampus immediately after training, returns to baseline by 30 min and then increases again 1–3 h after training and this second wave lasts for more than 20 h after training. Similar results were obtained on the front of hippocampal CREB requirement in a spatial task: While knocking down the expression of hippocampal CREB immediately after training led to an impairment in memory consolidation, its disruption had no effect if it occurred starting 1 day after training, implying that CREB function during memory consolidation is restricted to a limited temporal window.

Similar studies were used to investigate the endogenous regulation and role of CREB target genes. It was found that the expression of the transcription factor C/EBPβ temporally followed, in the same hippocampal cells, that of pCREB and specifically increased between 6 and 9 h after training, remained elevated for more than 28 h and returned to baseline by 48 h after training. Antisense-mediated blockade of the C/EBP increase between 5 and 24 h after training, but not earlier or later, completely disrupted memory consolidation. These studies for the first time revealed that memory consolidation recruits endogenous CREB and C/EBP-dependent gene expression for about 24–48 h in the hippocampus after learning but not at later times.

Important questions that remain to be addressed are: What target genes are endogenously regulated during this temporal window? What is the end-result of such regulation? Do other brain regions, such as for example cortical regions use similar mechanisms? Is the temporal evolution of cortical molecular mechanisms underlying memory consolidation similar or different?

2.4.2.2. BDNF: A putative target gene of the CREB-C/EBP pathway and its role in memory consolidation

One of the putative target genes of the CREB-C/EBP pathway, which has emerged as an important regulator of synaptic plasticity and memory is the BDNF. In neuronal cultures, BDNF activates CREB and its expression is in turn regulated by CREB. BDNF is an activity-regulated dimeric protein that is secreted both post- and presynaptically in response to neuronal activity. Once secreted, BDNF binds with high affinity to the tropomyosin related kinase B (TrkB) receptor, which is present on presynaptic, post-synaptic, and glial termini. BDNF binding induces ligand-receptor dimerisation and autophosphorylation of tyrosine residues in the intracellular kinase domain of the receptor leading to the recruitment of adaptor molecules subserving diverse signalling cascades. Three main intracellular signaling cascades are activated by the binding of these adaptor molecules to TrkB: the RAS-MAPK pathway, the phosphoinositide 3-kinase-Akt and the phospholipase C gamma-Ca^{2+} pathway.

A number of studies have established that BDNF is required for LTP as well as memory formation and its endogenous levels are regulated in line with the CREB and C/EBP-dependent activation. Indeed, BDNF has been shown to increase in the hippocampus following training in a variety of paradigms such as the Morris water maze and the radial maze, inhibitory avoidance, contextual fear conditioning and olfactory recognition. Additionally, blocking BDNF synthesis or TrkB, both genetically and pharmacologically, has been shown to impair memory formation. BDNF antisense-mediated knock down or anti-BDNF blocking antibody before training blocks acquisition in inhibitory avoidance, suggesting that its action precedes the activation of the CREB-C/EBP cascade. However,

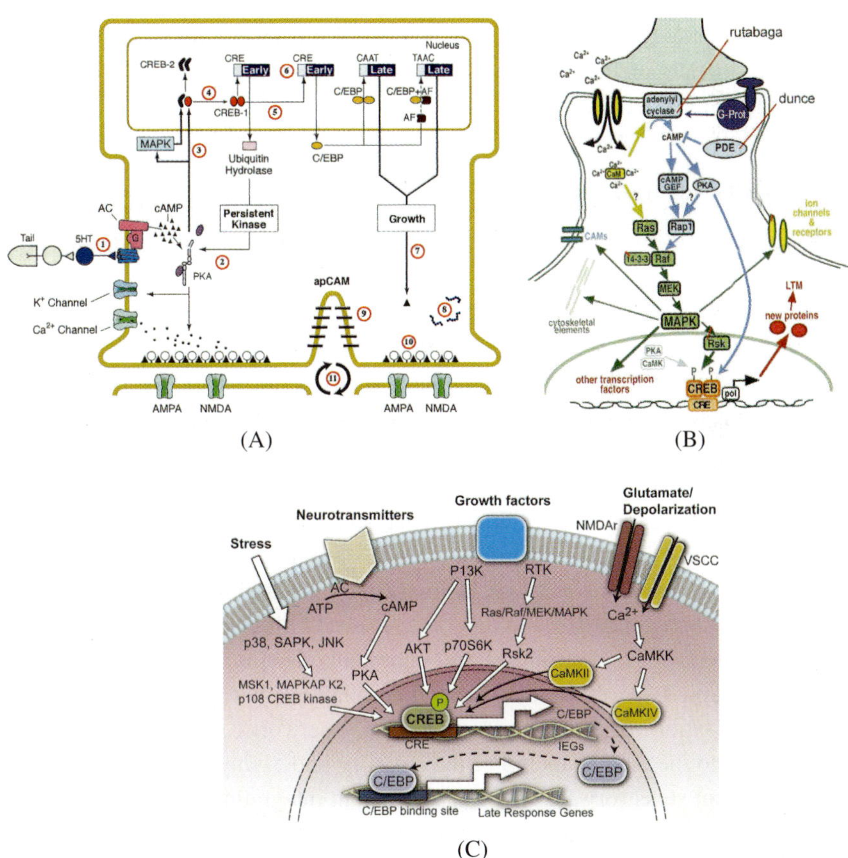

Figure 5.1. Evolutionarily conserved molecular mechanisms accompany long-term memory consolidation. (A) *Aplysia californica*. (B) *Drosophila melanogaster*. (C) Vertebrate systems.

like in neuronal cultures, BDNF actions also follow a multiphasic pattern similar to that seen for pCREB and protein synthesis during memory consolidation. Blocking BDNF with an anti-BDNF blocking antibody 1 or 4 h after inhibitory avoidance training impairs memory consolidation and the same manipulation as well as antisense injections at 10–12 h after training, although ineffective on memory retention at 2 days after training, lead to memory decay when retention is tested 7 days later (Bekinschtein *et al.*, 2008). This implies that the hippocampal expression of BDNF, like that of C/EBP and C/EBP target genes, needs to be maintained for more than 12 h after training in order for memory consolidation to complete.

In conclusion, these studies have begun to temporally and anatomically frame the cascade of events that are endogenously regulated following learning and play an essential role in memory consolidation. In Figure 5.1, we provide schematic models of the molecules found to have an evolutionarily conserved role in memory consolidation.

3. Conclusions and questions

In this chapter, we have highlighted the considerable progress that has been made towards the elucidation of the mechanisms of memory consolidation. It is easy to appreciate that several different techniques and approaches have uncovered a wealth of diverse data on the phenomenon of consolidation. However, several outstanding questions remain. Though much is now known about the time course and regional requirement of gene expression during memory consolidation, a comprehensive spatial and temporal profile of the specific genes endogenously regulated and involved will be invaluable in building a global view of memory consolidation. Whereas hippocampal mechanisms underlying memory consolidation are more understood, cortical mechanisms are not, and a formal molecular analysis of system consolidation is imperative. Additionally, the identity of mechanisms the brain employs to consolidate several different and related memories simultaneously is an intriguing and important question.

At the same time, the study of neural ensemble activity and its contribution to the stabilization of information into memory networks will be an important contribution to the understanding of the circuitry underlying memory consolidation. Important steps forward will be gained from integrating information across the disciplines of pharmacology, genetics, molecular biology, and electrophysiology to arrive at unified models of memory consolidation.

The study of memory consolidation promises to impact human society for the better. Understanding the mechanisms of consolidation may allow us to improve how knowledge is imparted in the classroom in the most efficient way. It will perhaps give us the information required to build artificial intelligence as well as neural prosthetics. But, perhaps most importantly, it may give us a deeper understanding of diseases of memory such as Alzheimer's disease, dementia, schizophrenia, depression, and post-traumatic stress disorder, so that we can find cures and prevention strategies.

References

Alberini, C.M. (2005). Mechanism of memory stabilization: Are consolidation and reconsolidation similar or distinct processes? *Trends Neurosci* **28**, 51–56.

Alberini, C.M. (2009). Transcription factors in long-term memory and synaptic plasticity. *Physiol Rev* **89**, 121–145.

Bekinschtein, P., Cammarota, M., Izquierdo, I. and Medina, J.H. (2008). BDNF and memory formation and storage. *Neuroscientist* **14**, 147–156.

Bourtchuladze, R., Frenguelli, B., Blendy, J., Cioffi, D., Schutz, G. and Silva, A.J. (1994). Deficient long-term memory in mice with a targeted mutation of the cAMP-responsive element-binding protein. *Cell* **79**, 59–68.

Brembs, B. (2003). Operant conditioning in invertebrates. *Curr Opin Neurobiol* **13**, 710–717.

Costa-Mattioli, M., Sossin, W.S., Klann, E. and Sonenberg, N. (2009). Translational control of long-lasting synaptic plasticity and memory. *Neuron* **61**, 10–26.

Davis, H.P. and Squire, L.R. (1984). Protein synthesis and memory: A review. *Psychol Bull* **96**, 518–559.

Davis, R.L. (2005). Olfactory memory formation in Drosophila: From molecular to systems neuroscience. *Annu Rev Neurosci* **28**, 275–302.

Dubnau, J. and Tully, T. (1998). Gene discovery in Drosophila: New insights for learning and memory. *Annu Rev Neurosci* **21**, 407–444.

Frankland, P.W. and Bontempi, B. (2005). The organization of recent and remote memories. *Nat Rev Neurosci* **6**, 119–30.

Gold, P.E. (2008). Protein synthesis inhibition and memory: Formation vs amnesia. *Neurobiol Learn Mem* **89**, 201–211.

Gkogkas, C., Sonenberg, N. and Costa-Mattioli, M. (2010) Translational control mechanisms in long-lasting synaptic plasticity and memory. *J Biol Chem* **285**, 31913–31917.

Guan, Z., Giustetto, M., Lomvardas, S., Kim, J.H., Miniaci, M.C., Schwartz, J.H., Thanos, D. and Kandel, E.R. (2002). Integration of long-term-memory-related synaptic plasticity involves bidirectional regulation of gene expression and chromatin structure. *Cell* **111**, 483–493.

Hawkins, R.D., Kandel, E.R., Bailey, C.H. (2006) Molecular mechanisms of memory storage in Aplysia. *Biol Bull* **210**, 174–191.

Kandel, E.R., Schwartz, J.H. and Jessell, T.M. (2000). *Principles of Neural Science* (Fourth ed.), New York: McGraw-Hill.

Kandel, E.R. (2001). The molecular biology of memory storage: A dialogue between genes and synapses. *Science* **294**, 1030–1038.

Keene, A.C. and Waddell, S. (2007) Drosophila olfactory memory: Single genes to complex neural ciruits. *Nat Rev Neurosci* **8**, 341–354.

Kuang, H., Wang, P.L. and Tsien, J.Z. (2009). Towards transgenic primates: What can we learn from mouse genetics? *Sci China C Life Sci* **52**, 506–514.

Lechner, H.A., Squire, L.R. and Byrne, J.H. (1999). 100 years of consolidation — remembering Müller and Pilzecker. *Learn Mem* **6**, 77–87.

Matynia, A., Kushner, S.A. and Silva, A.J. (2002). Genetic approaches to molecular and cellular cognition: A focus on LTP and learning and memory. *Annu Rev Genet* **36**, 687–720.

Mayford, M., Mansuy, I.M., Muller, R.U. and Kandel, E.R. (1997). Memory and behavior: A second generation of genetically modified mice. *Curr Biol* **7**, R580–9.

Mayr, B. and Montminy, M. (2001). Transcriptional regulation by the phosphorylation-dependent factor CREB. *Nat Rev Mol Cell Biol* **2**, 599–609.

McGaugh, J.L. (2000). Memory — a century of consolidation. *Science* **287**, 248–251.

McGaugh, J.L. and Roozendaal, B. (2002). Role of adrenal stress hormone in forming lasting memories in the brain. *Curr Opin Neurobiol* **12**, 205–210.

Müller G.E. and Pilzecker A. (1900) Experimentelle Beiträge zur Lehre vom Gedächtnis. *Z Psychol Ergänzungsband* **1**, 1–300.

Nadel, L., Samsonovich., A., Ryan., L. and Moscovitch., M. (2000). Multiple trace theory of human memory: Computional, neuroimaging and neuropsychological results. *Hippocampus* **10**, 352–368.

Nader, K. and Einarsson, E.O. (2010). Memory reconsolidation: an update. *Ann NY Acad Sci* **1191**, 27–41.

Nicolas, S. and Murray, D., 1999. Théodule Ribot (1839–1916), founder of French psychology: A biographical introduction. *History of Psychology* **2**, pp. 277–301.

O'Keefe, J. and Nadel, L. (1978). *The Hippocampus as a Cognitive Map*. London: Oxford University Press (Clarendon).

Squire, L.R., Stark, C.E. and Clark, R.E. (2004) The medial temporal lobe. *Annu Rev Neurosci* **27**, 279–306.

Wang, D.O., Martin, K.C. and Zukin, R.S. (2010). Spatially restricting gene expression by local translation at synapses. *Trends Neurosci* **33**, 173–182.

The Role of CREB and CREB Co-activators in Memory Formation

6

Melanie J. Sekeres*,†, Derya Sargin*, P. Joel Ross* and Sheena A. Josselyn*,†,‡

1. Introduction

The ability to learn from experience facilitates survival in a dynamic environment: accurate spatial representation of complex surroundings improves the likelihood of finding food and water; associative learning confers salience upon previously neutral cues that may predict the presence of a predator or a receptive mate. In this way, memories define the past and present, and help guide the cognitive construction of the future.

Memories can be subdivided into two general classes: short-term memory (STM) and long-term memory (LTM). These memory classes are operationally defined both in terms of temporal characteristics and underlying mechanism: STM lasts from minutes to hours and involves modification of pre-existing proteins, whereas LTM lasts from hours to a lifetime, and requires gene expression and protein synthesis. A key regulator of the stabilisation of LTM is the transcription factor CREB (cyclic adenosine-3′,5′-monophosphate (cAMP) response element-binding protein).

Below, we review the role of CREB and several CREB co-activators in memory formation.

2. Transcriptional regulation by CREB

The CREB family of transcription factors is encoded by three genes: *CREB*, cAMP response element modulator (*CREM*) and activating transcription factor 1 (*ATF-1*)

* Program in Neurosciences & Mental Health, Hospital for Sick Children, 555 University Ave. Toronto, ON, M5G 1X8

† Department of Physiology, University of Toronto, Toronto, ON, M5G 1X8, Canada

‡ Institute of Medical Sciences, University of Toronto, Toronto, ON, M5G 1X8, Canada

(Foulkes *et al.*, 1991; Gonzalez *et al.*, 1989; Rehfuss *et al.*, 1991). *CREB* encodes three alternatively spliced variants (CREB α, β, and σ), each of which stimulate transcription through promoters containing cyclic AMP response element (CRE) sequences. In contrast, the four major splice variants of CREM in the brain (CREM α, β, γ, and ICER (inducible cAMP early repressor)) repress CRE-dependent transcription (Mellstrom *et al.*, 1993).

CREB protein contains four conserved domains: the kinase inducible domain (KID), the basic helix-loop-helix (bHLH) domain, and glutamine (Q)-rich domains Q1 and Q2. The C-terminal bHLH domain mediates dimerisation and DNA binding. Q1 and Q2 promote constitutive, low-level transcription *in vitro*, whereas the KID is responsible for phosphorylation-dependent inducible transcription of CREB targets (Mayr and Montminy, 2001). Global analyses of CREB-DNA binding and transcriptional activation suggest that ~10–15% of mammalian genes are regulated, at least in part, by CREB (Conkright *et al.*, 2003b; Impey *et al.*, 2004; Zhang *et al.*, 2005). These studies identified ~4000–6300 unique genomic loci that were bound by CREB. Most CREB binding sites (~40%) are within 2 kb of an annotated transcription start site, either in the promoter region or within the first intron of target genes (Impey *et al.*, 2004).

At first glance, the CREB "regulon" seems vast. However, CREB-mediated transcription in specific cell types (e.g. neurons) is likely restricted to a much smaller set of target genes. Methylation of the CREB binding site, which is regulated in a cell- and tissue-specific manner (Jaenisch and Bird, 2003), impairs DNA binding (Iguchi-Ariga and Schaffner, 1989; Zhang *et al.*, 2005). Furthermore, CREB is unable to drive expression of genes that do not have a TATA box element within their core promoters [i.e. ~85% of mammalian genes (Conkright *et al.*, 2003b; Juven-Gershon and Kadonaga, 2010)]. Therefore, although CREB binds many loci within the genome, tissue-specific DNA methylation patterns and core promoter variability likely limit the range of relevant CREB target genes for mediating specific functions, such as learning and memory.

2.1 *Activation of CREB by extracellular signals*

CREB activity is regulated by numerous cellular signalling cascades that are initiated in response to neuronal activation, which primarily affect CREB activity via phosphorylation of serine 133 (S133) (West *et al.*, 2002). CREB phosphorylation is necessary (Gonzalez and Montminy, 1989) and sufficient (Cardinaux *et al.*, 2000; Du *et al.*, 2000) for CREB-dependent gene expression in most cell lines, although studies of primary hippocampal cultures suggest this may not be the case in neurons, where sustained CREB phosphorylation is necessary to promote efficient gene expression (Bito *et al.*, 1996; Deisseroth and Tsien, 2002; Wu *et al.*, 2001). Phosphorylation of CREB is dynamically regulated by kinases and phosphatases that respond to increased levels of cAMP and/or calcium (West *et al.*, 2002).

Behavioural training, neuronal depolarisation and synaptic stimulation result in increased intracellular levels of cAMP and calcium (Flavell and Greenberg, 2008). Neuronal activation results in calcium entry through L-type calcium channels and NMDA receptors (Greer and Greenberg, 2008); how cAMP signalling is initiated in response to these stimuli is not completely understood, although neurotransmitter-induced G-protein–coupled receptor signalling (Kandel, 2001) and calcium-induced adenylyl cyclase activity have been implicated (Poser and Storm, 2001). Calcium and cAMP initiate signalling cascades that converge on CREB, regulating its phosphorylation and transcriptional activity. Specifically, calcium/calmodulin-dependent protein kinases (CaMK) I and IV, mitogen and stress-activated protein kinase 1 (MSK1), ribosomal protein S6 kinase (RSK) 1 and 2, and protein kinase A (PKA) phosphorylate CREB at S133, and thereby promote activation of CREB target genes (Cohen and Greenberg, 2008; Flavell and Greenberg, 2008; West *et al.*, 2002).

Calcium-induced phosphorylation of CREB is primarily mediated via two pathways: by activating (1) calmodulin (Xia and Storm, 2005) and CaMKs (Wayman *et al.*, 2008), and (2) the Ras/mitogen-activated protein kinase (MAPK) pathway (Dolmetsch *et al.*, 2001; Tian and Feig, 2006). Calcium/calmodulin also activates adenylyl cyclases 1 and 8, resulting in enhanced synthesis of cAMP and activation of PKA (Poser and Storm, 2001). However, calcium and calmodulin also promote activity of the phosphatases calcineurin, protein phosphatase (PP) 2A, and PP1, which promote dephosphorylation of S133 (Alberts *et al.*, 1994; Bito *et al.*, 1996; Wadzinski *et al.*, 1993; Wu *et al.*, 2001). Therefore, although calcium/calmodulin signalling promotes CREB-dependent gene expression, negative feedback signals also limit calcium-dependent activation of CREB.

2.2. *The role of CREB phosphorylation in memory formation*

Behavioural training, in many different paradigms, phosphorylates CREB at the critical S133 residue [reviewed in Lonze and Ginty (2002)]. Treatments that disrupt function at various points along the pathways upstream of CREB phosphorylation induce deficits in LTM formation. For instance, systemic (Atkins *et al.*, 1998; Selcher *et al.*, 1999), intracerebroventricular (Bourtchouladze *et al.*, 1998; Schafe *et al.*, 1999) or intra-amygdala (Schafe and LeDoux, 2000) administration of inhibitors of PKA or MAPK activity impairs LTM. Disruption of CREB phosphorylation also impairs learning and memory in humans: Coffin-Lowry Syndrome is an X-linked disorder caused by mutations in the gene encoding RSK2, a protein kinase that phosphorylates CREB at the key S133 residue (Trivier *et al.*, 1996) and the IQ of patients with Coffin-Lowry Syndrome is directly correlated with the capacity of their mutated RSK2 to phosphorylate CREB (Harum *et al.*, 2001).

Genetically engineered mice with mutations upstream of CREB phosphorylation provide additional support for the important role of CREB phosphorylation in memory formation. For instance, the expression of a dominant negative PKA transgene in the forebrain of mice blocked hippocampus-dependent memory formation (Abel *et al.*, 1997). Deletion of the CREB kinases MSK1 (Chwang *et al.*, 2007) or CaMKIV (Wei *et al.*, 2002) not only decreased activity-dependent CREB phosphorylation but also impaired fear memory formation. RSK2-knockout mice also exhibited impaired spatial learning (Poirier *et al.*, 2007). On the other hand, inhibition of the CREB phosphatase PP1 during training increased CREB phosphorylation and improved both novel object recognition and spatial learning (Genoux *et al.*, 2002). Finally, double knockout animals missing the two calcium-dependent adenylyl cyclases (AC1 and AC8) failed to form LTM in the hippocampus-dependent passive avoidance task (Wong *et al.*, 1999), although this defect was overcome by delivery of the broad-spectrum adenylyl cyclase agonist forskolin to the CA1 region of the hippocampus. Together, these results suggest that CREB may act as a final central switch onto which various signalling pathways converge to regulate the likelihood of LTM formation.

2.3. *Identification of CREB as a key mediator of memory formation*

cAMP signalling was first implicated in memory formation in the late 1960s, when several studies revealed that depolarising agents and application of neurotransmitters stimulated cAMP production in mammalian brain slices (Kakiuchi and Rall, 1968a, b; Kakiuchi *et al.*, 1969). Studies of invertebrates revealed that cAMP signalling is essential for synaptic plasticity and STM. In the sea slug *Aplysia californica*, synaptic stimulation of motor neurons resulted in a doubling of cAMP levels (Cedar *et al.*, 1972), addition of exogenous cAMP enhanced synaptic transmission (Brunelli *et al.*, 1976), and inhibition of cAMP-dependent protein phosphorylation in sensory neurons blocked synaptic strengthening (Castellucci *et al.*, 1982). The first heritable modifiers of learning and memory were identified in the fruit fly *Drosophila melanogaster*. Initial screens for learning-impaired flies identified genes encoding dunce (Dudai *et al.*, 1976) and rutabaga (Aceves-Pina and Quinn, 1979). Further studies revealed that dunce and rutabaga encode key regulators of cAMP metabolism (Byers *et al.*, 1981; Livingstone *et al.*, 1984). Together, data from *Aplysia* and *Drosophila* clearly identified an important role for cAMP in memory acquisition.

Meanwhile, studies in several model systems revealed that gene expression and protein synthesis were necessary for LTM (Barondes and Jarvik, 1964; Davis and Squire, 1984; Flexner *et al.*, 1963). Kandel and Schwartz hypothesized that cAMP signalling may induce changes in gene expression that mediate LTM formation (Kandel and Schwartz, 1982). Cellular and molecular studies revealed that cAMP signalling indeed drove the expression of genes regulated by the CRE, with the consensus sequence TGACGTCA (Comb *et al.*, 1986; Montminy *et al.*, 1986; Short *et al.*, 1986). Montminy *et al.* soon identified the cellular protein bound to CRE, which they termed CRE binding protein (CREB) (Montminy and Bilezikjian, 1987). Subsequent studies revealed that CREB was phosphorylated in response to cAMP (Gonzalez *et al.*, 1989; Yamamoto *et al.*, 1988) and that phosphorylation drove transcription of CRE-containing genes (Gonzalez and Montminy, 1989; Yamamoto *et al.*, 1988).

Kandel *et al.* found that microinjection of oligonucleotides encoding consensus CREs into the nucleus of *Aplysia* sensory neuron blocked long-term facilitation, a strengthening of synaptic transmission thought to underlie learning in this system (Dash *et al.*, 1990). Observational analyses of CREB also revealed that neuronal activation resulted in cAMP- and calcium-dependent phosphorylation of CREB (Dash *et al.*, 1991; Sheng *et al.*, 1991), and that stimuli that produced long-term facilitation in *Aplysia* drove CREB-dependent transcription (Kaang *et al.*, 1993). By 1993, CREB was believed to represent a key mediator of the transition between STM and LTM, although definitive evidence awaited further technical developments.

The essential role of CREB in memory formation was shown in 1994, using reverse genetics to manipulate CREB activity during behavioral training paradigms in flies and rodents. Yin *et al.* generated transgenic *Drosophila* that conditionally-expressed an inhibitory isoform of CREB. These flies were unable to form LTM yet could learn and remember things for a short period of time (Yin *et al.*, 1994). Meanwhile, Bourtchuladze and Silva *et al.* found that genetically modified mice that lacked the two major isoforms of CREB [$CREB^{\alpha,\delta\ -/-}$ mice, CREB-deficient mice originally developed by Günther Schütz (Hummler *et al.*, 1994)], similarly exhibited a marked impairment in LTM formation but still showed intact STM (Bourtchuladze *et al.*, 1994). Together, the back-to-back publication of these findings clearly showed that CREB was necessary for the formation of long-term memories in a variety of species and laid the groundwork for our current understanding of the role of gene expression in memory consolidation.

After showing that CREB-induced transcription was necessary for LTM in *Drosophila*, Yin *et al.* performed an experiment in which they increased CREB function and trained flies in an odour discrimination task (Yin *et al.*, 1995). Typically flies require "spaced-training" to form long-lasting memories in this task (i.e. 15-min rest intervals between the 10 training sessions); massed training (i.e. no intervals between the 10 training sessions) fails to produce LTM. Upon conditional induction of CREB, however, flies were able to form LTM following massed training. These exciting data were the first to suggest that increasing CREB can facilitate memory formation.

Josselyn *et al.* found a similar phenotype in rats. Like flies, rats do not form a robust long-term memory of conditioned fear in the fear-potentiated startle paradigm when the training sessions are massed (Josselyn *et al.*, 2001). A critical site for plasticity in the encoding of these types of fear memories is the basolateral amygdala complex (BLA), particularly the lateral nucleus (LA) (Fanselow and LeDoux, 1999; LeDoux, 2003). Therefore, these researchers increased CREB levels only in BLA neurons. They found that rats with increased CREB levels in the BLA were able to form long-lasting fear memories even following massed training. These findings were the first to show that increasing CREB function enhances memory in mammals. Subsequent studies examining the role of CREB in memory formation have built on these initial findings.

2.4. *The role of CREB in hippocampus-mediated memory formation*

The hippocampus is important for episodic memory in humans. Several experiments have examined the role of CREB in hippocampus-dependent memory formation in rodents. Both global interference of CREB function throughout the brain, or local disruption of CREB strictly in the hippocampus (Bourtchuladze *et al.*, 1994; Florian *et al.*, 2006; Guzowski and McGaugh, 1997; Kogan *et al.*, 1997; Pittenger *et al.*, 2002) impairs hippocampal-dependent spatial memory formation in rodents as measured by the Morris water maze. These findings suggest that CREB is necessary for spatial memory formation. Interestingly, Sekeres *et al.* recently showed that replacing CREB only in the dorsal hippocampus of global CREB-deficient mice completely rescued their spatial memory deficit (Sekeres *et al.*, 2010). These findings clearly show the necessity of hippocampal CREB function for the formation of spatial memory. Additionally, these researchers found that increasing CREB levels in the dorsal hippocampus of wild-type mice promoted spatial memory formation under very weak training conditions that do not normally support memory formation, perhaps reminiscent of the initial findings in flies. Together, these result show CREB is both necessary and sufficient for spatial memory formation (Sekeres *et al.*, 2010).

Several other forms of hippocampus-dependent memory have also been shown to be sensitive to manipulations in CREB function. For instance, decreasing CREB function, either globally (Bourtchuladze *et al.*, 1994) or locally (Brightwell *et al.*, 2005), disrupted memory for the social transmission of a food preference. In addition, increasing CREB function by infusing a viral vector encoding a constitutively active form of CREB in the dorsal hippocampus of wild-type mice, Restivo *et al.* (2009) showed enhanced context fear memory (Restivo *et al.*, 2009). Together, these data indicate that CREB plays a conserved role in hippocampal memory formation.

The decline in memory, particularly hippocampus-dependent memory, is a well-known consequence of ageing (Albert, 1993; Craik, 1990). CREB protein and activity levels in the brain also seem to decline with age. For instance, Porte *et al.* found that phosphorylated CREB (pCREB) levels in CA1 and dentate gyrus of aged mice were lower than in young mice. Moreover, they reported a positive correlation between CA1 pCREB levels and spatial memory performance in old mice, suggesting that the age-related decrease in hippocampal pCREB may be partially responsible for age-related memory impairment (Porte *et al.*, 2008). This finding leads to the tantalizing hypothesis that at least some of the memory deficits associated with ageing may be overcome by enhancing CREB function.

So far, we have discussed experimental findings in which increasing CREB function enhanced memory formation. However, there are several reports that chronically increasing CREB function may, in fact, disrupt memory formation. For instance, Viosca *et al.* showed that in bitransgenic mice that express a constitutively active form of CREB in forebrain neurons, spatial learning and memory was impaired when the CREB transgene expression was "turned on" (Viosca *et al.*, 2009). Chronically increasing constitutively active CREB following acquisition of a spatial memory also interfered with the subsequent retrieval of previously acquired information (Viosca *et al.*, 2009). It is important to note that the constitutively active CREB expressed in these bitransgenic mice was 25-fold more active than endogenous CREB (Barco *et al.*, 2002). In contrast, more localised increases of wild-type CREB levels following spatial training did not interfere with memory retrieval (Sekeres *et al.*, 2010). These data suggest there may be a dose-dependent level at which CREB-mediated activity can either be beneficial or deleterious to both memory formation and retrieval. In line with previous findings in which spatial memory was found to be facilitated by a more localised activity-dependent CREB enhancement, there might be an inverted-U relationship between the CREB levels and memory formation.

3. CREB co-activators

The role of CREB in memory formation is now well established. While phosphorylation of CREB at S133 is necessary to initiate CREB-dependent gene transcription, other CREB co-activators, including CREB binding protein (CBP), p300, and the recently identified family of CREB-regulated transcriptional co-activators (CRTCs) also play a key role in regulating CREB-dependent gene transcription that may be important in memory formation.

3.1. *CBP/p300*

In response to phosphorylation at S133, CREB recruits the histone acetyltransferases CBP [CREB binding protein, encoded by the *CREBBP* gene (called *CBP* within this review)] and p300 (E1A-associated protein of 300 kDa, encoded by the *EP300* gene) (Note: according to the revised nomenclature, these proteins are now called lysine acetyltransferase (KAT) 3A and 3B, respectively (Allis *et al.*, 2007); however, we will use the more familiar nomenclature for this review). These proteins are thought to drive the transcription of CREB target genes by acetylating local histone tails and also by direct recruitment of RNA polymerase II (Vo and Goodman, 2001). Upon phosphorylation, the CREB KID (kinase inducible domain) serves as a docking-site for CBP and p300, via their KIX (KID interaction) domains. Recruitment of CBP and/or p300 is an important step in CREB-mediated transcription (Mayr and Montminy, 2001).

3.2. *The role of CBP in memory formation*

The fundamental importance of CREB-dependent gene expression for memory formation in humans was illustrated by the discovery that mutation of *CBP* or *EP300* causes Rubinstein-Taybi syndrome (RSTS; OMIM 180849) (Roelfsema and Peters, 2007). RSTS is a rare disorder (incidence is ~1 in 100,000) that is characterized by moderate to severe mental retardation and numerous physical abnormalities. Approximately 45% of cases of RSTS result from *de novo* loss mutations of *CBP*, with an additional ~5% resulting from mutations of *EP300*. The majority of RSTS mutations generate alleles that encode truncated proteins, which are hypothesised to act as dominant negative inhibitors of CREB-mediated transcription (Roelfsema and Peters, 2007). However, a missense point mutation of *CBP* (R1378P) that encoded an intact, but enzymatically inactive CBP also are associated with RSTS (Kalkhoven *et al.*, 2003; Murata *et al.*, 2001).

Upon identification of *CBP* mutations in patients with RSTS, many investigators sought to model the disease with genetically modified mice [see Josselyn (2005) for review]. The first model was a conventional knockout with a heterozygous null mutation of *CBP* (CBP+/–) (Kung *et al.*, 2000; Tanaka *et al.*, 1997). The second model had an engineered mutation in one *CBP* allele that produced CBP protein with a C-terminal truncation (CBP+/–C-term) (Oike *et al.*, 1999). This truncated protein lacked the histone acetyltransferase (HAT) domain and was thought to function as a dominant negative. Mice homozygous for either of these germline mutations died prenatally, indicating that CBP performed vital developmental functions. However, heterozygous mice were viable and showed phenotypic overlap with RSTS patients, including skeletal and cardiac abnormalities, growth retardation, and memory deficits (Alarcon *et al.*, 2004; Kung *et al.*, 2000; Oike *et al.*, 1999; Tanaka *et al.*, 1997). In both mouse models, however, CBP function was compromised throughout development and throughout the body. Therefore, the possible contributions of CBP to adult memory processes could not be adequately assessed.

Further analyses of the role of CBP in memory formation have been performed using more refined conditional-transgenic or knock-in mouse lines. To directly determine the effects of disrupting CBP function on synaptic plasticity and memory in adult brains, Wood *et al.* developed transgenic mice in which a truncated form of CBP (CBP Δ1) was expressed only in postnatal forebrain neurons (Wood *et al.*, 2005). The truncated form of CBP retained the N terminus (including the binding site for a subset of transcription factors that normally bind to CBP, such as CREB) but lacked the C terminus of full-length CBP (including the HAT domain that is essential for transcriptional activation). Therefore, this truncated protein should disrupt the function of a subset of CBP-binding transcription factors. The resulting CBP Δ1 transgenic mice showed no developmental abnormalities and had normal weights, locomotor activity, and anxiety levels. However, these mice showed deficits in some forms of synaptic potentiation, as well as marked impairment in spatial learning and contextual fear conditioning. These experiments established that CBP function is important for normal adult cognitive function.

To fully establish the role of the CREB/CBP interaction in memory formation, Wood *et al.* next employed mice encoding mutated CBP that was impaired in its ability to bind CREB (Wood *et al.*, 2006). This CBP allele, designated CBP^{KIX} (Kasper *et al.*, 2002), was generated by mutating three conserved residues within the KIX domain (i.e. Y650A, A654G, and Y658A), which coordinate the interaction between phosphorylated CREB and CBP. CBP binds many transcription factors via its KIX domain; however, CREB binds to a specific surface of the KIX domain that is thought to be unnecessary for interaction between CBP and other transcription factors (Campbell and Lumb, 2002; Kasper *et al.*, 2002; Liu *et al.*, 2003). CBP^{KIX} encodes a protein with an intact HAT domain that is impaired only in its ability to bind CREB. Homozygous CBP^{KIX} mice had normal STM that lasted a few minutes, but decreased LTM in novel object recognition and contextual fear conditioning. CREB-induced transcription was also decreased in CBP^{KIX} animals. Together, these data suggest that the CREB/CBP interaction is critical for memory formation.

Histone acetylation is associated with transcriptional activation (Jenuwein and Allis, 2001), and CBP's acetyltransferase activity has been implicated in CREB-induced transcription (Ogryzko *et al.*, 1996; Vo and Goodman, 2001). However, CBP also binds many proteins involved in transcription, including RNA polymerase II (Neish *et al.*, 1998). The large size of CBP, along with its wide range of interacting proteins, has led to the hypothesis that CBP serves as a molecular scaffold that coordinates protein–protein interactions to drive CREB-dependent transcription (Chan and La Thangue, 2001; Ford and Thanos, 2010). To directly test the role of CBP HAT activity in memory formation, Korzus *et al.* generated mice with forebrain-specific and inducible expression of a mutant form of CBP that had no HAT activity (*CBP* Y1540A, F1541A, called CBP^{HAT-}) (Korzus *et al.*, 2004). CBP^{HAT-} encoded a protein with an intact KIX domain that was unable to acetylate histones in the promoters of CREB target genes. CBP^{HAT-} blocked transcription of the CREB target gene *in vitro* and *in vivo*, confirming its ability to function as a dominant negative inhibitor of CREB. Upon induction of CBP^{HAT-} expression, the mice lost their ability to form new LTM in spatial and object recognition tasks. When transgene expression was suppressed prior to training, the mice were once again able to learn and form a normal memory. Finally, systemic administration of the histone deacetylase inhibitor trichostatin A increased hippocampal histone acetylation and overcame the memory deficits exhibited by the CBP^{HAT-} mice. These data convincingly showed that CREB requires CBP-induced histone acetylation to efficiently promote expression of its target genes and to mediate memory formation.

Together, the results described here clearly show that CBP is essential for normal cognitive function. In response to neuronal activation, CREB is phosphorylated and CBP is recruited via its KIX domain. Once recruited to CREB target genes, CBP facilitates transcriptional activation by acetylating local histone tails and counteracting the repressive effects of histone deacetylases (Jenuwein and Allis, 2001; Strahl and Allis, 2000). Newly transcribed CREB target genes then mediate LTM formation.

3.3. *The role of p300 in memory formation*

Loss of p300 function has also been implicated in RSTS. Therefore, similar approaches to those described above for CBP have been applied to elucidating the role of p300 in memory formation. As has been observed with deletion of *CBP* (Alarcon *et al.*, 2004; Kung *et al.*, 2000), homozygous mutation of *p300* is embryonically lethal (Yao *et al.*, 1998), but heterozygous null mice are viable. Thorough behavioural characterization of *p300*$^{+/-}$ animals revealed a mild cognitive phenotype (Viosca *et al.*, 2010): the mice exhibited normal memory in novel object recognition and spatial water maze, as well as showing no deficits in cued or contextual fear conditioning. The only memory deficit emerged in the transfer task of the water maze: the animals were able to navigate the water maze and find a hidden platform normally; however, when the platform was moved to a new quadrant of the water maze, *p300*$^{+/-}$ mice showed a modest, but significant, impairment in their ability to learn its new location. Brains from these animals also exhibited no changes in histone acetylation or gene expression. Together these data suggest that p300 activity may be less critical than CBP for memory formation.

Another approach to disrupt p300 function in the brain was the generation of a dominant negative *p300*Δ1 allele (Oliveira *et al.*, 2007). Similar to the *CBP*Δ1 mice described above, *p300*Δ1 mice conditionally-produced truncated p300 (which was able bind CREB but did not contain the HAT domain) within the forebrain. In contrast to the other studies of p300 in learning and memory, analysis of *p300*Δ1 animals revealed memory deficits that were similar to those observed in *CBP*Δ1 transgenic animals (Wood *et al.*, 2005). Specifically, expression of dominant negative p300 impaired memory in novel object recognition and contextual fear conditioning (Oliveira *et al.*, 2007). Although the goal of this study was to specifically examine the role of p300 in memory, by employing a dominant negative approach, the authors in effect tested the role of the KIX domain in blocking memory formation. The KIX domains of p300 and CBP are 90% identical (Georges *et al.*, 2003), so overexpression of dominant negative p300 and CBP should have similar effects. This is a likely explanation for remarkably similar memory deficits exhibited by *p300*Δ1 and *CBP*Δ1 transgenic animals (Oliveira *et al.*, 2007; Wood *et al.*, 2005). Therefore, in light of the modest cognitive deficits exhibited by specific abrogation of p300 function in the brain, *p300Δ1* animals likely exhibited memory deficits due (at least in part) to nonspecific inhibition of CREB/CBP complex formation.

What might explain the different roles of p300 and CBP in memory formation? The two proteins are similar in structure, HAT activity, and capacity to induce transcription. However, numerous studies have revealed subtle differences in function between the two co-activators [reviewed in Kalkhoven (2004), Roelfsema and Peters (2007), Vo and Goodman (2001)]. One intriguing explanation is differential phosphorylation of CBP and p300 in response to neuronal activation. NMDA receptor signalling induces CaMKIV-mediated phosphorylation of CBP at S301, which is necessary for activity-dependent CREB target gene expression in primary hippocampal cultures (Impey *et al.*, 2002). Although CaMKIV is necessary for CBP-mediated transcription, dominant negative CaMKIV has no effect on p300 target gene expression. This is likely due to the fact that S301 is not conserved between CBP and p300. Perhaps, phosphorylation of CBP ensures efficient detection of coincident elevation of cAMP and calcium in response to neuronal activation, thereby maximizing CREB-induced transcription within a discrete temporal window. Interestingly, *p300* mutations causing RSTS are rare, with only seven confirmed carriers (Bartsch *et al.*, 2010; Roelfsema and Peters, 2007; Viosca *et al.*, 2010). Mutation of *p300* results in milder manifestation of RSTS-associated cognitive impairment compared to *CBP* mutations (Bartsch *et al.*, 2010; Roelfsema and Peters, 2007). Therefore, although p300 is an important CREB co-activator in many biological systems, CBP plays a more critical role than p300 in CREB-mediated memory formation.

3.4. *CREB co-activators: CRTCs*

KID-mediated recruitment of CBP/p300 provides a mechanism for phosphorylation-dependent induction of CREB target genes. However, numerous studies revealed that CREB was able to induce transcription in a phosphorylation-independent manner. High-throughput genomics approaches to screen for additional CREB co-activators revealed a family of conserved proteins that bind unphosphorylated CREB and promote its transcriptional activity (Conkright *et al.*, 2003a; Iourgenko *et al.*, 2003). These proteins were originally termed TORCs (transducers of regulated CREB); however, to avoid confusion with the Target Of Rapamycin Complex, they have been renamed CREB-regulated transcriptional co-activator (CRTC).

The CRTC family consists of three genes: CRTC1, 2 and 3, which encode coiled–coil domain proteins that shuttle between the cytoplasm and nucleus (Bittinger *et al.*, 2004; Screaton *et al.*, 2004). The three proteins have similar sequences and functions, although they are expressed in distinct tissue and cell types; for example, CRTC1 is expressed in the brain at highest relative to CRTC2 or CTRC3 (Altarejos *et al.*, 2008; Conkright *et al.*, 2003a; Komiya *et al.*, 2010). CRTCs promote CREB-dependent gene expression by binding to TATA box binding protein-associated factor (TAF) 4 [which was previously called $TAF_{II}130/135$ in mammals and $TAF_{II}110$ in *Drosophila* (Tora, 2002)] (Conkright *et al.*, 2003a). CREB drives constitutive low-level transcription through binding of TAF4 to its Q2 domain (Felinski and Quinn, 1999, 2001). TAF4 is a component of the transcription factor II D (TFIID) complex, which is the core promoter recognition component of the RNA polymerase II basal transcription machinery (Dynlacht *et al.*, 1991). TAF4 is necessary for both constitutive and induced CREB-dependent gene expression (Mengus *et al.*, 2005). CRTCs presumably stabilise the interaction between TAF4 and CREB, thereby promoting CREB-dependent gene expression.

Subcellular localization of CRTCs is regulated by post-translational modifications, particularly phosphorylation. Under basal conditions, CRTC2 is phosphorylated at S171 and S275 (Jansson *et al.*, 2008; Screaton *et al.*, 2004), which promotes association with the 14-3-3 complex and cytoplasmic retention. In response to cAMP or calcium signalling, CRTC2 is dephosphorylated and swiftly shuttles to the nucleus where it binds the CREB bZIP domain and promotes CREB-dependent transcription. CRTCs are particularly well-suited to participate in memory-associated transcription, as they are maximally induced by coincident elevations of cAMP and calcium (Screaton *et al.*, 2004).

Several kinases and phosphatases have been implicated in the activity-dependent regulation of CRTC function. The first CRTC kinases to be identified were the salt-inducible kinases (SIK) 1–3, which are members of the 5′ adenosine monophosphate activated protein kinase (AMPK) family (Koo *et al.*, 2005; Screaton *et al.*, 2004). Several other AMPK family members directly phosphorylate CRTCs, including microtubule affinity-regulating kinase 2 (MARK2) and AMPK itself (Jansson *et al.*, 2008; Koo *et al.*, 2005). Genetic ablation of the upstream AMPK regulatory kinase LKB1 results in hypophosphorylation of CRTC2, suggesting AMPKs are the major CRTC2 kinases (Shaw *et al.*, 2005).

CRTCs contain at least two key phosphorylation sites that regulate transcriptional activity: S171 and S275 (Jansson *et al.*, 2008; Screaton *et al.*, 2004). Distinct signalling pathways differentially regulate the phosphorylation of these residues. Although AMPK, SIKs, and MARK2 phosphorylate both residues *in vitro*, AMPK and SIKs prefer S171, whereas MARK2 prefers S275 (Fu and Screaton, 2008). Dephosphorylation occurs in response to calcium or cAMP (Jansson *et al.*, 2008; Kovacs *et al.*, 2007; Screaton *et al.*, 2004). In response to elevated cAMP (with or without elevated calcium), PKA-mediated phosphorylation of SIKs disrupts the

SIK/CRTC complex, resulting in dephosphorylation of S171 (Screaton *et al.*, 2004). Calcium elevation alone has little effect on S171 phosphorylation, although it strongly potentiates dephosphorylation of S275 via activation of calcineurin (Fu and Screaton, 2008; Screaton *et al.*, 2004). CREB's transcriptional activity can be blocked by pharmacological induction of AMPK (Koo *et al.*, 2005) or inhibition of calcineurin (Kovacs *et al.*, 2007; Screaton *et al.*, 2004), overexpression of SIK1 (Koo *et al.*, 2005), or mutation of the calcineurin-binding domains of CRTC2 (Screaton *et al.*, 2004). Therefore, control of CRTC phosphorylation plays an important role in the regulation of CREB-dependent transcription. But do CRTCs play a role in memory formation?

3.5. *The role of CRTCs in memory formation?*

The role of CRTCs in memory has not yet been directly examined. However, recent findings suggest that CRTCs are important in CREB-dependent control of synaptic plasticity. Two groups have examined the role of CRTCs in synaptic plasticity in cultured neurons (Kovacs *et al.*, 2007; Li *et al.*, 2009; Zhou *et al.*, 2006). Both groups found that CRTC1 was (1) expressed in hippocampal and cortical pyramidal neurons, (2) translocated to the nucleus in response to neuronal activity, and (3) necessary for both activity-dependent expression of CREB target genes and late phase long-term potentiation (L-LTP). The necessity of CRTC1 for CREB-dependent transcription was determined using both dominant negative CRTC1 (truncation mutants that encode only the CREB-binding domain, without the transcriptional activation or nuclear export domains) and small interference RNA-mediated knockdown of CRTC1, which decreased both basal- and induced-expression of CREB target genes. In response to 3–4 trains of high frequency stimulation, L-LTP (3 h) was observed in control neurons; however, expression of dominant negative CRTC1 resulted in failure to maintain potentiation. Interestingly, CRTC1 overexpression facilitated synaptic potentiation: a single high-frequency train, which was unable to induce synaptic potentiation in control cells, promoted stable L-LTP in acute hippocampal slices transduced with a CRTC1-encoding viral vector (Zhou *et al.*, 2006). Together, these results suggest that CRTC1-mediated CREB-target gene transcription might be a critical component of LTP, and perhaps memory formation.

4. Conclusion

Several lines of evidence indicate that CREB is both necessary and sufficient for memory formation of in the adult brain. This conclusion is based on findings from different experimental models ranging in complexity from *Aplysia* to rodents. More recent studies have begun to examine the role of CREB co-activators in memory formation. The findings that several CREB co-activators are associated with human disease suggests not only that CREB is important for intact cognitive function in humans, but also that targeting CREB could be a novel treatment target for impaired memory in humans.

References

Abel, T., Nguyen, P.V., Barad, M., Deuel, T.A., Kandel, E.R. and Bourtchouladze, R. (1997). Genetic demonstration of a role for PKA in the late phase of LTP and in hippocampus-based long-term memory. *Cell* **88**, 615–626.

Aceves-Pina, E.O. and Quinn, W.G. (1979). Learning in normal and mutant Drosophila larvae. *Science* **206**, 93–96.

Alarcon, J.M., Malleret, G., Touzani, K., Vronskaya, S., Ishii, S., Kandel, E.R. and Barco, A. (2004). Chromatin acetylation, memory and LTP are impaired in CBP+/– mice: A model for the cognitive deficit in Rubinstein-Taybi syndrome and its amelioration. *Neuron* **42**, 947–959.

Albert, M. (1993). Neuropsychological and neurophysiological changes in healthy adult humans across the age range. *Neurobiol Aging* **14**, 623–625.

Alberts, A.S., Montminy, M., Shenolikar, S. and Feramisco, J.R. (1994). Expression of a peptide inhibitor of protein phosphatase 1 increases phosphorylation and activity of CREB in NIH 3T3 fibroblasts. *Mol Cell Biol* **14**, 4398–4407.

Allis, C.D., Berger, S.L., Cote, J., Dent, S., Jenuwien, T., Kouzarides, T., Pillus, L., Reinberg, D., Shi, Y., Shiekhattar, R., *et al.* (2007). New nomenclature for chromatin-modifying enzymes. *Cell* **131**, 633–636.

Altarejos, J.Y., Goebel, N., Conkright, M.D., Inoue, H., Xie, J., Arias, C.M., Sawchenko, P.E. and Montminy, M. (2008). The Creb1 coactivator Crtc1 is required for energy balance and fertility. *Nat Med* **14**, 1112–1117.

Atkins, C.M., Selcher, J.C., Petraitis, J.J., Trzaskos, J.M. and Sweatt, J.D. (1998). The MAPK cascade is required for mammalian associative learning. *Nat Neurosci* **1**, 602–609.

Barco, A., Alarcon, J.M. and Kandel, E.R. (2002). Expression of constitutively active CREB protein facilitates the late phase of long-term potentiation by enhancing synaptic capture. *Cell* **108**, 689–703.

Barondes, S.H. and Jarvik, M.E. (1964). The influence of actinomycin-D on brain RNA synthesis and on memory. *J Neurochem* **11**, 187–195.

Bartsch, O., Labonte, J., Albrecht, B., Wieczorek, D., Lechno, S., Zechner, U. and Haaf, T. (2010). Two patients with EP300 mutations and facial dysmorphism different from the classic Rubinstein-Taybi syndrome. *Am J Med Genet A* **152A**, 181–184.

Bito, H., Deisseroth, K. and Tsien, R.W. (1996). CREB phosphorylation and dephosphorylation: A Ca(2+)- and stimulus duration-dependent switch for hippocampal gene expression. *Cell* **87**, 1203–1214.

Bittinger, M.A., McWhinnie, E., Meltzer, J., Iourgenko, V., Latario, B., Liu, X., Chen, C.H., Song, C., Garza, D. and Labow, M. (2004). Activation of cAMP response element-mediated gene expression by regulated nuclear transport of TORC proteins. *Curr Biol* **14**, 2156–2161.

Bourtchouladze, R., Abel, T., Berman, N., Gordon, R., Lapidus, K. and Kandel, E.R. (1998). Different training procedures recruit either one or two critical periods for contextual memory consolidation, each of which requires protein synthesis and PKA. *Learn Mem* **5**, 365–374.

Bourtchuladze, R., Frenguelli, B., Blendy, J., Cioffi, D., Schutz, G. and Silva, A.J. (1994). Deficient long-term memory in mice with a targeted mutation of the cAMP-responsive element-binding protein. *Cell* **79**, 59–68.

Brightwell, J.J., Smith, C.A., Countryman, R.A., Neve, R.L. and Colombo, P.J. (2005). Hippocampal overexpression of mutant creb blocks long-term, but not short-term memory for a socially transmitted food preference. *Learn Mem* **12**, 12–17.

Brunelli, M., Castellucci, V. and Kandel, E.R. (1976). Synaptic facilitation and behavioral sensitization in Aplysia: possible role of serotonin and cyclic AMP. *Science* **194**, 1178–1181.

Byers, D., Davis, R.L. and Kiger, J.A., Jr. (1981). Defect in cyclic AMP phosphodiesterase due to the dunce mutation of learning in Drosophila melanogaster. *Nature* **289**, 79–81.

Campbell, K.M. and Lumb, K.J. (2002). Structurally distinct modes of recognition of the KIX domain of CBP by Jun and CREB. *Biochemistry* **41**, 13956–13964.

Cardinaux, J.R., Notis, J.C., Zhang, Q., Vo, N., Craig, J.C., Fass, D.M., Brennan, R.G. and Goodman, R.H. (2000). Recruitment of CREB binding protein is sufficient for CREB-mediated gene activation. *Mol Cell Biol* **20**, 1546–1552.

Castellucci, V.F., Nairn, A., Greengard, P., Schwartz, J.H. and Kandel, E.R. (1982). Inhibitor of adenosine 3′:5′-monophosphate-dependent protein kinase blocks presynaptic facilitation in Aplysia. *J Neurosci* **2**, 1673–1681.

Cedar, H., Kandel, E.R. and Schwartz, J.H. (1972). Cyclic adenosine monophosphate in the nervous system of Aplysia californica. I. Increased synthesis in response to synaptic stimulation. *J Gen Physiol* **60**, 558–569.

Chan, H.M. and La Thangue, N.B. (2001). p300/CBP proteins: HATs for transcriptional bridges and scaffolds. *J Cell Sci* **114**, 2363–2373.

Chwang, W.B., Arthur, J.S., Schumacher, A. and Sweatt, J.D. (2007). The nuclear kinase mitogen- and stress-activated protein kinase 1 regulates hippocampal chromatin remodeling in memory formation. *J Neurosci* **27**, 12732–12742.

Cohen, S. and Greenberg, M.E. (2008). Communication between the synapse and the nucleus in neuronal development, plasticity, and disease. *Annu Rev Cell Dev Biol* **24**, 183–209.

Comb, M., Birnberg, N.C., Seasholtz, A., Herbert, E. and Goodman, H.M. (1986). A cyclic AMP- and phorbol ester-inducible DNA element. *Nature* **323**, 353–356.

Conkright, M.D., Canettieri, G., Screaton, R., Guzman, E., Miraglia, L., Hogenesch, J.B. and Montminy, M. (2003a). TORCs: Transducers of regulated CREB activity. *Mol Cell* **12**, 413–423.

Conkright, M.D., Guzman, E., Flechner, L., Su, A.I., Hogenesch, J.B. and Montminy, M. (2003b). Genome-wide analysis of CREB target genes reveals a core promoter requirement for cAMP responsiveness. *Mol Cell* **11**, 1101–1108.

Craik, F.I. (1990). Changes in memory with normal aging: a functional view. *Adv Neurol* **51**, 201–205.

Dash, P.K., Hochner, B. and Kandel, E.R. (1990). Injection of the cAMP-responsive element into the nucleus of Aplysia sensory neurons blocks long-term facilitation. *Nature* **345**, 718–721.

Dash, P.K., Karl, K.A., Colicos, M.A., Prywes, R. and Kandel, E.R. (1991). cAMP response element-binding protein is activated by Ca^{2} /calmodulin- as well as cAMP-dependent protein kinase. *Proc Natl Acad Sci U S A* **88**, 5061–5065.

Davis, H.P. and Squire, L.R. (1984). Protein synthesis and memory: A review. *Psychol Bull* **96**, 518–559.

Deisseroth, K. and Tsien, R.W. (2002). Dynamic multiphosphorylation passwords for activity-dependent gene expression. *Neuron* **34**, 179–182.

Dolmetsch, R.E., Pajvani, U., Fife, K., Spotts, J.M. and Greenberg, M.E. (2001). Signalling to the nucleus by an L-type calcium channel-calmodulin complex through the MAP kinase pathway. *Science* **294**, 333–339.

Du, K., Asahara, H., Jhala, U.S., Wagner, B.L. and Montminy, M. (2000). Characterization of a CREB gain-of-function mutant with constitutive transcriptional activity *in vivo*. *Mol Cell Biol* **20**, 4320–4327.

Dudai, Y., Jan, Y.N., Byers, D., Quinn, W.G. and Benzer, S. (1976). Dunce, a mutant of Drosophila deficient in learning. *Proc Natl Acad Sci U S A* **73**, 1684–1688.

Dynlacht, B.D., Hoey, T. and Tjian, R. (1991). Isolation of coactivators associated with the TATA-binding protein that mediate transcriptional activation. *Cell* **66**, 563–576.

Fanselow, M.S. and LeDoux, J.E. (1999). Why we think plasticity underlying Pavlovian fear conditioning occurs in the basolateral amygdala. *Neuron* **23**, 229–232.

Felinski, E.A. and Quinn, P.G. (1999). The CREB constitutive activation domain interacts with TATA-binding protein-associated factor 110 (TAF110) through specific hydrophobic residues in one of the three subdomains required for both activation and TAF110 binding. *J Biol Chem* **274**, 11672–11678.

Felinski, E.A. and Quinn, P.G. (2001). The coactivator dTAF(II)110/hTAF(II)135 is sufficient to recruit a polymerase complex and activate basal transcription mediated by CREB. *Proc Natl Acad Sci U S A* **98**, 13078–13083.

Flavell, S.W. and Greenberg, M.E. (2008). Signaling mechanisms linking neuronal activity to gene expression and plasticity of the nervous system. *Annu Rev Neurosci* **31**, 563–590.

Flexner, J.B., Flexner, L.B. and Stellar, E. (1963). Memory in mice as affected by intracerebral puromycin. *Science* **141**, 57–59.

Florian, C., Mons, N. and Roullet, P. (2006). CREB antisense oligodeoxynucleotide administration into the dorsal hippocampal CA3 region impairs long- but not short-term spatial memory in mice. *Learn Mem* **13**, 465–472.

Ford, E. and Thanos, D. (2010). The transcriptional code of human IFN-beta gene expression. *Biochim Biophys Acta* **1799**, 328–336.

Foulkes, N.S., Borrelli, E. and Sassone-Corsi, P. (1991). CREM gene: Use of alternative DNA-binding domains generates multiple antagonists of cAMP-induced transcription. *Cell* **64**, 739–749.

Fu, A. and Screaton, R.A. (2008). Using kinomics to delineate signaling pathways: Control of CRTC2/TORC2 by the AMPK family. *Cell Cycle* **7**, 3823–3828.

Genoux, D., Haditsch, U., Knobloch, M., Michalon, A., Storm, D. and Mansuy, I.M. (2002). Protein phosphatase 1 is a molecular constraint on learning and memory. *Nature* **418**, 970–975.

Georges, S.A., Giebler, H.A., Cole, P.A., Luger, K., Laybourn, P.J. and Nyborg, J.K. (2003). Tax recruitment of CBP/p300, via the KIX domain, reveals a potent requirement for acetyltransferase activity that is chromatin dependent and histone tail independent. *Mol Cell Biol* **23**, 3392–3404.

Gonzalez, G.A. and Montminy, M.R. (1989). Cyclic AMP stimulates somatostatin gene transcription by phosphorylation of CREB at serine 133. *Cell* **59**, 675–680.

Gonzalez, G.A., Yamamoto, K.K., Fischer, W.H., Karr, D., Menzel, P., Biggs, W., 3rd, Vale, W.W. and Montminy, M.R. (1989). A cluster of phosphorylation sites on the cyclic AMP-regulated nuclear factor CREB predicted by its sequence. *Nature* **337**, 749–752.

Greer, P.L. and Greenberg, M.E. (2008). From synapse to nucleus: Calcium-dependent gene transcription in the control of synapse development and function. *Neuron* **59**, 846–860.

Guzowski, J.F. and McGaugh, J.L. (1997). Antisense oligodeoxynucleotide-mediated disruption of hippocampal cAMP response element binding protein levels impairs consolidation of memory for water maze training. *Proc Natl Acad Sci USA* **94**, 2693–2698.

Harum, K.H., Alemi, L. and Johnston, M.V. (2001). Cognitive impairment in Coffin-Lowry syndrome correlates with reduced RSK2 activation. *Neurology* **56**, 207–214.

Hummler, E., Cole, T.J., Blendy, J.A., Ganss, R., Aguzzi, A., Schmid, W., Beermann, F. and Schutz, G. (1994). Targeted mutation of the CREB gene: Compensation within the CREB/ATF family of transcription factors. *Proc Natl Acad Sci USA* **91**, 5647–5651.

Iguchi-Ariga, S.M. and Schaffner, W. (1989). CpG methylation of the cAMP-responsive enhancer/promoter sequence TGACGTCA abolishes specific factor binding as well as transcriptional activation. *Genes Dev* **3**, 612–619.

Impey, S., Fong, A.L., Wang, Y., Cardinaux, J.R., Fass, D.M., Obrietan, K., Wayman, G.A., Storm, D.R., Soderling, T.R. and Goodman, R.H. (2002). Phosphorylation of CBP mediates transcriptional activation by neural activity and CaM kinase IV. *Neuron* **34**, 235–244.

Impey, S., McCorkle, S.R., Cha-Molstad, H., Dwyer, J.M., Yochum, G.S., Boss, J.M., McWeeney, S., Dunn, J.J., Mandel, G. and Goodman, R.H. (2004). Defining the CREB regulon: A genome-wide analysis of transcription factor regulatory regions. *Cell* **119**, 1041–1054.

Iourgenko, V., Zhang, W., Mickanin, C., Daly, I., Jiang, C., Hexham, J.M., Orth, A.P., Miraglia, L., Meltzer, J., Garza, D., *et al.* (2003). Identification of a family of cAMP response element-binding protein coactivators by genome-scale functional analysis in mammalian cells. *Proc Natl Acad Sci U S A* **100**, 12147–12152.

Jaenisch, R. and Bird, A. (2003). Epigenetic regulation of gene expression: How the genome integrates intrinsic and environmental signals. *Nat Genet* **33** *Suppl*, 245–254.

Jansson, D., Ng, A.C., Fu, A., Depatie, C., Al Azzabi, M. and Screaton, R.A. (2008). Glucose controls CREB activity in islet cells via regulated phosphorylation of TORC2. *Proc Natl Acad Sci U S A* **105**, 10161–10166.

Jenuwein, T. and Allis, C.D. (2001). Translating the histone code. *Science* **293**, 1074–1080.

Josselyn, S.A. (2005). What's right with my mouse model? New insights into the molecular and cellular basis of cognition from mouse models of Rubinstein-Taybi Syndrome. *Learn Mem* **12**, 80–83.

Josselyn, S.A., Shi, C., Carlezon, W.A., Jr., Neve, R.L., Nestler, E.J. and Davis, M. (2001). Long-term memory is facilitated by cAMP response element-binding protein overexpression in the amygdala. *J Neurosci* **21**, 2404–2412.

Juven-Gershon, T. and Kadonaga, J.T. (2010). Regulation of gene expression via the core promoter and the basal transcriptional machinery. *Dev Biol* **339**, 225–229.

Kaang, B.K., Kandel, E.R. and Grant, S.G. (1993). Activation of cAMP-responsive genes by stimuli that produce long-term facilitation in Aplysia sensory neurons. *Neuron* **10**, 427–435.

Kakiuchi, S. and Rall, T.W. (1968a). The influence of chemical agents on the accumulation of adenosine 3′,5′-Phosphate in slices of rabbit cerebellum. *Mol Pharmacol* **4**, 367–378.

Kakiuchi, S. and Rall, T.W. (1968b). Studies on adenosine 3′,5′-phosphate in rabbit cerebral cortex. *Mol Pharmacol* **4**, 379–388.

Kakiuchi, S., Rall, T.W. and McIlwain, H. (1969). The effect of electrical stimulation upon the accumulation of adenosine 3′,5′-phosphate in isolated cerebral tissue. *J Neurochem* **16**, 485–491.

Kalkhoven, E. (2004). CBP and p300: HATs for different occasions. *Biochem Pharmacol* **68**, 1145–1155.

Kalkhoven, E., Roelfsema, J.H., Teunissen, H., den Boer, A., Ariyurek, Y., Zantema, A., Breuning, M.H., Hennekam, R.C. and Peters, D.J. (2003). Loss of CBP acetyltransferase activity by PHD finger mutations in Rubinstein-Taybi syndrome. *Hum Mol Genet* **12**, 441–450.

Kandel, E.R. (2001). The molecular biology of memory storage: A dialogue between genes and synapses. *Science* **294**, 1030–1038.

Kandel, E.R. and Schwartz, J.H. (1982). Molecular biology of learning: Modulation of transmitter release. *Science* **218**, 433–443.

Kasper, L.H., Boussouar, F., Ney, P.A., Jackson, C.W., Rehg, J., van Deursen, J.M. and Brindle, P.K. (2002). A transcription-factor-binding surface of coactivator p300 is required for haematopoiesis. *Nature* **419**, 738–743.

Kogan, J.H., Frankland, P.W., Blendy, J.A., Coblentz, J., Marowitz, Z., Schutz, G. and Silva, A.J. (1997). Spaced training induces normal long-term memory in CREB mutant mice. *Curr Biol* **7**, 1–11.

Komiya, T., Coxon, A., Park, Y., Chen, W.D., Zajac-Kaye, M., Meltzer, P., Karpova, T. and Kaye, F.J. (2010). Enhanced activity of the CREB co-activator Crtc1 in LKB1 null lung cancer. *Oncogene* **29**, 1672–1680.

Koo, S.H., Flechner, L., Qi, L., Zhang, X., Screaton, R.A., Jeffries, S., Hedrick, S., Xu, W., Boussouar, F., Brindle, P., *et al.* (2005). The CREB coactivator TORC2 is a key regulator of fasting glucose metabolism. *Nature* **437**, 1109–1111.

Korzus, E., Rosenfeld, M.G. and Mayford, M. (2004). CBP histone acetyltransferase activity is a critical component of memory consolidation. *Neuron* **42**, 961–972.

Kovacs, K.A., Steullet, P., Steinmann, M., Do, K.Q., Magistretti, P.J., Halfon, O. and Cardinaux, J.R. (2007). TORC1 is a calcium- and cAMP-sensitive coincidence detector involved in hippocampal long-term synaptic plasticity. *Proc Natl Acad Sci U S A* **104**, 4700–4705.

Kung, A.L., Rebel, V.I., Bronson, R.T., Ch'ng, L.E., Sieff, C.A., Livingston, D.M. and Yao, T.P. (2000). Gene dose-dependent control of hematopoiesis and hematologic tumor suppression by CBP. *Genes Dev* **14**, 272–277.

LeDoux, J. (2003). The emotional brain, fear, and the amygdala. *Cell Mol Neurobiol* **23**, 727–738.

Li, S., Zhang, C., Takemori, H., Zhou, Y. and Xiong, Z.Q. (2009). TORC1 regulates activity-dependent CREB-target gene transcription and dendritic growth of developing cortical neurons. *J Neurosci* **29**, 2334–2343.

Liu, Y.P., Chang, C.W. and Chang, K.Y. (2003). Mutational analysis of the KIX domain of CBP reveals residues critical for SREBP binding. *FEBS Lett* **554**, 403–409.

Livingstone, M.S., Sziber, P.P. and Quinn, W.G. (1984). Loss of calcium/calmodulin responsiveness in adenylate cyclase of rutabaga, a Drosophila learning mutant. *Cell* **37**, 205–215.

Lonze, B.E. and Ginty, D.D. (2002). Function and regulation of CREB family transcription factors in the nervous system. *Neuron* **35**, 605–623.

Mayr, B. and Montminy, M. (2001). Transcriptional regulation by the phosphorylation-dependent factor CREB. *Nat Rev Mol Cell Biol* **2**, 599–609.

Mellstrom, B., Naranjo, J.R., Foulkes, N.S., Lafarga, M. and Sassone-Corsi, P. (1993). Transcriptional response to cAMP in brain: Specific distribution and induction of CREM antagonists. *Neuron* **10**, 655–665.

Mengus, G., Fadloun, A., Kobi, D., Thibault, C., Perletti, L., Michel, I. and Davidson, I. (2005). TAF4 inactivation in embryonic fibroblasts activates TGF beta signalling and autocrine growth. *EMBO J* **24**, 2753–2767.

Montminy, M.R. and Bilezikjian, L.M. (1987). Binding of a nuclear protein to the cyclic-AMP response element of the somatostatin gene. *Nature* **328**, 175–178.

Montminy, M.R., Sevarino, K.A., Wagner, J.A., Mandel, G. and Goodman, R.H. (1986). Identification of a cyclic-AMP-responsive element within the rat somatostatin gene. *Proc Natl Acad Sci U S A* **83**, 6682–6686.

Murata, T., Kurokawa, R., Krones, A., Tatsumi, K., Ishii, M., Taki, T., Masuno, M., Ohashi, H., Yanagisawa, M., Rosenfeld, M.G.*, et al.* (2001). Defect of histone acetyltransferase activity of the nuclear transcriptional coactivator CBP in Rubinstein-Taybi syndrome. *Hum Mol Genet* **10**, 1071–1076.

Neish, A.S., Anderson, S.F., Schlegel, B.P., Wei, W. and Parvin, J.D. (1998). Factors associated with the mammalian RNA polymerase II holoenzyme. *Nucleic Acids Res* **26**, 847–853.

Ogryzko, V.V., Schiltz, R.L., Russanova, V., Howard, B.H. and Nakatani, Y. (1996). The transcriptional coactivators p300 and CBP are histone acetyltransferases. *Cell* **87**, 953–959.

Oike, Y., Hata, A., Mamiya, T., Kaname, T., Noda, Y., Suzuki, M., Yasue, H., Nabeshima, T., Araki, K. and Yamamura, K. (1999). Truncated CBP protein leads to classical Rubinstein-Taybi syndrome phenotypes in mice: Implications for a dominant-negative mechanism. *Hum Mol Genet* **8**, 387–396.

Oliveira, A.M., Wood, M.A., McDonough, C.B. and Abel, T. (2007). Transgenic mice expressing an inhibitory truncated form of p300 exhibit long-term memory deficits. *Learn Mem* **14**, 564–572.

Pittenger, C., Huang, Y.Y., Paletzki, R.F., Bourtchouladze, R., Scanlin, H., Vronskaya, S. and Kandel, E.R. (2002). Reversible inhibition of CREB/ATF transcription factors in region CA1 of the dorsal hippocampus disrupts hippocampus-dependent spatial memory. *Neuron* **34**, 447–462.

Poirier, R., Jacquot, S., Vaillend, C., Soutthiphong, A.A., Libbey, M., Davis, S., Laroche, S., Hanauer, A., Welzl, H., Lipp, H.P. and Wolfer, D.P. (2007). Deletion of the Coffin-Lowry syndrome gene Rsk2 in mice is associated with impaired spatial learning and reduced control of exploratory behavior. *Behav Genet* **37**, 31–50.

Porte, Y., Buhot, M.C. and Mons, N. (2008). Alteration of CREB phosphorylation and spatial memory deficits in aged 129T2/Sv mice. *Neurobiol Aging* **29**, 1533–1546.

Poser, S. and Storm, D.R. (2001). Role of Ca2+-stimulated adenylyl cyclases in LTP and memory formation. *Int J Dev Neurosci* **19**, 387–394.

Rehfuss, R.P., Walton, K.M., Loriaux, M.M. and Goodman, R.H. (1991). The cAMP-regulated enhancer-binding protein ATF-1 activates transcription in response to cAMP-dependent protein kinase A. *J Biol Chem* **266**, 18431–18434.

Restivo, L., Tafi, E., Ammassari-Teule, M. and Marie, H. (2009). Viral-mediated expression of a constitutively active form of CREB in hippocampal neurons increases memory. *Hippocampus* **19**, 228–234.

Roelfsema, J.H. and Peters, D.J. (2007). Rubinstein-Taybi syndrome: Clinical and molecular overview. *Expert Rev Mol Med* **9**, 1–16.

Schafe, G.E. and LeDoux, J.E. (2000). Memory consolidation of auditory pavlovian fear conditioning requires protein synthesis and protein kinase A in the amygdala. *J Neurosci* **20**, RC96.

Schafe, G.E., Nadel, N.V., Sullivan, G.M., Harris, A. and LeDoux, J.E. (1999). Memory consolidation for contextual and auditory fear conditioning is dependent on protein synthesis, PKA, and MAP kinase. *Learn Mem* **6**, 97–110.

Screaton, R.A., Conkright, M.D., Katoh, Y., Best, J.L., Canettieri, G., Jeffries, S., Guzman, E., Niessen, S., Yates, J.R., 3rd, Takemori, H., *et al.* (2004). The CREB coactivator TORC2 functions as a calcium- and cAMP-sensitive coincidence detector. *Cell* **119**, 61–74.

Sekeres, M.J., Neve, R.L., Frankland, P.W. and Josselyn, S.A. (2010). Dorsal hippocampal CREB is both necessary and sufficient for spatial memory. *Learn Mem* **17**, 280–283.

Selcher, J.C., Atkins, C.M., Trzaskos, J.M., Paylor, R. and Sweatt, J.D. (1999). A necessity for MAP kinase activation in mammalian spatial learning. *Learn Mem* **6**, 478–490.

Shaw, R.J., Lamia, K.A., Vasquez, D., Koo, S.H., Bardeesy, N., Depinho, R.A., Montminy, M. and Cantley, L.C. (2005). The kinase LKB1 mediates glucose homeostasis in liver and therapeutic effects of metformin. *Science* **310**, 1642–1646.

Sheng, M., Thompson, M.A. and Greenberg, M.E. (1991). CREB: a Ca(2+)-regulated transcription factor phosphorylated by calmodulin-dependent kinases. *Science* **252**, 1427–1430.

Short, J.M., Wynshaw-Boris, A., Short, H.P. and Hanson, R.W. (1986). Characterization of the phosphoenolpyruvate carboxykinase (GTP) promoter-regulatory region. II. Identification of cAMP and glucocorticoid regulatory domains. *J Biol Chem* **261**, 9721–9726.

Strahl, B.D. and Allis, C.D. (2000). The language of covalent histone modifications. *Nature* **403**, 41–45.

Tanaka, Y., Naruse, I., Maekawa, T., Masuya, H., Shiroishi, T. and Ishii, S. (1997). Abnormal skeletal patterning in embryos lacking a single Cbp allele: A partial similarity with Rubinstein-Taybi syndrome. *Proc Natl Acad Sci U S A* **94**, 10215–10220.

Tian, X. and Feig, L.A. (2006). Age-dependent participation of Ras-GRF proteins in coupling calcium-permeable AMPA glutamate receptors to Ras/Erk signaling in cortical neurons. *J Biol Chem* **281**, 7578–7582.

Tora, L. (2002). A unified nomenclature for TATA box binding protein (TBP)-associated factors (TAFs) involved in RNA polymerase II transcription. *Genes Dev* **16**, 673–675.

Trivier, E., De Cesare, D., Jacquot, S., Pannetier, S., Zackai, E., Young, I., Mandel, J.L., Sassone-Corsi, P. and Hanauer, A. (1996). Mutations in the kinase Rsk-2 associated with Coffin-Lowry syndrome. *Nature* **384**, 567–570.

Viosca, J., Lopez-Atalaya, J.P., Olivares, R., Eckner, R. and Barco, A. (2010). Syndromic features and mild cognitive impairment in mice with genetic reduction on p300 activity: Differential contribution of p300 and CBP to Rubinstein-Taybi syndrome etiology. *Neurobiol Dis* **37**, 186–194.

Viosca, J., Malleret, G., Bourtchouladze, R., Benito, E., Vronskava, S., Kandel, E.R. and Barco, A. (2009). Chronic enhancement of CREB activity in the hippocampus interferes with the retrieval of spatial information. *Learn Mem* **16**, 198–209.

Vo, N. and Goodman, R.H. (2001). CREB-binding protein and p300 in transcriptional regulation. *J Biol Chem* **276**, 13505–13508.

Wadzinski, B.E., Wheat, W.H., Jaspers, S., Peruski, L.F., Jr., Lickteig, R.L., Johnson, G.L. and Klemm, D.J. (1993). Nuclear protein phosphatase 2A dephosphorylates protein kinase A-phosphorylated CREB and regulates CREB transcriptional stimulation. *Mol Cell Biol* **13**, 2822–2834.

Wayman, G.A., Lee, Y.S., Tokumitsu, H., Silva, A.J. and Soderling, T.R. (2008). Calmodulin-kinases: Modulators of neuronal development and plasticity. *Neuron* **59**, 914–931.

Wei, F., Qiu, C.S., Liauw, J., Robinson, D.A., Ho, N., Chatila, T. and Zhuo, M. (2002). Calcium calmodulin-dependent protein kinase IV is required for fear memory. *Nat Neurosci* **5**, 573–579.

West, A.E., Griffith, E.C. and Greenberg, M.E. (2002). Regulation of transcription factors by neuronal activity. *Nat Rev Neurosci* **3**, 921–931.

Wong, S.T., Athos, J., Figueroa, X.A., Pineda, V.V., Schaefer, M.L., Chavkin, C.C., Muglia, L.J. and Storm, D.R. (1999). Calcium-stimulated adenylyl cyclase activity is critical for hippocampus-dependent long-term memory and late phase LTP. *Neuron* **23**, 787–798.

Wood, M.A., Attner, M.A., Oliveira, A.M., Brindle, P.K. and Abel, T. (2006). A transcription factor-binding domain of the coactivator CBP is essential for long-term memory and the expression of specific target genes. *Learn Mem* **13**, 609–617.

Wood, M.A., Kaplan, M.P., Park, A., Blanchard, E.J., Oliveira, A.M., Lombardi, T.L. and Abel, T. (2005). Transgenic mice expressing a truncated form of CREB-binding protein (CBP) exhibit deficits in hippocampal synaptic plasticity and memory storage. *Learn Mem* **12**, 111–119.

Wu, G.Y., Deisseroth, K. and Tsien, R.W. (2001). Activity-dependent CREB phosphorylation: convergence of a fast, sensitive calmodulin kinase pathway and a slow, less sensitive mitogen-activated protein kinase pathway. *Proc Natl Acad Sci U S A* **98**, 2808–2813.

Xia, Z. and Storm, D.R. (2005). The role of calmodulin as a signal integrator for synaptic plasticity. *Nat Rev Neurosci* **6**, 267–276.

Yamamoto, K.K., Gonzalez, G.A., Biggs, W.H., 3rd, and Montminy, M.R. (1988). Phosphorylation-induced binding and transcriptional efficacy of nuclear factor CREB. *Nature* **334**, 494–498.

Yao, T.P., Oh, S.P., Fuchs, M., Zhou, N.D., Ch'ng, L.E., Newsome, D., Bronson, R.T., Li, E., Livingston, D.M. and Eckner, R. (1998). Gene dosage-dependent embryonic development and proliferation defects in mice lacking the transcriptional integrator p300. *Cell* **93**, 361–372.

Yin, J.C., Del Vecchio, M., Zhou, H. and Tully, T. (1995). CREB as a memory modulator: Induced expression of a dCREB2 activator isoform enhances long-term memory in Drosophila. *Cell* **81**, 107–115.

Yin, J.C., Wallach, J.S., Del Vecchio, M., Wilder, E.L., Zhou, H., Quinn, W.G. and Tully, T. (1994). Induction of a dominant negative CREB transgene specifically blocks long-term memory in Drosophila. *Cell* **79**, 49–58.

Zhang, X., Odom, D.T., Koo, S.H., Conkright, M.D., Canettieri, G., Best, J., Chen, H., Jenner, R., Herbolsheimer, E., Jacobsen, E., *et al.* (2005). Genome-wide analysis of cAMP-response element binding protein occupancy, phosphorylation, and target gene activation in human tissues. *Proc Natl Acad Sci U S A* **102**, 4459–4464.

Zhou, Y., Wu, H., Li, S., Chen, Q., Cheng, X.W., Zheng, J., Takemori, H. and Xiong, Z.Q. (2006). Requirement of TORC1 for late-phase long-term potentiation in the hippocampus. *PLoS One* **1**, e16.

Sex Differences in Memory Formation

7

Keiko Mizuno* and Karl Peter Giese*

1. Introduction

Why study sex differences (see Text Box) in memory formation? There is sufficient evidence that there are sex differences in protein expression and protein signalling that impact on synaptic plasticity and learning/memory in adulthood (Cahill, 2006; Andreano and Cahill, 2009; Mizuno and Giese, 2010). Furthermore, neuropsychiatric disorders, such as schizophrenia, Alzheimer's disease and some major depressions, differ between the genders in incident, prevalence, symptoms, age at onset and severity (Cahill, 2006; Kim, 2010).These diseases are believed

Sex differences: *The term "sex differences" concerns genetic differences between males and females, whereas the term 'gender differences' considers both genetic and environmental differences. In the laboratory, the environmental conditions may be the same for both male and female animals. Therefore, differences between male and female animals in the laboratory are due to sex differences and they should not be referred to as "gender differences".*

Sex hormone oestradiol: *Oestrogen is a class of sex steroid hormones, predominantly synthesized by the ovary including oestrone, oestradiol and oestriol. Oestradiol (also described as E2 or 17β-oestradiol) is the most abundant circulating oestrogen in females, its concentration fluctuates during the oestrous cycle, and it is present in males at a lower level. Oestradiol is usually applied for investigation of the effect of oestrogen.*

---

*King's College London, Institute of Psychiatry, Department of Neuroscience, 125 Coldharbour Lane, London, SE5 9NU, UK.

to result from synaptic deficits. Thus, molecular mechanisms that bring about sex differences in synaptic function are candidate mechanisms that may contribute to memory disorders. Therefore, to comprehend memory formation in health and disease, sex differences should be considered.

2. Sex differences in memory formation

Males and females differ with regard to various behaviours including hippocampus-dependent memory formation (Andreano and Cahill, 2009). Such sex differences in memory formation in humans and rodents are quantitative and/or qualitative. Quantitative differences are found in the rate and accuracy of memory formation. For example, males perform better than females in some spatial memory tasks (Postma *et al.*, 2004; Jonasson, 2005), whereas in other spatial tasks, such as object location tasks, females outperform males (Andreano and Cahill, 2009). Further, there are qualitative differences in learning strategies. For example, in humans and rodents males and females use different strategies to solve spatial memory tasks, making use of either geometric or landmark cues (Sandstrom *et al.*, 1998; Roof and Stein, 1999). Additionally, depending on the stage of the oestrous cycle female rats appear to prefer either an allocentric place learning strategy or an egocentric response learning strategy (Korol *et al.*, 2004). Qualitative differences may be explained by the use of different neuronal circuits-within separate brain structures, such as hippocampus-dependent learning for environmental layout (geometry) and striatal-dependent learning for responses to individual landmark cues (Doeller *et al.*, 2008). Alternatively, qualitative differences may arise from distinct molecular signalling within the same brain structure (Mizuno and Giese, 2010). Additionally, sex differences in hippocampal anatomy have been described (Madeira and Lieberman, 1995). In principle, these anatomical differences, although small, may contribute to sex differences in hippocampus-dependent memory formation.

3. Sex differences in stress responses on memory formation

Acute and chronic stress impact on the ability to form memory. Stress effects on memory formation differ between the sexes. For example, acute stress has opposite effects on classical conditioning in male and female rats (Shors, 2004; Bangasser and Shors, 2010). After a threshold of stress is reached, performance is enhanced in males and impaired in females. The opposing effects of acute stress on memory formation depend on the basolateral nucleus of the amygdala (Waddell *et al.*, 2008) and also require the hippocampus in both sexes (Bangasser and Shors, 2007). However, the bed nucleus of the stria terminalis (BNST) is part of the circuitry for the enhancing effects of acute stress in male, but not female, rats (Bangasser and Shors, 2008). Furthermore, inactivation of the medial prefrontal cortex (mPFC), by infusion of GABA-A receptor agonist muscimol, prevents stress-induced impairment of memory formation in females, but has no effect in males (Maeng *et al.*, 2011). Disconnection of the communication between basolateral amygdala (BLA) and the mPFC also prevents the stress-induced impairments of memory formation in females.

Taken together, males and females use the same as well as different brain structures in response to the effect of stressful experience on memory formation. Within the same brain structure, acute stress can have opposite effects in males and females. Accordingly, acute stress increases spine density in CA1 pyramidal neurons in males whilst it decreases in females (Leuner and Shors, 2004). The effect on spine density correlates with the impact on memory formation, which is increased in males and decreased in females. Thus, it has been suggested that the number of spines (synapses) before training in the memory task determines how many synapses may be engaged during memory formation (Leuner and Shors, 2004).

4. Oestradiol and memory formation

A widely accepted view is that the sex hormone oestradiol (see Text Box) can affect synaptic plasticity and memory formation (McEwen, 2002). Most oestradiol effects have been studied in females, because they have higher concentration of this circulating sex hormone than males. Studies on the role of oestradiol in males have been neglected. However, in both males and females oestradiol can be synthesized in target tissues, such as in the hippocampus, by either aromatase from testosterone or by *de novo* synthesis from cholesterol. Therefore, oestradiol effects should not only be considered in females, but also in males, and additionally the effects of testosterone in males should be of consideration (Hojo *et al.*, 2008; Gillies and Mcarthur, 2010; Mukai *et al.*, 2010).

In females oestradiol affects synaptic plasticity. These effects are either slow or rapid (Woolley, 2007; Spencer *et al.*, 2008; Smejkalova and Woolley, 2010). Slow actions of oestradiol occur via classical nuclear receptors and they appear to induce formation of dendritic spines. Oestrogen receptor beta activation induces the cAMP responsive element-binding protein (CREB)-dependent transcription and spine formation (Liu *et al.*, 2008; Schwarz *et al.*, 2008). Higher hippocampal spine density improves memory formation, possibly because more synapses can be engaged during memory formation (Leuner and Shors, 2004). The rapid actions of oestradiol occur via synapse-localized estrogen receptor alpha or beta. Long-term potentiation (LTP) is enhanced via activation of oestrogen receptor beta (Liu *et al.*, 2008). The impact on LTP is thought to contribute to memory formation (see, Chapter 1).

Oestradiol enhances hippocampus-dependent memory formation (Romeo *et al.*, 2004; Rissman, 2008; Brinton, 2009). However, it does not enhance all types of memory formation and depending on the dose, oestradiol can even impair memory formation (Barha *et al.*, 2010; Gibbs, 2010). Thus, there is no simple relationship between oestradiol and memory formation. Oestradiol can also modulate the effects of stress on memory formation (Andreano and Chahill, 2009).

5. Qualitative sex differences for synaptic kinases in memory formation

In recent years it has emerged that there are sex differences in the functioning of synaptic kinases. The synaptic kinases Ca^{2+}/calmodulin (CaM) kinase kinase 1 (CaMKK1) and CaMKK2 are required for memory formation in male, but not in female mice (Mizuno and Giese, 2010). These two kinases activate two downstream kinases, CaMKI and CaMKIV, that have been implicated in synaptic signalling as well as the activation of the transcription factor CREB that is required for long-term memory formation (Wayman *et al.*, 2008). In males, but not in female mice, CaMKK2 is required for spatial memory formation, CREB activation after spatial training, and the late phase of LTP in hippocampal area CA1 (Peters *et al.*, 2003; Mizuno *et al.*, 2007). CaMKK1 is required for contextual fear memory formation and activation of brain-derived neurotrophic factor (BDNF) expression in males but not females (Blaeser *et al.*, 2006; Mizuno *et al.*, 2006). The sex-specific requirement for CaMKK1 and 2 for memory formation is not a consequence of sexually dimophic protein expression (KM and KPG, unpublished data). Distinct calcium signalling during memory formation may account for the sex-specific requirement of the CaMKKs.

Another example for a sex-dependent requirement of a synaptic kinase for memory formation comes from studies with p25 transgenic mice. p25 is a truncated form of p35, the main activator of cyclin-dependent kinase 5 (Cdk5). The calcium-sensitive protease calpain cleaves p35 into p25. Both p25 and p35 are activators of cyclin-dependent kinase 5 (Cdk5), but p25 is more stable because it cannot be degraded by the proteasome. Thus, p25 formation overactivates Cdk5. p25 formation was believed to be a pathological process but recently it was shown that p25 can be formed under physiological conditions as it occurs during memory formation in female mice (Engmann *et al.*, 2011). Transgenic overexpression of physiologic levels of p25 improves spatial memory formation and L-LTP at CA1 synapses in females, but not in males (Ris *et al.*, 2005).

Spatial memory formation requires the hippocampus both in males and females (Angelo *et al.*, 2003; Antunes-Martins *et al.*, 2007). As discussed, the synaptic kinases as CaMKK2 and p25/Cdk5 are required in a sex-dependent manner for hippocampal LTP and spatial memory formation. This suggests males and females use distinct mechanisms for synaptic plasticity underlying memory formation. However, it should be realized that some signalling processes are used by both sexes for memory formation. For example, αCaMKII autophosphorylation is required for CA1 LTP and spatial memory formation in both sexes (Irvine *et al.*, 2006). It remains unclear why some signalling processes are required for memory formation in both sexes, whilst others are required in only one sex.

6. Qualitative sex differences in gene transcription during memory formation

The previous section addressed sex-specific requirements for signalling processes for memory formation. These findings do not directly address the issue of whether sex-specific plasticity occurs during memory formation. This point was addressed when investigating the activation of synaptic kinases as well as the analysis of gene transcription induced during memory formation.

The extracellular signal-regulated kinase (ERK) cascade is essential for memory formation. After contextual fear conditioning, activation of ERK is higher in the ventral than in the dorsal hippocampus in male, but not in female, rats (Gresack *et al.*, 2009). Because the ERK cascade regulates the expression of plasticity-associated genes through activation of CREB, male-specific ERK activation may lead to sex-specific CREB activation, which was observed after contextual fear conditioning (Kudo *et al.*, 2004). Male-specific activation of CREB was also shown after spatial training in the Morris water maze (Mizuno *et al.*, 2007). Accordingly, in mutant mice with reduced CREB expression, spatial memory formation is impaired in adult males, but not in females (Hebda-Bauer *et al.*, 2007). Another transcription factor with a sexually dimorphic function appears to be nuclear factor κB (NF-κB). Studies with transgenic mice revealed that astroglial NF-κB plays a role in L-LTP, hippocampus-dependent memory formation and expression of neuronal proteins such as mGluR5 and PSD95 in female, but not in male, mice (Bracci-Ricard *et al.*, 2008).

Sex differences in gene transcription during memory formation have also been demonstrated. For example, the hippocampal expression of BDNF mRNA, a CREB target gene, is sexually dimorphic after contextual fear conditioning.

Contextual fear conditioning upregulates BDNF mRNA expression in male, but not in female mice (Mizuno *et al.*, 2006). Not only BDNF expression, but also BDNF effects on behaviours have shown sex differences. Male heterozygous BDNF knockouts and male inducible-forebrain BDNF knockout mice are impaired in contextual, but not cued, fear memory formation (Liu *et al.*, 2004, Monteggia *et al.*, 2004). However, female heterozygotes are not impaired in contextual fear memory formation (Chourbaji *et al.*, 2004). Female, but not male, BDNF forebrain knockout mice have increased depression-like behaviour (Monteggis *et al.*, 2007, Autry *et al.*, 2009). Since BDNF is an important plasticity gene, the sex-specific regulation of BDNF after contextual fear conditioning strongly suggests that the sexes use distinct forms of synaptic plasticity during contextual memory formation. Another example for a qualitative sex difference in mRNA expression during memory formation is for the splicing factor arginine/serine-rich 3 (SRp20) (Antunes-Martins *et al.*, 2007). The hippocampal SRp20 mRNA expression is upregulated in males, but not females, during spatial memory formation and after contextual fear conditioning. Furthermore, SRp20 expression regulation requires CaMKK2 signalling that is required for spatial memory formation in males, but not females (Mizuno *et al.*, 2007). Male-specific regulation of SRp20 expression implies that alternative splicing may contribute to sex differences in memory formation. Further studies are required to identify target genes of SRp20 for memory formation. However, *in vitro* studies suggest that Ca^{2+} signalling pathways, including CaMKIV activation, cause alternative splicing of ion channel transcripts. This includes the generation of NMDA receptor 1 transcripts containing exon 21 and effects on NMDA receptor-induced gene expression, which regulate neuronal excitation (Li *et al.*, 2007).

In addition to sexually dimorphic mRNA expression during memory formation within a given brain structure, there is also evidence that the sexes activate transcription in distinct neuronal circuits. This was shown for c-Fos expression during spatial memory formation in a working memory task (Mendez-Lopez *et al.*, 2009). c-Fos expression increases in the infralimbic cortex (IL) and dorsal hippocampal areas CA1 and CA3 only in adolescent female rats. In adult rats, c-Fos expression is induced in the IL only in males. In contrast, only in females, c-Fos expression is activated in the lateral mammillary region. These c-Fos expression patterns suggest that there are sex differences in neuronal network activation associated with spatial working memory formation at different ages. Involvement of different brain circuits in memory formation is consistent with the idea that males and females use different spatial learning strategies (Sandstrom *et al.*, 1998; Roof and Stein, 1999).

Transcriptions specifically occurring during memory formation are markers for particular types of plasticity that have been generated by training in a memory task. Thus, the fact that gene transcription qualitatively differs between males and females, indicates that the sexes induce distinct forms of plasticity despite receiving the same sensory inputs. An important question is why males and females use distinct molecular mechanisms to learn the same tasks. It is possible that the sexes use distinct forms of synaptic plasticity for memory formation. Alternatively, sex-dependent strategies in the same learning situation might activate distinct molecular processes.

7. Genetic and epigenetic mechanisms that may contribute to sex differences in memory formation

Genes on sex chromosomes may potentially contribute to differences in memory formation. For example, the testis-determining *Sry* gene is a Y chromosome gene that is expressed in male brain (Dewing *et al.*, 2006). Sry is expressed in adult dopaminergic neurons and knock-down of Sry expression causes decreased tyrosine hydroxylase expression and motor deficits in male rats. This shows that a Y chromosome gene contributes to brain function in males and, obviously, this mechanism does not exist in females as they do not have a Y chromosome. Furthermore, X chromosome genes contribute to memory formation. Some of these may escape X inactivation so that females express a higher dose than males. Indeed it is interesting to note that mutations in a large number of X-linked genes contribute to mental retardation (Ropers and Hamel, 2005).

Epigenetic mechanisms alter gene activity by modulating DNA–protein interactions without changing the genetic code. Gene expression can be regulated by post-translational modification of histones including acetylation, methylation, and/or ubiquitylation. There is growing evidence that histone modification and DNA methylation are involved in memory formation (Day and Sweatt, 2011). Epigenetic mechanisms underlying memory formation might differ between the sexes. For example, the expression level of the X chromosome gene *UTX* is higher in adult cortex and hippocampus in females than in males (Xu *et al.*, 2008a). UTX is a histone demethylase (HDM) that catalyzes the removal of a methyl group at lysine 27 of histone H3. Therefore, UTX may stimulate greater gene expression in females than in males. Another example is the X chromosome gene *Jarid1c* that encodes another HDM, which is specific for lysine 4 of histone H3. Mutations in this gene cause mental retardation (Jensen *et al.*, 2005). *Jarid1c* mRNA is expressed in adult mouse brain, including hippocampus, at higher levels in females than in males (Xu *et al.*, 2008b). The X chromosome gene *Usp9x* (ubiquitin-specific protease 9), which can act at synapses (Chen *et al.*, 2003), is also expressed at higher levels in adult female brains than in males (Xu *et al.*, 2005). Both *UTX* and *Jarid1c* escape X-inactivation in human and mouse females, and *Usp9x* escapes X-inactivation in humans, but not in mice (Xu and Andreassi, 2011).

8. What are the direct causes of sex differences in memory formation?

Sex differences in memory formation are caused by the genetic differences between males and females (XX vs. XY). For mechanistic understanding of these sex differences, it is important to know what the *direct* causes are. Sex hormones contribute directly to sex differences in memory formation. These hormones influence brain development and they have activational effects, which have been briefly discussed in the case of oestradiol. However, hormonal effects do not account for all of the sex differences in memory formation. There are also direct effects by sex chromosome genes. The genetic differences could be from Y chromosome genes in males, higher doses of X chromosome genes that have not been inactivated in females, and/or X-linked imprinted genes in females. An imprinted gene is expressed only from one parent allele. Males have only the maternal X chromosomal imprint, whereas females have a mixture of paternal and maternal X chromosomal imprints. These X-linked gene dosage and imprinting effects may cause sex differences in brain function. Finally, there are also sex differences for the function of autosome genes, as shown for CaMKK1 and 2. These proteins may act downstream of sex hormones or they function in a pathway with sex chromosome genes.

To dissociate the effects of sex chromosome genes from sex hormonal impact on memory formation, a powerful mouse model, referred to as the "four core genotypes (FCG) model" was developed. In the FCG model, the testis-determining gene *Sry* is deleted from the Y chromosome and a transgenic copy of *Sry* is inserted onto an autosome. By comparing XX and XY mice with ovaries or with testes, the effect of sex chromosome complement (XX versus XY) can be assessed independent of gonadal status (ovaries vs. testes). The FCG model has been used to show that sex chromosome genes regulate habit formation (Quinn *et al.*, 2007). Habit formation is a simple type of learning. Initially, a behavioural response is goal-directed to obtain a response outcome, such as food reward. With continued training, the behaviour becomes habitual. There are sex differences in habit formation and studies with the FCG model have shown that regardless of the gonadal phenotype XX mice have faster habit formation than XY mice.

However, not all types of memory formation seem to depend on sex chromosome genes, as shown in the one-way active avoidance task with the FCG model (McPhie-Lalmansingh *et al.*, 2008). In future, it will be important to use FCG mice to study more sophisticated types of memory formation and their dependence upon sex chromosome genes.

9. Sex differences in Alzheimer's disease

A key feature of Alzheimer's disease (AD) is memory dysfunction. The risk of getting AD is thought to be higher in women than in men. Accordingly, sex-specific phenotypes have been obtained in mouse models of AD. For example, in transgenic mice (rTg4510) that model tau aggregation and neurofibrillary tangle formation female-specific deficits were identified (Yue *et al.*, 2011). Female rTg4510 mutants have higher levels of tau hyperphosphorylation, a prerequisite for tau aggregation, and more severe impairments in spatial memory formation in comparison with male rTg4510 mice. This suggests that there are female-specific mechanisms that promote tau pathology and cognitive deficits.

Studies with other AD mouse models have suggested that there also female-specific factors that enhance amyloid precursor protein (APP) processing into toxic Aβ42 peptide. Using the 5xFAD mouse model, it was shown that 5-day restraint stress induces Aβ42 peptide formation in females, but not in males, at an age point when Aβ42 peptide formation is minor under non-stress conditions (Devi *et al.*, 2010). The Aβ42 peptide is believed to induce synaptic dysfunction in the early stages of AD (Selkoe, 2002). Thus, stress may promote in females, but not in males, Aβ42 peptide formation that leads to synaptic dysfunction in AD. Higher levels of Aβ42 peptide accumulation in females in comparison with males were also reported for other AD mouse models, such as Tg2576 (Callahan *et al.*, 2001), APP/PS1 (Wang, 2003), APPswe/PSEN1E9 (Halford and Russell, 2009) and APP23 mice (Sturchler-Pierrat and Staufenbiel, 2000). Aged female APP23 transgenic mice have more extracellular space (ECS) and more impaired memory formation than age-matched male mutants (Sykova *et al.*, 2005). The enlarged ECS is believed to cause a decrease in extracellular concentration of neurotransmitters, such as acetylcholine, and may change neuron–glial interactions, synaptic efficacy, and/or synaptic plasticity. Taken together, the earlier onset and higher levels of Aβ42 peptide accumulation in females than in males might be one of the reasons for sex differences in AD.

Sex differences in AD may reflect differences of the effects of sex hormones. Sex hormones may have either organisational effects during critical periods of development that induce permanent brain dimorphisms, or activational effects during adulthood that regulate adult brain function. The effect of sex hormones on Aβ42 peptide accumulation was studied in the 3xTg-AD mouse model (Carroll *et al.*, 2010). Female 3xTg-AD mice are more severely impaired than males in a hippocampus-dependent working memory task. Further, the female mutants have more Aβ42 peptide accumulation in frontal cortex, subiculum and hippocampal area CA1. When pharmacologically disrupting the organizational effects of sex hormones, demasculinized male 3xTg-AD mice have increased Aβ42 peptide accumulation in CA1 and subiculum. Defeminized female 3xTg-AD mice showed significantly lower Aβ levels in frontal cortex, but not in other brain areas. Therefore, organizational effects of sex hormones partially make the female brain more vulnerable to AD pathogenesis than the male brain.

Testosterone may play an important role in AD in men. A decreased testosterone level in aged men is associated with impaired cognitive function and an increased risk for getting AD. However, testosterone can be converted into oestradiol by aromatase. Thus, it is difficult to determine whether the link between testosterone and AD neuropathology is due to the aromatization of testosterone into oestradiol (which may be protective) or whether it is oestrogen-independent. To study this question, aromatase knockout (ArKO) mice were crossed with APP23 micegenerating APP23/$Ar^{+/-}$(McAllister *et al.*, 2010). These double mutants have less Aβ42 peptide accumulation and less impaired spatial memory formation in comparison with male APP23 mice. Thus, the endogenous testosterone protects from AD pathology. Testosterone was shown to downregulate β-secretase, which reduce APP cleavage into Aβ42 peptideand to upregulate neprilysin that degrades Aβ42 peptide.

BDNF Val66Met polymorphism has been associated with poor episodic memory in human (Egan *et al.*, 2003) and BDNF levels are reduced in AD. Thus, BDNF may be involved in the development of AD. Accordingly, there is a significant allelic association between the Val66Met polymorphism and AD in women, but not in men (Fukumoto *et al.*, 2009).

Qualitative sex differences in synaptic kinase function and gene transcription have not been studied in mouse models of AD. However, it is likely that these processes are affected in AD, as p25 formation is disrupted in the early stages of AD and p25 formation enhances memory formation primarily in female mice (Ris *et al.*, 2005; Engmann *et al.*, 2011).

10. Conclusion

In a variety of species there are sex differences in memory formation. However, the whys and wherefores of these sex differences are not well understood. So far, only a few mechanistic studies of memory formation have considered sex as a factor. The vast majority of these studies were conducted only in male rodents, due in part to the argument that females may exhibit greater variability due to the oestrous cycle. However, in the studies where females were used across different stages of the oestrous cycle the variability of the data appeared similar to the ones obtained with males (e.g. Mizuno *et al.*, 2007). With advances in molecular biology and the analysis of both sexes in parallel, the idea has emerged that there are sex differences at many levels, including plasticity mechanisms that contribute to memory formation (Figure 7.1). Currently, it is not known whether these differences are due to sex hormone effects and/or differences in sex chromosome gene expression in the brain. This chapter has summarized key findings from relevant

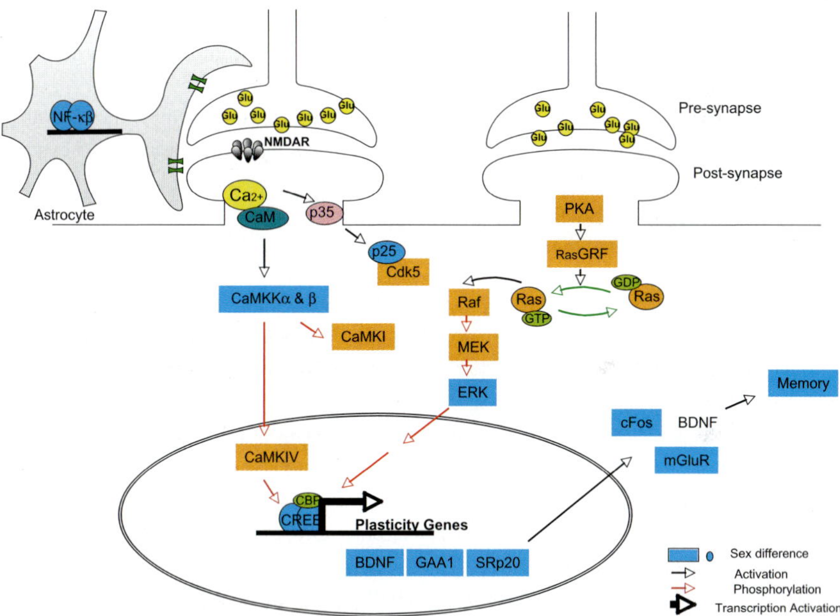

Figure 7.1. Summary of recently identified sex differences in molecular signalling underlying memory formation. Synaptic plasticity is thought to underlie memory formation. At glutamatergic synapses, calcium influx through the NMDA receptor triggers synaptic plasticity by activating various synaptic kinases and inducing gene transcription followed by protein synthesis. Furthermore, neurons and astrocytes communicate by way of tripartite synapses. Sex differences have been identified for the role of (and activation of)various synaptic kinases, transcription factors and gene transcriptions; these differences are highlighted in blue. Figure taken from Mizuno and Giese, 2010 with permission from Elsevier.

molecular biological studies, which include sex differences in molecular signalling that induce gene transcription for memory formation and also female-specific impairments in AD mouse models. The molecular understanding of sex differences in memory formation is fundamental not only for understanding memory processes. It is also essential for understanding and treating psychiatric diseases in which memory formation is affected and gender differences are observed. Many psychiatric diseases, including Alzheimer's disease, differ between males and females in incidence, time of onset, and/or degree of severity. Thus, a mechanistic understanding of sex differences in memory formation may provide new therapeutic possibilities for psychiatric diseases.

References

Andreano, J.M. and Cahill, L. (2009). Sex influences on the neurobiology of learning and memory. *Learn Mem* **16**, 248–266.

Angelo, M., Plattner, F., Irvine, E.E. and Giese, K.P. (2003). Improved reversal learning and altered fear conditioning in transgenic mice with regionally restricted p25 expression. *Eur J Neurosci* **18**, 423–431.

Antunes-Martins, A., Mizuno, K., Irvine, E.E., Lepicard, E.M. and Giese, K.P. (2007). Sex-dependent up-regulation of two splicing factors, Psf and Srp20, during hippocampal memory formation. *Learn Mem* **14**, 693–702.

Autry, A.E., Adachi, M., Cheng, P. and Monteggia, L.M. (2009). Gender-specific impact of brain-derived neurotrophic factor signaling on stress-induced depression-like behavior. *Biol Psychiatry* **66**, 84–90.

Bangasser, D.A. and Shors, T.J. (2007). The hippocampus is necessary for enhancements and impairments of learning following stress. *Nat Neurosci* **10**, 1401–1403.

Bangasser, D.A. and Shors, T.J. (2008). The bed nucleus of the stria terminalis modulates learning after stress in masculinized but not cycling females. *J Neurosci* **28**, 6383–6387.

Bangasser, D.A. and Shors, T.J. (2010). Critical brain circuits at the intersection between stress and learning. *Neurosci Biobehav Rev* **34**, 1223–1233.

Barha, C.K., Dalton, G.L. and Galea, L.A. (2010). Low doses of 17alpha-estradiol and 17beta-estradiol facilitate, whereas higher doses of estrone and 17alpha- and 17beta-estradiol impair, contextual fear conditioning in adult female rats. *Neuropsychopharmacology* **35**, 547–559.

Blaeser, F., Sanders, M.J., Truong, N., Ko, S., Wu, L.J., Wozniak, D.F., Fanselow, M.S., Zhuo, M. and Chatila, T.A. (2006). Long-term memory deficits in Pavlovian fear conditioning in Ca^{2+}/calmodulin kinase kinase alpha-deficient mice. *Mol Cell Biol* **26**, 9105–9115.

Bracchi-Ricard, V., Brambilla, R., Levenson, J., Hu, W.H., Bramwell, A., Sweatt, J.D., Green, E.J. and Bethea, J.R. (2008). Astroglial nuclear factor-kappaB regulates learning and memory and synaptic plasticity in female mice. *J Neurochem* **104**, 611–623.

Brinton, R.D. (2009). Estrogen-induced plasticity from cells to circuits: Predictions for cognitive function. *Trends Pharmacol Sci* **30**, 212–222.

Cahill, L. (2006). Why sex matters for neuroscience. *Nat Rev Neurosci* **7**, 477–484.

Callahan, M.J., Lipinski, W.J., Bian, F., Durham, R.A., Pack, A. and Walker, L.C. (2001). Augmented senile plaque load in aged female beta-amyloid precursor protein-transgenic mice. *Am J Pathol* **158**, 1173–1177.

Carroll, J.C., Rosario, E.R., Kreimer, S., Villamagna, A., Gentzschein, E., Stanczyk, F.Z. and Pike, C.J. (2010). Sex differences in beta-amyloid accumulation in 3xTg-AD mice: Role of neonatal sex steroid hormone exposure. *Brain Res* **1366**, 233–245.

Chen, H., Polo, S., Di Fiore, P.P. and De Camilli, P.V. (2003). Rapid Ca^{2+}-dependent decrease of protein ubiquitination at synapses. *Proc Natl Acad Sci USA* **100**, 14908–14913.

Chourbaji, S., Hellweg, R., Brandis, D., Zörner, B., Zacher, C., Lang, U.E., Henn, F.A., Hörtnagl, H. and Gass, P. (2004). Mice with reduced brain-derived neurotrophic factor expression show decreased choline acetyltransferase activity, but regular brain monoamine levels and unaltered emotional behavior. *Mol Brain Res* **121**, 28–36.

Day, J.J. and Sweatt, J.D. (2011). Epigenetic mechanisms in cognition. *Neuron* **70**, 813–829.

Devi, L., Alldred, M.J., Ginsberg, S.D. and Ohno, M. Sex- and brain region-specific acceleration of beta-amyloidogenesis following behavioral stress in a mouse model of Alzheimer's disease. *Mol Brain* **3**, 34.

Dewing, P., Chiang, C.W., Sinchak, K., Sim, H., Fernagut, P.O., Kelly, S., Chesselet, M.F., Micevych, P.E., Albrecht, K.H., Harley, V.R. and Vilain, E. (2006). Direct regulation of adult brain function by the male-specific factor SRY. *Curr Biol* **16**, 415–420.

Doeller, C.F., King, J.A. and Burgess, N. (2008). Parallel striatal and hippocampal systems for landmarks and boundaries in spatial memory. *Proc Natl Acad Sci USA* **105**, 5915–5920.

Egan, M,F., Kojima, M., Callicott, J.H., Goldberg, T.E., Kolachana, B.S., Bertolino. A., Zaitsev, E., Gold, B., Goldman, D., Dean, M., Lu, B. and Weinberger, D.R. (2003). The BDNF val66met polymorphism affects activity-dependent secretion of BDNF and human memory and hippocampal function. *Cell* **112**, 257–269.

Engmann, O., Hortobagyi, T., Thompson, A.J., Guadagno, J., Troakes, C., Soriano, S., Al-Sarraj, S., Kim, Y. and Giese, K.P. (2011). Cyclin-dependent kinase 5 activator p25 is generated during memory formation and is reduced at an early stage in Alzheimer's disease. *Biol Psychiatry* **70**, 159–168.

Fukumoto, N., Fujii, T., Combarros, O., Kamboh, M.I., Tsai, S.J., Matsushita, S., Nacmias, B., Comings, D.E., Arboleda, H., Ingelsson, M., Hyman, B.T., Akatsu, H., Grupe, A., Nishimura, A.L., Zatz, M., Mattila, K.M., Rinne, J., Goto, Y., Asada, T., Nakamura, S. and Kunugi, H. (2010). Sexually dimorphic effect of the Val66Met polymorphism of BDNF on susceptibility to Alzheimer's disease: New data and meta-analysis. *Am J Med Genet B Neuropsychiatr Genet* **153B**, 235–242.

Gibbs, R.B. (2010). Estrogen therapy and cognition: A review of the cholinergic hypothesis. *Endocr Rev* **31**, 224–253.

Gillies, G.E. and McArthur, S. (2010). Estrogen actions in the brain and the basis for differential action in men and women: A case for sex-specific medicines. *Pharmacol Rev* **62**, 155–198.

Gresack, J.E., Schafe, G.E., Orr, P.T. and Frick, K.M. (2009). Sex differences in contextual fear conditioning are associated with differential ventral hippocampal extracellular signal-regulated kinase activation. *Neuroscience* **159**, 451–467.

Halford, R.W. and Russell, D.W. (2009). Reduction of cholesterol synthesis in the mouse brain does not affect amyloid formation in Alzheimer's disease, but does extend lifespan. *Proc Natl Acad Sci USA* **106**, 3502–3506.

Hebda-Bauer, E.K., Luo, J., Watson, S.J. and Akil, H. (2007). Female CREBalphadelta-deficient mice show earlier age-related cognitive deficits than males. *Neuroscience* **150**, 260–272.

Hojo, Y., Murakami, G., Mukai, H., Higo, S., Hatanaka, Y., Ogiue-Ikeda, M., Ishii, H., Kimoto, T. and Kawato, S. (2008). Estrogen synthesis in the brain — role in synaptic plasticity and memory. *Mol Cell Endocrinol* **290**, 31–43.

Irvine, E.E., von Hertzen, L.S.J, Plattner, F. and Giese, K.P. (2006) alpha CaMKII autophosphorylation: A fast-track to memory. *Trends Neurosci* **29**, 459–465.

Jensen, L.R., Amende, M., Gurok, U., Moser, B., Gimmel, V., Tzschach, A., Janecke, A.R., Tariverdian, G., Chelly, J., Fryns, J.P., *et al.* (2005). Mutations in the JARID1C gene, which is involved in transcriptional regulation and chromatin remodeling, cause X-linked mental retardation. *Am J Hum Genet* **76**, 227–236.

Jonasson, Z. (2005). Meta-analysis of sex differences in rodent models of learning and memory: A review of behavioral and biological data. *Neurosci Biobehav Rev* **28**, 811–825.

Kim, A.M., Tingen, C.M. and Woodruff, T.K. (2010). Sex bias in trials and treatment must end. *Nature* **465**, 688–689.

Korol, D.L., Malin, E.L., Borden, K.A., Busby, R.A. and Couper-Leo, J. (2004). Shifts in preferred learning strategy across the estrous cycle in female rats. *Horm Behav* **45**, 330–338.

Kudo, K., Qiao, C.X., Kanba, S. and Arita, J. (2004). A selective increase in phosphorylation of cyclic AMP response element-binding protein in hippocampal CA1 region of male, but not female, rats following contextual fear and passive avoidance conditioning. *Brain Res* **1024**, 233–243.

Leuner, B. and Shors, T.J. (2004). New spines, new memories. *Mol Neurobiol* **29**, 117–130.

Li, Q., Lee, J.A. and Black, D.L. (2007). Neuronal regulation of alternative pre-mRNA splicing. *Nat Rev Neurosci* **8**, 819–831.

Liu, F., Day, M., Bitran, D., Arlas, R., Reviall-Sanchez, R., Grauer, S., Zhang, G., Kelley, C., Pulito, V., Sung, A., Mervis, R.F., Navarra, R., Hirst, W.D., Reinhart, P.H., Marquis, K.L., Moss, S.J., Pangalos, M.N., Brandon, N.J. (2008). Activation of estrogen receptor-beta regulates hippocampal synaptic plasticity and improves memory. *Nat Neurosci* **11**, 334–343.

Liu, I.Y., Lyons, W.E., Mamounas, L.A. and Thompson, R.F. (2004). Brain-derived neurotrophic factor plays a critical role in contextual fear conditioning. *J Neurosci* **24**, 7958–7963.

Madeira, M.D. and Lieberman, A.R. (1995). Sexual dimorphism in the mammalian limbic system. *Prog Neurobiol* **45**, 275–333.

Maeng, L.Y., Waddell, J. and Shors, T.J. (2011). The prefrontal cortex communicates with the amygdala to impair learning after acute stress in females but not in males. *J Neurosci* **30**, 16188–16196.

McAllister, C., Long, J., Bowers, A., Walker, A., Cao, P., Honda, S., Harada, N., Staufenbiel, M., Shen, Y. and Li, R. (2010). Genetic targeting aromatase in male amyloid precursor protein transgenic mice down-regulates beta-secretase (BACE1) and prevents Alzheimer-like pathology and cognitive impairment. *J Neurosci* **30**, 7326–7334.

McEwen, B. (2002). Estrogen actions throughout the brain. Recent *Prog Horm Res* **57**, 357–384.

McPhie-Lalmansingh, A.A., Tejada, L.D., Weaver, J.L., Rissman, E.F. (2008). Sex chromosome complement affects social interactions in mice. *Horm Behav* **54**, 565–570.

Mizuno, K., Antunes-Martins, A., Ris, L., Peters, M., Godaux, E. and Giese, K.P. (2007). Calcium/calmodulin kinase kinase beta has a male-specific role in memory formation. *Neuroscience* **145**, 393–402.

Mizuno, K. and Giese, K.P. (2010). Towards a molecular understanding of sex differences in memory formation. *Trends Neurosci* **33**, 285–291.

Mizuno, K., Ris, L., Sanchez-Capelo, A., Godaux, E. and Giese, K.P. (2006). Ca^{2+}/calmodulin kinase kinase alpha is dispensable for brain development but is required for distinct memories in male, though not in female, mice. *Mol Cell Biol* **26**, 9094–9104.

Monteggia, L.M., Barrot, M., Powell, C.M., Berton, O., Galanis, V., Gemelli, T., Meuth, S., Nagy, A., Greene, R.W. and Nestler, E.J. (2004). Essential role of brain-derived neurotrophic factor in adult hippocampal function. *Proc Natl Acad Sci USA* **101**, 10827–10832.

Monteggia, L.M., Luikart, B., Barrot, M., Theobold, D., Malkovska, I., Nef, S., Parada, L.F. and Nestler, E.J. (2007). Brain-derived neurotrophic factor conditional knockouts show gender differences in depression-related behaviors. *Biol Psychiatry* **61**, 187–197.

Mukai, H., Kimoto, T., Hojo, Y., Kawato, S., Murakami, G., Higo, S., Hatanaka, Y. and Ogiue-Ikeda, M. (2010). Modulation of synaptic plasticity by brain estrogen in the hippocampus. *Biochim Biophys Acta* **1800**, 1030–1044.

Peters, M., Mizuno, K., Ris, L., Angelo, M., Godaux, E. and Giese, K.P. (2003). Loss of Ca^{2+}/calmodulin kinase kinase beta affects the formation of some, but not all, types of hippocampus-dependent long-term memory. *J Neurosci* **23**, 9752–9760.

Postma, A., Jager, G., Kessels, R.P., Koppeschaar, H.P. and van Honk, J. (2004). Sex differences for selective forms of spatial memory. *Brain Cogn* **54**, 24–34.

Quinn, J.J., Hitchcott, P.K., Umeda, E.A., Arnold, A.P. and Taylor, J.R. (2007). Sex chromosome complement regulates habit formation. *Nat Neurosci* **10**, 1398–1400.

Ris, L., Angelo, M., Plattner, F., Capron, B., Errington, M.L., Bliss, T.V., Godaux, E. and Giese, K.P. (2005). Sexual dimorphisms in the effect of low-level p25 expression on synaptic plasticity and memory. *Eur J Neurosci* **21**, 3023–3033.

Rissman, E.F. (2008). Roles of oestrogen receptors alpha and beta in behavioural neuroendocrinology: Beyond Yin/Yang. *J Neuroendocrinol* **20**, 873–879.

Romeo, R.D., Waters, E.M. and McEwen, B.S. (2004). Steroid-induced hippocampal synaptic plasticity: Sex differences and similarities. *Neuron Glia Biol* **1**, 219–229.

Roof, R.L. and Stein, D.G. (1999). Gender differences in Morris water maze performance depend on task parameters. *Physiol Behav* **68**, 81–86.

Ropers, H.H. and Hamel, B.C. (2005). X-linked mental retardation. *Nat Rev Genet* **6**, 46–57.

Sandstrom, N.J., Kaufman, J. and Huettel, S.A. (1998). Males and females use different distal cues in a virtual environment navigation task. *Brain Res Cogn Brain Res* **6**, 351–360.

Schwarz, J.M., Liang, S.L., Thompson, S.M. and McCarthy, M.M. (2008). Estradiol induces hypothalamic dendritic spines by enhancing glutamate release: A mechanism for organizational sex difference. *Neuron* **58**, 584–598.

Selkoe, D.J. (2002). Alzheimer's disease is a synaptic failure. *Science* **298**, 789–791.

Shors, T.J. (2004). Learning during stressful times. *Learn Mem* **11**, 137–144.

Smejkalova, T. and Woolley, C.S. (2010). Estradiol acutely potentiates hippocampal excitatory synaptic transmission through a presynaptic mechanism. *J Neurosci* **30**, 16137–16148.

Spencer, J.L., Waters, E.M., Romeo, R.D., Wood, G.E., Milner, T.A. and McEwen, B.S. (2008). Uncovering the mechanisms of estrogen effects on hippocampal function. *Front Neuroendocrinol* **29**, 219–237.

Sturchler-Pierrat, C. and Staufenbiel, M. (2000). Pathogenic mechanisms of Alzheimer's disease analyzed in the APP23 transgenic mouse model. *Ann N Y Acad Sci* **920**, 134–139.

Sykova, E., Vorisek, I., Antonova, T., Mazel, T., Meyer-Luehmann, M., Jucker, M., Hajek, M., Ort, M. and Bures, J. (2005). Changes in extracellular space size and geometry in APP23 transgenic mice: A model of Alzheimer's disease. *Proc Natl Acad Sci USA* **102**, 479–484.

Waddell, J., Bangasser, D.A. and Shors, T.J. (2008). The basolateral nucleus of the amygdala is necessary to induce the opposing effects of stressful experience on learning in males and females. *J Neurosci* **28**, 5290–5294.

Wang, J., Tanila, H., Puolivali, J., Kadish, I. and van Groen, T. (2003). Gender differences in the amount and deposition of amyloidbeta in APPswe and PS1 double transgenic mice. *Neurobiol Dis* **14**, 318–327.

Wayman, G.A., Lee, Y.S., Tokumitsu, H., Silva, A.J. and Soderling, T.R. (2008). Calmodulin-kinases: Modulators of neuronal development and plasticity. *Neuron* **59**, 914–931.

Woolley, C.S. (2007). Acute effects of estrogen on neuronal physiology. *Annu Rev Pharmacol Toxicol* **47**, 657–680.

Xu, J. and Andreassi, M. (2011) Reversible histone methylation regulates brain gene expression and behavior. *Horm Behav* **59**, 383–392.

Xu, J., Deng, X., Watkins, R. and Disteche, C.M. (2008a). Sex-specific differences in expression of histone demethylases Utx and Uty in mouse brain and neurons. *J Neurosci* **28**, 4521–4527.

Xu, J., Deng, X. and Disteche, C.M. (2008b). Sex-specific expression of the X-linked histone demethylase gene Jarid1c in brain. *PLoS One* **3**, e2553.

Xu, J., Taya, S., Kaibuchi, K. and Arnold, A.P. (2005). Sexually dimorphic expression of Usp9x is related to sex chromosome complement in adult mouse brain. *Eur J Neurosci* **21**, 3017–3022.

Yue, M., Hanna, A., Wilson, J., Roder, H. and Janus, C. (2011). Sex difference in pathology and memory decline in rTg4510 mouse model of tauopathy. *Neurobiol Aging* **32**, 590–603.

Neocortical Learning, Memory and Experience-dependent Plasticity

8

Giorgia Albieri* and Gerald T. Finnerty*

1. Introduction

One of the most intriguing features of the brain is its ability to acquire, store, and recall information. This facility enables an organism to modify its behaviour based on previous experience. The neocortex is thought to be the final repository for many forms of long-term or remote memories. Despite more than a century of investigation, the cellular and circuits bases for learning and memory remain key unanswered questions in neuroscience. The complexity of mammalian neocortex has meant that many mechanistic studies have made slow progress. Instead, investigations have focused on the hippocampus and the cerebellum in mammals or on invertebrates with comparatively simple nervous systems, e.g. the sea mollusc, *Aplysia*, the fruit fly, *Drosophila melanogaster* (Dudai, 1989). More recently, however, technological advances have led to fresh attacks on the mechanisms underling learning and memory in the neocortex.

There are different types of learning and memory (Dudai, 1989; Squire, 1987). Many of the declarative forms of memory rely on the hippocampus. It has been proposed that memories are stored in the hippocampus for short periods of time and are then transferred to the neocortex for long-term storage. This is accompanied by some forms of declarative memory becoming hippocampus-independent whereas others, such as spatial memory, always require hippocampal activity. Neocortical lesioning studies have suggested that long-term or remote memories are not stored in discrete locations, but instead involve neural circuits that are distributed across the neocortex (Lashley, 1950). Therefore, a major problem for any study investigating this process is to ascertain the neurons that play a role in memory (Frankland and Bontempi, 2005; Silva *et al.*, 2009). Many recent studies have sought to make the problem tractable by studying memory that is

* MRC Centre for Neurodegeneration Research, King's College London, De Crespigny Park, London SE5 8AF, UK.

independent of the hippocampus. In this chapter, we consider the mechanisms underpinning learning and memory that is encoded directly into the neocortex, e.g. perceptual learning, motor skill learning. We explore how procedural learning and memory relates to experience-dependent plasticity in an effort to gain further insights into the underlying plasticity mechanisms.

The brain can only use a few broad categories of cellular plasticity mechanisms to change the output of neural circuits. The categories include: changes in synaptic strength (see Chapter 1), rewiring of connections between neurons (Table 8.1, see Chapter 4 for neurogenesis) and altered neuronal excitability (see Chapter 3). We refer to these categories as the plasticity toolbox. The limited number of plasticity categories may lead one to think that it should be relatively simple to determine the cellular basis for cortical plasticity. However, this is not the case. Neural circuits contain multiple connections and each connection can be modified by one or more mechanisms from the plasticity toolbox. Hence, it is a complex task to disentangle the cellular mechanisms that underpin changes in behaviour.

The difficulties are compounded further by the realisation that neither learning nor memory is one uniform process. Learning is frequently divided into different phases, e.g. fast and slow, to reflect the observation that tasks are not learnt at a constant rate (Karni *et al.*, 1998). Similarly, memory is divided into different stages, which include memory formation, storage, and retrieval. Memory formation is commonly subdivided into encoding and consolidation. Many investigations have focused on consolidation (Dudai, 2004).

Table 8.1. Strategies for rewiring. Adapted from Barnes & Finnerty, Neuroscientist (2010).

Existing Connections

- 'Unsilence' silent synapses
- Alter synapse number
  - Synaptogenesis/elimination
- Relocate synapses
  - Different dendritic branches
  - Synapse clustering

New Connections

- Synaptogenesis
- Growth of axon and dendrite
- Neurogenesis

2. Sensory learning and memory

Sensory systems exhibit several forms of learning and memory. Perceptual learning and sensory associative learning have been studied most frequently. Perceptual learning is an improvement in the ability to discriminate between sensory stimuli (Gilbert *et al.*, 2001; Goldstone, 1998). Associative learning is difficult to define simply (Dudai, 1989). However, a useful operational definition is that associative learning is a persistent change in behaviour that accompanies a stimulus or context. An example of associative sensory learning is conditioned taste aversion, in which rodents associate the taste of a food with having been sick previously and, therefore, do not eat that food (Shema *et al.*, 2007).

More recently, interest has increased in a very simple form of sensory memory in which animals show evidence of memory of previous sensory experience. Evidence for this type of memory has come from behaviour (Guic-Robles *et al.*, 1989), but has also been obtained from studies showing accelerated reorganization of cortical maps (Hofer *et al.*, 2006; Hofer *et al.*, 2009). Some may argue that accelerated reorganization of cortical maps does not really constitute a form of memory because there is no measurable change in behaviour. However, memory of previous sensory experience, whether investigated behaviourally or through cortical map reorganization, exhibits savings and, therefore, conforms to Ebbinghaus' original definition of memory (Ebbinghaus, 1885).

2.1. *Perceptual learning and memory*

Perceptual learning improves with practice over many trials (Gilbert *et al.*, 2001; Goldstone, 1998). It was originally thought that perceptual learning was characterized by the specificity to the stimulus feature that was learnt and the lack of transfer to other tasks and attention to the task (Ahissar and Hochstein, 1993). However, it has become clear recently that perceptual learning can occur as a result of repeated exposure to a sensory stimulus without any requirement for attention or need for the sensory stimulus to play a central role in the perceptual task (Dinse *et al.*, 2003; Frenkel *et al.*, 2006; Gutnisky *et al.*, 2009; Li *et al.*, 2006; Sasaki *et al.*, 2010; Watanabe *et al.*, 2001).

There has been great interest in the neural basis for perceptual learning. In the classical view of sensory perception, sensory inputs were thought to be processed linearly in a series of cortical areas. The broad idea was that sensory inputs were first passed to their respective primary sensory cortex and then on to higher sensory regions before reaching association cortex. As the hierarchy of cortical areas was ascended, neurons responded to more and more complex features of the sensory stimulus. Although it is not clear whether perceptual learning occurs in exactly the same way in all sensory systems, the principles overlap and will, therefore, be discussed together.

Perceptual learning typically involves training on a task that requires the subject to focus on a set of sensory inputs, which are presented repetitively. A key question is "Where does perceptual learning occur in the brain?". Three major ideas have been put expounded to answer this question. An early hypothesis arose from studies of the psychophysics of perceptual learning. It was argued that primary sensory cortex was the major site for perceptual learning because the learning was, first, specific to simple features of the sensory stimulus and, second, did not transfer to other tasks (Karni and Sagi, 1991). Essentially, primary sensory cortex reorganizes in response to changes in sensory input. This process may be referred to as "bottom up". This proposal received some support from electrophysiological recording of neuronal firing (Schoups *et al.*, 2001). However, there was debate about whether the changes in spiking were sufficient to explain the improved performance. Moreover, greater changes in neuronal firing were found in sensory brain regions outside primary sensory cortex (Ghose *et al.*, 2002; Yang and Maunsell, 2004).

A second theory is that perceptual learning occurred when processing of sensory features is moved from higher association cortices to primary sensory cortex (Ahissar and Hochstein, 1997; Ahissar and Hochstein, 2004). The premise is that the complex processing required for difficult perceptual tasks is initially performed in association cortex. However, as task performance improves, more and more of the sensory processing required for the task is moved from association cortex to primary sensory cortex. The proposal is termed the reverse hierarchy theory because it involved reversing the classical hierachical progression of sensory processing (Ahissar and Hochstein, 1997; Ahissar and Hochstein, 2004). This proposal may also be viewed as a top-down process because inputs from higher centres feedback and modify primary sensory cortex.

Both the bottom-up and top-down hypotheses suggests that primary sensory cortex is the key site in perceptual learning and memory. The third hypothesis states that it is not primary sensory cortex per se that is important. Rather, it is the output from primary sensory cortex that is processed differently in association cortex. The central idea is that the read-out from primary sensory cortex changes (Dosher and Lu, 1998). This can be accomplished by strengthening and weakening of specific outputs from primary sensory cortex, possibly through Hebbian synaptic plasticity (Dosher and Lu, 1998). In this theory, the key cellular changes are not found in primary sensory cortex, but in higher centres for sensory processing.

These theories have been tested in humans and animals, most intensively in the visual system. Typically, experiments combine psychophysics to measure task performance and functional magnetic resonance imaging (fMRI) to determine which parts of the brain network are activated when performing the task. A particular popular form of fMRI is based on the blood-oxygen-level–dependent (BOLD) signal (Logothetis and Wandell, 2004; Ogawa *et al.*, 1990). The advantage of functional imaging is that it enables the whole brain to be imaged at one sitting. Hence, the brain regions involved in perceptual learning and memory can be identified and the functional connectivity between those regions can be measured (Buchel *et al.*, 1999). Animal studies give further information because they can combine psychophysics and fMRI with direct measurements of neural activity, such as recordings of neural spiking in different brain regions.

Human fMRI studies have demonstrated increased BOLD signal in primary visual cortex after visual perceptual learning (Furmanski *et al.*, 2004; Schwartz *et al.*, 2002; Yotsumoto *et al.*, 2008). This is accompanied by greater functional connectivity between visual cortex, posterior parietal cortex, and prefrontal regions (Lewis *et al.*, 2009). Hence, the fMRI studies suggest that perceptual learning involves a network of regions in different lobes of the brain, which includes primary sensory cortex. Notably, the prefrontal and posterior parietal regions are part of the top-down attention network and reflect the recruitment of attention when difficult perceptual tasks are learnt (Corbetta and Shulman, 2002). However, changes in the functional connectivity between brain regions in resting state networks has limited power to discriminate between the any of the hypotheses for perceptual learning.

Electrophysiological studies have shown evidence of increased spiking in sensory cortex. Although changes in spiking have been reported in primary sensory cortex following perceptual learning, more pronounced changes in firing are found outside primary sensory cortex (Ghose *et al.*, 2002; Sasaki *et al.*, 2010; Yang and Maunsell, 2004). More recently, support for the read-out hypothesis has come from a study that reported increased spiking in posterior parietal cortex with unchanged spiking in medial temporal visual cortex after training on a visual discrimination task of motion direction (Law and Gold, 2008).

The theories of perceptual learning and associated experiments have tended to focus on either a general change in a brain region or on interactions between brain regions. However, the cellular mechanisms underpinning perceptual learning and memory remain unclear. It is widely believed that perceptual learning involves synaptic plasticity. This is supported by the finding that exposure-based perceptual learning in humans can be blocked by the NMDA receptor antagonist, memantine (Dinse *et al.*, 2003). However, there is little information on the role of synaptic plasticity in task-relevant perceptual learning. Furthermore, the contribution of altered neuronal excitability and rewiring to perceptual learning and memory are poorly understood.

2.2. *Sensory associative learning*

Sensory associative learning is a general feature of sensory systems. Interest has focused on the neural circuits supporting the association and the cellular and molecular mechanisms underlying the learning and memory. A popular hypothesis is that learning and memory requires changes in synaptic strength. A major issue, however, has been explaining how changes in synaptic strength are maintained in the face of molecular turnover. Persistently active kinases have been popular candidate molecules for this role. Recent evidence suggests that a constitutively-active isoform of protein kinase C termed PKMzeta may be involved in maintaining synaptic strengthening (Pastalkova *et al.*, 2006). PKMzeta can be blocked *in vivo* by infusing zeta inhibitory peptide (ZIP), which can cross cell membranes. This has led to ZIP being used as a tool to probe memory mechanisms. Infusion of ZIP into the insular cortex erased long-term memory of conditioned taste aversion (Shema *et al.*, 2007). Furthermore, injection of a lentivirus to overexpress PKMzeta in insular cortex enhanced long-term memory of conditioned taste aversion whereas a dominant-negative form of PKMzeta blocks long-term memory (Shema *et al.*, 2011).

Further supportive evidence for PKMzeta playing a role in long-term memory has come from studies of sensory associative learning with a strong emotional component. Injection of ZIP into secondary sensory cortex impaired remote, but not recent fear memories (Sacco and Sacchetti, 2010). This finding suggests that secondary sensory cortex contributes to long-term storage of sensory memories that are associated with a strong emotional charge.

2.3. *Memory of previous sensory experience*

Memory of previous sensory experience lies in the overlap between memory studies based on behaviour and experience-dependent plasticity. This area of research has highlighted the contribution of both structural and functional synaptic plasticity to neocortical learning and memory. A short period of monocular deprivation during adolescence results in reorganization of primary visual cortex that reverses when binocular vision is restored. *In vivo* imaging of dendritic spines in visual cortex revealed that monocular deprivation results in increased turnover of dendritic spines and a small increase in spine density on the apical tufts of layer 5 pyramidal neurons (Hofer *et al.*, 2009). A repeated episode of monocular deprivation during adulthood is accompanied by more rapid reorganization of the visual cortical map (Hofer *et al.*, 2006). Spine counts do not increase further. However, some of the new spines formed during the first episode of monocular deprivation become larger (Hofer *et al.*, 2009). Taken together, these findings suggest that the initial episode of visual deprivation led to the formation of new persistent synapses and the rewiring of the microcircuits in visual cortex. The second episode of deprivation, in which dendritic spines enlarged, was interpreted to signify strengthening of the new synapses.

3. Motor learning and memory

As for sensory systems, there are several types of motor learning and memory. Motor cortex has been implicated in motor skill learning. Motor associative learning has been studied intensively, but the circuits usually involve brainstem reflexes and the cerebellum, e.g. eyeblink conditioning and adaptation of the vestibulo-ocular reflex. Therefore, we will concentrate on motor skill learning in keeping with this chapter's focus on the neocortex.

Motor skill learning is assayed by performance on motor tasks that require relatively complex motor movements, e.g. reaching for food pellets in rodents, or tapping a sequence with the digits of one hand in humans. Learning is commonly divided into fast and slow phases (Karni *et al.*, 1995; Ungerleider *et al.*, 2002). The fast phase is initiated during training, and may involve cortico-striatal or cortico-cerebellar circuits. Fast learning leaves a fragile memory trace in primary motor cortex, which can be disrupted up to 6 h after training (Brashers-Krug *et al.*, 1996; Muellbacher *et al.*, 2002). This phase of the learning process is thought to be instrumental in setting up an optimized motor strategy for the task being learned. In contrast, slow learning is the incremental gain in performance that develops progressively over many training sessions. This phase of learning is accompanied by reorganization of primary motor cortex. Specifically, representations of the trained movements expand, as shown by fMRI (Karni *et al.*, 1995) or by electrophysiological mapping experiments (Nudo *et al.*, 1996). Finally, the transition from a fragile, short-term memory to a robust, enduring memory is termed consolidation. Evidence suggests that consolidation of motor skill memory occurs over hours rather than days to weeks, which constrains the plasticity mechanisms that could be operational (Krakauer and Shadmehr, 2006).

It is generally believed that motor skill learning and memory result in changes to the microcircuits in primary motor cortex (Komiyama *et al.*, 2010). However, it has proven difficult to establish precisely which mechanisms from the plasticity toolbox are invoked and when they are implemented. Much information about the cellular mechanisms underpinning memory traces in motor cortex has come from studies based on rodents reaching for small food pellets with one forepaw. This model has been particularly useful for probing the contribution of functional and structural synaptic plasticity to motor cortex reorganization. After 5 days of motor skill training, stimulation of horizontal intracortical pathways in layer 2/3 of trained motor cortex evokes larger amplitude local field potentials (Rioult-Pedotti *et al.*, 1998). Furthermore, long-term potentiation (LTP) is diminished and long-term depression (LTD) is enhanced (Rioult-Pedotti *et al.*, 2000). Hence, it is harder to potentiate "trained" synapses further, but it becomes easier to weaken them. Collectively, these findings suggest that motor skill learning induces LTP-like strengthening of synaptic connections in primary motor cortex.

The evidence that changes in synapse number in the neocortex mediate motor skill memory has been mixed. Electron microscopy (EM) studies indicated that synapse density increased in primary motor cortex after a few weeks of training on the rodent reaching task (Kleim *et al.*, 1996; 2004). In contrast, longitudinal *in vivo* imaging of dendritic spines, which act as a surrogate marker for excitatory synapses, suggests that synapse number undergoes minimal changes (Xu *et al.*, 2009; Yang *et al.*, 2009). Although training leads to the formation of new dendritic spine within hours, synapse elimination is also markedly increased. Hence, the major accompaniment of learning is increased synapse turnover rather than increased synapse number. Some of the new dendritic spines become persistent suggesting that the increased synapse turnover facilitates establishment of new synapses. Formation of new synapses enables the connections between cortical neurons to rewire (Barnes and Finnerty, 2010). Hence, the rapid appearance of new spines suggests that rewiring of circuits in motor cortex could contribute to both the fast and the slow phases of motor skill learning.

It has been proposed that disinhibiton contributes to the reorganization of primary motor cortex, either by revealing latent horizontal intracortical connections or through facilitation of synaptic potentiation (Jacobs and Donoghue, 1991). Support for this idea comes from experiments in which pharmacological disinhibition of motor cortex leads to reorganization of motor maps (Jacobs and Donoghue, 1991). It is surprising, therefore, that there is little direct evidence that disinhibition of primary motor cortex occurs in behaving animals during motor skill learning. There is a pressing need for these experiments to be done.

Motor learning and memory is facilitated by neuromodulatory inputs. In particular, lesioning the nucleus basalis, which removes cholinergic afferents to the neocortex, abolishes motor skill learning (Ramanathan *et al.*, 2009). It remains to be determined whether it is just cholinergic inputs to primary motor cortex that are central to motor skill learning or whether cholinergic inputs also mediate their effects through other brain regions.

4. Experience-dependent plasticity

Experience-dependent plasticity occurs when the brain adapts to changes in an organism's environment. Sensory systems provide a host of models for studying the mechanistic basis for experience-dependent plasticity. This is because the neocortex harbours orderly topographic representations of the sensory periphery, termed cortical maps, which reorganize when sensory experience is altered persistently. Typically, senses have more than one cortical map, each in a different subdivision of sensory cortex (Alonso *et al.*, 2008; Kaas, 1987). However, mechanistic investigations have usually concentrated on reorganization of the map in primary sensory cortex.

Two major factors, developmental status and experimental paradigm, should be considered carefully when comparing experience dependent plasticity with neocortical learning and memory. First, experience-dependent plasticity is greatest during developmental critical periods, when neural circuits undergo pronounced reconfiguration (Buonomano and Merzenich, 1998; Feldman, 2009; Fox and Wong, 2005). Plasticity persists in adulthood after critical periods have closed and neural circuits are established, although at a lower level. Therefore, cortical reorganization in adulthood has to address a different set of issues from those that drive critical periods. Essentially, adult cortical plasticity involves modifying neural circuits that are established whereas critical period plasticity refines neural circuits that are not fully developed with the goal of optimizing their output. Hence, it is not reasonable to assume that the mechanistic basis for adult cortical plasticity is the same as critical period plasticity, but just at lower levels. Second, the nature of the experimental paradigm is an important factor. Lesioning the sensory periphery and non-traumatic sensory deprivation both result in cortical reorganization. However, there are important differences in the mechanistic basis for cortical plasticity induced by deafferentation compared with nontraumatic sensory deprivation (Barnes and Finnerty, 2010; Jones, 2000). Therefore, in this chapter, we restrict ourselves to mechanisms that underlie experience-dependent plasticity induced non-traumatically after developmental critical periods have been completed.

Space allocation in the neocortex does not simply reflect the area of the sensory input. For instance, in sensorimotor cortex, the face has a large representation whereas the trunk and legs have small representations. A widely held belief is that greater use or salience of a sensory input results in expansion of its sensory representation in the neocortex. This is supported by the finding that repetitive tactile stimulation as part of a training protocol results in expansion of the cortical representation of the stimulated skin region (Recanzone *et al.*, 1992a; Recanzone *et al.*, 1992b). Similarly, loss of a sensory input, which renders that less salient, results in expansion of representations of neighbouring retained sensory inputs into the cortical territory previously used to process the sensory inputs that were lost (Buonomano and Merzenich, 1998; Feldman and Brecht, 2005; Fox and Wong, 2005). However, the view that bigger representations are better has been challenged recently (Frostig, 2006). In the somatosensory system, enhanced sensory experience can result in expanded or contracted representations depending on the behavioural context (Polley *et al.*, 1999; Polley *et al.*, 2004). With these issues in mind, we ask which mechanisms from the plasticity toolbox are implemented during experience-dependent plasticity.

4.1. *Synaptic plasticity in excitatory circuits*

There is good evidence that experience-dependent plasticity is accompanied by changes in the strength of excitatory connections. Rodents have a map of their whisker sensory input in primary somatosensory cortex (SI). Plasticity of the SI whisker map can be induced by trimming a subset of rodents' whiskers. Electrophysiological recordings from pairs of neurons in layer 2/3 of SI that has retained its whisker sensory input show that excitatory connections between neighbouring pyramidal neurons become progressively stronger over several weeks (Cheetham *et al.*, 2007). *In vivo* imaging suggests that spine turnover is increased by altering sensory input (Trachtenberg *et al.*, 2002). However, strengthening is not accompanied by an increase in synapse number suggesting that individual synapses are potentiated (Cheetham *et al.*, 2007).

4.2. *Rewiring of excitatory circuitry*

It has become clear recently that the experience-dependent plasticity leads to rewiring of the connections between neurons. The axons and dendrites of neurons can be thought of as the wiring in the brain (Cherniak, 1992; Chklovskii *et al.*, 2004). Hence, rewiring can be defined as any structural alteration to the axonal arbour or dendritic tree that leads to a change in the number of spatial pattern of functioning synapses between neurons (Barnes and Finnerty, 2010). In essence, rewiring is a reconfiguration of the connections between neurons.

In vivo imaging studies have shown that experience-dependent plasticity is associated with increased turnover of dendritic spines (Holtmaat *et al.*, 2005; Majewska *et al.*, 2006; Trachtenberg *et al.*, 2002; Zuo *et al.*, 2005) and remodelling of axonal arbours (De Paola *et al.*, 2006; Marik *et al.*, 2010; Stettler *et al.*, 2006). In adult cortex, new synapses are formed by a dendritic spine growing towards an axonal bouton (Knott *et al.*, 2006). Thus, increased turnover of dendritic spines and axonal boutons provides a physical substrate for rewiring through elimination of old synapses and formation of new synapses.

Growth of the axonal arbour has been proposed to contribute to cortical reorganization in two ways. First, axonal arbours could expand into new territory and form new connections (Marik *et al.*, 2010). Alternatively, local axonal remodelling could lead to rewiring of existing connections (Cheetham *et al.*, 2008). Both suggestions have been studied in SI after trimming a subset of rodents' whiskers. Whisker trimming causes part of SI to retain its normal whisker sensory input (spared) whereas part of SI loses its principal sensory input (deprived). *In vivo* imaging has shown that the axons of layer 2/3 pyramidal neurons in spared cortex invade deprived cortex over weeks (Marik *et al.*, 2010). This long-range axonal growth that crosses cortical columns could contribute to long-term expansion of spared whisker representations into deprived cortex.

Local remodelling of the axonal arbour has been postulated to be involved in memory of previous somatosensory experience (Cheetham *et al.*, 2008). Rodents can learn to perform discrimination tasks with only a subset of their whiskers after they have been trimmed. If their whiskers are allowed to regrow back to normal size and then trimmed again, then the rodent will relearn the same task faster than on the first occasion (Guic-Robles *et al.*, 1992; Harris *et al.*, 1999; Harris *et al.*, 2001; Knutsen *et al.*, 2006). The axonal remodelling of layer 2/3 pyramidal neurons that accompanies whisker trimming may contribute to memory of previous sensory experience because the remodelling is targeted (Cheetham *et al.*, 2008). Specifically, the axonal growth increases the number of regions where the axon is close to the postsynaptic dendrite of neighbouring pyramidal neurons (Figures. 8.1A, B). Hence, there are a greater number of sites where synapses could be formed between connected neurons (Figure. 8.1B). The outcome of axonal growth is not an increase synapse number per connection, which remains constant. Instead, the proposal is that the axonal remodelling enables some of the synapses at strengthened connections to relocate to different parts of the postsynaptic dendritic tree (Figure. 8.1B). Relocating the synapses on the dendritic tree is worthwhile because dendrites are active, not passive structures. Hence, relocating synapses to a different dendritic compartment enables synaptic responses to be modified. When the trimmed whiskers regrow, the connections

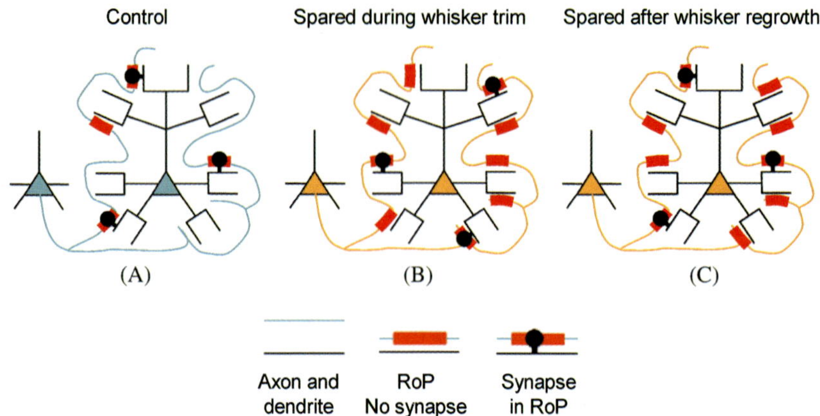

Figure 8.1. A model of synaptic rewiring after whisker trimming is shown. The model shows the synaptic connections and regions of proximity (RoPs) between two pyramidal neurons in somatosensory cortex before or after whisker trimming. This figure is reprinted from Cheetham *et al.*, 2008.

that were strengthened following whisker trimming weaken to regain their original strength and the pattern of synapses returns to the pre-trimmed configuration (Figure. 8.1C). Importantly, the axon that grew during the phase of axonal remodelling does not retract. This unretracted axon acts as an anatomical memory trace of previous experience. Hence, if whisker deprivation is repeated at a later time, new synapses can grow at the regions of proximity that harboured synapses during the first episode of whisker trimming. Thus, synapses can resume the "whisker trimmed" configuration more rapidly than during the first episode of whisker trimming. This manifests experimentally as a shorter time to adapt to whisker trimming. However, it remains unclear whether disinhibition lasting for hours to days is necessary for cortical reoganization in beharing animals. An alternative possibility is that brain olasticity *in vivo* only requires transient disinhibition and that this occurs on the background of other mechanisms that maintain the overall balance between excitation and inhibition.

4.3. *Inhibitory circuitry*

Much work has concentrated on changes to excitatory circuitry. In contrast, less attention has been paid to modifications to inhibitory circuitry in adult neocortex following experience-dependent plasticity, learning and memory. Interneurons represent the 20–30% of the cortical cells population. The majority are inhibitory neurons use GABA as their neurotransmitter (Markram *et al.*, 2004). Mature interneurons typically have aspiny dendrites and usually project inside the column or laterally within the same cortical lamina but rarely outside the neocortex. Hence, they are principally involved in cortical microcircuits. Interneurons are commonly classified into different subtypes on the basis of several features including their morphology, electrical properties, expression of molecular markers, and the site of inhibitory synapses formed by the interneuron with the postsynaptic neuron, e.g. somatic, dendritic (Kawaguchi, 1995; Markram *et al.*, 2004).

Disinhibition has been proposed to play a role in expansion of cortical maps (Jones, 1993). Early support for this idea came from experiments that measured anatomical markers of inhibition. Expression of the inhibitory neurotransmitter, GABA, and its synthesizing enzyme, glutamic acid decarboxylase, were decreased in deprived cortex (Akhtar and Land, 1991; Hendry and Jones, 1988). However, this takes several weeks to develop with nontraumatic sensory deprivation (Akhtar and Land, 1991; Hendry and Jones, 1988). Furthermore, metabolic imaging indicates that interneurons in both spared and deprived whisker barrel cortex remain highly active during the early stages of cortical map reorganization suggesting that intracortical inhibitory circuitry is not persistently weak (Maier *et al.*, 2003).

Trimming a subset of rodents' whiskers has been reported to increase neuronal spiking in spared SI of behaving animals (Kelly *et al.*, 1999). It has been suggested that the increased spiking may be attributable to disinhibition. This may be mediated by top-down inputs to SI (Hsiao *et al.*, 1993; Polley *et al.*, 2006). Neuromodulatory inputs may be a further contributory factor. Certainly, the cholinergic input to the neocortex from the nucleus basalis plays a role in cortical map reorganization (Juliano *et al.*, 1991; Kilgard and Merzenich, 1998). Electrophysiological recordings show that repetitive activation of the cholinergic inputs results in increased excitatory activity and decreased inhibitory activity (Froemke *et al.*, 2007).

Further supportive evidence for a role of disinhibition in cortical map plasticity comes from experiments that reactivate cortical plasticity in adulthood. Injection of an enzyme, chondroitinase ABC, into visual cortex degrades the extracellular matrix and breaks down the perineuronal nets that surround fast spiking interneurons. This is accompanied by the re-emergence of ocular dominance plasticity in adult animals (Pizzorusso *et al.*, 2002).

Similarly, the anti-depressant, fluoxetine, has been reported to reactivate ocular dominance plasticity in adulthood (Maya Vetencourt *et al.*, 2008). Fluoxetine-induced plasticity is associated with reduced extracellular GABA levels, increased expression of BDNF and can be completely abolished by treatment with diazepam (Maya Vetencourt *et al.*, 2008). Finally, infusion of a

$GABA_A$ receptor antagonist into the visual cortex partially reactivates ocular dominance plasticity (Harauzov *et al.*, 2010). Collectively, these data indicate that persistent disinhibition can promote experience-dependent plasticity in adult neocortex. However, it remains unclear whether disinhibition lasting for hours to days is necessary for cortical reorganization in behaving animals. An alternative possibility is that brain plasticity *in vivo* only requires transient disinhibition and that this occurs on the background of other mechanisms that maintain the overall balance between excitation and inhibition.

Far less attention has been paid to the effects of sensory overstimulation. Repetitive passive stimulation of a whisker for 24 h reduces evoked spiking in layer 4 of the whisker's barrel column and is associated with a persistent increase in dendritic spines in layer 4 with both an excitatory and an inhibitory synapse (Knott *et al.*, 2002).

5. Relationship between experience-dependent plasticity and neocortical learning and memory

Investigations of experience-dependent plasticity typically modify sensory inputs to induce cortical map plasticity. The behavioural ramifications of the cortical map plasticity are not usually studied. However, when behaviour is probed, it becomes evident that sensory deprivation leaves a memory trace in the brain. This may be thought of as similar to the bottom-up plasticity that occurs during exposure-based perceptual learning and memory. However, this view may be unnecessarily narrow. Many sensory systems are not simply passive acceptors of information from the environment. Instead, acquisition of sensory information is an active process. For instance, rats move their whiskers back and forth when using them to exploring objects or surfaces. Active sensing systems are likely to engage top-down inputs such as attention, particularly when sensory deprived, e.g. by whisker trimming. Hence, there may be greater overlap between adult experience-dependent plasticity and procedural learning and memory than is commonly thought.

There has been considerable debate about the role of cortical map expansion in procedural learning and memory. Cortical map expansion has been shown after training on a behavioural task (Nudo *et al.*, 1996; Polley *et al.*, 2006; Recanzone *et al.*, 1992a; Recanzone *et al.*, 1992b). However, some reports have indicated that learning persists even though the cortical map retracts back to normal dimensions over days to weeks (Molina-Luna *et al.*, 2008; Reed *et al.*, 2011; Yotsumoto *et al.*, 2008). This raises the question about the cortical map expansion is an epiphenomenon that is irrelevant to learning and memory. In contrast, others have argued that expansion of cortical representations is directly related to the strength of sensory memory (Bieszczad and Weinberger, 2010). These findings are not irreconcilable. Cortical map expansion may be a manifestation of reorganization of cortical microcircuits. Retraction of the map could occur when the new circuits have been reconfigured. However, there may be cases where a larger population of neurons is required to maintain a memory. Under those circumstances, cortical maps may be persistently enlarged.

6. Conclusion

Many studies of neocortical learning and memory have focused on the neocortical regions that are involved and the connections between those regions. This has advanced knowledge of learning and memory at the network level. Little is known about the cellular basis for neocortical learning and memory. In particular, the contributions of changes in synaptic strength, rewiring and altered neuronal excitability have yet to be teased apart. The overlap between adult experience-dependent plasticity and procedural learning and memory suggests that insights into the mechanisms underlying experience-dependent plasticity can guide future experiments into the cellular basis for neocortical learning and memory.

References

Ahissar, M., and Hochstein, S. (1993). Attentional control of early perceptual learning. *Proc. Natl Acad Sci USA* **90**, 5718–5722.

Ahissar, M., and Hochstein, S. (1997). Task difficulty and the specificity of perceptual learning. *Nature* **387**, 401–406.

Ahissar, M. and Hochstein, S. (2004). The reverse hierarchy theory of visual perceptual learning. *Trends Cogn Sci* **8**, 457–464.

Akhtar, N.D. and Land, P.W. (1991). Activity-dependent regulation of glutamic acid decarboxylase in the rat barrel cortex: Effects of neonatal versus adult sensory deprivation. *J Comp Neurol* **307**, 200–213.

Alonso, B.D., Lowe, A.S., Dear, J.P., Lee, K.C., Williams, S.C. and Finnerty, G.T. (2008). Sensory inputs from whisking movements modify cortical whisker maps visualized with functional magnetic resonance imaging. *Cereb Cortex* **18**, 1314–1325.

Barnes, S.J. and Finnerty, G.T. (2010). Sensory experience and cortical rewiring. *Neuroscientist* **16**, 186–198.

Bieszczad, K.M. and Weinberger, N.M. (2010). Representational gain in cortical area underlies increase of memory strength. *Proc Natl Acad Sci USA* **107**, 3793–3798.

Brashers-Krug, T., Shadmehr, R. and Bizzi, E. (1996). Consolidation in human motor memory. *Nature* **382**, 252–255.

Buchel, C., Coull, J.T. and Friston, K.J. (1999). The predictive value of changes in effective connectivity for human learning. *Science* **283**, 1538–1541.

Buonomano, D.V. and Merzenich, M.M. (1998). Cortical plasticity: From synapses to maps. *Annu Rev Neurosci* **21**, 149–186.

Cheetham, C.E., Hammond, M.S., Edwards, C.E. and Finnerty, G.T. (2007). Sensory experience alters cortical connectivity and synaptic function site specifically. *J Neurosci* **27**, 3456–3465.

Cheetham, C.E., Hammond, M.S., McFarlane, R. and Finnerty, G.T. (2008). Altered sensory experience induces targeted rewiring of local excitatory connections in mature neocortex. *J Neurosci* **28**, 9249–9260.

Cherniak, C. (1992). Local optimization of neuron arbors. *Biol Cybern* **66**, 503–510.

Chklovskii, D.B., Mel, B.W. and Svoboda, K. (2004). Cortical rewiring and information storage. *Nature* **431**, 782–788.

Corbetta, M. and Shulman, G.L. (2002). Control of goal-directed and stimulus-driven attention in the brain. *Nat Rev Neurosci* **3**, 201–215.

De Paola, V., Holtmaat, A., Knott, G., Song, S., Wilbrecht, L., Caroni, P. and Svoboda, K. (2006). Cell type-specific structural plasticity of axonal branches and boutons in the adult neocortex. *Neuron* **49**, 861–875.

Dinse, H.R., Ragert, P., Pleger, B., Schwenkreis, P. and Tegenthoff, M. (2003). Pharmacological modulation of perceptual learning and associated cortical reorganization. *Science* **301**, 91–94.

Dosher, B.A. and Lu, Z.L. (1998). Perceptual learning reflects external noise filtering and internal noise reduction through channel reweighting. *Proc Natl Acad Sci USA* **95**, 13988–13993.

Dudai, Y. (1989). *The Neurobiology of Memory: Concepts, Findings, Trends*. (Oxford, UK: Oxford University Press).

Dudai, Y. (2004). The neurobiology of consolidations, or, how stable is the engram? *Annu Rev Psychol* **55**, 51–86.

Ebbinghaus, H. (1885). Über das Gedächtnis. (English translation (1913) entitled '*Memory: A Contribution to Experimental Psychology*'). (New York, NY: Dover).

Feldman, D.E. (2009). Synaptic mechanisms for plasticity in neocortex. *Annu Rev Neurosci* **32**, 33–55.

Feldman, D.E. and Brecht, M. (2005). Map plasticty in somatosensory cortex. *Science* **310**, 810–815.

Fox, K. and Wong, R.O. (2005). A comparison of experience-dependent plasticity in the visual and somatosensory systems. *Neuron* **48**, 465–477.

Frankland, P.W. and Bontempi, B. (2005). The organization of recent and remote memories. *Nat Rev Neurosci* **6**, 119–130.

Frenkel, M.Y., Sawtell, N.B., Diogo, A.C., Yoon, B., Neve, R.L. and Bear, M.F. (2006). Instructive effect of visual experience in mouse visual cortex. *Neuron* **51**, 339–349.

Froemke, R.C., Merzenich, M.M. and Schreiner, C.E. (2007). A synaptic memory trace for cortical receptive field plasticity. *Nature* **450**, 425–429.

Frostig, R.D. (2006). Functional organization and plasticity in the adult rat barrel cortex: Moving out-of-the-box. *Curr Opin Neurobiol* **16**, 445–450.

Furmanski, C.S., Schluppeck, D. and Engel, S.A. (2004). Learning strengthens the response of primary visual cortex to simple patterns. *Curr Biol* **14**, 573–578.

Ghose, G.M., Yang, T. and Maunsell, J.H. (2002). Physiological correlates of perceptual learning in monkey V1 and V2. *J Neurophysiol* **87**, 1867–1888.

Gilbert, C.D., Sigman, M. and Crist, R.E. (2001). The neural basis of perceptual learning. *Neuron* **31**, 681–697.

Goldstone, R.L. (1998). Perceptual learning. *Annu Rev Psychol* **49**, 585–612.

Guic-Robles, E., Jenkins, W.M. and Bravo, H. (1992). Vibrissal roughness discrimination is barrelcortex-dependent. *Behav Brain Res* **48**, 145–152.

Guic-Robles, E., Valdivieso, C. and Guajardo, G. (1989). Rats can learn a roughness discrimination using only their vibrissal system. *Behav Brain Res* **31**, 285–289.

Gutnisky, D.A., Hansen, B.J., Iliescu, B.F. and Dragoi, V. (2009). Attention alters visual plasticity during exposure-based learning. *Curr Biol* **19**, 555–560.

Harauzov, A., Spolidoro, M., DiCristo, G., De, P.R., Cancedda, L., Pizzorusso, T., Viegi, A., Berardi, N. and Maffei, L. (2010). Reducing intracortical inhibition in the adult visual cortex promotes ocular dominance plasticity. *J Neurosci* **30**, 361–371.

Harris, J.A., Petersen, R.S. and Diamond, M.E. (1999). Distribution of tactile learning and its neural basis. *Proc Natl Acad Sci USA* **96**, 7587–7591.

Harris, J.A., Petersen, R.S. and Diamond, M.E. (2001). The cortical distribution of sensory memories. *Neuron* **30**, 315–318.

Hendry, S.H. and Jones, E.G. (1988). Activity-dependent regulation of GABA expression in the visual cortex of adult monkeys. *Neuron* **1**, 701–712.

Hofer, S.B., Mrsic-Flogel, T.D., Bonhoeffer, T. and Hubener, M. (2006). Prior experience enhances plasticity in adult visual cortex. *Nat Neurosci* **9**, 127–132.

Hofer, S.B., Mrsic-Flogel, T.D., Bonhoeffer, T. and Hubener, M. (2009). Experience leaves a lasting structural trace in cortical circuits. *Nature* **457**, 313–317.

Holtmaat, A.J.G.D., Trachtenberg, J.T., Wilbrecht, L., Shepherd, G.M., Zhang, X.Q., Knott, G.W. and Svoboda, K. (2005). Transient and persistent dendritic spines in the neocortex *in vivo*. *Neuron* **45**, 279–291.

Hsiao, S.S., O'Shaughnessy, D.M. and Johnson, K.O. (1993). Effects of selective attention on spatial form processing in monkey primary and secondary somatosensory cortex. *J Neurophysiol* **70**, 444–447.

Jacobs, K.M. and Donoghue, J.P. (1991). Reshaping the cortical motor map by unmasking latent intracortical connections. *Science* **251**, 944–947.

Jones, E.G. (1993). GABAergic neurons and their role in cortical plasticity in primates. Cereb *Cortex* **3**, 361–372.

Jones, E.G. (2000). Cortical and subcortical contributions to activity-dependent plasticity in primate somatosensory cortex. *Annu Rev Neurosci* **23**, 1–37.

Juliano, S.L., Ma, W. and Eslin, D. (1991). Cholinergic depletion prevents expansion of topographic maps in somatosensory cortex. *Proc Natl Acad Sci USA* **88**, 780–784.

Kaas, J.H. (1987). The organization of cortex in mammals: Implications for theories of brain function. *Annu Rev Psychol* **38**, 124–151.

Karni, A., Meyer, G., Jezzard, P., Adams, M.M., Turner, R. and Ungerleider, L.G. (1995). Functional MRI evidence for adult motor cortex plasticity during motor skill learning. *Nature* **377**, 155–158.

Karni, A., Meyer, G., Rey-Hipolito, C., Jezzard, P., Adams, M.M., Turner, R. and Ungerleider, L.G. (1998). The acquisition of skilled motor performance: Fast and slow experience-driven changes in primary motor cortex. *Proc Natl Acad Sci USA* **95**, 861–868.

Karni, A. and Sagi, D. (1991). Where practice makes perfect in texture discrimination: Evidence for primary visual cortex plasticity. *Proc Natl Acad Sci USA* **88**, 4966–4970.

Kawaguchi, Y. (1995). Physiological subgroups of nonpyramidal cells with specific morphological characteristics in layer II/III of rat frontal cortex. *J Neurosci* **15**, 2638–2655.

Kelly, M.K., Carvell, G.E., Kodger, J.M. and Simons, D.J. (1999). Sensory loss by selected whisker removal produces immediate disinhibition in the somatosensory cortex of behaving rats. *J Neurosci* **19**, 9117–9125.

Kilgard, M.P. and Merzenich, M.M. (1998). Cortical map reorganization enabled by nucleus basalis activity. *Science* **279**, 1714–1718.

Kleim, J.A., Hogg, T.M., VandenBerg, P.M., Cooper, N.R., Bruneau, R. and Remple, M. (2004). Cortical synaptogenesis and motor map reorganization occur during late, but not early, phase of motor skill learning. *J Neurosci* **24**, 628–633.

Kleim, J.A., Lussnig, E., Schwarz, E.R., Comery, T.A. and Greenough, W.T. (1996). Synaptogenesis and Fos expression in the motor cortex of the adult rat after motor skill learning. *J Neurosci* **16**, 4529–4535.

Knott, G.W., Holtmaat, A., Wilbrecht, L., Welker, E. and Svoboda, K. (2006). Spine growth precedes synapse formation in the adult neocortex *in vivo*. *Nat Neurosci* **9**, 1117–1124.

Knott, G.W., Quairiaux, C., Genoud, C. and Welker, E. (2002). Formation of dendritic spines with GABAergic synapses induced by whisker stimulation in adult mice. *Neuron* **34**, 265–273.

Knutsen, P.M., Pietr, M. and Ahissar, E. (2006). Haptic object localization in the vibrissal system: Behavior and performance. *J Neurosci* **26**, 8451–8464.

Komiyama, T., Sato, T.R., O'Connor, D.H., Zhang, Y.X., Huber, D., Hooks, B.M., Gabitto, M. and Svoboda, K. (2010). Learning-related fine-scale specificity imaged in motor cortex circuits of behaving mice. *Nature* **464**, 1182–1186.

Krakauer, J.W. and Shadmehr, R. (2006). Consolidation of motor memory. *Trends Neurosci* **29**, 58–64.

Lashley, K.S. (1950). In search of the engram. *Soc Exp Biol Symp* **4**, 454–482.

Law, C.T. and Gold, J.I. (2008). Neural correlates of perceptual learning in a sensory-motor, but not a sensory, cortical area. *Nat Neurosci* **11**, 505–513.

Lewis, C.M., Baldassarre, A., Committeri, G., Romani, G.L. and Corbetta, M. (2009). Learning sculpts the spontaneous activity of the resting human brain. *Proc Natl Acad Sci USA* **106**, 17558–17563.

Li, W., Luxenberg, E., Parrish, T. and Gottfried, J.A. (2006). Learning to smell the roses: Experience-dependent neural plasticity in human piriform and orbitofrontal cortices. *Neuron* **52**, 1097–1108.

Logothetis, N.K. and Wandell, B.A. (2004). Interpreting the BOLD signal. *Annu Rev Physiol* **66**, 735–769.

Maier, D.L., Grieb, G.M., Stelzner, D.J. and McCasland, J.S. (2003). Large-scale plasticity in barrel cortex following repeated whisker trimming in young adult hamsters. *Exp Neurol* **184**, 737–745.

Majewska, A.K., Newton, J.R. and Sur, M. (2006). Remodeling of synaptic structure in sensory cortical areas *in vivo*. *J Neurosci* **26**, 3021–3029.

Marik, S.A., Yamahachi, H., McManus, J.N., Szabo, G. and Gilbert, C.D. (2010). Axonal dynamics of excitatory and inhibitory neurons in somatosensory cortex. *PLoS Biol* **8**, e1000395.

Markram, H., Toledo-Rodriguez, M., Wang, Y., Gupta, A., Silberberg, G. and Wu, C.Z. (2004). Interneurons of the neocortical inhibitory system. *Nat Rev Neurosci* **5**, 793–807.

Maya Vetencourt, J.F., Sale, A., Viegi, A., Baroncelli, L., De, P.R., O'Leary, O.F., Castren, E. and Maffei, L. (2008). The antidepressant fluoxetine restores plasticity in the adult visual cortex. *Science* **320**, 385–388.

Molina-Luna, K., Hertler, B., Buitrago, M.M. and Luft, A.R. (2008). Motor learning transiently changes cortical somatotopy. *Neuroimage* **40**, 1748–1754.

Muellbacher, W., Ziemann, U., Wissel, J., Dang, N., Kofler, M., Facchini, S., Boroojerdi, B., Poewe, W. and Hallett, M. (2002). Early consolidation in human primary motor cortex. *Nature* **415**, 640–644.

Nudo, R.J., Milliken, G.W., Jenkins, W.M. and Merzenich, M.M. (1996). Use-dependent alterations of movement representations in primary motor cortex of adult squirrel monkeys. *J Neurosci* **16**, 785–807.

Ogawa, S., Lee, T.M., Kay, A.R. and Tank, D.W. (1990). Brain magnetic resonance imaging with contrast dependent on blood oxygenation. *Proc Natl Acad Sci USA* **87**, 9868–9872.

Pastalkova, E., Serrano, P., Pinkhasova, D., Wallace, E., Fenton, A.A. and Sacktor, T.C. (2006). Storage of spatial information by the maintenance mechanism of LTP. *Science* **313**, 1141–1144.

Pizzorusso, T., Medini, P., Berardi, N., Chierzi, S., Fawcett, J.W. and Maffei, L. (2002). Reactivation of ocular dominance plasticity in the adult visual cortex. *Science* **298**, 1248–1251.

Polley, D.B., Chen-Bee, C.H. and Frostig, R. (1999). Two directions of plasticity in sensory-deprived adult cortex. *Neuron* **24**, 623–637.

Polley, D.B., Kvasnak, E. and Frostig, R.D. (2004). Naturalistic experience transforms sensory maps in the adult cortex of caged animals. *Nature* **429**, 67–71.

Polley, D.B., Steinberg, E.E. and Merzenich, M.M. (2006). Perceptual learning directs auditory cortical map reorganization through top-down influences. *J Neurosci* **26**, 4970–4982.

Ramanathan, D., Tuszynski, M.H. and Conner, J.M. (2009). The basal forebrain cholinergic system is required specifically for behaviorally mediated cortical map plasticity. *J Neurosci* **29**, 5992–6000.

Recanzone, G.H., Merzenich, M.M. and Jenkins, W.M. (1992a). Frequency discrimination training engaging a restricted skin surface results in an emergence of a cutaneous response zone in cortical area 3a. *J Neurophysiol* **67**, 1057–1070.

Recanzone, G.H., Merzenich, M.M., Jenkins, W.M., Grajski, K.A. and Dinse, H.R. (1992b). Topographic reorganization of the hand representation in cortical area 3b owl monkeys trained in a frequency-discrimination task. *J Neurophysiol* **67**, 1031–1056.

Reed, A., Riley, J., Carraway, R., Carrasco, A., Perez, C., Jakkamsetti, V. and Kilgard, M.P. (2011). Cortical map plasticity improves learning but is not necessary for improved performance. *Neuron* **70**, 121–131.

Rioult-Pedotti, M.S., Friedman, D. and Donoghue, J.P. (2000). Learning-induced LTP in neocortex. *Science* **290**, 533–536.

Rioult-Pedotti, M.S., Friedman, D., Hess, G. and Donoghue, J.P. (1998). Strengthening of horizontal cortical connections following skill learning. *Nat Neurosci* **1**, 230–234.

Sacco, T. and Sacchetti, B. (2010). Role of secondary sensory cortices in emotional memory storage and retrieval in rats. *Science* **329**, 649–656.

Sasaki, Y., Nanez, J.E. and Watanabe, T. (2010). Advances in visual perceptual learning and plasticity. *Nat Rev Neurosci* **11**, 53–60.

Schoups, A., Vogels, R., Qian, N. and Orban, G. (2001). Practising orientation identification improves orientation coding in V1 neurons. *Nature* **412**, 549–553.

Schwartz, S., Maquet, P. and Frith, C. (2002). Neural correlates of perceptual learning: A functional MRI study of visual texture discrimination. *Proc Natl Acad Sci USA* **99**, 17137–17142.

Shema, R., Haramati, S., Ron, S., Hazvi, S., Chen, A., Sacktor, T.C. and Dudai, Y. (2011). Enhancement of consolidated long-term memory by over expression of protein kinase Mzeta in the neocortex. *Science* **331**, 1207–1210.

Shema, R., Sacktor, T.C. and Dudai, Y. (2007). Rapid erasure of long-term memory associations in the cortex by an inhibitor of PKM zeta. *Science* **317**, 951–953.

Silva, A.J., Zhou, Y., Rogerson, T., Shobe, J. and Balaji, J. (2009). Molecular and cellular approaches to memory allocation in neural circuits. *Science* **326**, 391–395.

Squire, L.R. (1987). *Memory and Brain*. (New York, USA: Oxford University Press).

Stettler, D.D., Yamahachi, H., Li, W., Denk, W. and Gilbert, C.D. (2006). Axons and synaptic boutons are highly dynamic in adult visual cortex. *Neuron* **49**, 877–887.

Trachtenberg, J.T., Chen, B.E., Knott, G.W., Feng, G., Sanes, J.R., Welker, E. and Svoboda, K. (2002). Long-term *in vivo* imaging of experience-dependent synaptic plasticity in adult cortex. *Nature* **420**, 788–794.

Ungerleider, L.G., Doyon, J. and Karni, A. (2002). Imaging brain plasticity during motor skill learning. *Neurobiol Learn Mem* **78**, 553–564.

Watanabe, T., Nanez, J.E. and Sasaki, Y. (2001). Perceptual learning without perception. *Nature* **413**, 844–848.

Xu, T., Yu, X., Perlik, A.J., Tobin, W.F., Zweig, J.A., Tennant, K., Jones, T. and Zuo, Y. (2009). Rapid formation and selective stabilization of synapses for enduring motor memories. *Nature* **462**, 915–919.

Yang, G., Pan, F. and Gan, W.B. (2009). Stably maintained dendritic spines are associated with lifelong memories. *Nature* **462**, 920–924.

Yang, T. and Maunsell, J.H. (2004). The effect of perceptual learning on neuronal responses in monkey visual area V4. *J Neurosci* **24**, 1617–1626.

Yotsumoto, Y., Watanabe, T. and Sasaki, Y. (2008). Different dynamics of performance and brain activation in the time course of perceptual learning. *Neuron* **57**, 827–833.

Zuo, Y., Lin, A., Chang, P. and Gan, W.B. (2005). Development of long-term dendritic spine stability in diverse regions of cerebral cortex. *Neuron* **46**, 181–189.

Mechanisms of Memory Storage 9

Kimberly Lackenby* and Karl Peter Giese*

1. Introduction

Long-term memories can last a lifetime. How are such memories stored? Protein synthesis-dependent memory consolidation (Chapter 5) is not sufficient for storing long-term memory. This is because consolidation is a transient, and not a persistent, process that is essential for memory storage. This chapter will discuss candidate persistent mechanisms that are thought to mediate memory storage.

To understand memory storage, knowledge of the cellular mechanism(s) underlying the formation of memory is helpful. For example, if the establishment of new synaptic connections between neurons were critical for memory formation (Chapters 2 and 10), then these connections should be maintained, if the memory is to last long time. In principle, no memory-specific maintenance mechanism is required for this, as many non-plastic synaptic connections in the nervous system are maintained throughout the lifespan. Thus, basic maintenance mechanisms may be sufficient for the persistence of newly formed synapses. Nonetheless, studies with cultured neurons from *Aplysia* have suggested that a prion-like mechanism may maintain newly formed synapses and memory (Bailey *et al.*, 2004). The prion-like protein is the cytoplasmic polyadenylation element-binding protein (CPEB) that promotes local translation of mRNA encoding structural proteins, such as actin and tubulin, among other things. CPEB exists in an active or inactive conformational state. Active CPEB can activate inactive CPEB (like a prion protein that converts one conformation into another; however, in the case of prion protein, the converted form is inactive or pathological) providing a perpetuating mechanism for local mRNA translation of structural proteins at the synapse.

However, long-term memory formation involves strengthening of existing synapses (Chapter 1). To maintain synaptic strengthening, molecular signalling is required. This poses the problem that molecules have only a short lifetime

*King's College London, Institute of Psychiatry, Department of Neuroscience, 125 Coldharbour Lane, London, SE5 9NU, UK

(minutes to weeks), as molecules are turned over. Therefore, Francis Crick and John Lisman independently proposed that feed-forward signalling switches at existing synapses are critical for memory storage (Crick, 1984; Lisman, 1985). In this chapter we introduce candidate feed-forward mechanisms and discuss their importance in memory storage.

2. CaMKII and memory storage

In the mid 1980s, the calcium/calmodulin-dependent kinase II (CaMKII), a major synaptic kinase, was shown to switch its activity from Ca^{2+}-dependent to-independent activity upon autophosphorylation (Saitoh and Schwartz, 1985; Miller and Kennedy, 1986). Ca^{2+} activates CaMKII and strong Ca^{2+} signals induce the autophosphorylation. Because of this switch-like property and because the autophosphorylation represents a feed-forward signal that may overcome protein turnover (Figure 9.1), CaMKII was suggested to maintain synaptic strength and to mediate memory storage (Lisman *et al.*, 1994, 2002).

CaMKII is the major protein in the postsynaptic density of excitatory synapses (Lucchesi *et al.*, 2011). It has different subunits, of which the α- and β-subunits are most abundant in brain. Each subunit is composed of a catalytic domain, an association domain, and an inhibitory domain. Individual subunits assemble to form a holoenzyme of 12 subunits, arranged as two hexameric rings lying on top of each other. In the absence of Ca^{2+}, the inhibitory domain binds to the catalytic domain, preventing phosphorylation of substrates. Binding of Ca^{2+}/calmodulin to the inhibitory domain activates CaMKII by inducing a conformational change that exposes the catalytic domain for interaction with substrates. Active CaMKII can autophosphorylate within the inhibitory domain, at threonine (T)-286 in case of αCaMKII and at T287 of βCaMKII. These autophosphorylations occur within a holoenzyme between individual Ca^{2+}/calmodulin-bound subunits (inter-subunit reaction) (Lisman *et al.*, 2002). T286/T287 autophosphorylations entrap bound Ca^{2+}/calmodulin and they switch the kinase into an autonomous activity mode that persists when Ca^{2+} levels drop to baseline. Thus, it was suggested that CaMKII autophosphorylation remembers Ca^{2+} stimuli at synapses.

Hippocampus-dependent memory formation requires NMDA receptors (see Chapter 1). Activation of the NMDA receptor results in Ca^{2+} influx at glutamatergic synapses, which in turn activates CaMKII. Active CaMKII phosphorylates several proteins, but most importantly CaMKII phosphorylations induce trafficking of AMPA receptors. Consequently, the density of AMPA receptors increases at the synapse, leading to enhanced glutamatergic transmission. The Ca^{2+} stimulus

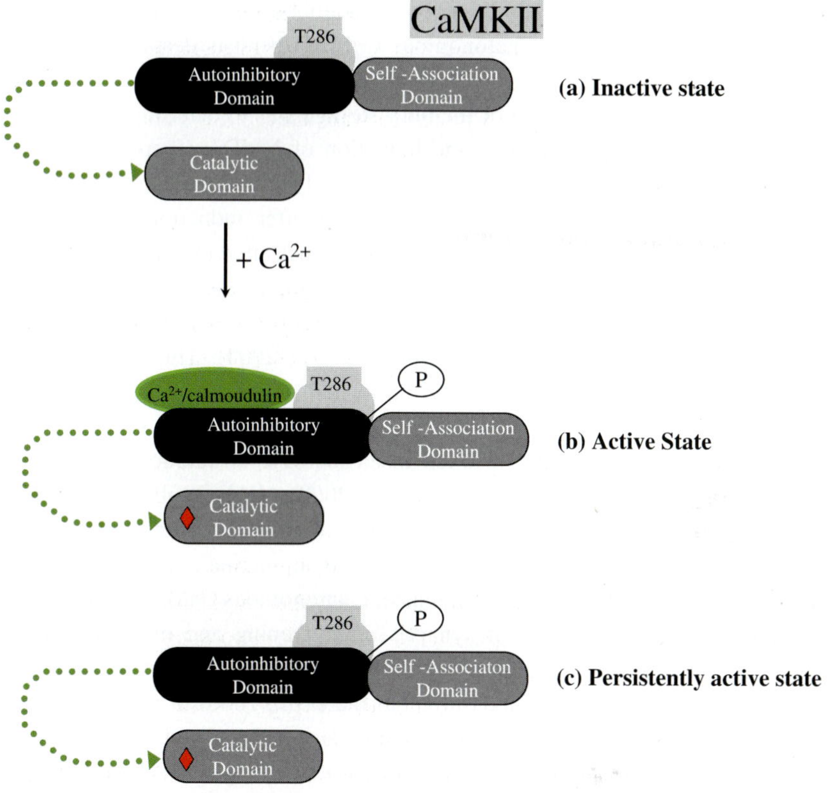

Figure 9.1. The model of how CaMKII activity persists. Ca^{2+}/calmodulin binds to the autoinhibitory domain; this induces a conformational change thus exposing the catalytic domain. In the active state, T286 can become autophosphorylated by neighbouring subunits, this enables CaMKII to remain active even after Ca^{2+} levels return to baseline and Ca^{2+}/calmodulin dissociates. The kinase is switched to autonomous activity that is Ca^{2+} independent and persists due to subunits re-phosphorylating each other.

through the NMDA receptor also leads to CaMKII autophosphorylation, switching CaMKII into an autonomous activity mode. It has been suggested that autonomous CaMKII activity persists as long as the synapse is strengthened (Lisman *et al.*, 2002). Thus, persistent CaMKII signalling was proposed to be critical for the maintenance of synaptic strengthening. Autonomous CaMKII activity might last for a lifetime, as it was suggested to overcome the molecular turnover problem (Lisman, 1994): Once an autophosphorylated subunit is turned over, a naïve subunit may be incorporated into the holoenzyme. Afterwards, the naïve isoform may be autophosphorylated by a neighbouring autonomously active subunit. Thus, the activity of autonomous CaMKII persists despite turnover of individual subunits.

The first test of this model of memory storage was to determine whether or not CaMKII activity persists after the induction of NMDA receptor-dependent synaptic strengthening in the hippocampus. An early study found that the autonomous CaMKII activity is induced immediately after induction of long-term potentiation (LTP) and that it persisted as long as measured (1 h after LTP induction) (Fukunaga *et al.*, 1993). However, two follow-up studies could not replicate this finding. In one of these studies, synaptic strengthening was induced by pharmacological block of K^+ channels (Lengyel *et al.*, 2004). This resulted in an increase autonomous CaMKII activity that lasted only up to 2 min. The other study monitored CaMKII activity with a reporter construct at a single synapse and found that LTP induction resulted in an increase of autonomous CaMKII activity for about one minute (Lee *et al.*, 2009). Additionally, autonomous CaMKII activity in the hippocampus was shown to be only transiently upregulated after training in a passive avoidance task (Cammarota *et al.*, 1998). Taken together, these findings suggest that persistent, autonomous CaMKII activity does not underlie the maintenance of synaptic strengthening and memory storage. Interestingly however, CaMKII autophosphorylation appears to persist after LTP induction as long as it can be measured (Irvine *et al.*, 2006). It remains unclear why persistent CaMKII autophosphorylation is not translated into persistent, autonomous CaMKII activity. One possibility is that endogenous CaMKII inhibitors block CaMKII activity whilst sparing the autophosphorylation (Lucchesi *et al.*, 2011).

αCaMKII autophosphorylation-deficient knock-in (T286A) mutants have been generated to study the function of the autophosphorylation in learning and memory (Giese *et al.*, 1998). These mutants express a normal level of αCaMKII that cannot be autophosphorylated. The T286A mutants have fully blocked NMDA receptor-dependent LTP at hippocampal CA1 synapses (Giese *et al.*, 1998; Cooke *et al.*, 2006), indicating that αCaMKII autophosphorylation is important for the induction of synaptic strengthening. However, since the T286A mice are not inducible mutants, these findings do not address whether αCaMKII autophosphorylation is also required for the maintenance of synaptic strengthening. Behavioural studies with the T286A mutants in aversive conditioning tasks were more informative about the role of αCaMKII autophosphorylation in memory storage. These studies showed that the αCaMKII autophosphorylation is required for memory formation after a single training trial, but not after multiple training trials (Irvine *et al.*, 2005, 2006, 2011). The memory after repeated training lasts for at least 4 weeks. Therefore, memory storage does not require αCaMKII autophosphorylation (and autonomous activity that requires the autophosphorylation). Instead, the prime function of αCaMKII autophosphorylation appears to be the enabling of memory formation after a single training trial, possibly by prolonging signalling at the synapse that may be needed for the induction of consolidation-specific gene transcription (Irvine *et al.*, 2006).

Pharmacological approaches were also used to address the hypothesis that autonomously, active CaMKII may mediate memory storage. Most of the previously used CaMKII blockers were unspecific so that no *bona fide* conclusion could be reached. However, recently a sufficiently specific CaMKII blocker, tatCN21, was developed. The peptide inhibitor tatCN21 blocks hippocampal LTP induction and contextual fear memory formation (Buard *et al.*, 2010). However, it does not block LTP maintenance or established contextual fear memory. The conclusions from these findings are in agreement with the analysis of the T286A mutants: autonomously active CaMKII is not required for memory storage.

Two alternative models have emerged of how CaMKII might contribute to memory storage. Both of these models focus rather on a long-lasting change in CaMKII signalling as the mechanism underlying memory storage. However, they do not take protein turnover into consideration. The first model suggests that upon synaptic stimulation, αCaMKII becomes autophosphorylated that would lead to a long-lasting translocation of CaMKII into the postsynaptic density (PSD). CaMKII translocation may be a biochemical memory that changes signalling at the synapse (Shen *et al.*, 2000). The experimental evidence for this model derives from studies with cultured neurons, a system that can be prone to artifacts. Furthermore, αCaMKII autophosphorylation-deficient mice can store memory, suggesting that CaMKII translocation is not required for memory storage. The second model proposes that synaptic stimulation leads to binding of CaMKII to NMDA receptor subunits and that this alters CaMKII signalling at the synapse. CaMKII binding to the NMDA receptor is potentiated by autophosphorylation, but it does not require the autophosphorylation (Leonard *et al.*, 2002). A high dose of the CaMKII inhibitor tatCN21 blocks CaMKII/NMDA receptor interaction and impairs LTP maintenance (Sanhueza *et al.*, 2011). Accordingly, there is evidence that the blocking of the CaMKII/NMDA receptor interaction impairs hippocampus-dependent memory formation (Zhou *et al.*, 2007). These findings suggest that CaMKII/NMDA receptor interaction may be important for memory formation. However, the critical behavioural studies to address the requirement for memory storage have not yet been performed, and it is unclear how CaMKII/NMDA receptor complexes outlast molecular turnover.

3. PKMζ and memory storage

There is overwhelming evidence that PKMζ, an atypical isoform of protein kinase C (PKC), is critical for memory storage (Sacktor, 2011). Normally, PKC isoforms have a regulatory domain that contains an autoinhibitory pseudosubstrate region to inhibit the catalytic domain. However, PKMζ lacks this inhibitory domain making it constitutively active. PKMζ is encoded by its own mRNA that is transported into dendrites of neurons. Local translation of the dendritic mRNA leads to PKMζ synthesis, providing the constitutively active kinase (Figure 9.2).

The development of a specific PKMζ inhibitory peptide, termed zeta inhibitory peptide (ZIP), was instrumental in demonstrating that PKMζ is essential for memory storage. In a landmark study, Pastalkova *et al.* showed that injection of ZIP into the hippocampus erases a consolidated, spatial long-term memory (Pastalkova *et al.*, 2006). The authors used an active, spatial avoidance task and they injected ZIP or saline 1 day or 30 days after training into the hippocampus. In both cases, ZIP treatment erased the spatial long-term memory. Additionally, the ZIP treatment also blocked LTP maintenance. Taken together, these findings suggest that persistent, synaptic strengthening underlies memory storage in the hippocampus (see also Chapter 1). A follow-up investigation showed that ZIP also blocks consolidated, long-term taste aversion memory that is dependent on neocortex (Shema *et al.*, 2007). Further, there was no recovery of long-term memory after ZIP treatment. Afterwards, the same group established that the effect of ZIP treatment is specific for PKMζ, as overexpression of a dominant-negative form of PKMζ also blocks neocortical long-term memory (Shema *et al.*, 2011).

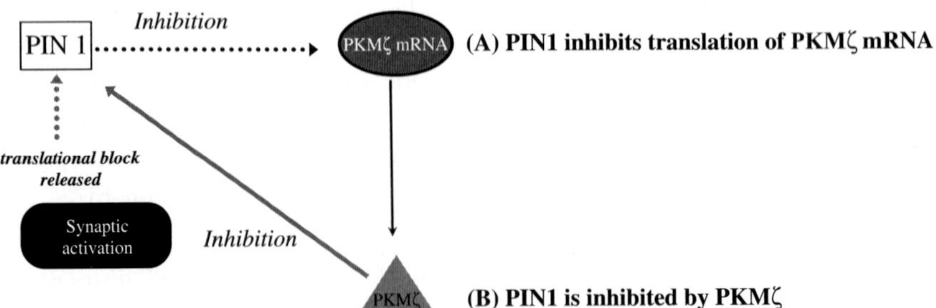

Figure 9.2. The model of how PKMζ activity persists. (A) PIN1 inhibits local translation of PKMζ mRNA in dendrites. Synaptic activation releases the translational block, allowing synthesis of PKMζ in a dendritic shaft. PKMζ translocates into tagged synapses and is delivered to the strengthened synapse. (B) Active PKMζ phosphorylates PIN1, preventing block of local translation so that synthesis of PKMζ continues.

Additionally, overexpression of PKMζ enhances neocortical long-term memory. Overexpression of PKMζ even enhances two distinct consolidated, long-term memories. Taken together, these molecular biological and pharmacological studies have established that PKMζ is essential for the maintenance of spatial and taste aversion long-term memory.

Do all memory types require PKMζ for storage? ZIP injection into the baso-lateral amygdala blocks inhibitory avoidance memory, memory after tone fear conditioning, contextual fear memory, and drug-induced conditioned place preference memory (Serrano *et al.*, 2008; Kwapis *et al.*, 2009; He *et al.*, 2011). The inhibition of PKMζ in nucleus accumbens core abolishes long-term drug reward memory (Li *et al.*, 2011). Injection of ZIP into somatosensory cortex blocks sensorimotor memory (von Kraus *et al.*, 2010). Administration of ZIP into the hippocampus blocks spatial reference memory, trace eyeblink memory and object location memory (Serrano *et al.*, 2008; Hardt *et al.*, 2010; Madronal *et al.*, 2010). However, injection of ZIP into the dorsal hippocampus does not impair contextual fear memory, although this part of the hippocampus is essential for this type of memory (Serrano *et al.*, 2008; Kwapis *et al.*, 2009). Furthermore, injection of ZIP into insular cortex does not block attenuation of neophobia and taste latent inhibition (Shema *et al.*, 2007). Recently, it was suggested that ZIP administration into the baso-lateral amygdala would only erase memory when a memory test is given within a few days after treatment (Parsons and Davis, 2011). In this study, a memory test at least 10 days after ZIP treatment did not reveal a memory deficit. However, the experimental design of this study has been criticized (Nader, 2011). In summary, most, but not all, memories require PKMz for storage.

How is PKMζ activated and how is its activity maintained? Under basal conditions, local translation of dendritic PKMζ mRNA is repressed by protein interacting with NIMA1 (PIN1) (Westmark *et al.*, 2010). Synaptic activation reduces PIN1 activity so that PKMζ mRNA is translated. PKMζ phosphorylates PIN1, which inhibits PIN1 function. Thus, once PKMζ is locally synthesised it promotes its continued synthesis by blocking PIN1. Under basal conditions, local translation occurs primarily in the dendritic shaft. After synaptic activation, polyribosomes, which are fundamental for local translation, redistribute into spines, suggesting that local translation occurs at activated synapses (Ostroff *et al.*, 2002). Tagging, a mechanism that allows for the entry of molecules into strengthened but not other synapses (Redondo and Morris, 2011), may be important for the redistribution of polyribosomes, PKMζ mRNA, and other molecules needed for local synthesis of PKMζ at strengthened synapses.

What is the function of PKMζ? The key function of newly synthesized PKMζ is thought to be the promotion of trafficking of GluA2 (GluR2)-containing

AMPA receptors. Trafficking of AMPA receptors enhances glutamatergic transmission (see Chapter 1). Blocking the PKMζ-induced GluR2 trafficking blocks LTP maintenance in the hippocampus (Yao *et al.*, 2008). This was also shown for LTP maintenance in the amygdala (Migues *et al.*, 2010). Further, fear memory storage requires continued trafficking of GluR2 by PKMζ in the amygdala (Migues *et al.*, 2010).

4. DNA methylation and memory storage

Recently, DNA methylation has been suggested to be a candidate mechanism of memory storage (Miller *et al.*, 2010; Day and Sweatt, 2011) (Figure 9.3). Changes in DNA methylation decrease or increase the expression of genes, or they change the production of isoforms deriving from alternative splicing, depending on their position within the genome (Day and Sweatt, 2011). DNA methylation occurs at cytosine residues within a CpG dinucleotide. DNA methyltransfrases (DNMTs) catalyze the methylation. There are two types of DNMTs: *de novo* DNMTs that methylate non-methylated cytosines, and maintenance DNMTs that detect hemimethylated CpGs and proceed to methylate the non-methylated cytosine. Maintenance DNMTs, as the name suggests, maintain once established DNA methylation by overcoming spontaneous demethylation of a cytosine. In principle, this maintenance mechanism enables DNA methylation to last as long as the lifetime of the animal. Thus, training-induced DNA methylation may be important for memory storage. If so, training-induced DNA methylation should persist and it should be associated with continuous alteration in mRNA and protein expression. Further, block of persistent, methylation-induced expression should impair memory storage. A question arising is whether the nuclear expression changes impact at the synaptic level. A lasting synaptic tag may be required to persistently deliver the alter expression of plasticity-related proteins at memory storing synapses. However, an alternative scenario is that persistent, methylation-induced expression changes do not act at the synaptic level, and that they rather alter the excitability of memory storing neurons (see Chapter 3).

So far, only one experiment supports the model that DNA methylation is required for memory storage (Miller *et al.*, 2010). This study showed that contextual fear conditioning induces a lasting change for in DNA methylation, mRNA and protein expression of the calcineurin gene. Calcineurin is a phosphatase that is considered to be a memory suppressor gene (Abel *et al.*, 1998). Thirty days after contextual fear conditioning methylation of the calcineurin gene was increased and the mRNA/protein expression was reduced in prefrontal cortex. Further, inhibiting DMNTs 30 days after contextual fear conditioning impaired

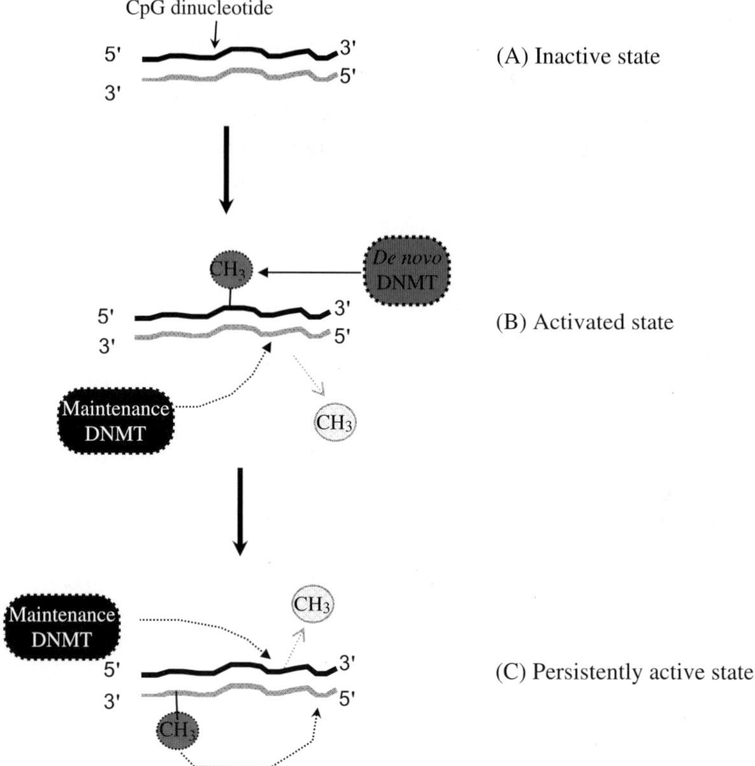

Figure 9.3. The model of how DNA methylation persists. DNA methylation occurs at cytosine residues within a CpG dinucleotide. DNA methyltransfrases (DNMTs) catalyse the methylation: *de novo* DNMTs methylate non-methylated cytosines, and maintenance DNMTs detect hemi-methylated CpGs and methylate the non-methylated cytosine. Maintenance DNMTs, maintain once established DNA methylation by overcoming spontaneous demethylation of a cytosine.

the memory. Taken together, these findings suggest that a persistent, DNA methylation-mediated suppression of calcineurin expression may be involved in memory storage.

5. Conclusion

In this chapter, we have discussed candidate mechanisms of memory storage. After decades of research, it has emerged that CaMKII autophosphorylation is unlikely to be important for memory storage. Instead, this biochemical switch appears to enable long-term memory formation after a single training trial. Recently, it has emerged that PKMζ is at the heart of memory storage. However, there appear to be additional processes that are required for memory storage, as not all memories require PKMζ. For example, it is conceivable that DNA methylation-mediated changes in protein expression contribute to memory storage. More studies are needed to investigate this possibility.

References

Abel, T., Martin, K.C., Bartsch, D. and Kandel, E.R. (1998). Memory suppressor genes: Inhibitory constraints on the storage of long-term memory. *Science* **279**, 338–341.

Bailey, C.H., Kandel, E.R. and Si, K. (2004). The persistence of long-term memory: A molecular approach to self-sustaining changes in learning-induced synaptic growth. *Neuron* **44**, 49–57.

Buard, I., Coultrap, S.J., Freund, R.K., Lee, Y.S., Dell'Acqua, M.L., Silva, A.J. and Bayer, K.U. (2010). CaMKII "autonomy" is required for initiating but not for maintaining neuronal long-term information storage. *J Neurosci* **30**, 8214–8220.

Cammarota, M., Bernabeu, R., Levi De Stein, M., Izquierdo, I. and Medina, J.H. (1998). Learning-specific, time-dependent increases in hippocampal Ca^{2+}/calmodulin-dependent protein kinase II activity and AMPA GluR1 subunit immunoreactivity. *Eur J Neurosci* **10**, 2669–2676.

Cooke, S.F., Wu, J., Plattner, F., Errington, M., Rowan, M., Peters, M., Hirano, A., Bradshaw, K.D., Anwyl, R., Bliss, T.V. and Giese, K.P. (2006). Autophosphorylation of alphaCaMKII is not a general requirement for NMDA receptor-dependent LTP in the adult mouse. *J Physiol* **574**, 805–818.

Crick, F. (1984). Memory and molecular turnover. *Nature* **312**, 101.

Day, J.J. and Sweatt, J.D. (2011). Cognitive neuroepigenetics: A role for epigenetic mechanisms in learning and memory. *Neurobiol Learn Mem* **96**, 2–12.

Fukunaga, K., Stoppini, L., Miyamoto, E. and Muller, D. (1993). Long-term potentiation is associated with an increased activity of Ca2+/calmodulin-dependent protein kinase II. *J Biol Chem* **268**, 7863–7867.

Giese, K.P., Fedorov, N.B., Filipkowski, R.K. and Silva, A.J. (1998). Autophosphorylation at Thr286 of the alpha calcium-calmodulin kinase II in LTP and learning. *Science* **279**, 870–873.

Hardt, O., Migues, P.V., Hastings, M., Wong, J. and Nader, K. (2010). PKMzeta maintains 1-day- and 6-day-old long-term object location but not object identity memory in dorsal hippocampus. *Hippocampus* **20**, 691–695.

He, Y.Y., Xue, Y.X., Wang, J.S., Fang, Q., Liu, J.F., Xue, L.F. and Lu, L. (2011). PKMzeta maintains drug reward and aversion memory in the basolateral amygdala and extinction memory in the infralimbic cortex. *Neuropsychopharmacology*, in press

Irvine, E.E., Danhiez, A., Radwanska, K., Nassim, C., Lucchesi, W., Godaux, E., Ris, L. and Giese, K.P. (2011). Properties of contextual memory formed in the absence of alphaCaMKII autophosphorylation. *Mol Brain* **4**, 8.

Irvine, E.E., Vernon, J. and Giese, K.P. (2005). AlphaCaMKII autophosphorylation contributes to rapid learning but is not necessary for memory. *Nat Neurosci* **8**, 411–412.

Irvine, E.E., von Hertzen, L.S., Plattner, F. and Giese, K.P. (2006). alphaCaMKII autophosphorylation: a fast track to memory. *Trends Neurosci* **29**, 459–465.

Kwapis, J.L., Jarome, T.J., Lonergan, M.E. and Helmstetter, F.J. (2009). Protein kinase Mzeta maintains fear memory in the amygdala but not in the hippocampus. *Behav Neurosci* **123**, 844–850.

Lee, S.J., Escobedo-Lozoya, Y., Szatmari, E.M. and Yasuda, R. (2009). Activation of CaMKII in single dendritic spines during long-term potentiation. *Nature* **458**, 299–304.

Lengyel, I., Voss, K., Cammarota, M., Bradshaw, K., Brent, V., Murphy, K.P., Giese, K.P., Rostas, J.A. and Bliss, T.V. (2004). Autonomous activity of CaMKII is only transiently increased following the induction of long-term potentiation in the rat hippocampus. *Eur J Neurosci* **20**, 3063–3072.

Leonard, A.S., Bayer, K.U., Merrill, M.A., Lim, I.A., Shea, M.A., Schulman, H. and Hell, J.W. (2002). Regulation of calcium/calmodulin-dependent protein kinase II docking to N-methyl-D-aspartate receptors by calcium/calmodulin and alpha-actinin. *J Biol Chem* **277**, 48441–48448.

Li, Y.Q., Xue, Y.X., He, Y.Y., Li, F.Q., Xue, L.F., Xu, C.M., Sacktor, T.C., Shaham, Y., Lu, L. (2011). Inhibition of PKMzeta in nucleus accumbens core abolishes long-term drug reward memory. *J Neurosci* **31**, 5436–5446.

Lisman, J. (1994). The CaM kinase II hypothesis for the storage of synaptic memory. *Trends Neurosci* **17**, 406–412.

Lisman, J., Schulman, H. and Cline, H. (2002). The molecular basis of CaMKII function in synaptic and behavioural memory. *Nat Rev Neurosci* **3**, 175–190.

Lisman, J.E. (1985). A mechanism for memory storage insensitive to molecular turnover: A bistable autophosphorylating kinase. *Proc Natl Acad Sci U S A* **82**, 3055–3057.

Lucchesi, W., Mizuno, K. and Giese, K.P. (2011). Novel insights into CaMKII function and regulation during memory formation. *Brain Res Bull* **85**, 2–8.

Madronal, N., Gruart, A., Sacktor, T.C. and Delgado-Garcia, J.M. (2010). PKMzeta inhibition reverses learning-induced increases in hippocampal synaptic strength and memory during trace eyeblink conditioning. *PLoS One* **5**, e10400.

Migues, P.V., Hardt, O., Wu, D.C., Gamache, K., Sacktor, T.C., Wang, Y.T. and Nader, K. (2010). PKMzeta maintains memories by regulating GluR2-dependent AMPA receptor trafficking. *Nat Neurosci* **13**, 630–634.

Miller, C.A., Gavin, C.F., White, J.A., Parrish, R.R., Honasoge, A., Yancey, C.R., Rivera, I.M., Rubio, M.D., Rumbaugh, G. and Sweatt, J.D. (2010). Cortical DNA methylation maintains remote memory. *Nat Neurosci* **13**, 664–666.

Miller, S.G. and Kennedy, M.B. (1986). Regulation of brain type II Ca^{2+}/calmodulin-dependent protein kinase by autophosphorylation: A Ca^{2+}-triggered molecular switch. *Cell* **44**, 861–870.

Nader, K. (2011). On the temporary nature of disruption of fear-potentiated startle following PKMzeta inhibition in the amygdale. *Front Behav Neurosci* **5**, 29.

Ostroff, L.E., Fiala, J.C., Allwardt, B. and Harris, K.M. (2002). Polyribosomes redistribute form dendritic shafts into spines with enlarged synapses during LTP in developing rat hippocampal slices. *Neuron* **35**, 535–545.

Parsons, R.G. and Davis, M. (2011). Temporary disruption of fear-potentiated startle following PKMzeta inhibition in the amygdala. *Nat Neurosci* **14**, 295–296.

Pastalkova, E., Serrano, P., Pinkhasova, D., Wallace, E., Fenton, A.A. and Sacktor, T.C. (2006). Storage of spatial information by the maintenance mechanism of LTP. *Science* **313**, 1141–1144.

Redondo, R.L. and Morris, R.G. (2011). Making memories last: The synaptic tagging and capture hypothesis. *Nat Rev Neurosci* **12**, 17–30.

Sacktor, T.C. (2011). How does PKMzeta maintain long-term memory? *Nat Rev Neurosci* **12**, 9–15.

Saitoh, T. and Schwartz, J.H. (1985). Phosphorylation-dependent subcellular translocation of a Ca^{2+}/calmodulin-dependent protein kinase produces an autonomous enzyme in Aplysia neurons. *J Cell Biol* **100**, 835–842.

Sanhueza, M., Fernandez-Villalobos, G., Stein, I.S., Kasumova, G., Zhang, P., Bayer, K.U., Otmakhov, N., Hell, J.W. and Lisman, J. (2011). Role of the CaMKII/NMDA receptor complex in the maintenance of synaptic strength. *J Neurosci* **31**, 9170–9178.

Sanhueza, M., McIntyre, C.C. and Lisman, J.E. (2007). Reversal of synaptic memory by Ca^{2+}/calmodulin-dependent protein kinase II inhibitor. *J Neurosci* **27**, 5190–5199.

Serrano, P., Friedman, E.L., Kenney, J., Taubenfeld, S.M., Zimmerman, J.M., Hanna, J., Alberini, C., Kelley, A.E., Maren, S., Rudy, J.W., *et al.* (2008). PKMzeta maintains spatial, instrumental and classically conditioned long-term memories. *PLoS Biol* **6**, 2698–2706.

Shema, R., Haramati, S., Ron, S., Hazvi, S., Chen, A., Sacktor, T.C. and Dudai, Y. (2011). Enhancement of consolidated long-term memory by overexpression of protein kinase Mzeta in the neocortex. *Science* **331**, 1207–1210.

Shema, R., Sacktor, T.C. and Dudai, Y. (2007). Rapid erasure of long-term memory associations in the cortex by an inhibitor of PKM zeta. *Science* **317**, 951–953.

Shen, K., Teruel, M.N., Connor, J.H., Shenolikar, S. and Meyer, T. (2000). Molecular memory by reversible translocation of calcium/calmodulin-dependent protein kinase II. *Nat Neurosci* **3**, 881–886.

von Kraus, L.M., Sacktor, T.C. and Francis, J.T. (2010). Erasing sensorimotor memories via PKMzeta inhibition. *PLoS One* **5**, e11125.

Westmark, P.R., Westmark, C.J., Wang, S., Levenson, J., O'Riordan, K.J., Burger, C. and Malter, J.S. (2010). Pin1 and PKMzeta sequentially control dendritic protein synthesis. *Sci Signal* **3**, ra18.

Yao, Y., Kelly, M.T., Sajikumar, S., Serrano, P., Tian, D., Bergold, P.J., Frey, J.U. and Sacktor, T.C. (2008). PKM zeta maintains late long-term potentiation by N-ethylmaleimide-sensitive factor/GluR2-dependent trafficking of postsynaptic AMPA receptors. *J Neurosci* **28**, 7820–7827.

Zhou, Y., Takahashi, E., Li, W., Halt, A., Wiltgen, B., Ehninger, D., Li, G.D., Hell, J.W., Kennedy, M.B. and Silva, A.J. (2007). Interactions between the NR2B receptor and CaMKII modulate synaptic plasticity and spatial learning. *J Neurosci* **27**, 13843–13853.

Memory Reconsolidation and Extinction

10

Satoshi Kida*

1. Introduction

Memory consolidation is often thought of as a process of fixation or storage; that is, memories become increasingly immune to disruption as they mature (McGaugh, 1966). The important biochemical feature of memory consolidation is its requiment for gene expression (Abel *et al.*, 1997; Davis and Squire, 1984; Flexner *et al.*, 1965; McGaugh, 2000). Studies show that the pharmacological inactivation of gene expression blocks formation of long-term memory (LTM) (Abel *et al.*, 1997; Kida *et al.*, 2002; Suzuki *et al.*, 2004).

Although it was previously thought that the consolidation of a new memory occurs just once, there is growing evidence that memory retrieval is a dynamic process that either reinforces or alters consolidated memories (Anokhin *et al.*, 2002; Judge and Quartermain, 1982; Lewis, 1979; Mactutus *et al.*, 1979; Misanin *et al.*, 1968; Nader *et al.*, 2000a; Pedreira *et al.*, 2002; Przybyslawski *et al.*, 1997, 1999; Sara, 2000; Schneider and Sherman, 1968). Similar to the observations of initial consolidation, inhibition of protein synthesis before or immediately following memory retrieval disrupts the subsequent expression of that memory (Debiec *et al.*, 2002; Kida *et al.*, 2002; Nader *et al.*, 2000a; Taubenfeld *et al.*, 2001). This experimental evidence suggests that following retrieval, the retrieved memory returns to a labile state similar to short-term memory (STM) and requires gene expression-dependent memory reconsolidation for storage (Debiec *et al.*, 2002; Kida *et al.*, 2002; Nader *et al.*, 2000a; Taubenfeld *et al.*, 2001). Thus, memory reconsolidation may serve to update or integrate new information into long-term memories (Dudai, 2002; Nader *et al.*, 2000b; Sara, 2000).

In Pavlovian fear conditioning, a conditioned stimulus (CS; such as a context) is paired with an unconditioned stimulus (US; such as footshock). After such

*Department of Bioscience, Faculty of Applied Bioscience, Tokyo University of Agriculture, 1-1-1 Sakuragaoka, Setagaya-ku, Tokyo 156-8502, Japan; E-mail: kida@nodai.ac.jp

conditioning, when the animal is placed back in the training context, it exhibits conditioned fear responses such as freezing, in the absence of the US. Experimentally, cued recall typically involves re-exposing subjects to the CS without the US. This reminder initiates not only reconsolidation, but also extinction in which the CS comes to predict no US and loses its ability to evoke a conditioned response (Baum, 1988; Bouton, 1993; Myers and Davis, 2002; Pavlov, 1927). Importantly, experimental extinction is thought to reflect new learning of a CS-no US association that inhibits the conditioned response (Konorksi 1967; Rescorla and Heth, 1975; Rescorla, 2001; Robbins, 1990).

Thus, memory retrieval initiates two potentially dissociable and opposite processes: reconsolidation and extinction. Reconsolidation acts to stabilise, whereas extinction tends to weaken, the expression of the original memory. In this chapter, I will review our current understanding of the regulations and mechanisms of reconsolidation and extinction and their relationships to one another.

2. Memory reconsolidation

STM is labile. To generate LTM, STM is stabilised via memory consolidation (Davis and Squire, 1984; Flexner *et al.*, 1965; McGaugh, 2000; McGaugh, 1966). Studies show that inhibiting protein synthesis around the time of learning blocks the formation of LTM without affecting STM, indicating that the formation of LTM but not STM depends on gene expression (Abel *et al.*, 1997; Kida *et al.*, 2002; Suzuki *et al.*, 2004). Thus, a critical biochemical feature of memory consolidation is a requirement for new gene expression.

Similar to the biochemical signatures of memory consolidation, retrieved memories undergo gene expression-dependent reconsolidation. Studies show that blocking gene expression using inhibitors of translation or transcription disrupt long-term post-retrieved memory without affecting short-term post-retrieved memories (Debiec *et al.*, 2002; Kida *et al.*, 2002; Nader *et al.*, 2000a; Taubenfeld *et al.*, 2001; Suzuki *et al.*, 2004). Therefore, when memory is reactivated by retrieval, the memory is first destabilised and then re-stabilised though reconsolidation.

The majority of these studies used conditioned fear to examine reconsolidation (Anokhin *et al.*, 2002; Debiec *et al.*, 2002; Kida *et al.*, 2002; Kraus *et al.*, 2002; Milekic and Alberini, 2002; Nader *et al.*, 2000a; Pedreira *et al.*, 2002; Taubenfeld *et al.*, 2001). Further studies extended these observations and showed that almost all kinds of memory, including spatial memory, object recognition memory, cocaine-seeking behaviour, incentive memory, and conditioned place preference, undergo reconsolidation following retrieval (Eisenberg *et al.*, 2003; Lee *et al.*, 2005; Miller *et al.*, 2005; Milton *et al.*, 2008; Rossato *et al.*, 2007; Suzuki *et al.*, 2004; Tronel *et al.*, 2005; Wang *et al.*, 2005; but see Blum *et al.*, 2006; Morris *et al.*, 2006). Moreover, the phenomenon of memory reconsolidation has been observed in all levels of animals from *Caenorhabditis elegans* to crab, medaka fish, chicken, and mammals, including rodents and human (Anokin *et al.*, 2002; Eisenberg *et al.*, 2003; Pedreira *et al.*, 2002; Rose *et al.*, 2006; Sangha *et al.*, Walker *et al.*, 2002).

Additional studies identified some of the molecular mechanisms underlying the restabilisation of retrieved memory (Duvarci *et al.*, 2005; Kida *et al.*, 2002; Lee *et al.*, 2004; Miller and Marshall 2005; Taubenfeld *et al.*, 2001; Tronson *et al.*, 2006; von Hertzen and Giese 2005). Restabilisation engages molecular processes that are similar to, but distinct from, those of the initial consolidation. Activation of CREB-mediated transcription is required for both consolidation and reconsolidation of fear memory (Kida *et al.*, 2002). On the other hand, BDNF and Zif268 are required in the hippocampus for consolidation and reconsolidation, respectively, of contextual fear memory (Lee *et al.*, 2004). Interestingly, CAAT/enhancer-binding protein b(C/EBPb) mediated transcription in the hippocampus is required for consolidation, but not for reconsolidation, of inhibitory avoidance (IA) memory (Taubenfeld *et al.* 2001); in contrast, reconsolidation of such memory requires C/EBPb-mediated transcription in the amygdala, while consolidation does not (Tronel *et al.*, 2005). These findings suggest that mechanisms of memory consolidation and reconsolidation are dissociable at the molecular and anatomical levels.

3. Memory destabilisation after retrieval

Abundant studies have focused on investigating the mechanisms for restabilising (reconsolidating) retrieved memories (Taubenfeld *et al.*, 2001; Kida *et al.*, 2002; Lee *et al.*, 2004; Duvarci *et al.*, 2005; Miller and Marshall, 2005; Tronson *et al.*, 2006). In contrast, the molecular mechanisms underlying the initial destabilisation remain poorly understood. A recent report demonstrated the importance of activating the NMDA-glutamate receptors (NMDAR) in the destabilisation of retrieved cued fear memory (Ben Mamou *et al.*, 2006). Ben Mamou *et al.* showed that blocking the NMDAR but not the AMPA-glutamate receptor (AMPAR) in the amygdala inhibited the disruption of retrieved cued fear memory induced by protein synthesis inhibitors. This evidence indicates that activating the NMDAR triggers the destabilisation of retrieved memory. Interestingly, another study showed that the destabilisation of retrieved contextual fear memory requires the activation of protein degradation via the proteasome (Lee *et al.*, 2008), suggesting that retrieved memory is destabilised via protein degradation, thereby requiring protein synthesis-dependent restabilisation (reconsolidation). Studies also indicated that activation of the L-type voltage-gated calcium channel (LVGCC) and the cannabinoid receptor 1 (CB1) are required for destabilising contextual fear memory following its retrieval (Suzuki *et al.*, 2008). Similar to the blockade of NMDARs or proteasome-dependent protein degradation, pharmacological blockade of these molecules protected retrieved memories from the amnestic effects of protein synthesis inhibition (Suzuki *et al.*, 2008). Collectively, these findings indicate the existence of an active process inducing the destabilisation of retrieved memory.

Interestingly, several studies have shown that infusing CB1 agonists into the amygdala or insular cortex following retrieval of cued fear or conditioned taste aversion memories, respectively, leads to subsequent deficits in the expressions of those memories (Lin *et al.*, 2006, Kobilo *et al.*, 2007; de Oliveira *et al.*, 2008). Together with the finding that inactivation of CB1 blocks the destabilisation of retrieved memory (Suzuki *et al.*, 2008), these results raises the possibility that strong activation of CB1-dependent signalling may facilitate destabilisation to such a degree that restabilisation is no longer possible. Thus CB1 is thought to play essential roles in regulating memory destabilisation following its retrieval.

How do NMDARs, CB1s, LVGCCs, and proteasome-mediated protein degradation cooperatively regulate the destabilisation of retrieved memory? LVGCCs are expressed in postsynaptic sites of excitatory neurons in the hippocampus, and mediate increases in postsynaptic Ca^{2+} concentrations in response to depolarisation following the activation of NMDARs. Endocannabinoids are released in response to increases in postsynaptic Ca^{2+} concentrations and then act as a retrograde signal to activate presynaptic CB1 receptors, leading to decreased presynaptic neurotransmitter release. Therefore, it is possible that CB1 receptors are activated via the activation of postsynaptic LVGCCs. Further studies are necessary to investigate the molecular cascades linking NMDAR, CB1, LVGCCs, and proteasome-dependent protein degradation to destabilise retrieved memory.

As mentioned above, previous study has shown that the pharmacological blockade of NMDAR by pre-, but not post-, re-exposure infusion of a NMDAR blocker into the amygdala impaired the destabilisation of retrieved cued fear memory (Ben Mamou *et al.*, 2006). On the other hand, this blockade by post-re-exposure infusion into the hippocampus disrupted the retrieved spatial memory (Kim *et al.*, 2011). It is possible that the function of NMDAR on destabilisation and restabilisation differs in brain regions and types of memory. Moreover, there is another possibility that the requirements of NMDAR for destabilisation and restabilisation of retrieved memory display distinct critical time windows although NMDAR is required for both destabilisation and restabilisation of retrieved memory.

4. Boundary conditions affecting the induction of memory reconsolidation

The phenomenon of reconsolidation raises a fundamental question: does memory always become labile following retrieval? Several studies have examined this question. Twenty-four hours after contextual fear conditioning, inhibiting protein synthesis along with re-exposure to the context (3 min) disrupted retrieved fear memory, indicating that this duration of fear memory retrieval induces memory reconsolidation (Kida *et al.*, 2002; Suzuki *et al.*, 2004). In contrast, inhibiting protein synthesis along with a shorter contextual re-exposure (1 min) failed to disrupt retrieved fear memory, even though the mice displayed high levels of freezing behaviour (Suzuki *et al.*, 2004). These observations strongly suggest that memory does not always become labile when retrieved; brief retrieval leaves the original memory unaffected. These findings also indicate that the important variable that determines the stability of a retrieved memory is duration of its retrieval.

While the initial encoding and storage of fear memories depends upon the hippocampus, permanent storage is thought to depend on the neocortex (Anagnostaras *et al.*, 1999, 2001; Eichenbaum *et al.*, 1994; Frankland *et al.*, 2001; Kim and Fanselow, 1992; LeDoux, 2000; McGaugh, 2000; McClelland *et al.*, 1995; Quevedo *et al.*, 1999; Squire and Alvarez 1995). Based on these data, memories are thought to be reorganized over the course of their life. Therefore, it is interesting to ask whether remote memory undergoes memory reconsolidation. Studies demonstrated that retrieved remote memories (~28 days) of IA and contextual fear in rodents were resistant to the inhibition of protein synthesis (Milekic and Alberini, 2002; Suzuki *et al.*, 2004, but see Debiec *et al.*, 2002). Similar observations were made in medaka fish (Eisenberg and Dudai, 2004). These findings indicate that another important variable that determines the stability of a retrieved memory is its age.

Similar to the features of remote memory, stronger memories are more resistant to reconsolidation than weaker memories. For example, retrieved contextual fear memory induced by three footshocks, but not by only one footshock, was immune to the disruptive effects of protein synthesis inhibition (Suzuki *et al.*, 2004). A similar observation was made when strong cued fear memory, formed after 10 pairings of the CS-US, was tested (Wang *et al.*, 2009). These findings suggest that an additional important variable that determines the stability of a retrieved memory is its strength.

Further investigations of the boundary conditions for inducing reconsolidation demonstrated that even strong or old memories of contextual fear undergo reconsolidation with longer re-exposure (Suzuki *et al.*, 2004). The retrieval of strong or old memories induced memory reconsolidation when mice were re-exposed to a CS (context) for 10 min, but not for 3. Re-exposure to the context for 10 min, but not for 3, was required for the disruption of retrieved strong or remote contextual fear memory by the inhibition of protein synthesis, while the same re-exposure for only 3 min was sufficient to disrupt retrieved weak or recent memory by protein synthesis inhibition.

Similarly, stronger and older object memories are more resistant to reconsolidation than weaker and younger memories. However, the presence of new information such as salient novel contextual information enables the induction of the reconsolidation of stronger and older memories (Winters *et al.*, 2009).

On the other hand, even re-exposure to 10 CSs (tone) 24 h after training with 10 CS-US pairings did not induce reconsolidation of the strong cued fear memory, while the same re-exposure did induce reconsolidation when the rats were re-exposed to only one CS 56 days after the training (Wang *et al.*, 2009). In this study, the authors suggested that this resistance to induction of reconsolidation of recent strong cued fear memory is due to decrease in expression level of NMDAR NR2B subunit.

In summary, boundary conditions affecting the induction of reconsolidation have been identified; re-exposure duration, the age of the memory, and the strength of the memory influence the stability of retrieved memory in tasks that model declarative memory.

5. Memory extinction

Extinction, during which a CS comes to predict no US and loses its ability to evoke a conditioned response, is thought to depend upon the medial prefrontal cortex (mPFC; Morgan *et al.*, 1993; Morgan and LeDoux, 1995, 1999; Morrow *et al.*, 1999; Quirk *et al.*, 2000; Teich *et al.*, 1989) and amygdala (Herry *et al.*, 2008; Myers and Davis, 2002; Quirk *et al.*, 2000). The amygdala plays a significant role in the acquisition of tone fear extinction, whereas the mPFC, especially the prelimbic cortex, is implicated in its consolidation (Quirk *et al.*, 2000; Santini *et al.*, 2004). Furthermore, the extinction of contextual fear memory was shown to be consolidated through new gene expression in the mPFC and amygdala (Mamiya *et al.*, 2009). Strikingly, a recent study identified a population of "extinction" neurons in the basolateral amygdala (BLA) whose activation decreases high fear behaviour (Herry *et al.*, 2008).

Recent pharmacology and mouse genetics studies indicate that the downstream signalling transduction pathways involving LVGCCs, CB1, cdk5 signaling, NMDAR, activation of the ubiquitin–proteasomal pathway, actin rearrangement, and BDNF are required for extinction of fear memory (Fischer *et al.*, 2004; Marsciano *et al.*, 2002; Sananbenesi *et al.*, 2007; de Oliveira Alvares *et al.*, 2008; Lee *et al.*, 2008; Peters *et al.*, 2010). Molecular cascades among those molecules should be investigated to understand the molecular mechanisms regulating the extinction of fear memory.

6. Clinical application of reconsolidation and extinction processes for the treatment of emotional disorders

The finding that long-term memory may be more dynamic and plastic than previously thought may have important clinical implications for the treatment of emotional disorders. Understanding the circumstances under which maladaptive memories become plastic (and modifiable) is of clear clinical relevance. As described above, previous studies showed that LVGCCs, CB1, and protein degradation via proteasome-dependent processes are all important in weakening (destabilising and extinguishing) fear memories (Lee *et al.*, 2008; Marsciano *et al.*, 2002; Suzuki *et al.*, 2004, 2008). Thus, they might serve as useful therapeutic targets for the treatment of conditions such as post-traumatic stress disorder (PTSD) and phobias.

Exposure therapy for the treatment of such emotional disorders as PTSD is thought to reflect some aspect of the biological basis of memory extinction (Delgado *et al.*, 2008; Phelps and LeDoux 2005; Rauch *et al.*, 2006). Therefore, blocking the reconsolidation and facilitation of extinction are thought to be candidates for reducing acquired fear that leads to emotional disorders. However, their clinical efficacies in human are limited: blocking reconsolidation requires toxic drugs, and extinction is transient, because spontaneous recovery of fear memory is observed (Myers and Davis, 2002). Interestingly, recent studies demonstrated that a fear memory in rats could be destabilised and reinterpreted as safe by presenting a retrieval trial just before an extinction session (Monfils *et al.*, 2009; Schiller *et al.*, 2010). This finding demonstrates the role of reconsolidation as a window of chance for updating emotional memories, suggesting a non-invasive treatment that can be safely used in humans to prevent the recovery of fear.

7. Relationship between reconsolidation and extinction: Behavioural level

As described above, memory retrieval initiates two opposite processes: reconsolidation and extinction. Reconsolidation may serve to strengthen or modify the existing memory trace, whereas extinction involves the formation of an inhibitory memory that competes with the original memory. Because the temporal dynamics of these two processes differ, which comes to gain control over behaviour depends, in part, on the duration of the retrieval episode; reconsolidation is the dominant process following shorter duration re-exposures, whereas extinction is the dominant process following longer duration re-exposures (Debiec *et al.*, 2002; Eisenberg *et al.*, 2003; Nader, 2003; Pedreira and Maldonado, 2003; Suzuki *et al.*, 2004). Moreover, blocking protein synthesis during brief re-exposure disrupts the original memory, whereas blocking protein synthesis during longer re-exposure blocks the formation of a new extinction memory (Eisenberg *et al.*, 2003; Pedreira and Maldonado, 2003; Suzuki *et al.*, 2004). The requirement for protein synthesis in the formation of a new extinction memory is perhaps not surprising. However, the fact that the inhibition of protein synthesis selectively affects the formation of the extinction memory and leaves the original memory unaffected is surprising, because it is expected that both reconsolidation and extinction are initiated under these conditions. Therefore, there might be an interaction between the phases of extinction and reconsolidation at the behavioural level; the acquisition of a new extinction memory may transiently alter the stability of the original memory.

It is important to note that inhibition of protein synthesis in the BLA disrupted both the briefly reactivated and extinguished cued fear memories after retrieval. This suggests that extinction does not always interact with reconsolidation (Duvarci *et al.*, 2006).

Interestingly, studies using the crab's contextual memory model have shown that CS-offset (termination of re-exposure to the CS) is required for the induction of reconsolidation and extinction following the re-exposure to the CS for a short or longer time, respectively (Pedreira *et al.*, 2004; Perez-Cuesta *et al.*, 2007). These results suggest that CS-offset as well as the duration of re-exposure to the CS act as switches determining reconsolidation or extinction.

8. Relationship between reconsolidation and extinction: Anatomical level

The consolidation of tone fear extinction requires protein synthesis and NMDA receptor-dependent bursting in the mPFC (Santini *et al.*, 2004; Burgos-Robles *et al.*, 2006). On the other hand, reconsolidation of fear memories is mediated by the amygdala (Nader *et al.*, 2000a) and the hippocampus (Debiec *et al.*, 2002). To compare the mechanisms responsible for reconsolidation and extinction of contextual fear memory at the anatomical level, brain regions activated during these two phases was analysed (Mamiya *et al.*, 2009). As CREB-mediated transcription is required for reconsolidation and extinction of contextual fear memory (Mamiya *et al.*, 2009), the activity of CREB was measured following either brief (3 min) or prolonged (30 min) re-exposure to a conditioning context (CS) that induced reconsolidation or extinction, respectively (Figure 10.1). When the short context re-exposures were used, CREB activity was increased in the hippocampus and the amygdala, but not in the mPFC. Consistent with this, pharmacological studies using the microinfusion of anysomycin into these brain regions indicate that reconsolidation of the contextual fear memory depends on new gene expression in the hippocampus and amygdala but not in the mPFC (Mamiya *et al.*, 2009). In contrast, following prolonged re-exposure to the context (30 min), CREB-activity was increased in the amygdala and the mPFC, but not in the hippocampus. Likewise, pharmacological studies indicated that long-term extinction of contextual fear memory depends on new gene expression in the mPFC and amygdala, but not the hippocampus (Mamiya *et al.*, 2009). These observations extend behavioural observations that re-exposure to the CS triggers two distinct, time-dependent processes, reconsolidation and extinction, and indicate that these processes are dissociable at the anatomical level.

Restabilisation of retrieved contextual fear memory requires new gene expression in the hippocampus (Debiec *et al.*, 2002; Mamiya *et al.*, 2009; Suzuku *et al.*, 2008). In contrast, inhibition of protein synthesis in the hippocampus did not affect long-term extinction (Mamiya *et al.*, 2009), suggesting that the new protein synthesis in the hippocampus does not contribute to the consolidation of extinction memory. These contrasting observations suggest that the hippocampus plays distinct roles in reconsolidation and extinction of contextual fear memory. Interestingly, the hippocampus plays critical roles in the extinction of contextual fear. Indeed, cdk5 signalling, activation of proteasome-dependent protein degradation, activation of CB1 receptors, and actin rearrangement in the hippocampus are required for extinction of contextual fear memory (Fischer *et al.*, 2004; Sananbenesi *et al.*, 2007; de Oliveira Alvares *et al.*, 2008; Lee *et al.*, 2008). Therefore, it is possible that the hippocampus regulates contextual fear extinction through activation of these pathways without activating gene expression. Suppression of gene expression in the hippocampus might be a critical step that leads to long-term extinction. Further studies are required to investigate the regulatory roles of the hippocampus in memory extinction.

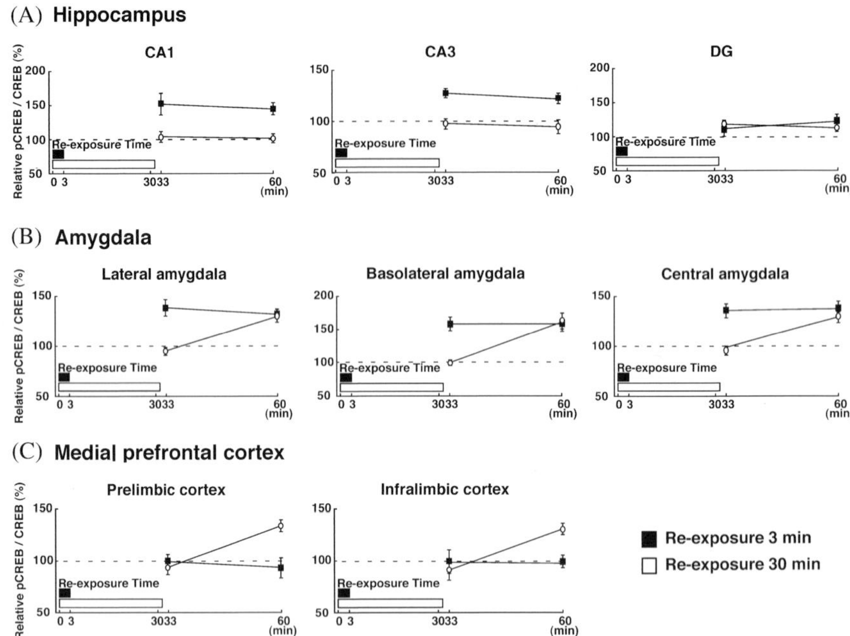

Figure 10.1. Regulation of CREB activity in the hippocampus, amygdala, and mPFC during the reconsolidation and extinction phases of contextual fear memory. Time course of CREB activation in the (A) hippocampus, (B) amygdala, and (C) mPFC from the beginning of the reconsolidation (3 min) and extinction (30 min) re-exposure to the context (using the identical time schedule) 24 h after the contextual fear conditioning (Mamiya *et al.*, 2009). Error bars are SEM; CREB activation (relative pCREB/CREB levels) was calculated by normalizing the number of phosphorylated CREB (pCREB)-positive cells to the total number of CREB-positive cells. Data for CREB activation for each group were then expressed as percentages of the averaged values in the control group (dotted lines at 100%) that were exposed and re-exposed to the training context without receiving the electrical footshocks (no-US).

Studies showed that gene expression was activated in the amygdala during both the reconsolidation and extinction phases (Mamiya *et al.*, 2009; Herry and Mons, 2004; Herry *et al.*, 2008); however, the observation that the time course of amygdaloid CREB-activation differed during the reconsolidation and extinction phases of contextual fear memory (Figure 10.1B; Mamiya *et al.*, 2009) raises the possibility that distinct populations of amygdaloid neurons are activated during these phases. Consistent with this, Reijimers *et al.* (2008) demonstrated that retrieval of a contextual fear memory reactivates BLA-neurons that are activated during contextual fear conditioning, whereas extinction learning prevents their activation. Furthermore, two distinct populations of BLA neurons (fear and extinction neurons) were identified, whose activities were reversely correlated with low and high fear behaviour (Herry *et al.*, 2008). Therefore, it is important to examine whether identical or different population of neurons are activated in the reconsolidation and extinction phases of contextual fear memory.

9. Relationship between reconsolidation and extinction: Molecular level

The observation that there is an interaction between the reconsolidation and extinction phases of contextual fear memory at the behavioural level raises the possibility that molecular processes engaged during extinction may also regulate the stability of the original trace after the retrieval (but see Duvarci *et al.*, 2006). As described above, the activation of LVGCCs, CB1s, and proteasome-dependent protein degradation are required for extinction of contextual fear memory (Cain *et al.*, 2002; Marsciano *et al.*, 2002; Suzuki *et al.* 2004; Lee *et al.*, 2008). Importantly, these signal transduction pathways are required in the hippocampus also for the destabilisation of retrieved contextual fear memory (Lee *et al.*, 2008; Suzuki *et al.*, 2004, 2008), suggesting that molecular processes involved in fear extinction in the hippocampus also regulate the stability of the original fear memory after retrieval. These findings suggest that memory retrieval triggers the activation of LVGCCs, CB1 receptors, and proteosome-dependent protein degradation, leading to destabilisation and/or extinction of contextual fear memory, and support the interaction between reconsolidation and extinction at the molecular levels. Therefore, it is important to evaluate whether acquisition of contextual fear extinction prevents the destabilisation of retrieved memory or whether the original contextual fear memory is destabilised and then restabilised (inactivated) in the extinction phase in a gene expression-independent manner (such as the redistribution or relocalisation of pre-existing proteins).

Furthermore, as described above, differences in CREB activity during the reconsolidation and extinction phases identified the hippocampus and the amygdala as candidate anatomical loci for the interaction between these two processes at the molecular level (Figure 10.1; Mamiya *et al.*, 2009). First, an increase in CREB activity was observed in the hippocampus following short, but not longer, re-exposure to the CS (Figure 10.1A), suggesting that gene expression in the hippocampus undergoes distinct regulation in the reconsolidation and extinction phases. The observation that CREB-mediated gene expression was not induced in the extinction phase, even though contextual fear was retrieved, indicates that such induction must be suppressed when within-session extinction occurs. Second, differences in the time courses of CREB activation in the amygdala were observed in the reconsolidation and extinction phases (Figure 10.1B), indicating that amygdaloid CREB activation is differentially regulated in these two memory phases, and furthermore, that amygdaloid CREB activation observed in the extinction phase is not simply due to fear memory reactivation. Moreover, inhibition of protein synthesis in the amygdala following prolonged re-exposure blocked long-term extinction without affecting the subsequent expression of the original fear memory, while such inhibition following brief re-exposure disrupted the fear memory. Collectively, this evidence suggests that the interaction between the reconsolidation and extinction phases occurs in both the hippocampus and the amygdala at the molecular level.

References

Abel, T., Nguye, P.V., Barad, M., Deuel, T.A., Kandel, E.R. and Bourtchouladze, R. (1997). Genetic demonstration of a role for PKA in the late phase of LTP and in hippocampus-based long-term memory. *Cell* **88**, 615–626.

Anagnostaras, S.G., Maren. S. and Fanselow, M.S. (1999). Temporally graded retrograde amnesia of contextual fear after hippocampus damage in rat: Within-subjects examination. *J Neurosci* **19**, 1106–1114.

Anagnostaras, S.G., Gale, G.D. and Fanselow, M.S. (2001). Hippocampus and contextual fear conditioning: Recent controversies and advances. *Hippocampus* **11**, 8–17.

Anokhin, K.V., Tiunova, A.A. and Rose, S.P.R. (2002). Reminder effects-reconsolidation or retrieval deficit? Pharmachological dissection with protein synthesis inhibitors following reminder for a passive-avoidance task in young chicks. *Eur J Neurosci* **15**, 1759–1765.

Bailey, C.H. and Kandel, E.R. (1993). Structural changes accompanying memory storage. *Annu Rev Physiol* **55**, 397–426.

Bauer, E.P., Schafe, G.E. and LeDoux, J.E. (2002). NMDA receptors and L-type voltage-gated calcium channels contribute to long-term potentiation and different components of fear memory formation in the lateral amygdala. *J Neurosci* **22**, 5239–5249.

Baum, M. (1988). Spontaneous recovery from the effects of flooding (exposure) in animals. *Behav Res Ther* **26**, 185–186.

Berman, D.E. and Dudai, Y. (2001). Memory extinction, Learning anew, and learning the new: Dissociation in the molecular machinery of learning in cortex. *Science* **291**, 2417–2419.

Blanchard, R.J. and Blanchard, D.C. (1969). Passive and active reactions to fear-eliciting stimuli. *J Comp Physiol Psychol* **68**, 129–135.

Blum, S., Runyan, J.D. and Dash, P.K. (2006). Inhibition of prefrontal protein synthesis following recall does not disrupt memory for trace fear conditioning. *BMC Neurosci* **7**, 67.

Bontempi, B., Laurent-Demir, C., Destrade, C. and Jaffard, R. (1999). Time-dependent reorganization of brain circuitry underlying long-term memory strage. *Nature* **400**, 671–675.

Bourtchuladze, R., Frenguelli, B., Blendy, J., Cioffi, D., Schutz, G. and Silva, A.J. (1994). Deficient long-term memory in mice with a targeted mutation of the cAMP-responsive element-binding protein. *Cell* **79**, 59–68.

Bouton, M.E. (1993). Context, time, and memory retrieval in the interference paradigms of Pavlovian learning. *Psychol Bull* **114**, 80–99.

Burgos-Robles, A., Vidal-Gonzalez, I., Santini, E. and Quirk, G.J. (2007). Consolidation of fear extinction requires NMDA receptor-dependent bursting in the ventromedial prefrontal cortex. *Neuron* **53**, 871–880.

Cain, C., Blouin, A. and Barad, M.G. (2002). L-type voltage-gated calcium channels are required for extinction, but not for acquisition or expression, of conditioned fear in mice. *J Neurosci* **22**, 9113–9121.

Davis, H.P. and Squire, L.R. (1984). Protein synthesis and memory. *Psychol Bull* **96**, 518–559.

Debiec, J., LeDoux, J.E. and Nader, K. (2002). Cellular and systems reconsolidation in the hippocampus. *Neuron* **36**, 527–538.

Delgado, M.R., Nearing, K.I., Ledoux, J.E. and Phelps, E.A. (2008). Neural circuitry underlying the regulation of conditioned fear and its relation to extinction. *Neuron* **59**, 829–838.

de Oliveira Alvares, L., Pasqualini Genro, B., Diehl, F., Molina, V.A. and Quillfeldt, J.A. (2008). Opposite action of hippocampal CB1 receptors in memory reconsolidation and extinction. *Neuroscience* **154**, 1648–1655.

Dudai, Y. (2002). Molecular bases of long-term memories: A question of persistence. *Curr Opin Neurobiol* **12**, 211–216.

Duvarci, S., Mamou, C.B. and Nader, K. (2006). Extinction is not a sufficient condition to prevent fear memories from undergoing reconsolidation in the basolateral amygdala. *Eur J Neurosci* **24**, 249–260.

Eichenbaum, H., Otto, T. and Cohen, N.J. (1994). Two functional components of the hippocampal memory system. *Behav Brain Sci* **17**, 449–518.

Eisenberg, M., Kobilo, T., Berman, D.E. and Dudai, Y. (2003). Stability of retrieved memory: Inverse correlation with trace dominance. *Science* **301**, 1102–1104.

Eisenberg, M. and Dudai, Y. (2004). Reconsolidation of fresh, remote, and extinguished fear memory in Medaka: Old fears don't die. *Eur J Neurosci* **20**, 3397–3403.

Fanselow, M.S. (1980). Conditioned and unconditional components of post-shock freezing. *Pavlov J Biol Sci* **15**, 177–182.

Fischer, A., Sananbenesi, F., Schrick, C., Spiess, J. and Rdulovic, J. (2004). Distinct roles of hippocampal *de novo* protein synthesis and actin rearrangement in extinction of contextual fear. *J Neurosci* **24**, 1962–1966.

Flexner, L.B., Flexner, J.B. and Stellar, E. (1965). Memory and cerebral protein synthesis in mice as affected by graded amounts of puromycin. *Exp Neurol* **13**, 264–272.

Flood, J.F., Rosenzweig, M.R., Bennett, E.L. and Orme, A.E. (1973). The influence of duration of protein synthesis inhibition on memory. *Physiol Behav* **10**, 555–562.

Frankland, P.W., O'Brien, C., Ohno, M., Kirkwood, A. and Silva, A.J. (2001). Alpha-CaMK-dependent plasticity in the cortex is required for permanent memory. *Nature* **411**, 309–313.

Frankland, P.W., Bontempi, B., Talton, L.E., Kaczmarek, L. and Silva, A.J. (2004). The involvement of the anterior cingulated cortex in remote contextual fear memory. *Science* **304**, 881–883.

Frankland, P.W., Ding, H.K., Takahashi, E., Suzuki, A., Kida, S. and Silva, A.J. (2006). Stability of recent and remote contextual fear memory. *Learn Mem* **13**, 451–457.

Flood, J.F., Rosenzweig, M.R., Bennett, E.L. and Orme, A.E. (1973). The influence of duration of protein synthesis inhibition on memory. *Physiol Behav* **10**, 555–562.

Guzowski, J.F., McNaughton, B.L., Barnes, C.A. and Worley, P.E. (1999). Environment-specific expression of the immediate-early gene Arc in hippocampal neuronal ensembles. *Nat Neurosci* **2**, 1120–1124.

Hall, J., Thomas, K.L. and Everitt, B.J. (2001). Fear memory retrieval induces CREB phosphorylation and Fos expression within the amydgala. *Eur J Neurosci* **13**, 1453–1458.

Herry, C. and Mons, N. (2004). Resistance to extinction is associated with impaired immediate early gene induction in medial prefrontal cortex and amygdala. *Eur J Neurosci* **20**, 781–790.

Herry, C., Ciocchi, S., Senn, V., Demmou, L., Muller, C. and Luthi, A. (2008). Switching on and off fear by distinct neuronal circuits. *Nature* **454**, 600–606.

Ji, J., Maren, S. (2007). Hippocampal involvement in contextual modulation of fear extinction. *Hippocampus* **17**, 749–758.

Judge, M.E. and Quartermain, D. (1982). Alleviation of anisomycin-induced amnesia by pre-test treatment with lysine-vasopressin. *Pharmacol Biochem Behav* **16**, 463–466.

Kida, S., Josselyn, S.A., deOrtiz, S.P., Kogan, J.H., Chevere, I., Masushige, S. and Silva, A.J. (2002). CREB required for the stability of new and reactivated fear memories. *Nat Neurosci* 5, 348–355.

Kim, J.J. and Fanselow, M.S. (1992). Modality-specific retrograde amnesia of fear. *Science* **256**, 675–677.

Kim, R., Moki, R. and Kida, S. (2011). Molecular mechanisms for the destabilisation and restabilisation of reactivated spatial memory in the Morris water maze. *Mol Brain* **4**, 9.

Kobilo, T., Hazvi, S. and Dudai, Y. (2007). Role of cortical cannabinoid CB1 receptor in conditioned taste aversion memory. *Eur J Neurosci* **25**, 3417–3421.

Konorski, J. (1967). Some new ideas concerning the physiological mechanisms of perception. *Acta Biol Exp (Warsz)* **27**, 147–161.

Kraus, M., Schicknick, H., Wetzel, W., Ohl, F., Staak, S. and Tischmeyer, W. (2002). Memory consolidation for the discrimination of frequency-modulated tones in mongolian gerbils is sensitive to protein-synthesis inhibitors applied to the auditory cortex. *Learn Mem* **9**, 293–303.

Lattal, K.M. and Abel, T. (2001). Different requirements for protein synthesis in acquisition and extinction of spatial preferences and context-evoked fear. *J Neurosci* **21**, 5773–5780.

LeDoux, J.E. (2000). Emotion circuits in the brain. *Annu Rev Neurosci* **23**, 155–184.

Lee, J.L., Di Ciano, P., Thomas, K.L. and Everitt, B.J. (2005). Disrupting reconsolidation of drug memories reduces cocaine-seeking behaviour. *Neuron* **47**, 795–801.

Lee, J.L., Thomas, K.L. and Everitt, B.J. (2001). Cellular Imaging of zif268 Expression in the hippocampus and amygdala during contextual and cued fear memory retrieval: Selective activation of hippocampal CA1 neurons during the recall of contextual memories. *J Neurosci* **21**, 2186–2193.

Lee, J.L., Milton, A.L. and Everitt, B.J. (2006). Reconsolidation and extinction of conditioned fear: Inhibition and potentiation. *J Neurosci* **26**, 10051–10056.

Lee, S.H., Choi, J.H., Lee, N., Lee, H.R., Kim, J.I., Yu, N.K., Choi, S.L., Lee, S.H., Kim, H. and Kaang, B.K. (2008). Synaptic protein degradation underlies destabilisation of retrieved fear memory. *Science* **319**, 1253–1256.

Lewis, D.J. (1979). Psychobiology of active and inactive memory. *Psychol Bull* **86**, 1054–1083.

Lin, H.C., Mao, S.C. and Gean, P.W. (2006). Effects of intra-amygdala infusion of CB1 receptor agonists on the reconsolidation of fear-potentiated startle. *Learn Mem* **3**, 316–321.

Lu, K.T., Walker, D.L. and Davis, M. (2001). Mitogen-activated protein kinase cascade in the basolateral nucleus of amygdala is involved in extinction of fear-potentiated startle. *J Neurosci* **21**, RC162.

Mactutus, C.F., Riccio, D.C. and Ferek, J.M. (1979). Retrograde amnesia for old (reactivated) memory: Some anomalous characteristics. *Science* **204**, 1319–1320.

Mamiya, N., Fukushima, H., Suzuki, A., Matsuyama, Z., Homma, S., Frankland, P.W. and Kida, S.J. (2009). Brain region-specific gene expression activation required for reconsolidation and extinction of contextual fear memory. *J Neurosci* **29**, 402–413.

Marsicano, G., Wotjak, C.T., Azad, S.C., Bisogno, T., Rammes, G., Cascio, M.G., Hermann, H., Tang, J., Hofmann, C. and Zieglgansberger, W., *et al.* (2002). The endogenous cannabinoid system controls extinction of aversive memories. *Nature* **418**, 530–534.

McClelland, J.L., McNaughton, B.L. and O'Reilly, R.C. (1995). Why there are complementary learning systems in the hippocampus and neocortex: Insights from the successes and failures of connectionist models of learning and memory. *Psychol Rev* **102**, 419–457.

McGaugh, J.L. (2000). Memory — a century of consolidation. *Science* **287**, 248–251.

McGaugh, J.L. (1966). Time-dependent processes in memory storage. *Science* **153**, 1351–1358.

Milekic, M.H. and Alberini, C.M. (2002). Temporally graded requirement for protein synthesis following memory reactivation. *Neuron* **36**, 521–525.

Miller, C.A. and Marshall, J.F. (2005). Molecular substrates for retrieval and reconsolidation of cocaine-associated contextual memory. *Neuron* **47**, 873–884.

Milton, A.L., Lee, J.L. and Everitt, B.J. (2008). Reconsolidation of appetitive memories for both natural and drug reinforcement is dependent on {beta}-adrenergic receptors. *Learn Mem* **15**, 88–92.

Misanin, J.R., Miller, R.R. and Lewis, D.J. (1968). Retrograde amnesia produced by electroconvulsive shock after reactivation of a consolidated memory trace. *Science* **160**, 554–555.

Monfils, M.H., Cowansage, K.K., Klann, E. and LeDoux, J.E. (2009). Extinction-reconsolidation boundaries: Key to persistent attenuation of fear memories. *Science* **324**, 951–955.

Morgan, M.A. and LeDoux, J.E. (1995). Differential contribution of dorsal and ventral medial prefraontal cortex to the acquisition and extinction of conditioned fear in rats. *Behav Neurosci* **109**, 681–688.

Morgan, M.A. and LeDoux, J.E. (1999). Contribution of ventrolateral prefrontal cortex to the acquisition and extinction of conditioned fear in rats. *Neurobiol Learn Mem* **72**, 244–251.

Morgan, M.A., Romanski, L.M. and LeDoux, J.E. (1993). Extinction of emotional learning: Contribution of medial prefrontal cortex. *Neurosci Lett* **163**, 109–113.

Morris, R.G., Inglis, J., Ainge, J.A., Olverman, H.J., Tulloch, J., Dudai, Y. and Kelly, P.A. (2006). Memory reconsolidation: Sensitivity of spatial memory to inhibition of protein synthesis in dorsal hippocampus during encoding and retrieval. *Neuron* **50**, 479–489.

Morrow, B.A., Elsworth, J.D., Rasmusson, A.M. and Roth, R.H. (1999). The role of mesoprefrontal dopamine neurons in the acquisition and expression of conditioned fear in the rat. *Neuroscience* **92**, 553–564.

Myers, K.M. and Davis, M. (2002). Behavioural and neural analysis of extinction. *Neuron* **36**, 567–584.

Myers, K.M. and RDavis, M. (2007). Mechanisms of fear extinction. *Mol Psychiatry* **12**, 120–150.

Myers, K.M. and Davis, M. (2002). Behavioural and neural analysis of extinction. *Neuron* **36**, 567–584.

Nader, K. (2003). Memory traces unbound. *Trends Neurosci* **26**, 65–72.

Nader, K., Schafe, G.E. and Le Doux, J.E. (2000a). Fear memories require protein synthesis in the amygdala for reconsolidation after retrieval. *Nature* **406**, 722–726.

Nader, K., Schafe, G.E. and Le Doux, J.E. (2000b). The labile nature of consolidation theory. *Nat Rev Neurosci* **1**, 216–219.

Pavlov, I.P. (1927). *Conditioned Reflexes* (London: Oxford UP).

Pedreira, M.E., Perez-Cuesta, L.M. and Maldonado, H. (2002). Reactivation and reconsolidation of long-term memory in the crab Chasmagnathus: Protein synthesis requirement and mediation by NMDA-type glutamatergic receptors. *J Neurosci* **22**, 8305–8311.

Pedreira, M.E. and Maldonado, H. (2003). Protein synthesis subserves reconsolidation or extinction depending on reminder duration. *Neuron* **38**, 863–869.

Pedreira, M.E., Pérez-Cuesta, L.M. and Maldonado, H. (2004). Mismatch between what is expected and what actually occurs triggers memory reconsolidation or extinction. *Learn Mem* **11**, 579–585.

Pérez-Cuesta, L.M., Hepp, Y., Pedreira, M.E. and Maldonado, H. (2007). Memory is not extinguished along with CS presentation but within a few seconds after CS-offset. *Learn Mem* **14**, 101–108.

Peters, J., Dieppa-Perea, L.M., Melendez, L.M. and Quirk, G.J. (2010). Induction of fear extinction with hippocampal-infralimbic BDNF. *Science* **328**, 1288–1290.

Phelps, E.A. and LeDoux, J.E. (2005). Contributions of the amygdala to emotion processing: From animal models to human behaviour. *Neuron* **48**, 175–187.

Przybyslawski, J. and Sara, S.J. (1997). Reconsolidation of memory after its reactivation. *Behav Brain Res* **84**, 241–246.

Przybyslawski, J., Roullet, P. and Sara, S.J. (1999). Attenuation of emotional and nonemotional memories after their reactivation: Role of beta adrenergic receptors. *J Neurosci* **19**, 6623–6628.

Quevedo, J., Vianna, M.R., Roesler, R., de-Paris, F., Izquierdo, I. and Rose, S.P. (1999). Two time windows of anisomycin-induced amnesia for inhibitory avoidance training in rats: Protection from amnesia by pretraining but not pre-exposure to the task apparatus. *Learn Mem* **6**, 600–607.

Quirk, G.J., Gracia, R., Gonzalez-Lima, F. (2006). Prefrontal mechanisms in extinction of conditioned fear. *Biol Psychiatry* **60**, 337–343.

Quirk, G.J., Russo, G.K., Barron, J.L. and Lebron, K. (2000). The role of ventromedial prefrontal cortex in the recovery of extinguished fear. *J Neurosci* **20**, 6225–6231.

Rauch, S.L., Shin, L.M. and Phelps, E.A. (2006). Neurocircuitry models of posttraumatic stress disorder and extinction: Human neuroimaging research — past, present, and future. *Biol Psychiatry* **60**, 376–382.

Reijmers, L.G., Perkins, B.L., Matsuo, N. and Mayford, M. (2007). Localization of a stable neuronal correlate of associative memory. *Science* **317**, 1230–1233.

Rescorla, R.A. and Heth, C.D. (1975). Reinstatement of fear to an extinguished conditioned stimulus. *J Exp Psychol Anim Behav Process* **1**, 88–96.

Rescorla, R.A. (2001). Experimental extinction. In R.R. Mowrer and S. Klein, eds. *Handbook of Contemporary Learning Theories*. pp. 119–154.

Robbins, S.J. (1990). Mechanisms underlying spontaneous recovery in autoshaping. *J Exp Psychol Anim Behav Process* **16**, 235–249.

Rose, J.K. and Rankin, C.H. (2006). Blocking memory reconsolidation reverses memory-associated changes in glutamate receptor expression. *J Neurosci* **8**, 11582–11587.

Rosenblum, K., Meiri, N. and Dudai, Y. (1993) Taste memory: The role of protein synthesis in gustatory cortex. *Behav Neural Biol* **59**, 49–56.

Rossato, J.I., Bevilaqua, L.R., Myskiw, J.C., Medina, J.H., Izquierdo, I. and Cammarota, M. (2007). On the role of hippocampal protein synthesis in the consolidation and reconsolidation of object recognition memory. *Learn Mem* **14**, 36–46.

Sananbenesi, F., Fischer, A., Wang, X., Schrick, C., Neve, R., Radulovic, J. and Tsai, L.H. (2007). A hippocampal Cdk5 pathway regulates extinction of contextual fear. *Nat Neurosci* **10**, 1012–1019.

Sangha, S., Scheibenstock, A. and Lukowiak, K. (2003a). Reconsolidation of a long-term memory in Lymnaea requires new protein and RNA synthesis and the soma of right pedal dorsal 1. *J Neurosci* **23**, 8034–8040.

Santini, E., Ge, H., Ren, K., deOrtiz, S.P. and Quirk, G.J. (2004). Consolidation of fear extinction reuires protein synthesis in the medial prefrontal cortex. *J Neurosci* **24**, 5704–5710.

Sangha, S., Scheibenstock, A., Morrow, R. and Lukowiak, K. (2003b). Extinction requires new RNA and protein synthesis and the soma of the cell right pedal dorsal 1 in Lymnaea stagnalis. *J Neurosci* **23**, 9842–9851.

Santini, E., Muller, R.U. and Quirk, G.J. (2001). Consolidation of extinction learning involves transfer from NMDA-independent to NMDA-dependent memory. *J Neurosci* **21**, 9009–9017.

Sara, S.J. (2000). Retrieval and reconsolidation: Toward a neurobiology of remembering. *Learn Mem* **7**, 73–84.

Schafe, G.E., Nadel, N.V., Sullivan, G.M., Harris, A. and LeDoux, J.E. (1999). Memory consolidation for contextual and auditory fear conditioning is dependent on protein synthesis, PKA, and MAP kinase. *Learn Mem* **6**, 97–110.

Schiller, D., Monfils, M.H., Raio, C.M., Johnson, D.C., Ledoux, J.E. and Phelps, E.A. (2010). Preventing the return of fear in humans using reconsolidation update mechanisms. *Nature* **463**, 49–53.

Schneider, A.M. and Sherman, W. (1968). Amnesia: A function of the temporal relation of footshock to electroconvulsive shock. *Science* **159**, 219–221.

Squire, L.R. and Alvarez, P. (1995). Retrograde amnesia and memory consolidation: A neurobiological perspective. *Curr Opin Neurobiol* **5**, 169–177.

Suzuki, A., Josselyn, S.A., Frankland, P.W., Masushige, S., Silva, A.J. and Kida, S. (2004). Memory reconsolidation and extinction have distinct temporal and biochemical signatures. *J Neurosci* **24**, 4787–4795.

Suzuki, A., Mukawa, T., Tsukagoshi, A., Frankland, P.W. and Kida, S. (2008). Activation of LVGCCs and CB1 receptors required for destabilisation of reactivated contextual fear memories. *Learn Mem* **15**, 426–433.

Taubenfeld, S.M., Milekic, M.H., Monti, B. and Alberini, C.M. (2001). The consolidation of new but not reactivated memory requires hippocampal C/EBPbeta. *Nat Neurosci* **4**, 813–818.

Tronel, S., Milekic, M.H. and Alberini, C.M. (2005). Linking new information to a reactivated memory requires consolidation and not reconsolidation mechanisms. *PLoS Biol* **3**, 1630–1638.

Tronson, N.C., Wiseman, S.L., Olausson, P. and Taylor, J.R. (2006). Bidirectional behavioural plasticity of memory reconsolidation depends on amygdalar protein kinase A. *Nat Neurosci* **9**, 167–169.

Teich, A.H., McCabe, P.M., Gentile, C.C., Schneiderman, L.S., Winters, R.W., Liskowsky, D.R. and Schneiderman N. (1989). Auditory cortex lesions prevent the extinction of Pavlovian differential heart rate conditioning to tonal stimuli in rabbits. *Brain Res* **480**, 210–218.

Vianna, M.R., Szapiro, G., McGaugh, J.L., Medina, J.H. and Izquierdo, I. (2001). Retrieval of memory for fear-motivated training initiates extinction requiring protein synthesis in the rat hippocampus. *Proc Natl Acad Sci USA* **98**, 12251–12254.

Viola, H., Furman, M., Izquierdo, L.A., Alonso, M., Barros, D.M., de Souza, M.M., Izquierdo, I. and Medina, J.H. (2000). Phosphorylated camp response element-binding protein as a marker of memory processing in rat hippocampus: Effect of novelty. *J Neurosci* **20**, RC112.

von Hertzen, L.S. and Giese, K.P. (2005). Memory reconsolidation engages only a subset of immediate-early genes induced during consolidation. *J Neurosci* **25**, 1935–1942.

Walker, M.P., Brakefield, T., Hobson, J.A. and Stickgold, R. (2003). Dissociable stages of human memory consolidation and reconsolidation. *Nature* **425**, 616–620.

Wang, S.H., de Oliveira Alvares, L. and Nader, K. (2009). Cellular and systems mechanisms of memory strength as a constraint on auditory fear reconsolidation. *Nat Neurosci* **12**, 905–912.

Wang, S.H., Ostlund, S.B., Nader, K. and Balleine, B.W. (2005). Consolidation and reconsolidation of incentive learning in the amygdala. *J Neurosci* **25**, 830–835.

Winters, B.D., Tucci, M.C. and DaCosta-Furtado, M. (2009). Older and stronger object memories are selectively destabilized by reactivation in the presence of new information. *Learn Mem* **16**, 545–553.

Memory Circuits in *Drosophila* 11

Katsuo Furukubo-Tokunaga*, Zoe N. Ludlow† and Frank Hirth†

1. Introduction

Learning and memory are neuronal mechanisms by which animal and human behaviour changes in response to experience. Learning refers to the neuronal process of memory acquisition, whereas memory itself is the maintenance of experience-induced changes in neuronal activities over time. These changes can be related to single or multiple sensory modalities, such as vision, olfaction and taste, and are either associative or non-associative with regard to the combination of the cognate modalities. Based on retention time, memory can be classified into either short- or long-term; it is thought to be processed and stored in distinct regions of the brain depending on its forms (Squire and Kandel, 2009). Although a wealth of knowledge is accumulating about the biochemical pathways involved in memory formation, the neurocircuits underlying learning and memory are rather enigmatic and have been the focus of intense research aiming to elucidate their functional anatomy. One of the foremost model organisms to address this issue is the fruitfly *Drosophila melanogaster*.

Drosophila provides a powerful genetic model system for the analysis of the molecular and cellular basis of behaviours that are relevant not only to insects but also to humans; the *Drosophila* brain is complex enough to mediate a range of sophisticated behaviours, yet it is still small enough for in depth structure–function analyses as compared to the more complex brains of vertebrates. Indeed, many genes and molecular processes first discovered in *Drosophila* have proven to be conserved in other organisms, including humans. In addition, genome projects in the past decade have successfully sequenced and assembled the entire coding regions in flies and humans, demonstrating that two-thirds of the human genes have homologous counterparts in the *Drosophila* genome

* Department of Life and Environmental Sciences, University of Tsukuba, Japan.
† Department of Neuroscience, MRC Centre for Neurodegeneration Research, Institute of Psychiatry, King's College London, UK.

(Adams *et al.*, 2000; Venter *et al.*, 2001). These data provide compelling evidence for the structural conservation of genes due to common origin, and suggest a deep homology underlying cell biological mechanisms that extends beyond gene structures to patterned protein expressions and functions. This notion is further supported by experiments demonstrating that genes involved in *Drosophila* and human brain development can functionally substitute each other, highlighting evolutionary conserved mechanisms underlying brain development in insects and mammals (Leuzinger *et al.*, 1998; Nagao *et al.*, 1998; reviewed in Hirth and Reichert, 1999, 2007).

Drosophila as a genetically tractable model system also offers unique opportunities in the study of learning and memory; fundamental cellular processes related to neurobiology are similar in *Drosophila* and humans, including synapse formation, neuronal communication, membrane trafficking, and cell death. Moreover, flies and mammals have common neurobiological bases of elementary behaviours, including sensory perception, as well as aspects of sensory integration and motor output. Most significantly, *Drosophila* displays various forms of associative and non-associative memory that are based on the plasticity of sensory modalities similar to mammals. Studies for more than 4 decades have identified a large number of genes and signalling pathways that underlie various forms of learning and memory. These include cAMP signalling involving cyclic-AMP response-element binding protein (CREB) and calcium–calmodulin kinase II (CaMKII) signalling, both of which play essential roles in long-term memory formation not only in insects but also in vertebrates [for reviews, see Margulies *et al.* (2005); McGuire *et al.* (2005); Keene and Waddell (2007); Davis (2011)].

In addition to these insights into the biochemical and intracellular signalling mechanisms underlying memory formation, *Drosophila* memory studies have identified central circuits that alter their neuronal activities upon associative olfactory learning and may harbor memory traces [for a recent review, see Davis (2011)]. Given the importance of the cellular and molecular organisation of memory circuitry, knowledge about the developmental neuroanatomy of the brain circuits involved in learning and memory is indispensible. Investigations into the development of memory circuits provide important insights not only into the cytoarchitectures of the memory networks but also into physiological functions — a prerequisite for comprehensive understanding of the neuronal mechanisms of memory. Here, we review recent progress on the development and specification of two major memory circuits in *Drosophila*, the mushroom bodies (MBs) and the central complex (CCX). Apart from these two brain centres, pharmacological and molecular studies in honeybees (Hammer and Menzel, 1998) and fruit flies (Davis, 2005; Thum *et al.*, 2007; Yu *et al.*, 2004) signify the antennal lobe (AL) as another important neuronal structure involved in the formation of short-term memory. For the anatomy and development of the AL, the reader is referred to excellent reviews that complement the focus of this review (Masse *et al.*, 2009; Vosshall and Stocker, 2007).

2. *Drosophila* brain development

Similar to mammalian brain development, the *Drosophila* brain derives from a monolayered epithelium called the neuroectoderm. During germ layer formation, the antagonistic interaction between the extracellular signalling proteins Decapentaplegic (Dpp) and Short-gastrulation (Sog) establish the region of the embryo that attains neurogenic potential and forms neuroectoderm, in which Sog is expressed and inhibits the action of invading Dpp signals (Hirth and Reichert, 2007). Subsequent neurogenesis is characterised by two neurogenic periods: one during embryogenesis and another during larval and pupal stages. The precursor cells of the developing brain, termed neuroblasts (NBs), derive from the embryonic procephalic neurogenic region to form proliferative clusters, and undergo a characteristic number of cell divisions, thereby generating neurons required for larval life. Towards the end of embryogenesis, most NBs stop proliferating and enter a period of quiescence. During the first larval instar stage, quiescent NBs re-enter the cell cycle in a characteristic spatiotemporal pattern to perform a second round of neurogenesis, resulting in the postembryonic neural progeny. These two phases of brain development are followed by an extensive morphological transformation during metamorphosis that ultimately leads to the adult fly brain (Figure 11.1) (Truman *et al.*, 1993; Maurange and Gould, 2005; Kim and Hirth, 2009).

The central brain of *Drosophila* derives from 106 embryonic brain neuroblasts (NBs) which were identified based on their position relationships and the NB-specific expression of segment polarity genes as well as dorsoventral patterning genes (Urbach and Technau, 2004). During the transition from immobile embryonic to crawling larval life, the 106 embryonic NBs undergo quiescence, and around 85 postembryonic NBs re-enter mitosis after larval hatching (Ito and Hotta, 1992; Truman *et al.*, 1993). These include postembryonic NB lineages whose neurons differentiate into specific functional domains of the adult brain such as the CCX, ALs, and MBs (Ito and Awasaki, 2008).

Another lineage-specific regulation of self-renewal has been observed for atypical postembryonic NBs recently identified in the dorsomedial region of the larval central brain (Bello *et al.*, 2008; Bowman *et al.*, 2008; Boone and Doe, 2008). Molecular genetic analyses indicate their mode of division is

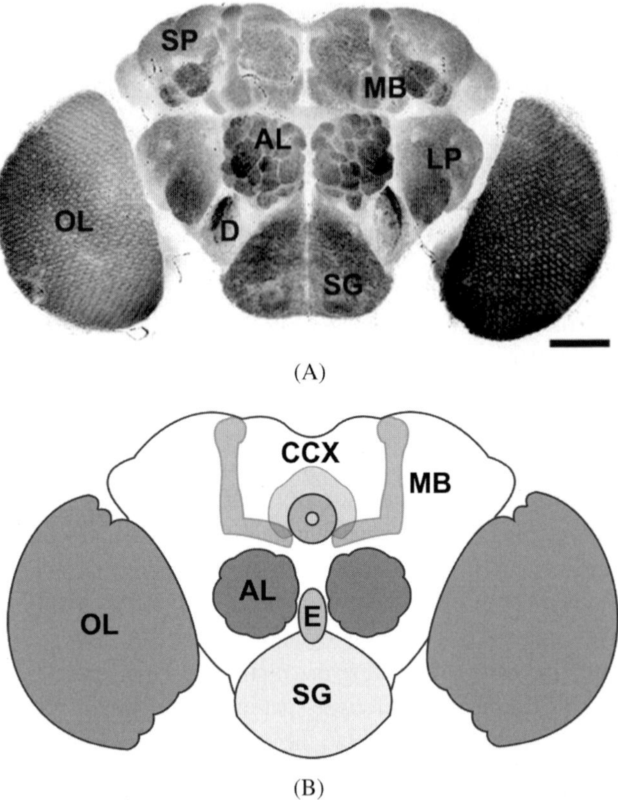

Figure 11.1. The adult brain of *Drosophila* and its major memory circuits. (A) Confocal image of a dissected adult brain immuno-labelled with an antibody that specifically recognizes synaptic terminals. This allows the visualisation of cortical areas in the fly brain, including optic lobes (OL), antennal lobes (AL), superior protocerebrum (SP), lateral protocerebrum (LP), mushroom bodies (MB), deuterocerebrum (D), and suboesophageal ganglion (SG). (B) Schematic representation of major neuropil structures and areas in the adult fly brain. Abbreviations as in A; CCX, central complex; E, oesophagus. The MBs have been identified as neural circuits involved in olfactory memory processing and courtship conditioning, whereas the CCX and its substructures are involved in the processing of spatial orientation and visual place memory. Scale bar: 50 μm.

morphologically symmetrical, but molecularly asymmetrical in that key cell fate determinants are segregated into only one of the two daughter cells (Bello *et al.*, 2008). These neural stem cells thus generate secondary intermediate precursor cells that undergo multiple rounds of self-renewing transit-amplifying divisions (Bowman *et al.*, 2008). Based on these morphological and molecular features, dorso-medial central brain postembryonic NB lineages and the resulting intermediate transit amplifying GMCs are described as type II lineages (Boone and Doe, 2008). The predominant, type I, NB lineages undergo repeated asymmetric cell divisions leading to GMCs that usually complete one terminal symmetric division to generate two differentiating neurons (Jan and Jan, 1998). Recent clonal analysis indicated that larval type II lineages contribute to the development of the adult CCX, although the exact contribution of each type II postembryonic NB is not clear yet (Bayraktar *et al.*, 2010).

Although approximately 95% of the 200,000 neurons in the adult *Drosophila* brain are generated postembryonically, the main architecture of the adult brain is already laid down during embryonic neurogenesis (Hirth and Reichert, 1999; Urbach and Technau, 2004). Thus, the *Drosophila* brain is composed of an anterior supraesophageal ganglion and a posterior subesophageal ganglion, both of which are interconnected by an axon scaffold surrounding the gut (Nassif *et al.*, 1998; Younossi-Hartenstein *et al.*, 2006). The supraoesophageal ganglion comprises the protocerebrum, the deutocerebrum, and the tritocerebrum, whereas the subesophageal ganglion comprises the mandibular, maxillary, and labial neuromeres (Boyan *et al.*, 2007). The major neural circuits underlying learning and memory formation in the *Drosophila* brain are located in the protocerebrum, including the MBs and the CCX, while the ALs are located in the deutocerebrum (Figure 11.1B).

3. The mushroom bodies — Centres for associative olfactory learning and memory

A large number of studies have demonstrated that the MBs are centres for higher-order functions, including olfactory learning (Heisenberg, 2003; Davis, 2005; Berry *et al.*, 2008), courtship behaviour (Villella and Hall, 2008), sleep regulation (Sehgal *et al.*, 2007), and cognitive functions such as visual context generalisation and attention (Guo *et al.*, 2009; Liu *et al.*, 1999; van Swinderen *et al.*, 2009). Historically, Erber and Menzel (1980) found that localised cooling of the MBs disrupt memory in the honeybee after conditioning in an olfactory reward paradigm. By analysing *Drosophila* mutants with defects in various regions of the brain, Heisenberg *et al.* (1985) identified the MBs as an indispensible structure for olfactory memory. After these initial discoveries, Davis *et al.* found that a memory mutant gene, *dunce*, which encodes cAMP phosphodiesterase, was preferentially expressed in the MBs (Nighorn *et al.*, 1991). Subsequent molecular genetic studies of memory genes demonstrated that many of the genes required for *Drosophila* olfactory memory are indeed preferentially expressed in the MBs (Berry *et al.*, 2008; Davis, 2005).

The functional importance of the MBs to memory formation was also shown in parallel by hydroxyurea-mediated pharmacological ablation of the MBs in the *Drosophila* brain (de Belle and Heisenberg, 1994), targeted expression of a constitutively active Gαs subunit that specifically inactivates the intracellular signalling cascade in the intrinsic MB neurons (Kenyon cells) (Connolly *et al.*, 1996), and targeted rescue of a memory mutant, *rutabaga*, which encodes an adenylate cyclase, via specific expression of a *rutabaga* transgene in a subset of the Kenyon cells (Zars, 2000). Behavioural studies with a temperature-sensitive form of dynamin (Shibire^{ts}) (Kitamoto, 2002) demonstrated that synaptic output of MB neurons is required for retrieval but not acquisition of olfactory memory (Dubnau *et al.*, 2001; McGuire *et al.*, 2001). Moreover, by careful characterisation of GAL4 drivers expressed in MB neurons, Krashes *et al.* (2007) found that different subsets of MB neurons are sequentially involved in forming, stabilising and retrieving memory during *Drosophila* odour memory processing. Finally, recent imaging studies uncovered the dynamics of cAMP signalling and PKA activities in MB neurons, further demonstrating the importance of MB neurons in associative olfactory learning (Gervasi *et al.*, 2010; Tomchik and Davis, 2009).

3.1. *Structure of the mushroom bodies in the adult Drosophila brain*

In the adult *Drosophila* brain, the MBs consist of approximately 2,500 neurons (Kenyon cells), which send densely packed parallel fibers through the peduncle to the lobes (Figure 11.2). The cell bodies of the Kenyon cells are located at the dorsal cortex, extending their dendrites into the calyx, where they receive olfactory information from projection neurons of the ALs. More distally, the bundles of MB axons bifurcate into the vertical and horizontal lobes. In adult *Drosophila*, Kenyon cells are grouped into one of the three classes, g, α′/β′ or α/β, based on their axonal trajectory (Crittenden *et al.*, 1998). While axons of α′/β′ and α/β neurons bifurcate distally, axons of the g neurons extend only horizontally. Accordingly, the MB lobes comprise two intertwined vertical lobes, each consisting of the fibers of α′ and α neurons, and three types of parallel horizontal lobes, each consisting of fibers of β, β′, and α neurons (Crittenden *et al.*, 1998). In addition to this apparently complex external morphology, the calyx of the MB seems to have an intricate internal organisation or microglomerular structures

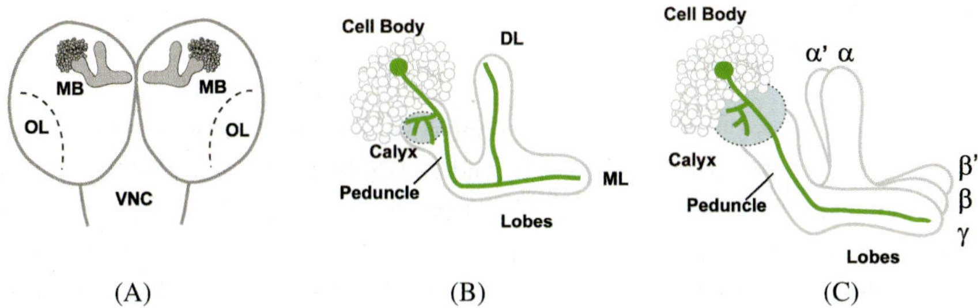

Figure 11.2. Anatomy of the *Drosophila* mushroom bodies. (A) Dorsal view of the larval central nervous system showing the developing (MB), the optic lobe (OL) primordium, and the anterior part of the ventral nerve cord (VNC). (B) Structure of the larval MBs. The larval MBs consist of a single dorsal lobe (DL) and a single medial lobe (ML). Each of the MB neurons branches dendrites in the calyx, extends its axon through the peduncle and bifurcates into the DL and ML. (C) Structure of the adult MB. The adult MB consists of two vertical and three horizontal lobes. The cell bodies of the MB neurons are located at the dorsal cortex and extend their dendrites into the calyx. The bundles of MB axons bifurcate into the vertical and horizontal lobes. The Kenyon cells are grouped into one of three classes, γ, α′/β′ or α/β, based on their axonal trajectory. While each of the axons of the α′/β′ or α/β neurons bifurcate into the distal and horizontal lobe, the axons of the γ neurons extend only horizontally (shown in green).

(Jefferis *et al.*, 2007; Lin *et al.*, 2007; Tanaka *et al.*, 2008), consistent with the notion that odour information transmitted to the calyx may first be processed there, and then transmitted to the output neurons that synapse on the axons of MB neurons on the lobes. Olfactory memory information is also transferred and modulated between different sets of lobes, particularly during acquisition and consolidation of associative olfactory memory (Gervasi *et al.*, 2010; Keene and Waddell, 2007; Krashes *et al.*, 2007). Neurocircuitry blocking experiments suggest that synaptic outputs of α′/β′ neurons are required for the acquisition and stabilisation of odour memory, whereas the activities of α/β neurons are required for memory retrieval (Gervasi *et al.*, 2010; Keene and Waddell, 2007; Krashes *et al.*, 2007; Pascual and Preat, 2001). Functional neuroimaging experiments further suggest that short-term memory trace is formed in α′/β′ neurons while long-term memory trace is formed in α/β neurons. Late-phase long-term memory that persists for days forms in the g neurons (Akalal *et al.*, 2010, 2011; Yu *et al.*, 2006a) [reviewed in Davis (2011)]. These results demonstrate functional subdivisions of the *Drosophila* MBs, and indicate that distinctive temporal forms of olfactory memory are processed and stored in different sets of neuronal circuitries in the MB.

3.2. *Development and structure of the embryonic and larval mushroom bodies*

Kenyon cells derive from four NBs born during embryonic stages (Ito and Hotta, 1992; Noveen *et al.*, 2000; Tettamanti, 1997). The MB primordia can be identified with a specific marker at the most anterior region (according to neuraxis) of the embryonic brain shortly after the protocerebral anlagen are formed (Kurusu *et al.*, 2000; Noveen *et al.*, 2000; Tettamanti, 1997). By late embryonic stages, the major axonal tracts of the MB are established with a peduncle projection that splits into the medial and dorsal lobes (Figure 11.3). The embryonic MB neurons achieve this bifurcated axonal structure by projecting to the medial lobe first and then sending collaterals to the dorsal lobe. The original architecture of the embryonic MB is basically maintained during the postembryonic growth in the subsequent larval instars with increasing numbers of Kenyon cells and their axonal projections. Thus, the larval MB involves only a single set of the horizontal and vertical lobes. Internally, the axonal projections of the larval Kenyon cells are organised in concentric layers surrounding a core that consists of fine axonal projections, rich in actin filaments (Figure 11.3). Mosaic analyses revealed sequential generation of the MB layers, in which axons of newly born Kenyon cells firstly project into the core and shift to more distal layers as they undergo further differentiation (Kurusu *et al.*, 2002). Regulatory proteins involved in growth cone guidance such as an immunoglobulin family molecule, Dscam (Zhan *et al.*, 2004) and receptor tyrosine phosphatases (Kurusu and Zinn, 2008) are preferentially expressed in the core fibers. Recently, Mochizuki *et al.* (2011) showed preferential expression of UNC-51, a conserved Ser/Thr kinase, and kinesin motor proteins in the MB core fibers, suggesting active intracellular transport along the newly elongating axons. Interestingly, the *Dscam* gene, which can potentially generate 38,016 different isoforms, is required for the proper segregation of axonal branches to the different lobes in the adult MBs (Wang *et al.*, 2002). Moreover, Dscam plays an essential role in promoting the sorting and selective fasciculation of young axons in the peduncle of the developing MBs (Zhan *et al.*, 2004).

Although the adult MBs have been shown as centres for olfactory memory, physiological functions of the larval MBs had not been addressed until recently. Honjo and Furukubo-Tokunaga (2005) showed that synaptic output of larval MB neurons is required for retrieval but not for acquisition and retention of olfactory memory in larvae. Comparative studies of memory stability in larvae revealed differential retention of olfactory memory formed with aversive and appetitive stimuli (Honjo and Furukubo-Tokunaga, 2005, 2009); appetitive memory

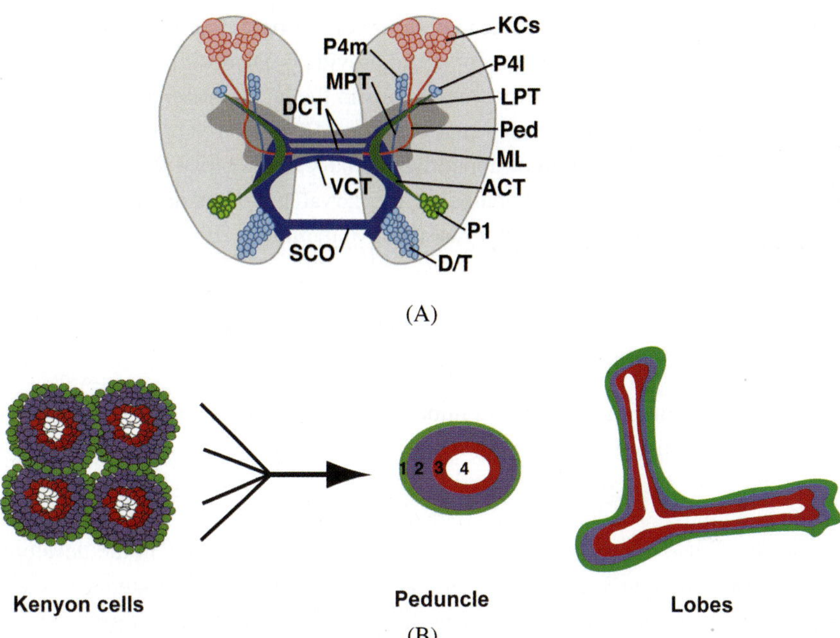

Figure 11.3. Development of the embryonic and larval mushroom bodies in *Drosophila*. (A) Schematic representation of the MB primordia in the embryonic brain. Frontal view of the embryonic brain at late stage 16. ACT, antenno-cerebral tract; DCT, dorsal commissural tract; KC, Kenyon cells; LPT, lateral protocerbral tract; MPT, medial protocerbral tract; Ped, peduncle; VCT, ventral commissural tract. P1, P4l, P4m and D/T, fibre tract founder clusters (Nassif *et al.*, 1998). Only major tracts are shown; optic lobes are excluded. (B) Concentric layer organisation of the larval mushroom bodies. Dorsal images of Kenyon cell clusters and cross sections of the peduncle and lobes. Corresponding subdivisions are shown in same colours. Note that the concentric subdivisions of each of the four Kenyon cell clusters topologically correspond to the unified concentric subdivisions in the peduncle and the lobes. The core consists of a bundle of newly formed axon fibres containing densely packed actin filaments. Axons of newly born Kenyon cells project into the core to shift to more distal layers as they undergo further differentiation.

generated with sucrose is maintained for 120 min, but aversive memory induced with quinine is lost in 20 min, even with a same odour and under an identical training paradigm. Neurocircuitry analyses showed that synaptic output of octopaminergic (OA) and dopaminergic (DA) neurons, which exhibit distinctive innervation patterns in the MB and AL, is differentially required for the acquisition of appetitive and aversive memory, respectively (Honjo and Furukubo-Tokunaga,

2005, 2009). These results suggest that genetically programmed memory circuitries constructed in early development might provide predisposition in the efficacy of inducing longer-lived memory components in the larval brain. Recently, Pauls *et al.* (2010) further demonstrated that intrinsic MB neurons born in the embryonic stage mediate formation of appetitive memory at the larval stage. They also identified 10 distinctive subdomains of the larval MB that are important for acquisition and retrieval of olfactory associative memory.

3.3. *Drosophila Pax6 genes are required for mushroom body development*

A large number of genes have been implicated in the regionalisation of the vertebrate forebrain during development. Transcription factors such as Otx, Emx, Dlx, Nkx, and Pax6 have essential functions in setting up the initial patterning and subdivision of the developing brain (Hoch *et al.*, 2009). Molecular genetic studies in flies have also demonstrated that *Pax6* genes play major roles in the development of the *Drosophila* MBs (Callaerts *et al.*, 2001; Furukubo-Tokunaga *et al.*, 2009; Kurusu *et al.*, 2000; Noveen *et al.*, 2000). The vertebrate *Pax6* gene is expressed in various regions of the telencephalon anlagen including those of the olfactory bulb, piriform cortex and amygdala, which play crucial roles in olfactory perception and memory formation. Mutations of *Pax6* result in profound defects in vertebrate brain development affecting a variety of these forebrain structures. In contrast to a single *Pax6* gene in vertebrates, the genome of *Drosophila* melanogaster encodes two *Pax6* genes: *eyeless* (*ey*) and *twin of eyeless* (*toy*) (Czerny *et al.*, 1999; Quiring *et al.*, 1994), both are expressed in various brain regions during embryonic and larval stages, and have important functions in the formation of major brain structures including the MBs (Adachi *et al.*, 2003; Callaerts *et al.*, 2001; Furukubo-Tokunaga *et al.*, 2009; Kurusu *et al.*, 2000; Noveen *et al.*, 2000).

Loss-of-function of either *ey* or *toy* causes profound structural disorganisation not only in the MBs but also in other brain structures, including the ellipsoid body (EB) of the CCX (Callaerts *et al.*, 2001; Furukubo-Tokunaga *et al.*, 2009; Kurusu *et al.*, 2000). In addition, a genome-wide survey of the *Drosophila* MB transcripts identified a large number of genes that are preferentially expressed in MBs and thus may represent candidates of the downstream targets of the *Pax6* genes in brain development (Kobayashi *et al.*, 2006). Indeed, RNA interference analyses showed that many of the identified genes are required for the development of normal MB lobes, while the rest of the genes might be required for the plasticity and physiological functions of the MB neurons.

3.4. *Four neuroblasts sequentially generate different types of mushroom body neurons*

Systematic clonal analysis demonstrated that each of the four MB-NBs produces an almost indistinguishable set of MB neurons during the course of development (Ito *et al.*, 1997). Subsequent studies further demonstrated that single MB-NBs sequentially generate the three types of morphologically distinct neurons in a temporally regulated manner (Lee *et al.*, 1999); neurons projecting into the γ lobe of the adult MB are born first, prior to the mid-third instar larval stage, and neurons projecting into the α′/β′ lobes are born between the mid-third instar larval stage and puparium formation. Finally, neurons projecting into the α/β lobes are born after puparium formation.

Several studies suggest that temporal progression of NB competence is defined by sequential expression of transcription factors that are maintained in GMCs and postmitotic neurons when they are produced by the cognate NBs (Brody and Odenwald, 2005; Egger *et al.*, 2008; Isshiki *et al.*, 2001; Jacob *et al.*, 2008; Pearson and Doe, 2003). By genetically labelling single MB neurons at specific developmental times, Zhu *et al.* (2006) identified a BTB-Zinc finger protein, Chinmo (Chronologically inappropriate morphogenesis), which controls the generation of the multiple neuronal identities by the MB-NBs over development. Intriguingly, Chinmo mutant neurons born at early developmental stages often adopted the fates of late-born neurons of the same lineage. In addition, early-born MB neurons seem to contain higher amount of Chinmo than their later-born siblings, providing a temporal gradient of the regulatory protein within each NB lineage. A similar mechanism has also been found in the sequential generation of different neuronal identities of the AL projection neurons (Jefferis *et al.*, 2001; Zhu *et al.*, 2006).

3.5. *Mushroom body neuroblasts proliferate throughout development*

As noted above, the proliferation mode of the *Drosophila* NBs can be subdivided into embryonic and postembryonic phases (Truman and Bate, 1988). Intriguingly, while most of the *Drosophila* NBs pause proliferation between embryonic and postembryonic neurogenesis, the four MB-NBs remain active along with a progenitor of the lateral AL neurons (Ito and Hotta, 1992; Prokop and Technau, 1994; Stocker *et al.*, 1997; Truman and Bate, 1988). Thus, only five NBs are active in each brain hemisphere for the first 8–12 h after larval hatching. As for the non-MB NBs, glia cells play key roles in regulating the initiation and termination of their quiescence by secreting the glycoprotein Anachronism (Ebens *et al.*, 1993), while an extracellular matrix molecule of the Perlecan family, participates in the activation of quiescent NBs (Park *et al.*, 2003; Voigt *et al.*, 2002). In addition, molecules involved in various signalling networks, such as Branchless/FGF, Hedgehog, and the target of rapamycin (TOR) pathway are required to regulate the initiation of NB division (Barrett *et al.*, 2008; Sousa-Nunes *et al.*, 2011). On the other hand, little is know about the molecular mechanism that maintains the proliferative activity of the MB-NBs during the embryonic and postembryonic transition.

The continued mitotic activity of the *Drosophila* NBs in the postembryonic period is regulated by numerous genes, including a histone acetyltransferase (*enok*) (Scott *et al.*, 2001), *Lis1* (Liu *et al.*, 2000), *RhoA* (Lee *et al.*, 2000b), as well as *Dhc64C* and *Roadblock*, which encode Dynein heavy and light chains, respectively (Reuter *et al.*, 2003). Genes controlling asymmetric cell division of NBs are also required for efficient NB proliferation (Caussinus and Hirth, 2007; Sousa-Nunes *et al.*, 2010; Yu *et al.*, 2006b). Although most NBs in the *Drosophila* brain cease postembryonic neurogenesis by the early pupal stage, only the four MB-NBs maintain exceptional proliferation activity until the end of the pupa stage (Ito and Hotta, 1992; Kurusu *et al.*, 2009; Truman and Bate, 1988). Recent studies (Kim *et al.*, 2009; Kurusu *et al.*, 2009) demonstrated that a conserved orphan nuclear receptor, Tailless (TLL), and a leucine-zipper protein, Bunched, are required for prolonged cell division of the MB-NBs at the pupal stages. Intriguingly, the *Drosophila* TLL exhibits high sequence similarity with the vertebrate homologue TLX (NR2E1) in the DNA binding domain (Monaghan *et al.*, 1995; Yu *et al.*, 1994). *Tlx* mutant mice exhibit a reduction of neuron numbers in rhinencephalon and limbic structures with emotional and learning defects (Monaghan *et al.*, 1997; Roy *et al.*, 2004; Roy *et al.*, 2002). Postnatally, TLX is localised to the adult neurogenic regions including the subgranular layer of the dentate gyrus to maintain stem cells in a proliferative and undifferentiated state (Monaghan *et al.*, 1997; Shi *et al.*, 2004; Zhang *et al.*, 2008). These results, as a whole, suggest an intriguing commonality of the genetic programs between flies and mammals that control the prolonged proliferation of neural stem cells in brain centres involved in memory formation and storage.

3.6. *Larval mushroom body axons are pruned during metamorphosis*

As in the vertebrate brain, selective synapse elimination and axon pruning is a vital process for the refinement of functional neural circuits in the fly brain.

Excessive neuronal processes formed during early neural development are selectively pruned to ensure the precise connectivity of the functional circuit in the adult brain. As a holometabolic insect, *Drosophila* undergoes complete metamorphosis. This drastic alteration applies not only to the various body parts but also to the peripheral and central nervous systems. Consequently, a large-scale reorganisation of neural networks is initiated in the fly brain by the steroid hormone 20-hydroxyecdysone (hereafter referred to as ecdysone) to construct the complex adult brain (Armstrong *et al.*, 1998; Lee *et al.*, 2000a).

At the beginning of the pupal stage, MBs undergo massive degeneration and reorganisation to form the adult architecture characterised by the three intertwined-projecting groups: α/β, α′/β′ and γ (Crittenden *et al.*, 1998). Molecular genetic studies demonstrated that the steroid hormone ecdysone signals MB neurons to recruit active glial engulfment and phagocytosis of the larval axons for pruning. During the onset of pupal metamorphosis, a spike of ecdysone induces local axon degeneration in the γ neurons, which express ecdysone receptors (Lee *et al.*, 2000a; Zhu *et al.*, 2003). While the initiation of axonal pruning requires cell-autonomous activation of the ecdysone receptor signalling cascades, local axonal degeneration is also mediated by intrinsic activity of the ubiqutin-proteasome system (Hoopfer *et al.*, 2008; Watts *et al.*, 2003) as well as postmitotic function of cohesin, which is known to maintain sister-chromatid cohesion during cell division (Schuldiner *et al.*, 2008). In addition, soon after the initiation of metamorphosis, a population of glial cells actively invades the MB lobes and engulfs axonal varicosities (Awasaki and Ito, 2004; Watts *et al.*, 2004). Moreover, glial cells seem to play an active role in axonal pruning since the loss of axonal varicosities and axonal pruning are both severely suppressed when the endocytosis and phagocytosis activities of glial cells are suppressed by temporally blocking dynamin function (Awasaki and Ito, 2004). Interestingly, these axonal degeneration mechanisms observed in the *Drosophila* MB parallels the processes involved in Wallerian degeneration in mammals, in which, as in *Drosophila*, both the ubiqutin proteasome system and glia engulfment are involved (Awasaki *et al.*, 2006; Ehlers, 2004; Hoopfer *et al.*, 2006; Martini *et al.*, 2008). The end result of these developmental processes is the intricate structure of the adult MB, which serves as one of the critical focal points of memory acquisition and consolidation in the fly brain (Heisenberg, 2003; Davis, 2005; Berry *et al.*, 2008; Davis, 2011).

3.7. *Mushroom body-related modulatory memory circuits*

As the central memory circuitry for olfactory associative memory, neuronal activities of MB neurons are modulated by a number of extrinsic neurons that synapse on distinctive parts of the MB lobes and the calyx. Studies in *Drosophila* and other insects provided insights into the neuroanatomy and functional properties of modulatory memory circuits, and suggest critical functions of monoaminergic neurons in mediating reinforcing stimuli (Hammer, 1993; Hammer and Menzel, 1998; Kim *et al.*, 2007; Schwaerzel *et al.*, 2003; Unoki *et al.*, 2005, 2006; Vergoz *et al.*, 2007). As in the mammalian brain, the *Drosophila* DA system is involved in a variety of neuronal activities including sleep and arousal, context generalisation, decision-making, behavioural flexibility, and memory [reviewed in Waddell (2010)]. The adult fly brain comprises ~600 DA neurons in 15 clusters that are found throughout the central nervous system (Mao and Davis, 2009). However, only a subset of DA neurons innervates the MBs, representing converging points in which the internal information of the odour stimuli and the reinforcing stimuli are integrated. Thus, a set of DA neurons extensively innervates the MB α/α'-lobes, the heel, parts of the β/γ-lobes, and less densely the calyx (Honjo and Furukubo-Tokunaga, 2009; Riemensperger *et al.*, 2005; Zhang *et al.*, 2007). Clonal studies demonstrate the existence of eight different types of DA neurons that innervate distinctive MB regions in the adult brain, suggesting functional differentiation among the DA neurons and MB target regions (Mao and Davis, 2009). Moreover, Selcho *et al.* (2009) showed similar anatomical differentiation of the larval DA neurons, which are grouped into three clusters in the central brain, six clusters along the midline of the ventral nerve, and three clusters in lateral regions of the ventral nerve. Behavioural studies in both adult flies and larvae suggest that synaptic output of DA neurons is required for acquisition, but not retrieval, of aversive memory (Honjo and Furukubo-Tokunaga, 2009; Schwaerzel *et al.*, 2003; Selcho *et al.*, 2009). By optically controlling neuronal activity with channel rhodopsin, Schroll *et al.* (2006) showed that light stimulation of DA neurons triggers aversive learning in larvae. Claridge-Chang *et al.* (2009) further showed that selective optogenetic stimulation of 12 DA neurons in the PPL1 cluster substitute aversive reinforcement signals in living adult flies. In addition, using a heat-inducible TRP channel, dTRPA1, Aso *et al.* (2010) showed that MB-MP1 type DA neurons in the PPL1 cluster could induce aversive odour memory. They also showed that activation of another type of DA neurons, MB-M3, in the PAM cluster could substitute for aversive reinforcement. Together, these studies suggest functional differentiation among DA neuron clusters and the importance of DA circuitry for mediating the reinforcing stimuli in aversive memory formation in the fly brain.

Whereas specific DA neurons are involved in aversive memory, synaptic output of OA (corresponding to vertebrate noradrenalin) neurons is necessary for the formation of sugar reward memory (Honjo and Furukubo-Tokunaga,

2009; Schroll *et al.*, 2006; Schwaerzel *et al.*, 2003). OA neurons project to multiple parts of the adult MBs, innervating the heel, parts of the γ-lobe, the calyx as well as the AL glomeruli (Sinakevitch and Strausfeld, 2006; Honjo and Furukubo-Tokunaga, 2009). Clonal analyses of OA neurons identified 27 types OA neurons that innervate multiple structures in the brain and subesophageal ganglia (Busch *et al.*, 2009). Unlike aversive memory, appetitive memory in the adult fly is regulated by the relative satiety state and promoted by hunger. Hence, appetitive memory is thought to be controlled by motivational inputs, but underlying neurocircuitry mechanisms remain elusive. Krashes *et al.* (2009) showed that stimulation of neurons that express neuropeptide F (dNPF), a mammalian NPY orthologue, mimics food deprivation and promotes memory performance in satiated flies. dNPF seems to act on distinctive set of DA neurons that express the receptor for dNPF and innervate the MB to suppress appetitive memory performance. Six DA neurons are identified as a key module of dNPF-regulated circuitry, through which the internal motivational states of hunger and satiety are represented in the MB. These results suggest that neurocircuitry of dNPF and DA provides the internal motivational switch in the MBs that controls the output of appetitive memory (Krashes *et al.*, 2009).

In addition to monoaminergic neurons, a large number of extrinsic neurons are described to innervate MB lobes (Ito *et al.*, 1998; Pitman *et al.*, 2011; Sejourne *et al.*, 2011; Wu *et al.*, 2011). Among them, DPM (dorsal paired medial) neurons project exclusively to the MB, and express the putative neuropeptide encoded by the *amnesiac* gene, which is similar to the vertebrate pituitary adenylyl cyclase activating peptide (PACAP). DPM neurons innervate the entire MB lobes and appear to be pre- and postsynaptic to MB neurons. Synaptic output of both DPM and the α′/β′ lobes is required for the formation of medium-term memory in both appetitive and aversive conditioning (Feany and Quinn, 1995; Keene *et al.*, 2006; Keene *et al.*, 2004; Waddell *et al.*, 2000). Recently, another paired (APL, anterior paired lateral) neuron groups, which are GABAergic, were identified to have a regulatory role in the formation of an anesthesia-sensitive labile component of mid-term memory (Pitman *et al.*, 2011; Wu *et al.*, 2011). APL neurons contact DPM neurons via gap junctions most densely in the α′/β′ lobes, and blocking synaptic output of APL neurons after training disrupts labile memory without compromising long-term memory. These results imply the importance of an inhibitory network in the stabilisation of labile memory traces in the recurrent DPM-α′/β′ neuron circuit. Whether additional neurons contribute to the modulation of the stability and consolidation of distinctive memory components awaits further behavioural and anatomical examinations.

4. The central complex — An integration centre for learning and memory processing

In addition to the MBs, research over the last decade identified the central complex (CCX) and its substructures as another brain region involved in learning and memory. The CCX is located at the midline of the posterior protocerebrum and composed of four neuropil structures: the fan-shaped body (FB), the ellipsoid body (EB), the protocerbral bridge and the paired noduli (Figure 11.4A) (Hanesch *et al.*, 1989; Young and Armstrong, 2010a). The caudal-most part is the protocerebral bridge, which is located between the MB calyces and is composed of 16 glomeruli (8 on each side of the brain). Rostral to the protocerebral bridge is the FB, the largest of the CCX neuropils; it is organised into six vertical segments and eight horizontal layers. The EB nestles in the rostral surface of the FB and forms a circular neuropil with a canal in the centre. There are two distinct rings of neuropil, organised in five concentric layers. The nodulis lie ventral to the FB and EB, and are subdivided into three horizontal layers. The CCX is the brain structure most commonly associated with higher control of locomotor behaviour in the fly (Strauss, 2002), and has been found in all other arthropod species studied so far (Loesel *et al.*, 2002), though with some variation; outside diptera, the EB is not a ring, but rather appears kidney-bean shaped (Hanesch *et al.*, 1989; Young and Armstrong, 2010a).

Associated with the CCX but not classified as part of it, are the accessory neuropils: the lateral triangles (also known as the bulbs) and the lateral accessory lobes (also called ventral bodies) (Hanesch *et al.*, 1989). The lateral accessory lobes lie laterally and rostrally to the EB, between the MBs and ALs and are connected by an interhemispheric commissure. The lateral accessory lobes are believed to be the main output nucleus of the CCX, and are connected to several brain regions, but they do not project to the MBs, ALs and optic lobes. The lateral triangles lie adjacent to the FB and lateral to the EB, and are connected directly by axon tracts to the EB, to the lateral accessory lobes and the optic foci as well as to other protocerebral regions (Hanesch *et al.*, 1989).

4.1. *The central complex is composed of large- and small-field neurons*

The CCX comprises more than 50 neuron types, the majority of which are large- and small-field neurons. Large-field neurons typically innervate only one of the CCX neuropils, and also one or both accessory neuropils. Small-field neurons interconnect parts of the CCX neuropils and some connect them

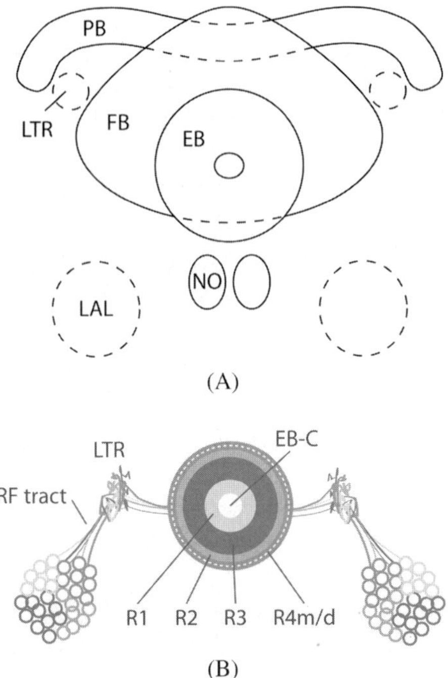

Figure 11.4. Structure of the *Drosophila* central complex. (A) The *Drosophila* CCX is composed of four neuropil structures: the fan-shaped body (FB), the ellipsoid body (EB), the protocerebral bridge (PB) and the paired noduli (NO). Associated with the CCX are the accessory neuropils: the lateral triangles (LTRs, also known as the bulbs) and the lateral accessory lobes (LALs, also called ventral bodies). (B) Schematic presentation of the *Drosophila* EB. The EB is composed of groups of extrinsic and intrinsic large-field ring neurons. Intrinsic ring neurons project their axons along the ring fiber tract (RF tract) and the lateral triangle (LTR) into the EB canal (EB-C) and finally arborise their terminals in a subtype-specific pattern. Based on their terminal arborisation within the EB ring, ring neurons are subdivided into R1-4 sub-groups, where R4 is further subdivided into R4m and R4d. Targeted genetic manipulation of individual classes of ring neurons revealed subtype-specific memory processing: R1 neurons are involved in visual place learning, R3 and R4d neurons appear to be required for spatial orientation memory, whereas R4m neurons are involved in consolidation of olfactory long-term memory.

to accessory neuropils (Hanesch *et al.*, 1989). Large-field neurons have been found to innervate the EB, FB or protocerebral bridge. Large-field neurons of EB are known as ring neurons, since they synapse in a ring within the neuropil. In the fan-shaped body, large-field neurons (F neurons) innervate distinct layers and arborise in the lateral accessory lobes, lateral triangles or extrinsic central brain regions. Large-field protocerebral bridge neurons arborise within all segments of the protocerebral bridge and connect the two hemispheres, but do not innervate any other brain areas (Hanesch *et al.*, 1989; Young and Armstrong, 2010a).

Pontine neurons are an example of small-field neurons that are intrinsic to one neuropil; they arborise only within the FB, connecting contralateral segments, adjacent segments or adjacent layers. The majority of small-field neurons connect subsections of different CCX neuropils; for example, bidirectional neurons connect FB and EB, and projecting neurons convey information from the EB to the ipsilateral protocerebral bridge and the ipsilateral or contralateral accessory lobe. Other small-field neurons have been identified that connect the CCX neuropils in seemingly every possible combination (Hanesch *et al.*, 1989). Some small-field neurons are likely to carry the main outputs of the CCX, via connections to the lateral accessory lobe, while others presumably operate feedbacks between the neuropils (Hanesch *et al.*, 1989; Young and Armstrong, 2010a).

The CCX receives innervations from much of the brain, including the optic lobes and ALs (Hanesch *et al.*, 1989) as well as indirect sensory input from the protocerebrum; in addition, the protocerebral bridge receives input from the thoracic ganglia (Strausfeld, 1999). Based upon its connection profile, the CCX appears to have a role in sensory integration and the modulation of motor output; its position spanning the midline of the brain suggests it functions in transfer of information between hemispheres (Bouhouche *et al.*, 1993), or processing information coming from both sides of the brain (bilateral signals), reminiscent of the coordinating functions of the basal ganglia and cerebellum in the mammalian brain. Indeed, structure–function analyses of various mutant strains and targeted genetic manipulation of specific substructures, identified the CCX as a higher centre for the control of motor behaviour (Strauss and Heisenberg, 1993) and memory processing, including courtship conditioning where a male fly learns to depress courtship toward a virgin female after an unproductive sexual experience with an already-mated female (Joiner and Griffith, 2000; Popov *et al.*, 2005; Sakai and Kitamoto, 2006) and long-term memory after conditioning to an odourant (Wu *et al.*, 2007). The CCX also plays a prominent role in visual pattern memory (Liu *et al.*, 2006; Pan *et al.*, 2009; Wang *et al.*, 2008), spatial orientation memory (Neuser *et al.*, 2008), and visual place memory (Ofstad *et al.*, 2011) which are explained in further detail below.

4.2. *Development of the central complex*

In contrast to the wealth of knowledge available for MB development, insights into the development of the CCX and its substructures are only starting to emerge. In locust, eight embryonic NBs have been identified as precursors of the larval CCX (Boyan and Williams, 1997). The *Drosophila* CCX is not evident at the larval stage, and only becomes identifiable as a neuropil group at the late larval/early pupal stages (Renn *et al.*, 1999; Young and Armstrong, 2010b). In the late larva, undifferentiated axon tracts of the CCX neurons form a primordium, in which all of the CCX compartments can be recognised as discrete entities (Pereanu *et al.*, 2011). A lineage-based study shows that most CCX neurons belong to as many as 10 clonally related groups that are produced during the larval stage (Pereanu *et al.*, 2011). Young and Armstrong (2010b) carried out an in-depth study of the temporal sequence of CCX formation, following from the earlier work by Renn *et al.* (1999). They found that the fan-shaped body and protocerebral bridge are the first structures to take shape, appearing in the third instar larva and becoming fully formed by 48 h after puparium formation (APF). The EB is not visible in the larva, becoming recognizable at 24 h APF and acquiring its final shape by 32 h APF. The noduli appear at 12 h APF and are fully developed by 48 h, at which time the CCX has acquired its adult appearance. The CCX neuropils subsequently differentiate from the interhemispheric commissure, and appear to use the transient interhemispheric fibrous ring (TIFR) as a scaffold. Ablation of the glial TIFR causes severe CCX defects (Boquet *et al.*, 2000a; Hitier *et al.*, 2000). In brains lacking the TIFR, the protocerebral bridge is severely fragmented or absent, the FB is flattened and the EB has a cleft on its ventral side (Hitier *et al.*, 2000).

Analysis of structural mutants affecting the CCX identified several genes that are involved in the development and specification of CCX substructures (Strauss *et al.*, 1992; Strauss and Heisenberg, 1993; Ilius *et al.*, 1994). *orthodenticle*, which encodes a homeodomain transcription factor essential for anterior brain development (Hirth *et al.*, 1995), *tay-bridge*, a novel gene with homology to a human autism-susceptibility candidate gene, *AUTS2* (Poeck *et al.*, 2008; Schumann *et al.*, 2011), and *no-bridge* (*nob*), a mutation as yet unrelated to a specific gene (Strauss and Heisenberg, 1993), were all found to function in protocerebral bridge development. *central-body-defect* (*cbd*), another mutation as yet unrelated to a specific gene, was found to affect the majority of the CCX; in most cases the FB and EB are fused to a single neuropil which is cleft along the midline, and noduli are often absent. *Central-complex* mutations (affected gene unknown) were found to cause defective protocerebral bridge and FB and EB development, whereas mutations in *ellipsoid body open* and *central-complex-deranged* (affected genes unknown) specifically affect the EB, as do *central-body-defect*, *central-complex-broad*, *central-brain-deranged* (affected genes unknown) (Strauss and Heisenberg, 1993). More recently, the activin receptor Baboon involved in TGFβ signalling (Zheng *et al.*, 2006), Ciboulot, which regulates actin assembly during brain metamorphosis (Boquet *et al.*, 2000b) and the Guanine exchange factor Vav (Malartre *et al.*, 2010) were all found to be involved in EB ring formation.

Mosaic analysis with a repressible cell marker (Lee and Luo, 2001) revealed that the majority of CCX substructures derive from a restricted number of larval NBs, and hence seem to represent lineages- or hemi-lineages of individual neural stem cells (Ito and Awasaki, 2008; Bayraktar *et al.*, 2010; Pereanu *et al.*, 2011). The embryonic origin and identity of some of these NB lineages is currently being uncovered (Ludlow *et al.*, submitted) and initial lineage analyses indicate that at least some CCX substructures are specified by a temporal cascade of gene activity comparable to neural lineage specification in the embryonic CNS (for review, see Brody and Odenwald, 2005; Jacob *et al.*, 2008). Lineage analysis of NB clones induced during larval life revealed that adult CCX neurons acquire different trajectories based on birth order, thereby leading to the innervation of distinct subcompartments of the noduli and the EB (Yu *et al.*, 2009b). The temporal fate and hence axonal projection pattern of these neurons appears to be dependent on *Chinmo*, which has been shown to regulate temporal identity in MB neurons (Zhu *et al.*, 2006). Thus, knocking out the *Chinmo* activity in an intermediate progenitor cell of a CCX sublineage caused neurons to acquire a temporal fate and axonal projections that are characteristic of a later-born neuron type (Yu *et al.*, 2009b). These data indicate that similar to MB lineages (Brody and Odenwald, 2005; Kao and Lee, 2010), CCX neuron identity appears to be specified by birth order and hence temporal fate that are defined by sequential expression of transcription factors.

4.3. *The central complex harbours memory traces*

Flies have the ability to discriminate between patterns in the environment, and to remember visual landmarks. They discriminate visual features, including size, color, contour direction, elevation and vertical compactness (Ernst and Heisenberg, 1999; Tang and Guo, 2001; Liu *et al.*, 2006). Visual pattern memory can be tested in a flight simulator, in which a fly is tethered but can change its orientation relative to artificial scenery. Traditional associative conditioning can be used to create an association between a particular pattern or landmark and an aversive stimulus;

mutant flies are subsequently tested to assess their memory. Using this experimental paradigm, Liu *et al.* (2006) showed that the ability to remember patterns and landmarks depends on the activity of *rutabaga* (*rut*), which is also involved in olfactory memory formation in the MBs (Zars *et al.*, 2000). Gal4/UAS-mediated, selective restoration of *rut* expression in specific sets of brain neurons in *rut* mutants, revealed that the upper part of the FB (in particular layer F5) is required for memory of elevation and the lower part (layer F1) is required for memory of contour orientation (Liu *et al.*, 2006).

A subsequent study identified *foraging* (*for*) as an essential gene in the formation of visual pattern memory, and that *sitter* (*for*^{S}) variants have impaired memory (Wang *et al.*, 2008). Overexpression of *for* in all neurons raised protein kinase G (PKG) levels in *for*^{S} flies and rescued the memory defects. Specific overexpression of *for* in the F5 layer of the FB in *for*^{S} flies rescued the memory for elevation but not contour orientation (Wang *et al.*, 2008). These data are in line with the findings of Liu *et al.* (2006), suggesting that the F5 layer of the FB plays an essential part in memorising elevation. Moreover, overexpression of *for* in R2 and R4m neurons of the EB in *for*^{S} flies rescued both elevation and contour orientation, whereas *for* expression in R3 and R4d neurons of the EB, was insufficient to restore memory in *for*^{S} flies (Wang *et al.*, 2008). Furthermore, Pan *et al.* (2009) showed that rescue of the *rut* memory phenotype can be achieved by overexpression in R2 and R4m neurons, and that all parameters are in fact restored (size, contour orientation, elevation and vertical compactness). Together these data suggest EB and FB cooperate in visual memory formation, but each seems to function differently; the FB appears to operate in a parameter-dependent manner (for specific visual features only), whereas memory formation in the EB appears to be parameter-independent (Pan *et al.*, 2009).

4.4. *The EB is required for memory processing*

In addition to the FB, several studies identified the EB as a specific relay for memory processing, and potentially also for memory storage. Initial experiments showed that, although the CCX is not involved in short-term olfactory memory, a subset of ellipsoid body neurons participate in the consolidation of olfactory long-term memory (Wu *et al.*, 2007; Zhao *et al.*, 2009). Thus, RNAi-mediated knockdown of the NMDA receptors Nmdar1 and Nmdar2 specifically in R4m neurons of the EB, but not in other neurons of the EB, disrupted long-term memory consolidation of olfactory information, but not its retrieval (Wu *et al.*, 2007). These data suggest that a specific cell type, R4m neurons of the EB, require NMDA receptor expression at physiological levels for memory formation.

However, it remains to be shown whether these parameters already identify components of the actual memory trace of a certain type of olfactory information.

More recently, several studies investigated spatial orientation and visual place memory in flies as the ability to remember the location of a landmark and to continue orienting towards it after it was removed (Neuser *et al.*, 2008) or moved (Ofstad *et al.*, 2011). Neuser *et al.* used a modified Buridan's paradigm (Götz, 1980; Strauss and Heisenberg, 1993) which tests object fixation behaviour, by presenting two targets opposite each other, out of reach of the fly. Flies then continuously walk backwards and forwards between the landmarks, attempting but being unable to reach them. From video recordings of the flies, several parameters including activity, velocity and straightness of walking can be measured. In the adapted Burdidan's paradigm, individual flies were allowed to walk backwards and forwards on the platform, but when they crossed an invisible midline, the target disappeared and reappeared at a 90° angle (Neuser *et al.*, 2008). After the flies had oriented to the new target, this also disappeared, leaving the fly with no visible landmark. Neuser *et al.* showed that wild-type flies usually reorient to the original target, which is invisible in a featureless environment. Flies must therefore have remembered the position of their bodies and the angle towards the former target, and thus showed spatial orientation memory. In addition, the group demonstrated that the EB was necessary for spatial orientation memory; persistence towards the original target was abolished in *ellipsoid body open* mutants (Neuser *et al.*, 2008). Moreover, Gal4/UAS-mediated cell type-specific silencing by tetanus toxin overexpression identified EB ring neurons required for memory of the original target and orientation towards it. In particular, GABAergic R3 and R4d neurons were found to be required for spatial orientation memory (Neuser *et al.*, 2008).

A recent study extended the findings of Neuser *et al.* and showed that *Drosophila* is capable of forming and retaining visual place memories that are used to guide selective navigation (Ofstad *et al.*, 2011). Visual place learning was examined by using a thermal–visual arena inspired by the Morris water maze (Morris, 1981), where flies must find a hidden "safe" target in an otherwise threatening environment. The hidden target is a tile cooled to standard room temperature of 25°C in a surrounding arena heated up to 36°C. There are no local cues available to identify the cool tile, however spatial cues are given with a surrounding electronic panorama that displays a pattern of evenly spaced bars in three orientations (Ofstad *et al.*, 2011). To challenge visual place memory in flies, the cool tile and the corresponding visual panorama were shifted to a new location (rotated by either 90° clockwise or 90° anticlockwise, chosen at random) but both target and visual panorama were coupled so that although the absolute position of the cool tile changed, its location relative to the visual panorama remained constant. During the trials, flies were tested for their navigation to the cool tile and the elapsed duration was measured. After several trials, flies took less time to locate the cool tile, and took shorter as well as more direct routes to the target, suggesting they had learned where and how to find the cool tile. Significantly, flies retained these visual place memories for at least 2 h (Ofstad *et al.*, 2011). Moreover, conditional silencing of subsets of neurons in adult flies by targeted expression of the inward-rectifying potassium channel Kir2.1 (Baines *et al.*, 2001) identified R1 EB ring neurons as a neuroanatomical substrate for processing visual place memory (Ofstad *et al.*, 2011).

Together, these data identify specific subsets of EB ring neurons as anatomical substrates for the selective processing and integration of visual information, and hence the formation of memory underlying spatial orientation and navigation. It is not clear, however, whether these neurons and their arborisations also store the actual visual information and thereby code for memory traces in the strict sense, as has been suggested for presynaptic plasticity (e.g. Christiansen *et al.*, 2011) and altered calcium influx (e.g. Akalal *et al.*, 2011) in Kenyon cells of the MBs.

5. Conclusions and perspective

Molecular genetic studies in *Drosophila* have revealed important insights not only into the genetic mechanisms but also the neural circuit principles underlying learning and memory in the brain. These studies identified the MBs and the CCX as essential memory circuits for the acquisition, consolidation and storage of multimodal information related to various types of memory. Studies investigating the development and specification of distinct subregions of the MBs and the CCX provide critical insights into the cytoarchitecture and functional organisation of these memory networks and thus are essential for the comprehensive analysis of physiological functions. Moreover, the development and functional anatomy of these two brain structures identifies several common principles that might be of general interest; clonal unit architecture, temporal identity and functional compartmentalisation.

5.1. *Clonal unit architecture*

A small number of neural stem cells generate identifiable neuronal substructures within the MBs and CCX. These substructures comprise clonally related cells and share a common lineage in functionally organised groups, which are involved in memory acquisition and storage. Hence, a genealogical relationship of clonally related cells, also referred to as ontogenetic clones, seems to underlie neural circuit architecture that appears to have functional relevance (Cardona *et al.*, 2010; Ito and Awasaki, 2008; Pereanu *et al.*, 2010). Intriguingly, this cytoarchitectonic principle underlying functional anatomy may also apply to mammals. A recent study in mice showed that clonally related neurons of the somatosensory cortex develop unidirectional chemical synapses with each other rather than with neighbouring nonsiblings; the synapses are characterised by the same interlaminar directional preference as those observed in the mature neocortex, suggesting that specific microcircuits develop preferentially within ontogenetic clones and contribute to the emergence of functional columns in the neocortex (Yu *et al.*, 2009a). It is not clear yet whether and how such ontogenetic clones code for/contribute to specific neuronal activities and thus behaviours, but their existence in both *Drosophila* and mouse suggests that clonal units are evolutionary conserved cytoarchitectonic modules underlying neural circuit formation and function in both insect and mammalian brains.

5.2. *Temporal identity*

As with vertebrate neural progenitor cells, *Drosophila* NBs generate diverse types of neurons and glia by spatially and temporally controlled mechanisms (Isshiki *et al.*, 2001; Pearson and Doe, 2003; Truman and Bate, 1988). Clonal analysis of neural stem cell lineages that generate the developing MBs and parts of the CCX revealed that lineage-specific progeny acquire a neuronal identity also in the central brain due to their birth order and the spatio-temporal activity of progenitor transcription factors. These progenitor transcription factors, in turn, regulate postmitotic transcription factors that specify temporal cell fates of neurons and glia. In the *Drosophila* brain, *Chinmo* specifies temporal fate and axonal projection patterns of neurons that are involved in olfactory memory processing in the MBs and visual memory processing in the CCX (Zhu *et al.*, 2006). *Chinmo* is also implicated in the generation of distinct type of neurons in the ALs (Jefferis *et al.*, 2001; Zhu *et al.*, 2006). Thus, along with lineage cues, time-dependent cell fate determinants appear to act as temporal codes to specify neuronal progeny that constitute memory circuits in the *Drosophila* brain. Comparable regulatory programs for the temporal guidance of neuronal fates have been identified in mammals (Jacob *et al.*, 2008; Kao and Lee, 2010). For example, Olig2-expressing precursor cells in the developing mouse telencephalon sequentially generate at least 10 subtypes of cortical interneurons in a time-dependent manner (Myoshi *et al.*, 2007). These data suggest that the mechanisms linking birth order to neuronal identity are conserved in insects and mammals. Given the conservation of genetic programs of brain development, we anticipate that, coupled with lineage-specific cues, temporal codes contribute to cell fate determination and memory circuit formation in both insect and mammalian brains.

5.3. *Functional compartmentalisation*

Targeted synaptic inactivation of MB and CCX substructures identified specific neuronal subtypes as essential anatomical substrates for olfactory memory processing in MBs (Davis, 2011; Keene and Waddell, 2007) and for visual memory processing in the CCX (Liu *et al.*, 2006; Neuser *et al.*, 2008; Ofstad *et al.*, 2011). In MBs, α'/β' neurons are required for the formation of short-term memory, whereas α/β neurons are essential for the formation of long-term memory traces (Davis, 2011). Visual pattern memory including size, contour orientation, elevation and vertical compactness, requires the EB and FB of the CCX. F5 layer neurons in the FB thus process memory of elevation, and F1 layer neurons process memory of contour orientation (Liu *et al.*, 2006). Spatial orientation memory on landmark location requires synaptic activity of EB R3 and R4d neurons (Neuser *et al.*, 2008), whereas R1 neurons process visual place memory (Ofstad *et al.*, 2011). These data suggest that, during the course of learning and memory, distinctive aspects of external information is acquired and stored in functionally compartmentalised cell clusters in *Drosophila* memory circuits.

Taken together, studies in *Drosophila* over the past decades demonstrate that systematic analyses of the developmental principles underlying memory circuitries provide an important framework for the analysis of the functional organisation and modular organisation of memory circuits in the fly brain. Moreover, recent structural and functional studies of memory networks are providing essential insights into the functional compartmentalisation of memory modules. Previous and current progress is furthered by new technology that allows optical imaging of physiologically active memory circuits in living animals (e.g. Martin *et al.*, 2007; Wang *et al.*, 2008). Similarly, recent applications of serial transmission electron microscopy (e.g. Anderson *et al.*, 2009; Cardona *et al.*, 2010) pave the way for ultra-structural analyses of subtle changes in neural circuitry associated with experienced-dependent behavioural plasticity. Combined with these promising new techniques, insights into the developmental principles of memory centres in *Drosophila* may hold a key to a comprehensive understanding of how memories are coded in the brain.

6. Acknowledgements

Supported by Grants-in-Aid for Scientific Research from MEXT of Japan to K.F.T., and the UK Medical Research Council (G070149), the Royal Society (Hirth/2007/R2), Parkinson's UK (G-0714), the Motor Neurone Disease Association (Hirth/Oct07/6233), and the Fondation-Thierry-Latran (2/2011/DrosALS) to F.H.

References

Adachi, Y., Hauck, B., Clements, J., Kawauchi, H., Kurusu, M., Totani, Y., Kang, Y.Y., Eggert, T., Walldorf, U., Furukubo-Tokunaga, K., *et al.* (2003). Conserved cis-regulatory modules mediate complex neural expression patterns of the eyeless gene in the Drosophila brain. *Mech Dev* **120**, 1113–1126.

Adams, M.D., Celniker, S.E., Holt, R.A., Evans, C.A., Gocayne, J.D., Amanatides, P.G., Scherer, S.E., Li, P.W., Hoskins, R.A., Galle, R.F., *et al.* (2000). The genome sequence of Drosophila melanogaster. *Science* **287**, 2185–2195.

Akalal, D.B., Yu, D. and Davis, R.L. (2010). A late-phase, long-term memory trace forms in the gamma Neurons of Drosophila mushroom bodies after olfactory classical conditioning. *J Neurosci* **30**, 16699–16708.

Akalal, D.B., Yu, D. and Davis, R.L. (2011). The long-term memory trace formed in the Drosophila alpha/beta mushroom body Neurons is abolished in long-term memory mutants. *J Neurosci* **31**, 5643–5647.

Anderson, J.R., Jones, B.W., Yang, J.H., Shaw, M.V., Watt, C.B., Koshevoy, P., Spaltenstein, J., Jurrus, E.U.V.K., Whitaker, R.T., Mastronarde, D., Tasdizen, T. and Marc, R.E. (2009). A computational framework for ultrastructural mapping of neural circuitry. *PLoS Biol* **7**, e1000074.

Armstrong, J.D., de Belle, J.S., Wang, Z. and Kaiser, K. (1998). Metamorphosis of the mushroom bodies; large-scale rearrangements of the neural substrates for associative learning and memory in Drosophila. *Learn Mem* **5**, 102–114.

Aso, Y., Siwanowicz, I., Bracker, L., Ito, K., Kitamoto, T. and Tanimoto, H. (2010). Specific dopaminergic Neurons for the formation of labile aversive memory. *Curr Biol* **20**, 1445–1451.

Awasaki, T. and Ito, K. (2004). Engulfing action of glial cells is required for programmed axon pruning during Drosophila metamorphosis. *Curr Biol* **14**, 668–677.

Awasaki, T., Tatsumi, R., Takahashi, K., Arai, K., Nakanishi, Y., Ueda, R. and Ito, K. (2006). Essential role of the apoptotic cell engulfment genes draper and ced-6 in programmed axon pruning during Drosophila metamorphosis. *Neuron* **50**, 855–867.

Baines, R.A., Uhler, J.P., Thompson, A., Sweeney, S.T. and Bate, M. (2001). Altered electrical properties in *Drosophila* neurons developing without synaptic transmission. *J Neurosci* **21**, 1523–1531.

Barrett, A.L., Krueger, S. and Datta, S. (2008). Branchless and Hedgehog operate in a positive feedback loop to regulate the initiation of neuroblast division in the Drosophila larval brain. *Dev Biol* **317**, 234–245.

Bayraktar, O.A., Boone, J.Q., Drummond, M.L. and Doe, C.Q. (2010). Drosophila type II neuroblast lineages keep Prospero levels low to generate large clones that contribute to the adult brain central complex. *Neural Dev* **5**, 26.

Bello, B.C., Izergina, N., Caussinus, E. (2008). Amplification of neural stem cell proliferation by intermediate progenitor cells in Drosophila brain development. *Neural Dev* **3**, 5.

Berry, J., Krause, W.C. and Davis, R.L. (2008). Olfactory memory traces in Drosophila. *Prog Brain Res* **169**, 293–304.

Boone, J.Q. and Doe, C.Q. (2008). Identification of Drosophila type II neuroblast lineages containing transit amplifying ganglion mother cells. *Dev Neurobiol* **68**, 1185–1195.

Boquet, I., Hitier, R., Dumas, M., Chaminade, M. and Préat, T. (2000a). Central brain postembryonic development in Drosophila: Implication of genes expressed at the interhemispheric junction. *J Neurobiol* **42**, 33048.

Boquet, I., Boujemaa, R., Carlier, M.-F. and Préat, T. (2000b). Ciboulot regulates actin assembly during Drosophila brain metamorphosis. *Cell* **102**, 797–808.

Bouhouche, A., Vaysse, G. and Corbière, M. (1993). Immunocytochemical and learning studies of a Drosophila melanogaster neurological mutant, *No-Bridge*^{KS49} as an approach to the possible role of the central complex. *J Neurogenet* **9**, 105–121.

Bowman, S.K., Rolland, V., Betschinger, J., Kinsey, K.A., Emery, G. and Knoblich, J.A. (2008). The tumor suppressors Brat and Numb regulate transit-amplifying neuroblast lineages in Drosophila. *Dev Cell* **14**, 535–546.

Boyan, G.S. and Williams, J.L.D. (1997). Embryonic development of the pars intercerebralis/central complex of the grasshopper. *Dev Genes Evol* **207**, 317–329.

Boyan, G., Williams, L. and Hirth, F. (2005). Commissural organisation and brain segmentation in insects. In T.H. Bullock, L.A. Krubitzer, T.M. Preuss, J.L.R Rubenstein, N.J. Strausfeld and G.F. Striedter, ed. *Evolution of Nervous Systems. Vol II.* (London, San Diego: Elsevier), pp. 349–359.

Brody, T. and Odenwald, W.F. (2005). Regulation of temporal identities during Drosophila neuroblast lineage development. *Curr Opin Cell Biol* **17**, 672–675.

Busch, S., Selcho, M., Ito, K. and Tanimoto, H. (2009). A map of octopaminergic neurons in the Drosophila brain. *J Comp Neurol* **513**, 643–667.

Callaerts, P., Leng, S., Clements, J., Benassayag, C., Cribbs, D., Kang, Y.Y., Walldorf, U., Fischbach, K.F. and Strauss, R. (2001). Drosophila Pax-6/eyeless is essential for normal adult brain structure and function. *J Neurobiol* **46**, 73–88.

Cardona, A., Saalfeld, S., Arganda, I., Pereanu, W., Schindelin, J. and Hartenstein, V. (2010). Identifying neuronal lineages of Drosophila by sequence analysis of axon tracts. *J Neurosci* **30**, 7538–7553.

Caussinus, E. and Hirth, F. (2007). Asymmetric stem cell division in development and cancer. *Prog Mol Subcell Biol* **45**, 205–225.

Christiansen, F., Zube, C., Andlauer, T.F., Wichmann, C., Fouquet, W., Owald, D., Mertel, S., Leiss, F., Tavosanis, G., Farca Luna, A.J., Fiala, A. and Sigrist, S.J. (2011). Presynapses in kenyon cell dendrites in the mushroom body calyx of Drosophila. *J Neurosci* **31**, 9696–9707.

Claridge-Chang, A., Roorda, R.D., Vrontou, E., Sjulson, L., Li, H., Hirsh, J. and Miesenbock, G. (2009). Writing memories with light-addressable reinforcement circuitry. *Cell* **139**, 405–415.

Connolly, J.B., Roberts, I.J., Armstrong, J.D., Kaiser, K., Forte, M., Tully, T. and O'Kane, C.J. (1996). Associative learning disrupted by impaired Gs signalling in Drosophila mushroom bodies. *Science* **274**, 2104–2107.

Crittenden, J.R., Skoulakis, E.M., Han, K.A., Kalderon, D. and Davis, R.L. (1998). Tripartite mushroom body architecture revealed by antigenic markers. *Learn Mem* **5**, 38–51.

Czerny, T., Halder, G., Kloter, U., Souabni, A., Gehring, W.J. and Busslinger, M. (1999). Twin of eyeless, a second Pax-6 gene of Drosophila, acts upstream of eyeless in the control of eye development. *Mol Cell* **3**, 297–307.

Davis, R.L. (2005). Olfactory memory formation in Drosophila: From molecular to systems neuroscience. *Annu Rev Neurosci* **28**, 275–302.

Davis, R.L. (2011). Traces of Drosophila memory. *Neuron* **70**, 8–19.

de Belle, J.S. and Heisenberg, M. (1994). Associative odour learning in Drosophila abolished by chemical ablation of mushroom bodies. *Science* **263**, 692–695.

Dubnau, J., Grady, L., Kitamoto, T. and Tully, T. (2001). Disruption of neurotransmission in Drosophila mushroom body blocks retrieval but not acquisition of memory. *Nature* **411**, 476–480.

Ebens, A.J., Garren, H., Cheyette, B.N. and Zipursky, S.L. (1993). The Drosophila anachronism locus: A glycoprotein secreted by glia inhibits neuroblast proliferation. *Cell* **74**, 15–27.

Egger, B., Chell, J.M. and Brand, A.H. (2008). Insights into neural stem cell biology from flies. *Philos Trans R Soc Lond B Biol Sci* **363**, 39–56.

Ehlers, M.D. (2004). Deconstructing the axon: Wallerian degeneration and the ubiquitin-proteasome system. *Trends Neurosci* **27**, 3–6.

Erber, J., Menzel, R. (1980). Localisation of short-term memory in the brain of the bee, *Apis mellifera*. *Physiol Entomol* **5**, 343–358.

Ernst, R. and Heisenberg, M. (1999). The memory template in Drosophila pattern vision at the flight simulator. *Vision Res* **39**, 3920–33.

Feany, M.B. and Quinn, W.G. (1995). A neuropeptide gene defined by the Drosophila memory mutant amnesiac. *Science* **268**, 869–873.

Furukubo-Tokunaga, K., Adachi, Y., Kurusu, M. and Walldorf, U. (2009). Brain patterning defects caused by mutations of the twin of eyeless gene in Drosophila melanogaster. *Fly (Austin)* **3**, 263–269.

Gervasi, N., Tchenio, P. and Preat, T. (2010). PKA dynamics in a Drosophila learning centre: Coincidence detection by rutabaga adenylyl cyclase and spatial regulation by dunce phosphodiesterase. *Neuron* **65**, 516–529.

Götz, K.G. (1980). Visual guidance in Drosophila. In O. Siddiqui, P. Babu, L.M. Hall, and J.C. Hall, eds. *Development and Neurobiology of Drosophila*. (New York: Plenum Press).

Greenspan, R.J. and Ferveur, J.F. (2000). Courtship in Drosophila. *Annu Rev Genet* **34**, 205–232.

Guo, A., Zhang, K., Peng, Y. and Xi, W. (2009). Heisenberg's roadmap guides our journey to the small cognitive world of Drosophila. *J Neurogenet* **23**, 100–103.

Hammer, M. (1993). An unidentifed Neuron mediates the unconditioned stimulus in associative olfactory learning in honey bees. *Nature* **366**, 59–63.

Hammer, M. and Menzel, R. (1998). Multiple sites of associative odour learning as revealed by local brain microinjections of octopamine in honeybees. *Learn Mem* **5**, 146–156.

Hanesch, U., Fischbach, K.-F. and Heisenberg, M. (1989). Neuronal architecture of the central complex in Drosophila melanogaster. *Cell Tissue Res* **257**, 343–366.

Heisenberg, M., Borst, A., Wagner, S. and Byers, D. (1985). Drosophila mushroom body mutants are deficient in olfactory learning. *J Neurogenet* **2**, 1–30.

Heisenberg, M. (2003). Mushroom body memoir: From maps to models. *Nat Rev Neurosci* **4**, 266–275.

Hirth, F., Therianos, S., Loop, T., Gehring, W.J., Reichert, H. and Furukubo-Tokunaga, K. (1995). Developmental defects in brain segmentation caused by mutations of the homeobox genes orthodenticle and empty spiracles in Drosophila. *Neuron* **15**, 769–778.

Hirth, F. and Reichert, H. (1999). Conserved genetic programs in insect and mammalian brain development. *Bioessays* **21**, 677–684.

Hirth, F. and Reichert, H. (2007). Basic nervous system types: One or many? In T.H. Bullock, L.A. Krubitzer, T.M. Preuss, J.L.R. Rubenstein, N.J. Strausfeld and G.F. Striedter, eds. *Evolution of Nervous Systems. Vol II*. (London, San Diego: Elsevier), pp. 349–359.

Hitier, R., Simon, A.F., Savarit, F. and Préat, T. (2000). No-bridge and linotte act jointly at the interhemispheric junction to build up the adult central brain of Drosophila melanogaster. *Mech Dev* **99**, 93–100.

Hoch, R.V., Rubenstein, J.L. and Pleasure, S. (2009). Genes and signalling events that establish regional patterning of the mammalian forebrain. *Semin Cell Dev Biol* **20**, 378–386.

Honjo, K. and Furukubo-Tokunaga, K. (2005). Induction of cAMP response element-binding protein-dependent medium-term memory by appetitive gustatory reinforcement in Drosophila larvae. *J Neurosci* **25**, 7905–7913.

Honjo, K. and Furukubo-Tokunaga, K. (2009). Distinctive Neuronal networks and biochemical pathways for appetitive and aversive memory in Drosophila larvae. *J Neurosci* **29**, 852–862.

Hoopfer, E.D., McLaughlin, T., Watts, R.J., Schuldiner, O., O'Leary, D.D. and Luo, L. (2006). Wlds protection distinguishes axon degeneration following injury from naturally occurring developmental pruning. *Neuron* **50**, 883–895.

Hoopfer, E.D., Penton, A., Watts, R.J. and Luo, L. (2008). Genomic analysis of Drosophila Neuronal remodeling: A role for the RNA-binding protein Boule as a negative regulator of axon pruning. *J Neurosci* **28**, 6092–6103.

Ilius, M., Wolf, R. and Heisenberg, M. (1994). The central complex of Drosophila melanogaster is involved in flight control: Studies on mutants and mosaics of the gene ellipsoid body open. *J Neurogenet* **9**, 189–206.

Isshiki, T., Pearson, B., Holbrook, S. and Doe, C.Q. (2001). Drosophila neuroblasts sequentially express transcription factors which specify the temporal identity of their Neuronal progeny. *Cell* **106**, 511–521.

Ito, K., Awano, W., Suzuki, K., Hiromi, Y. and Yamamoto, D. (1997). The Drosophila mushroom body is a quadruple structure of clonal units each of which contains a virtually identical set of Neurones and glial cells. *Development* **124**, 761–771.

Ito, K. and Awasaki, T. (2008). Clonal unit architecture of the adult fly brain. *Adv Exp Med Biol* **628**, 137–158.

Ito, K. and Hotta, Y. (1992). Proliferation pattern of postembryonic neuroblasts in the brain of Drosophila melanogaster. *Dev Biol* **149**, 134–148.

Ito, K., Suzuki, K., Estes, P., Ramaswami, M., Yamamoto, D. and Strausfeld, N.J. (1998). The organisation of extrinsic neurons and their implications in the functional roles of the mushroom bodies in Drosophila melanogaster Meigen. *Learn Mem* **5**, 52–77.

Jacob, J., Maurange, C. and Gould, A.P. (2008). Temporal control of Neuronal diversity: Common regulatory principles in insects and vertebrates? *Development* **135**, 3481–3489.

Jan, Y.N. and Jan, L.Y. (1998). Asymmetric cell division. *Nature* **392**, 775–778.

Jefferis, G.S., Marin, E.C., Stocker, R.F. and Luo, L. (2001). Target Neuron prespecification in the olfactory map of Drosophila. *Nature* **414**, 204–208.

Jefferis, G.S., Potter, C.J., Chan, A.M., Marin, E.C., Rohlfing, T., Maurer, C.R., Jr. and Luo, L. (2007). Comprehensive maps of Drosophila higher olfactory centres: Spatially segregated fruit and pheromone representation. *Cell* **128**, 1187–1203.

Joiner, M.A. and Griffith, L.C. (2000). Visual input regulates circuit configuration in courtship conditioning of Drosophila melanogaster. *Learn Mem* **7**, 32–42.

Kao, C.F. and Lee, T. (2010). Birth time/order-dependent Neuron type specification. *Curr Opin Neurobiol* **20**, 14–21.

Keene, A.C., Krashes, M.J., Leung, B., Bernard, J.A. and Waddell, S. (2006). Drosophila dorsal paired medial neurons provide a general mechanism for memory consolidation. *Curr Biol* **16**, 1524–1530.

Keene, A.C., Stratmann, M., Keller, A., Perrat, P.N., Vosshall, L.B. and Waddell, S. (2004). Diverse odour-conditioned memories require uniquely timed dorsal paired medial neuron output. *Neuron* **44**, 521–533.

Keene, A.C. and Waddell, S. (2007). Drosophila olfactory memory: Single genes to complex neural circuits. *Nat Rev Neurosci* **8**, 341–354.

Kim, Y.C., Lee, H.G. and Han, K.A. (2007). D1 dopamine receptor dDA1 is required in the mushroom body Neurons for aversive and appetitive learning in Drosophila. *J Neurosci* **27**, 7640–7647.

Kim, J., Lee, S., Hwang, M., Ko, S., Min, C. and Kim-Ha, J. (2009). Bunched specifically regulates alpha/beta mushroom body Neuronal cell proliferation during metamorphosis. *Neuroscience* **161**, 46–52.

Kim, D.W. and Hirth, F. (2009). Genetic mechanisms regulating stem cell self-renewal and differentiation in the central nervous system of Drosophila. *Cell Adh Migr* **3**, 402–411.

Kitamoto, T. (2002). Targeted expression of temperature-sensitive dynamin to study neural mechanisms of complex behaviour in Drosophila. *J Neurogenet* **16**, 205–228.

Kobayashi, M., Michaut, L., Ino, A., Honjo, K., Nakajima, T., Maruyama, Y., Mochizuki, H., Ando, M., Ghangrekar, I., Takahashi, K., *et al.* (2006). Differential microarray analysis of Drosophila mushroom body transcripts using chemical ablation. *Proc Natl Acad Sci USA* **103**, 14417–14422.

Krashes, M.J., DasGupta, S., Vreede, A., White, B., Armstrong, J.D. and Waddell, S. (2009). A neural circuit mechanism integrating motivational state with memory expression in Drosophila. *Cell* **139**, 416–427.

Krashes, M.J., Keene, A.C., Leung, B., Armstrong, J.D. and Waddell, S. (2007). Sequential use of mushroom body Neuron subsets during drosophila odour memory processing. *Neuron* **53**, 103–115.

Kurusu, M., Awasaki, T., Masuda-Nakagawa, L.M., Kawauchi, H., Ito, K. and Furukubo-Tokunaga, K. (2002). Embryonic and larval development of the Drosophila mushroom bodies: Concentric layer subdivisions and the role of fasciclin II. *Development* **129**, 409–419.

Kurusu, M., Maruyama, Y., Adachi, Y., Okabe, M., Suzuki, E. and Furukubo-Tokunaga, K. (2009). A conserved nuclear receptor, Tailless, is required for efficient proliferation and prolonged maintenance of mushroom body progenitors in the Drosophila brain. *Dev Biol* **326**, 224–236.

Kurusu, M., Nagao, T., Walldorf, U., Flister, S., Gehring, W.J. and Furukubo-Tokunaga, K. (2000). Genetic control of development of the mushroom bodies, the associative learning centres in the Drosophila brain, by the eyeless, twin of eyeless and Dachshund genes. *Proc Natl Acad Sci USA* **97**, 2140–2144.

Kurusu, M. and Zinn, K. (2008). Receptor tyrosine phosphatases regulate birth order-dependent axonal fasciculation and midline repulsion during development of the Drosophila mushroom body. *Mol Cell Neurosci* **38**, 53–65.

Lee, T., Lee, A. and Luo, L. (1999). Development of the Drosophila mushroom bodies: Sequential generation of three distinct types of Neurons from a neuroblast. *Development* **126**, 4065–4076.

Lee, T. and Luo, L. (2001). Mosaic analysis with a repressible cell marker (MARCM) for Drosophila neural development. *Trends Neurosci* **24**, 251–254.

Lee, T., Marticke, S., Sung, C., Robinow, S. and Luo, L. (2000a). Cell-autonomous requirement of the USP/EcR-B ecdysone receptor for mushroom body neuronal remodeling in Drosophila. *Neuron* **28**, 807–818.

Lee, T., Winter, C., Marticke, S.S., Lee, A. and Luo, L. (2000b). Essential roles of Drosophila RhoA in the regulation of neuroblast proliferation and dendritic but not axonal morphogenesis. *Neuron* **25**, 307–316.

Leuzinger, S., Hirth, F., Gerlich, D., Acampora, D., Simeone, A., Gehring, W.J., Finkelstein, R., Furukubo-Tokunaga, K. and Reichert, H. (1998). Equivalence of the fly orthodenticle gene and the human OTX genes in embryonic brain development of Drosophila. *Development* **125**, 1703–1710.

Lin, H.H., Lai, J.S., Chin, A.L., Chen, Y.C. and Chiang, A.S. (2007). A map of olfactory representation in the Drosophila mushroom body. *Cell* **128**, 1205–1217.

Liu, G., Seiler, H., Wen, A., Zars, T., Ito, K., Wolf, R., Heisenberg, M. and Liu, L. (2006). Distinct memory traces for two visual features in the Drosophila brain. *Nature* **439**, 551–556.

Liu, L., Wolf, R., Ernst, R. and Heisenberg, M. (1999). Context generalisation in Drosophila visual learning requires the mushroom bodies. *Nature* **400**, 753–756.

Liu, Z., Steward, R. and Luo, L. (2000). Drosophila Lis1 is required for neuroblast proliferation, dendritic elaboration and axonal transport. *Nat Cell Biol* **2**, 776–783.

Loesel, R., Nässel, D.R. and Strausfeld, N.J. (2002). Common design in a unique midline neuropil in the brains of arthropods. *Arthropod Struct Dev* **31**, 77–91.

Malartre, M., Ayaz, D., Amador, F.F. and Martín-Bermudo, M.D. (2010). The guanine exchange factor *vav* controls axon growth and guidance during Drosophila development. *J Neurosci* **30**, 2257–2267.

Mao, Z. and Davis, R.L. (2009). Eight different types of dopaminergic neurons innervate the Drosophila mushroom body neuropil: Anatomical and physiological heterogeneity. *Front Neural Circuits* **3**, 5.

Margulies, C., Tully, T. and Dubnau, J. (2005). Deconstructing memory in Drosophila. *Curr Biol* **15**, R700–R713.

Martin, J.R., Rogers, K.L., Chagneau, C. and Brûlet, P. (2007). *In vivo* bioluminescence imaging of Ca signalling in the brain of Drosophila. *PLoS One* **2**, e275.

Martini, R., Fischer, S., Lopez-Vales, R. and David, S. (2008). Interactions between Schwann cells and macrophages in injury and inherited demyelinating disease. *Glia* **56**, 1566–1577.

Masse, N.Y., Turner, G.C. and Jefferis, G.S. (2009). Olfactory information processing in Drosophila. *Curr Biol* **19**, R700–713.

Maurange, C. and Gould, A.P. (2005). Brainy but not too brainy: Starting and stopping neuroblast divisions in Drosophila. *Trends Neurosci* **28**, 30–36.

McGuire, S.E., Le, P.T. and Davis, R.L. (2001). The role of Drosophila mushroom body signalling in olfactory memory. *Science* **293**, 1330–1333.

McGuire, S.E., Deshazer, M. and Davis, R.L. (2005). Thirty years of olfactory learning and memory research in Drosophila melanogaster. *Prog Neurobiol* **76**, 328–347.

Mochizuki, H., Toda, H., Ando, M., Kurusu, M., Tomoda, T. and Furukubo-Tokunaga, K. (2011). Unc-51/ATG1 controls axonal and dendritic development via kinesin-mediated vesicle transport in the Drosophila brain. *PLoS ONE* **6**, e19632.

Monaghan, A.P., Bock, D., Gass, P., Schwager, A., Wolfer, D.P., Lipp, H.P. and Schutz, G. (1997). Defective limbic system in mice lacking the tailless gene. *Nature* **390**, 515–517.

Monaghan, A.P., Grau, E., Bock, D. and Schutz, G. (1995). The mouse homolog of the orphan nuclear receptor tailless is expressed in the developing forebrain. *Development* **121**, 839–853.

Nagao, T., Leuzinger, S., Acampora, D., Simeone, A., Finkelstein, R., Reichert, H. and Furukubo-Tokunaga, K. (1998). Developmental rescue of Drosophila cephalic defects by the human Otx genes. *Proc Natl Acad Sci USA* **95**, 3737–3742.

Morris, R.G.M. (1981). Spatial localisation does not require the presence of local cues. *Learning and Motivation* **12**, 239–260.

Miyoshi, G., Butt, S.J., Takebayashi, H. and Fishell, G. (2007). Physiologically distinct temporal cohorts of cortical interneurons arise from telencephalic Olig2-expressing precursors. *J Neurosci* **27**, 7786–7798.

Nassif, C., Noveen, A. and Hartenstein, V. (1998). Embryonic development of the Drosophila brain. I. Pattern of pioneer tracts. *J Comp Neurol* **402**, 10–31.

Neuser, K., Triphan, T., Mronz, M., Poeck, B. and Strauss, R. (2008). Analysis of a spatial orientation memory in Drosophila. *Nature* **453**, 1244–1247.

Nighorn, A., Healy, M.J. and Davis, R.L. (1991). The cyclic AMP phosphodiesterase encoded by the Drosophila dunce gene is concentrated in the mushroom body neuropil. *Neuron* **6**, 455–467.

Noveen, A., Daniel, A. and Hartenstein, V. (2000). Early development of the Drosophila mushroom body: The roles of eyeless and dachshund. *Development* **127**, 3475–3488.

Ofstad, T.A., Zuker, C.S. and Reiser, M.B. (2011). Visual place learning in Drosophila melanogaster. *Nature* **474**, 204–207.

Pan, Y., Zhou, Y., Guo, C., Gong, H., Gong, Z. and Liu, L. (2009). Differential roles of the fan-shaped body and the ellipsoid body in Drosophila visual pattern memory. *Learn Mem* **16**, 289–295.

Park, Y., Rangel, C., Reynolds, M.M., Caldwell, M.C., Johns, M., Nayak, M., Welsh, C.J., McDermott, S. and Datta, S. (2003). Drosophila perlecan modulates FGF and hedgehog signals to activate neural stem cell division. *Dev Biol* **253**, 247–257.

Pascual, A. and Preat, T. (2001). Localisation of long-term memory within the Drosophila mushroom body. *Science* **294**, 1115–1117.

Pauls, D., Selcho, M., Gendre, N., Stocker, R.F. and Thum, A.S. (2010). Drosophila larvae establish appetitive olfactory memories via mushroom body neurons of embryonic origin. *J Neurosci* **30**, 10655–10666.

Pearson, B.J. and Doe, C.Q. (2003). Regulation of neuroblast competence in Drosophila. *Nature* **425**, 624–628.

Pereanu, W., Kumar, A., Jennett, A., Reichert, H. and Hartenstein, V. (2010). Development-based compartmentalisation of the Drosophila central brain. *J Comp Neurol* **518**, 2996–3023.

Pereanu, W., Younossi-Hartenstein, A., Lovick, J., Spindler, S. and Hartenstein, V. (2011). Lineage-based analysis of the development of the central complex of the Drosophila brain. *J Comp Neurol* **519**, 661–689.

Pitman, J.L., Huetteroth, W., Burke, C.J., Krashes, M.J., Lai, S.L., Lee, T. and Waddell, S. (2011). A pair of inhibitory neurons are required to sustain labile memory in the Drosophila mushroom body. *Curr Biol* **21**, 855–861.

Poeck, B., Triphan, T., Neuser, K. and Strauss, R. (2008). Locomotor control by the central complex in Drosophila-An analysis of the tay bridge mutant. *Dev Neurobiol* **68**, 1046–1058.

Popov, A.V., Peresleni, A.I., Ozerskii, P.V., Shchekanov, E.E. and Savvateeva-Popova, E.V. (2005). The role of the flabellar and ellipsoid bodies of the central complex of the brain of Drosophila melanogaster in the control of courtship behaviour and communicative sound production in males. *Neurosci Behav Physiol* **35**, 741–750.

Prokop, A. and Technau, G.M. (1994). Normal function of the mushroom body defect gene of Drosophila is required for the regulation of the number and proliferation of neuroblasts. *Dev Biol* **161**, 321–337.

Quiring, R., Walldorf, U., Kloter, U. and Gehring, W.J. (1994). Homology of the eyeless gene of Drosophila to the Small eye gene in mice and Aniridia in humans. *Science* **265**, 785–789.

Renn, S.C.P., Armstrong, J.D., Yang, M., Wang, Z., An, X., Kaiser, K. and Taghert, P.H. (1999). Genetic analysis of the Drosophila ellipsoid body neuropil: Organisation and development of the central complex. *J Neurobiol* **41**, 189–207.

Reuter, J.E., Nardine, T.M., Penton, A., Billuart, P., Scott, E.K., Usui, T., Uemura, T. and Luo, L. (2003). A mosaic genetic screen for genes necessary for Drosophila mushroom body neuronal morphogenesis. *Development* **130**, 1203–1213.

Riemensperger, T., Voller, T., Stock, P., Buchner, E. and Fiala, A. (2005). Punishment prediction by dopaminergic neurons in Drosophila. *Curr Biol* **15**, 1953–1960.

Roy, K., Kuznicki, K., Wu, Q., Sun, Z., Bock, D., Schutz, G., Vranich, N. and Monaghan, A.P. (2004). The Tlx gene regulates the timing of neurogenesis in the cortex. *J Neurosci* **24**, 8333–8345.

Roy, K., Thiels, E. and Monaghan, A.P. (2002). Loss of the tailless gene affects forebrain development and emotional behaviour. *Physiol Behav* **77**, 595–600.

Sakai, T. and Kitamoto, T. (2006). Differential roles of two major brain structures, mushroom bodies and central complex, for Drosophila male courtship behaviour. *J Neurobiol* **66**, 821–834.

Schroll, C., Riemensperger, T., Bucher, D., Ehmer, J., Voller, T., Erbguth, K., Gerber, B., Hendel, T., Nagel, G., Buchner, E., *et al.* (2006). Light-induced activation of distinct modulatory Neurons triggers appetitive or aversive learning in Drosophila larvae. *Curr Biol* **16**, 1741–1747.

Schuldiner, O., Berdnik, D., Levy, J.M., Wu, J.S., Luginbuhl, D., Gontang, A.C. and Luo, L. (2008). PiggyBac-based mosaic screen identifies a postmitotic function for cohesin in regulating developmental axon pruning. *Dev Cell* **14**, 227–238.

Schumann, G., Coin, L.J., Lourdusamy, A., Charoen, P., Berger, K.H., Stacey, D., Desrivieres, S., Aliev, F.A., Khan, A.A., Amin, N., *et al.* (2011). Genome-wide association and genetic functional studies identify autism susceptibility candidate 2 gene (AUTS2) in the regulation of alcohol consumption. *Proc Natl Acad Sci USA* **108**, 7119–7124.

Schwaerzel, M., Monastirioti, M., Scholz, H., Friggi-Grelin, F., Birman, S. and Heisenberg, M. (2003). Dopamine and octopamine differentiate between aversive and appetitive olfactory memories in Drosophila. *J Neurosci* **23**, 10495–10502.

Scott, E.K., Lee, T. and Luo, L. (2001). Enok encodes a Drosophila putative histone acetyltransferase required for mushroom body neuroblast proliferation. *Curr Biol* **11**, 99–104.

Sehgal, A., Joiner, W., Crocker, A., Koh, K., Sathyanarayanan, S., Fang, Y., Wu, M., Williams, J.A. and Zheng, X. (2007). Molecular analysis of sleep: Wake cycles in Drosophila. *Cold Spring Harb Symp Quant Biol* **72**, 557–564.

Sejourne, J., Placais, P.Y., Aso, Y., Siwanowicz, I., Trannoy, S., Thoma, V., Tedjakumala, S.R., Rubin, G.M., Tchenio, P., Ito, K., *et al.* (2011). Mushroom body efferent Neurons responsible for aversive olfactory memory retrieval in Drosophila. *Nat Neurosci* **14**, 903–910.

Selcho, M., Pauls, D., Han, K.A., Stocker, R.F. and Thum, A.S. (2009). The role of dopamine in Drosophila larval classical olfactory conditioning. *PLoS ONE* **4**, e5897.

Shi, Y., Chichung Lie, D., Taupin, P., Nakashima, K., Ray, J., Yu, R.T., Gage, F.H. and Evans, R.M. (2004). Expression and function of orphan nuclear receptor TLX in adult neural stem cells. *Nature* **427**, 78–83.

Sinakevitch, I. and Strausfeld, N.J. (2006) Comparison of octopamine-like immunoreactivity in the brains of the fruit fly and blow fly. *J Comp Neurol* **494**, 460–475.

Sousa-Nunes, R., Cheng, L.Y. and Gould, A.P. (2010). Regulating neural proliferation in the Drosophila CNS. *Curr Opin Neurobiol* **20**, 50–57.

Sousa-Nunes, R., Yee, L.L. and Gould, A.P. (2011). Fat cells reactivate quiescent neuroblasts via TOR and glial insulin relays in Drosophila. *Nature* **471**, 508–512.

Squire, L.R. and Kandel, E.R. (2009). *Memory, From Mind to Molecules*, 2nd ed. (Roberts & Company, Roberts and Company Publisher).

Stocker, R.F., Heimbeck, G., Gendre, N. and de Belle, J.S. (1997). Neuroblast ablation in Drosophila P[GAL4] lines reveals origins of olfactory interneurons. *J Neurobiol* **32**, 443–456.

Strausfeld, N.J. (1999). A brain region in insects that supervises walking. *Prog Brain Res* **123**, 273–284.

Strauss, R. and Heisenberg, M. (1993). A higher control centre of locomotor behaviour in the Drosophila brain. *J Neurosci* **13**, 1852–1861.

Strauss, R., Hanesch, U., Kinklein, M., Wolf, R. and Heisenberg, M. (1992). No-bridge of Drosophila melanogaster: Portrait of a structural brain mutant of the central complex. *J Neurogenet* **8**, 125–155.

Strauss, R. (2002). The central complex and the genetic dissection of locomotor behaviour. *Curr Opin Neurobiol* **12**, 633–638.

Tanaka, N.K., Tanimoto, H. and Ito, K. (2008). Neuronal assemblies of the Drosophila mushroom body. *J Comp Neurol* **508**, 711–755.

Tang, S. and Guo, A. (2001). Choice behaviour of Drosophila facing contradictory visual cues. *Science* **294**, 1543–1547.

Tettamanti, M., Armstrong, J.D., Endo, K., Yang, M.Y., Furukubo-Tokunaga, K., Kaiser, K., Reichert, H. (1997). Early development of the Drosophila mushroom bodies, brain centres for associative learning and memory. *Dev Genes Evol* **207**, 242–252.

Thum, A.S., Jenett, A., Ito, K., Heisenberg, M. and Tanimoto, H. (2007). Multiple memory traces for olfactory reward learning in Drosophila. *J Neurosci* **27**, 11132–11138.

Tomchik, S.M. and Davis, R.L. (2009). Dynamics of learning-related cAMP signalling and stimulus integration in the Drosophila olfactory pathway. *Neuron* **64**, 510–521.

Truman, J.W. and Bate, M. (1988). Spatial and temporal patterns of neurogenesis in the central nervous system of Drosophila melanogaster. *Dev Biol* **125**, 145–157.

Truman, J.W., Taylor, B.J. and Awad, T.A. (1993). Formation of the adult nervous system. In M. Bate and A. Martinez-Arias, eds. *The Development of Drosophila melanogaster*. (New York: Cold Spring Harbor Laboratory Press), 1245–1275.

Unoki, S., Matsumoto, Y. and Mizunami, M. (2005). Participation of octopaminergic reward system and dopaminergic punishment system in insect olfactory learning revealed by pharmacological study. *Eur J Neurosci* **22**, 1409–1416.

Unoki, S., Matsumoto, Y. and Mizunami, M. (2006). Roles of octopaminergic and dopaminergic *Neuron*s in mediating reward and punishment signals in insect visual learning. *Eur J Neurosci* **24**, 2031–2038.

Urbach, R. and Technau, G.M. (2004). Neuroblast formation and patterning during early brain development in Drosophila. *BioEssays* **26**, 739–751.

van Swinderen, B., McCartney, A., Kauffman, S., Flores, K., Agrawal, K., Wagner, J. and Paulk, A. (2009). Shared visual attention and memory systems in the Drosophila brain. *PLoS ONE* **4**, e5989.

Venter, J.C., Adams, M.D., Myers, E.W., Li, P.W., Mural, R.J., Sutton, G.G., Smith, H.O., Yandell, M., Evans, C.A., Holt, R.A., *et al.* (2001). The sequence of the human genome. *Science* **291**, 1304–1351.

Vergoz, V., Roussel, E., Sandoz, J.C. and Giurfa, M. (2007). Aversive learning in honeybees revealed by the olfactory conditioning of the sting extension reflex. *PLoS ONE* **2**, e288.

Villella, A. and Hall, J.C. (2008). Neurogenetics of courtship and mating in Drosophila. *Adv Genet* **62**, 67–184.

Voigt, A., Pflanz, R., Schafer, U. and Jackle, H. (2002). Perlecan participates in proliferation activation of quiescent Drosophila neuroblasts. *Dev Dyn* **224**, 403–412.

Vosshall, L.B. and Stocker, R.F. (2007). Molecular architecture of smell and taste in Drosophila. *Annu Rev Neurosci* **30**, 505–533.

Waddell, S., Armstrong, J.D., Kitamoto, T., Kaiser, K. and Quinn, W.G. (2000). The amnesiac gene product is expressed in two neurons in the Drosophila brain that are critical for memory. *Cell* **103**, 805–813.

Waddell, S. (2010). Dopamine reveals neural circuit mechanisms of fly memory. *Trends Neurosci* **33**, 457–464.

Wang, J., Zugates, C.T., Liang, I.H., Lee, C.H. and Lee, T. (2002). Drosophila Dscam is required for divergent segregation of sister branches and suppresses ectopic bifurcation of axons. *Neuron* **33**, 559–571.

Wang, Z., Pan, Y., Li, W., Jiang, H., Chatzimanolis, L., Chang, J., Gong, Z. and Liu, L. (2008). Visual pattern memory requires foraging function in the central complex of Drosophila. *Learn Mem* **15**, 133–142.

Watts, R.J., Hoopfer, E.D. and Luo, L. (2003). Axon pruning during Drosophila metamorphosis: Evidence for local degeneration and requirement of the ubiquitin-proteasome system. *Neuron* **38**, 871–885.

Watts, R.J., Schuldiner, O., Perrino, J., Larsen, C. and Luo, L. (2004). Glia engulf degenerating axons during developmental axon pruning. *Curr Biol* **14**, 678–684.

Wu, C.L., Shih, M.F., Lai, J.S., Yang, H.T., Turner, G.C., Chen, L. and Chiang, A.S. (2011). Heterotypic gap junctions between two neurons in the drosophila brain are critical for memory. *Curr Biol* **21**, 848–854.

Wu, C.L., Xia, S., Fu, T.F., Wang, H., Chen, Y.H., Leong, D., Chiang, A.S. and Tully, T. (2007). Specific requirement of NMDA receptors for long-term memory consolidation in Drosophila ellipsoid body. *Nat Neurosci* **10**, 1578–1586.

Young, J.M. and Armstrong, J.D. (2010a). Building the central complex in Drosophila: Organisation of distinct neuronal subsets. *J Comp Neurol* **518**, 1500–1524.

Young, J.M. and Armstrong, J.D. Building the central complex in Drosophila: The generation and development of distinct neural subsets. *J Comp Neurol* **518**, 1525–1541.

Younossi-Hartenstein, A., Nguyen, B., Shy, D. and Hartenstein, V. (2006). Embryonic origin of the Drosophila brain neuropile. *J Comp Neurol* **497**, 981–998.

Yu, D., Ponomarev, A. and Davis, R.L. (2004). Altered representation of the spatial code for odours after olfactory classical conditioning; memory trace formation by synaptic recruitment. *Neuron* **42**, 437–449.

Yu, D., Akalal, D.B. and Davis, R.L. (2006a). Drosophila alpha/beta mushroom body neurons form a branch-specific, long-term cellular memory trace after spaced olfactory conditioning. *Neuron* **52**, 845–855.

Yu, F., Kuo, C.T. and Jan, Y.N. (2006b). Drosophila neuroblast asymmetric cell division: Recent advances and implications for stem cell biology. *Neuron* **51**, 13–20.

Yu, R.T., McKeown, M., Evans, R.M. and Umesono, K. (1994). Relationship between Drosophila gap gene tailless and a vertebrate nuclear receptor Tlx. *Nature* **370**, 375–379.

Yu, Y.-C., Bultje, R.S., Wang, X. and Shi, S.-H. (2009a). Specific synapses develop preferentially among sister excitatory Neurons in the neocortex. *Nature* **458**, 501–504.

Yu, H.-H., Chen, C.-H., Shi, L., Huang, Y. and Lee, T. (2009b). Twin-spot MARCM to reveal the developmental origin and identity of Neurons. *Nat Neurosci* **12**, 947–953.

Zars, T. (2000). Behavioural functions of the insect mushroom bodies. *Curr Opin Neurobiol* **10**, 790–795.

Zars, T., Fischer, M., Schulz, R. and Heisenberg, M. (2000). Localisation of a short-term memory in Drosophila. *Science* **288**, 672–675.

Zhan, X.L., Clemens, J.C., Neves, G., Hattori, D., Flanagan, J.J., Hummel, T., Vasconcelos, M.L., Chess, A. and Zipursky, S.L. (2004). Analysis of Dscam diversity in regulating axon guidance in Drosophila mushroom bodies. *Neuron* **43**, 673–686.

Zhang, C.L., Zou, Y., He, W., Gage, F.H. and Evans, R.M. (2008). A role for adult TLX-positive neural stem cells in learning and behaviour. *Nature* **451**, 1004–1007.

Zhang, K., Guo, J.Z., Peng, Y., Xi, W. and Guo, A. (2007). Dopamine-mushroom body circuit regulates saliency-based decision-making in Drosophila. *Science* **316**, 1901–1904.

Zhao, H., Zheng, X., Yuan, X., Wang, L., Wang, X., Zhong, Y., Xie, Z. and Tully, T. (2009). Ben Functions with scamp during synaptic transmission and long-term memory formation in Drosophila. *J Neurosci* **29**, 414–424.

Zheng, X., Zugates, C.T., Lu, Z., Shi, L., Bai, J.M. and Lee, T. (2006). Baboon/dSmad2 TGF-B signalling is required during late larval stage for development of adult-specific Neurons. *EMBO J* 615–627.

Zhu, S., Chiang, A.S. and Lee, T. (2003). Development of the Drosophila mushroom bodies: Elaboration, remodeling and spatial organisation of dendrites in the calyx. *Development* **130**, 2603–2610.

Zhu, S., Lin, S., Kao, C.F., Awasaki, T., Chiang, A.S. and Lee, T. (2006). Gradients of the Drosophila Chinmo BTB-zinc finger protein govern neuronal temporal identity. *Cell* **127**, 409–422.

Memory Circuits in the Hippocampus

12

Thomas J. McHugh*

1. Introduction

In 1957, Scoville and Milner conclude that bilateral medial temporal lobe resection "results in a persistent impairment of recent memory whenever the removal is carried far enough posteriorly to damage portions of the anterior hippocampus and hippocampal gyrus" (Scoville and Milner, 1957). In the half century since, a large body of neuropsychological and neurobiological studies have supported this finding — the hippocampus is crucial for learning and memory of events and facts both in humans and animals (Jarrard, 1993; Rempel-Clower *et al.*, 1996; Schmolck *et al.*, 2002; Scoville and Milner, 1957; Zola and Squire, 2001). In humans, the hippocampus plays a critical role in episodic (event) memory (Gabrieli *et al.*, 1988; Tulving, 1972, 1983). Similarly, in animals, the structure also seem to play a crucial role in the memory of what, where and when aspects of an event (episodic-like memory) (Clayton and Dickinson, 1998; Nakazawa *et al.*, 2003; Zentall *et al.*, 2001). The job of storing the contents of specific episodes requires the hippocampus to form and store its representations rapidly, in order to bind together temporally coincident events, and but also distinguish similar episodes or contexts (pattern separation). Furthermore, the hippocampus must be capable of using partial cues to retrieve previously stored patterns of representations due to the ever-changing nature of real life (pattern completion). Together, these requirements demand a reliable, yet flexible circuit.

For reasons of experimental tractability, the hippocampus has often been characterised as a "tri-synaptic" circuit, with information flowing from layer II stellate cells of the entorhinal cortex (EC) to the granule cells of the dentate gyrus

*Laboratory for Circuit and Behavioural Physiology, RIKEN Brain Science Institute, 2-1 Hirosawa, Wako City, Saitama, 351-0198 Japan.

(DG) through the perforant path (PP), from the granule cells to CA3 pyramidal cells through mossy fibers (MF), from CA3 pyramidal cells to CA1 pyramidal cells via the Schaffer collaterals (SC) and the loop closes itself by direct information flow from CA1 pyramidal cells to cells in the deep layers of EC or by indirect flow via the subiculum. Further, the connectivity among CA3 pyramidal cells is robust, a given cell being directly connected by recurrent collaterals (RC) with at least 2% of the other cells (MacVicar and Dudek, 1980; Miles and Wong, 1986), consequently forming an "autoassociative" network. Given this well-defined and tractable circuit, the hippocampus has been the focus of many computational models of its function in memory formation. Starting with the seminal work of Marr (1971), many have related the presumed mnemonic functions of the circuit to its underlying anatomical and physiological properties. The fundamental assumption has been that each of multiple hippocampal regions and their synaptic plasticity contributes to the overall role of the hippocampus in learning and memory by providing distinct and complementary functions (Buzsaki, 1996; McClelland and Goddard, 1996; McNaughton and Morris, 1987; O'Reilly and McClelland, 1994). The prevailing hypotheses have been that the hippocampus has two basic computational structures: the feed-forward pathway from EC to DG and to CA3, which is primarily responsible for pattern separation, and recurrent connectivity with synaptic modifiabilities within CA3, which is primarily responsible for pattern completion. In addition, CA1 is thought to "translate" the pattern-separated CA3 representation into the pattern-overlapping EC representation (O'Reilly and Rudy, 2001). Furthermore, it has been suggested that CA1 may be instrumental in recognizing novelty or familiarity of an event or context (Lisman and Otmakhova, 2001). This is thought to be accomplished by comparing the relatively raw sensory input arriving directly from the EC with the memory engram-driven input from CA3 via SC ("comparator" hypothesis (O'Reilly and Rudy, 2001; Vinogradova, 2001).

However, it is quite clear the hippocampus consists of many unique circuits, both convergent and parallel, that lead to a wider and more integrative view of information processing in the structure (Figure 12.1). These include the projections from entorhinal cortical layer II pyramidal cells directly to: (1) the CA3 pyramidal cells; hence, these pyramidal cells receive two inputs from EC cells, one processed by DG and the other unprocessed; (2) projections from EC layers II and III to CA2 pyramidal cells, where the EC inputs are paradoxically strong and excitatory and seemingly compete with strong feed-forward inhibition arriving from the CA3 via the SC (Chevaleyre and Siegelbaum, 2010); and (3) finally axons from EC layer III directly to CA1 pyramidal cells through the temporoammonic pathway (TA), where EC input has been suggested to influence plasticity

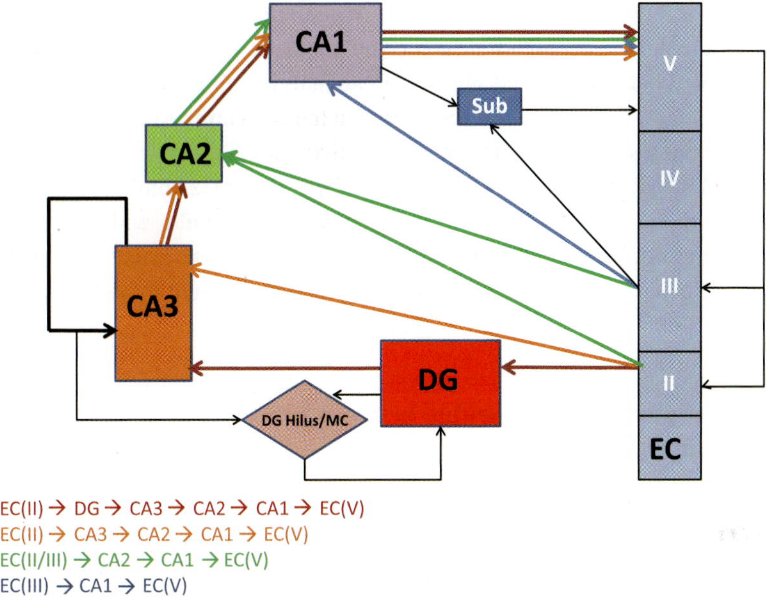

Figure 12.1. Multiple excitatory circuits in the hippocampus.

at the SC inputs (Ito and Schuman, 2007; Levy *et al.*, 1998; Remondes and Schuman, 2002, 2003). Thus, CA1 pyramidal cells receive multiple inputs from the EC, one processed by CA3 with or without DG involvement, one via CA2 without DG/CA3 and finally one unprocessed by all these areas. Because of the direct connections of the EC to the CA areas, the information received by EC from the association cortex undergoes hippocampal processing through three alternative and overlapping hippocampal loops, the trisynaptic loop involving DG, CA3 and CA1 (red in Figure 12.1), the bisynaptic loop involving CA3 and CA1 (green) and the monosynaptic loop involving only CA1 (purple). Further, other lesser characterised pathways also exist; for example, the excitatory pathways involving mossy cells (MC) of the hilus region of DG, which composes a "heteroassociative" network (Myers and Scharfman, 2010; Scharfman, 2007). One great challenge is to understand the contribution of these overlapping circuits to hippocampal function.

The growing physiological, computational and anatomical data have undoubtedly muddied what had once seemed like solid experimental conclusions, yet they now allow us a new found power to address the fundamental function of this structure known to play a key role in the learning and memory essential for our day to day life. Over the last decade, one successful approach has been the generation of mutant mice, both transgenic and knockouts, that when combined with sophisticated behavioural and physiological analysis, allow a finer dissection of the contribution of the major glutamatergic circuits to hippocampal function and the behaviour it supports (Bannerman *et al.*, 2003; McHugh *et al.*, 1996; McHugh *et al.*, 2007; Nakashiba *et al.*, 2008; Nakazawa *et al.*, 2002; Romberg *et al.*, 2009; Tsien *et al.*, 1996 among many). By analysing these mice with multifaceted technologies, including molecular and cellular biology, *in vitro* and *in vivo* physiology and behavioural paradigms, some of the long-standing hypotheses regarding hippocampal function have now been experimentally tested.

2. *In vivo* physiology: A bridge between plasticity and behaviour

One key to the multi-level examination of hippocampal function has been the discovery and characterisation of the "place cell". In 1971, O'Keefe and Dostrovsky (1971) reported that single hippocampal neurons recording in moving rats increased their firing rate whenever the rat traversed a particular region of an environment, dubbed the cell's "place field". This triggered the proposal that these cells underlie the ability of the hippocampus to serve as the neural substrate of a "cognitive map" (O'Keefe and Nadel, 1978) and in the 40 years following this initial description, hundreds of studies have attempted to characterise the properties and dependencies of place cell activity in rodents [reviewed in Best and White (1999); Best *et al.* (2001); Moser *et al.* (2008); Muller (1996)]. Studies using freely behaving rats have demonstrated that pyramidal cells show stable, long lasting, environmentally specific place fields, with between 30% and 50% of the CA1 and CA3 cell layer showing place specific activity in any given environment (Thompson and Best, 1990; Wilson and McNaughton, 1993; Wilson and McNaughton, 1994). As the field has matured, the techniques for recording and analysing place cells have evolved, allowing for the direct observation that any given place is encoded not by the activity of single neurons, but rather, by a population of co-activate cells (Wilson and McNaughton, 1993). By recording large numbers of cells simultaneously, researchers can look at the interactions between cells during and after learning. Using these approaches, numerous studies have provided evidence that place cells can be viewed as memory traces at the neuronal ensemble level (Eichenbaum *et al.*, 1999; Lee and Wilson, 2002; Louie and Wilson, 2001; Moser and Paulson, 2001; O'Keefe and Nadel, 1978; Pastalkova *et al.*, 2008; Poucet *et al.*, 2000; Wilson and McNaughton, 1993; Wilson and McNaughton, 1994). Among the most convincing are the findings that place cells display spatially characteristic stable firing patterns across temporally separated recording sessions and that firing patterns can be subsequently reactivated independently of behaviour in an experience-dependent manner (Kudrimoti *et al.*, 1999; Lee and Wilson, 2002; Louie and Wilson, 2001; Nadasdy *et al.*, 1999; Skaggs and McNaughton, 1996; Wilson and McNaughton, 1994). Given these properties, place cells represent a strong electrophysiological correlate of a natural form of learning in a freely behaving animal and an ideal system to examine the relationships among synaptic plasticity, *in vivo* electrophysiology, and behaviour. Finally, very recent advances in physiological and optical technology have permitted a look inside the membranes of place cells in behaving animals, offering the first glimpse of the mechanism underlying place coding (Dombeck *et al.*, 2010; Harvey *et al.*, 2009; Epsztein, 2010; Lee *et al.*, 2009; Lee *et al.*, 2006). This important work promises an ever deeper understanding of the input/output function of these cells and the circuit in the near future.

In addition to contributing to our understanding of how the hippocampus may underlie spatial learning, place cell recordings have also shed light on the ability of an animal to form and maintain representations for similar yet distinct contexts. Recent studies from several laboratories have investigated if pattern completion and pattern separation can be detected in the hippocampal circuit using the place cell recording technique in both the CA3 and CA1 regions of rats (Colgin *et al.*, 2008; Jeffery and Hayman, 2004; Lee and Kesner, 2004; Leutgeb *et al.*, 2004; Lever *et al.*, 2002; Wills *et al.*, 2005). The general approach in these experiments has been to modify a familiar environment and ask if the spatial representation persists (completion) or alters substantially (separation). However, there have been a variety of specific approaches to this question, with the some laboratories using distinct boxes and rooms, others using a "morphing" spatial environment that can assume multiple geometric shapes, and still others using the rotation of local and distal cues to alter the context via cue mismatch (Lee *et al.*, 2004; Leutgeb *et al.*, 2005a; Leutgeb *et al.*, 2005b; Lever *et al.*, 2002; Wills *et al.*, 2005). While this variation in protocol has made it difficult to synthesise these findings into a firm set of rules, there have been some significant and well-accepted observations. One particularly significant finding was that a pattern-separated ensemble representation can be detected in CA3 prior to CA1 (Leutgeb *et al.*, 2004), which agrees with the theory that DG input to CA3 may support pattern separation. A second is that relatively small environmental manipulations do not alter the stability of the place cell ensembles in CA1 and CA3, but when these contextual changes become more significant, the responses in the regions diverge, with CA1 remapping and CA3 remaining coherent, consistent with pattern completion in CA3

(Lee *et al.*, 2004). Although these results are at first glance seemingly conflicting, they can be reconciled in terms of the tradeoff between the pull of completion and the push of separation. The dichotomy of the CA3 response may be an accurate readout of the magnitude of the environmental changes, and of the response of the DG-CA3 circuit.

Hippocampal *in vivo* recording has also revealed that the local field potential (LFP) recorded in the structure can broadly define two major states of hippocampal activity: large irregular or sharp-wave/ripple activity (SWR), which correlates with both quiet wakefulness and slow wave sleep, and theta or regular slow wave activity, which correlates with motion, attentiveness and rapid-eye movement (REM) sleep (Vanderwolf, 1969). These two distinct states of the hippocampus have been hypothesized to underlie two distinct roles of the structure, memory encoding (theta) and memory consolidation (SWR) (Buzsaki, 1996). SWRs are thought to typically result from the synchronous depolarisation of a large number of pyramidal cells in area CA1 due to synchronous firing of CA3 pyramidal cells (Ylinen *et al.*, 1995a). During slow wave sleep or awake immobility, the EEG record characteristically contains short (40–120 ms) monophasic sharp waves that can occur at frequencies ranging from 0.02 to 3/s (Buzsaki, 1986). Between these bursts of activity the LFP often remains quite silent. An attractive hypothesis for the function of this form of population activity has been first proposed by Buzsaki (Buzsaki, 1989). He points out that during sharp-wave associated bursts, hippocampal neurons fire at high rates and with a high degree of synchrony. These two properties have been shown to be sufficient for inputs to strengthen their synaptic connections to their postsynaptic targets (Bliss and Collingridge, 1993; Bliss and Lømo, 1973). Therefore, it has been suggested that SWRs may serve a memory consolidation role in the hippocampal network, facilitating the transfer of information out to the subiculum, neocortex and possibly other brain regions. Studies have shown that sharp-wave discharges are correlated with population discharges in the cortical areas thought to be important for memory organisation or storage (Ji and Wilson, 2007; Siapas and Wilson, 1998) and recent work has demonstrated interference with hippocampal circuit activity specifically during this SWRs can disrupt memory formation (Ego-Stengel and Wilson, 2010; Girardeau *et al.*, 2009). These data support the hypothesis that this phenomenon may serve to facilitate information transfer and possibly memory consolidation.

The second major type of LFP signal detected in the rodent hippocampus is the theta rhythm. This is a sinusoidal oscillation at a frequency between 4 and 12 Hz in the hippocampal field potential that occurs during locomotion, sensory stimulation, urethane-induced anesthesia, and REM sleep (Buzsaki *et al.*, 1985; Vanderwolf, 1969). The theta rhythm is thought to have multiple generators, but primarily results from dipoles created by excitatory entorhinal input to the dentate granule cells and the pyramidal cells and inhibitory currents on the cell bodies of pyramidal cells generated via input from the medial septum (Buzsaki *et al.*, 1986; Buzsaki and Eidelberg, 1983; Lee *et al.*, 1994; Ylinen *et al.*, 1995b). The rhythm demonstrates a high coherence across much of the hippocampus, as well as both deep and superficial entorhinal cortex (Alonso and Garcia-Austt, 1987a, b; Alonso *et al.*, 1987; Buzsaki *et al.*, 1986). It has been suggested that the theta rhythm plays a vital role in memory encoding in the hippocampus, providing a periodic subthreshold excitation that allows inputs to trigger cycles of action potential firings. In fact, it has long been observed that the pyramidal cells of the hippocampus fire in correlation with the theta rhythm (Buzsaki and Eidelberg, 1983; Sinclair *et al.*, 1982). This may serve to increase the "signal to noise" ratio by silencing most of the pyramidal cells by keeping their membrane voltage close, but below, the firing threshold (Buzsaki *et al.*, 1986). Additionally, this regular periodic activity during behaviour may allow the coordination of neuronal activity across the various regions of the hippocampus (Buzsaki *et al.*, 1981; Lubenov and Siapas, 2009; Pavlides *et al.*, 1988; Rudell *et al.*, 1980). The precise role of theta in establishing the temporal structure of hippocampal memory formation continues to be investigated; however, it has been observed that it can modulate cellular properties such as firing rate, excitability, and membrane potential (Buzsaki and Eidelberg, 1983; Rudell and Fox, 1984; Wyble *et al.*, 2000). As a result, when a rodent first enters a cell's place field, the pyramidal cell will fire near the peak of theta, when inhibition is at its weakest. However, the cell will discharge earlier in reference to the theta cycle as the animal moves through the field. This phenomenon has been termed 'theta precession' (O'Keefe and Recce, 1993). A consequence of phase precession is an ordered discharge of cells with nearby fields during every theta cycle occurring in the overlapping region, a situation that may permit downstream targets to encode sequences of activity via LTP-associated mechanisms (Dragoi and Buzsaki, 2006; Skaggs *et al.*, 1996).

The remainder of this chapter will focus primarily on the use of the genetically modified mouse to understand how changes in the function of the hippocampal circuit are manifest on the level of behaviour and physiology. Much of our understanding of how the hippocampus encodes and uses spatial information derives from place cell recordings and, when combined with genetic technology, this technique has proved valuable in relating molecule to circuit to physiology to behaviour.

3. Genetic approaches to understanding plasticity and hippocampal circuit function

3.1. *The dentate gyrus*

While the numbers of pyramidal cells in areas CA1 and CA3 of a rodent have been estimated to be 100,000–250,000 each, with a similar number of stellate cells in the EC, the number of DG granule cells (GC) has been estimated to be close to 1 million (Boss *et al.*, 1985; Boss *et al.*, 1987; Squire, 1989). Each of these GCs receives about 5000 weak synaptic inputs from the EC, and of these, only 5–10% are active at any given time during behaviour (Barnes *et al.*, 1990). Additionally, each GC then forms synapses to only 10–15 CA3 pyramidal cells (Squire, 1989). This 4- to 10-fold increase in cell number from EC to DG, combined with the diffuse connectivity of the perforant and mossy fiber pathways, provides the ideal anatomical substrate for a pattern separation computation; allowing the DG to convert similar or overlapping EC inputs into unique patterns of CA3 ensemble activity. In addition, the extremely low level of activity in the DG (< 0.4%) can augment this pattern separation (i.e. "orthogonalisation") (Marr, 1971; McNaughton and Morris, 1987; O'Reilly and McClelland, 1994; Rolls, 1996).

Most behavioural studies employing physical or chemical lesions of the DG report that damage to the GC layer leads to severe deficits in spatial reference memory and spatial working memory (WM) (Czurko *et al.*, 1997; Gilbert *et al.*, 2001; Jeltsch *et al.*, 2001; McLamb *et al.*, 1988; McNaughton *et al.*, 1989; Sutherland *et al.*, 1983; Xavier *et al.*, 1999). Additionally, it has been reported that DG lesions lead to impairments in differentiation of cues in spatially similar locations, a phenotype attributed to a loss of DG-based pattern separation (Lee and Kesner, 2002). However, the lesion techniques employed limited the interpretation of these results and these experiments do not address the role of DG synaptic plasticity, a mechanism proposed to aid in the computational role of the DG in pattern separation (McClelland and Goddard, 1996). The encoding of novel information has been hypothesised to trigger long-term changes at the PP-DG synapse (Straube and Frey, 2003; Straube *et al.*, 2003; Xu *et al.*, 1998) which may underlie the formation of unique spatial representation (Dragoi *et al.*, 2003). Lee and Kesner attempted to address the potential role of plasticity at PP-DG synapses in memory by infusing the NR antagonist AP5 locally into the DG, but it was difficult to distinguish CA1- and DG-specific roles (Lee and Kesner, 2002). Several genetically engineered mice that displayed DG LTP deficits have been subjected to various memory tasks, but most of these studies were complicated by the lack of restriction of the genetic alterations which made it difficult to draw any definite conclusion regarding the relationship between the specific physiological and behavioural phenotypes (Bordi *et al.*, 1997; Brandon *et al.*, 1995; Chen *et al.*, 2001; Tecott *et al.*, 1998).

Using genetic techniques [Cre/loxP system (Hoess *et al.*, 1982; Orban *et al.*, 1992)] which allow the targeting of genetic deletion to a defined group of neurons, two groups independently generated mice which lack NMDA receptor-mediated plasticity specifically in the granule cells of the DG. Niewoehner *et al.* (2007) reported that a mouse with a deletion of the NR1 gene, encoding the NMDA receptor 1 subunit which is critical for receptor function, specifically in the granule cells of the DG lacked synaptic plasticity at the perforant path-GC synapse (DG-NR1 KO mouse). Behavioural characterisation of this line of mice was used to address the specific learning deficits that result from a loss of plasticity at this primary input synapse to the hippocampal circuit. The acquisition of spatial reference memory was tested using a 3/6 version of the radial arm maze, in which the mouse learns the location of three fixed reward locations on an 6-arm radial maze and the animal is explicitly prevented from re-entering the previously visited arm, thus eliminating the possibility of working memory errors. The performance of this task is vulnerable to chemical or physical lesions of the DG, however, the authors found no deficits in the DG-NR1 KO mouse. They then used variations in the training protocol to address the role of DG plasticity in spatial working memory. Animals were tested in a similar task, only now the animals were allowed to reenter previously visited arms, an indicator of working memory error, following a 5- or 15-s delay period. Under these protocols a significant deficit in spatial working memory was revealed at the longer delay time of 15 s. The authors undertook a closer examination of the data to ask if there was evidence that the working memory errors indicated a deficit in spatial separation, i.e. did the errors reflect the tendency of the mice to reenter previously visited locations adjacent to unvisited goal location, as opposed to revisiting arms spatial distant; however, there was no indication of a spatial pattern separation deficit in their performance.

A second line of mice with a granule cell-specific deletion of the NR1 gene was generated concurrently by second group and used to test the role of DG plasticity in its proposed function in contextual discrimination/pattern separation (McHugh *et al.*, 2007). Similar to the experiments above, the DG-NR1 KO mouse was found to have intact spatial memory, as tested both in the Morris water maze, a test of spatial learning and navigation (Morris *et al.*, 1982b), and in standard contextual fear conditioning (Fanselow, 1990), a paradigm in which the mouse learns to associate a training context with an aversive foot-shock. Together with the Niewoehner data, this suggests plasticity at the PP-GC synapse is dispensable for spatial reference memory.

The models of hippocampal function discussed above suggest that while the specific abolition of PP-DG NR-mediated plasticity may not affect spatial reference memory, it may lead to deficits in memory tasks which require distinction of similar sensory inputs, i.e. pattern separation. Building on the results from the contextual fear experiments, the authors then subjected the DG-NR1 KO mouse to a behavioural paradigm that required the discrimination of two highly similar chambers (A and B), one of which was always paired with an aversive shock and the second which never was. In this protocol, conditioning took place incrementally over several days, allowing the effects of repeated experiences to be investigated (Fanselow, 1981). On the first 3 days of the experiment, the mice were placed only into chamber A where they received a single footshock, after which both genotypes demonstrated fear acquisition but extensive generalisation between contexts. During the subsequent discrimination phase of the task, mice visited the two chambers daily for 12 days, always receiving a shock in A, but never in B. Control mice quickly learned to distinguish the chambers; however, the DG-NR1 KO mice exhibited a transient, yet very significant deficit during the acquisition of the discrimination task. This deficit in the mutant mice was exhibited as elevated freezing in the shock-free Chamber B, suggesting a deficit in their ability to reliably discriminate the training contexts.

These same DG-NR1 KO mice were then subject to mulitelectrode recordings to monitor hippocampal ensemble activity as they explored two distinct contexts. As reported in rats (Leutgeb *et al.*, 2005b), the exposure of animals to two similar, but distinct, contexts generates place cell firing rates in the two boxes that overlap significantly less in CA3 than in CA1, with many CA3 neurons undergoing what is termed “rate remapping”, a change not in the location of their firing field, but a significant and context-specific modulation of their firing rate. Rate remapping was robustly observed in rats when the contextual changes were relatively minor, i.e. changes in the local color and shape of the recording box, but no change in the distal spatial cues. Using a similar protocol, CA3 and CA1 place cell activity was recorded as DG-NR1 KO mice explored for scattered food rewards in a white circular box and a black square box, located in the identical position in the room relative to the distal cues. Rate remapping in both CA1 and CA3 was analysed by comparing the average firing rates of each neuron during exploration of the two boxes. As in rats, the remapping in the CA3 region in control mice was significantly greater than that in the CA1 region. In contrast, DG-NR1 KO mice showed similar remapping in CA3 and CA1, with the rate difference in CA3 significantly different from controls, thus providing a strong physiological correlate of the observed behavioural phenotypes.Together with the behavioural data, this suggests that the DG, and specifically plasticity at the

PP-GC synapse, does contribute to contextual discrimination and is important for the rapid separation of similar, yet distinct contexts.

A property unique to the dentate gyrus is the presence of postnatal neurogenesis (see Chapter 4). The last decade has seen intense interest in defining the contribution of adult-born granule cells in the DG to hippocampal information processing and memory (Deng *et al.*, 2010). These adult born GCs are transiently more excitable than the embryonically derived GCs (van Praag *et al.*, 2002); however, if this impacts their ability to participate in memory formation remains unclear (Kee *et al.*, 2007; Stone *et al.*, 2010). Recently, experiments to address the contribution of adult born neurons to pattern separation were reported. Neurogenesis in the DG was inhibited either through the application of γ-irradiation of the hippocampus or a genetic approach that used a lentiviral vector to inhibit the function of the Wnt-1 protein, a factor necessary for neurogenesis (Clelland *et al.*, 2009). Behavioural testing of these mice employed two novel tests of pattern separation, a delayed non-matching to place version of the radial arm maze and a two-choice spatial discrimination task. Both these paradigms allowed the experimenters to present the mice with highly separated (easy) or closely spaced (difficult) choice versions of the task. Mice with decreased neurogenesis were found to be impaired in both tasks only when the stimuli or locations to be discriminated were close together in space. These data suggest that adult-born neurons may play a particularly important role in spatial pattern separation computation of the dentate gyrus.

3.2. *Area CA3*

The CA3 region, and in particular, its robust and highly plastic recurrent collaterals, has received a great deal of attention for its role in hippocampal memory processes. Two prominent suggestions have been that CA3 is required for the rapid encoding of hippocampal memories and is necessary for the pattern completion function of the circuit. Pattern completion refers to the ability of the circuit to recall a memory based on partial or noisy cues. Marr and others have suggested that a recurrent neuronal network with robust synaptic plasticity, like the one present in the CA3, may be ideally suited to this computation (Marr, 1971; Rolls, 1996; Treves, 2004).

While data from traditional methods of circuit intervention, such as chemical or physical lesion, are less numerous in CA3, there are some suggestive results in the literature concerning CA3's contribution to hippocampal processing. A study by Kesner *et al.* showed that rats with ibotenic acid-mediated CA3 lesion exhibited normal 24-h memory in multitrial contextual fear conditioning, although they were slower than the control mice in the acquisition (Lee and Kesner, 2004). Moser *et al.* reported that rats with ibotenic acid-mediated lesions of the CA3 area were normal in spatial learning in an annular water maze task and exhibited only a very light deficit in the standard Morris water maze task (Steffenach *et al.*, 2002). The same group also introduced knife cuts between CA3 and CA1 and showed that the place fields recorded in the CA1 region of these "fiber lesioned" animals in a familiar open arena were only slightly larger than those of the control rats (Brun *et al.*, 2002). These studies suggest animals do not entirely depend on the integrity of the trisynaptic loop and SC input for acquisition of spatial reference memory or formation and reactivation of CA1 representations of space and, in turn, suggest a role of the direct EC input to CA1 in these processes.

Further, rats with CA3 lesion exhibited specific deficits in spatial learning tasks requiring pattern completion (Gold and Kesner, 2005). Interestingly, in a second task in rats with similar lesions, this same group also detected deficits in a task requiring spatial separation (Gilbert and Kesner, 2006). Although it is difficult to interpret this deficit because the CA3 lesion also limits the contribution of the DG to hippocampal processing, it suggests these two processes may share common mechanisms. However, one must take these suggestions with caution because of the limitations of lesion techniques.

Perhaps the most useful tools in testing Marr's hypothesis regarding auto-associative plasticity comes from work analysing a conditional knockout mouse in which the NMDA receptor gene is ablated specifically in the CA3 pyramidal cells of adult mice (CA3 NR1 KO) (Nakazawa *et al.*, 2002). Slice physiology data demonstrated functional ablation of NMDA receptors at the recurrent CA3 synapses (C/A → CA3) with respect to both the NMDA current and LTP. Extensive behavioural and physiological characterisation of this line has provided a deeper understanding of the role of CA3 plasticity in memory function.

Much like the DG-NR1 KO mice discussed above, the CA3-NR1 KO mutant mice do not display any impairment in the acquisition or retrieval of spatial reference memory in the standard, hidden platform version of the Morris water maze. However, when the mice were first trained to navigate to the hidden platform using four salient distal cues, then later challenged to navigate to the goal with only one randomly selected distal cue of the four present, the mutants exhibited a significantly lower search preference for the target area compared to the control mice (Nakazawa *et al.*, 2002). These results strongly suggested that NMDA receptors in CA3 played a crucial role specifically in pattern completion-based recall of spatial reference memory. Next, Nakazawa *et al.* determined the effect of partial cue removal on the activity of CA1 place cells. Using a paradigm similar to the behavioural test, CA1 place cells were recorded as mice explored a familiar context defined by four prominent distal cues, then again later in the day with only a single cue present. In the control mice, there was no effect of partial cue condition on the spatial tuning properties, including the average firing rate of the cells and the place field size. In contrast, cells in the mutants showed a significant reduction in their spatial tuning properties (Nakazawa *et al.*, 2002). These physiological results were compatible with the behavioural results, suggesting that reductions in CA1 output as the consequence of reduced CA3 drive resulting from cue removal may make it more difficult for mutants to retrieve spatial memories. This impairment may underlie the inability of mutants to solve spatial memory tasks, such as the water maze, when only partial distal cues are available.

A second set of physiological experiments in the CA3-NR1 KO mice found that CA1 place cells in the mutants were slower to mature when the animal first explored a novel environment, suggesting a deficit in the speed of the formation of a contextual representation. Indeed, the mice displayed deficits in rapidly acquiring the memory of a novel hidden platform location after one trial but were normal when tested with a previously experienced platform location (Nakazawa *et al.*, 2003). Recently, this encoding deficit was further examined using a variation of the contextual fear task (McHugh and Tonegawa, 2009). Hippocampus-dependent contextual fear conditioning has properties which make it a useful tool in assessing an animal's ability to form and recall contextual memory. If the foot shock [unconditioned stimulus (US)] is applied immediately after the rodent is placed in a novel context, conditioning is ineffective; at least a 20- to 40-s placement-to-shock interval (PSI) is needed (Fanselow, 1990). Further, under asymptotic values of 1–2 min, as the PSI is increased, the amount of fear also increased, indicating a strong correlation between the time in the context, the salience of the contextual CS, and the strength of the CS-US association. Thus, the task can provide a behavioural measure of the kinetics of the formation of a "usable" representation of the context. Mice were placed in a novel conditioning chamber and received a single unsignalled foot shock at one of four PSIs (0, 20, 40 and 60 s). When fear memory was assessed the next day, the mutants froze significantly less at the intermediate PSIs, although by 60 s of exposure the genotypes exhibited indistinguishable 24-h fear memory. The presence of a deficit only at intermediate PSIs suggests that the loss of NRs in CA3 does not abolish a hippocampal-dependent representation of the contextual CS, but does cause a significant change in the rate at which the context is processed and encoded. The relationship between the amount of time in the context and the saliency of the contextual CS is presumably dependent of many aspects of hippocampal processing; however, the specificity of the genetic manipulation revealed a quantifiable contribution of CA3 NR-mediated plasticity to the rapid formation of the representation. In a similar set of experiments, Cravens *et al.* also found deficits in the rapid encoding of context in these mice, as well as transient (3–24 h) deficit in pattern separation, again suggesting interplay between the processes of pattern completion and separation (Cravens *et al.*, 2006).

The 2002 Brun *et al.*, "CA3 knife cut" study cited above did not address a number of interesting questions, such as the potential role of CA3 output in rapid-learning or the contribution of CA3 to memory consolidation. The observation that CA1 place cells are normal in a familiar environment in the rats with a cut of the SCs suggests that the TA input may eventually be sufficient to form CA1 place cells with nearly normal fields. A thorough test of these issues would ideally use an inducible, reversible and complete blockade of CA3 synaptic transmission. Recently, this was achieved via genetic means using a triple transgenic mouse in which synaptic transmission is blocked by CA3 pyramidal cell-restricted and temporally controlled expression of the tetanus toxin (TeTX) light chain (Nakashiba *et al.*, 2008). TeTX is an endopeptidase specific for VAMP2, a protein essential for activity-dependent neurotransmitter release from the presynaptic terminal. Hippocampal slices prepared from mice expressing TeTX in CA3 pyramidal cells demonstrate a complete loss of synaptic transmission at the CA3-CA1 synapse. These mice were subject to the reference memory version of the Morris water maze and once more there was no observable difference in behaviour between the mutants and controls, suggesting this type of learning can be done in the absence of any CA3/DG contribution to CA1 activity. However, when the mice were tested in the contextual fear task, significantly poorer memory was observed one day after learning, suggesting CA3 to CA1 transmission is dispensable for the acquisition of slowly acquired spatial learning, as in the water maze, but is necessary for the rapid encoding of a novel context, as is required in the contextual fear task.

To examine the physiological consequences of a loss of SC input into CA1, place field recordings from CA1 in this line was conducted. The experiments indentified a strong physiological correlate of this behavioural phenotype; when CA-TeTX mice were allowed to explore a novel environment CA1 place fields were present however, the spatial specificity of individual cells was significantly poorer than control neurons. Surprisingly, when the mutant mice were returned to the same environment 24 and 48 h later, the spatial coding in CA1 showed a dramatic improvement. Although the spatial representation of CA1 pyramidal cells never quite reached the level of control mice, this slow experience-dependent improvement may reflect the ability of these mice to acquire tasks across many days of training and explain the data from the SC "knife cut" experiments of Brun *et al.* (2002).

The inducible nature of the genetic blockade in the CA3-TeTX mice is based on the tetracycline-inducible system (Gossen *et al.*, 1994). This method of drug-modulated transcriptional control of a transgene allows a slow, but complete, blockade to be induced in the adult mouse, with a time line of 2–3 weeks. Taking advantage of this, the CA3-TeTX mice were used to address the role of CA3 to CA1 transmission in the consolidation of spatial memory. Mutant mice were removed from the drug and subject to contextual fear conditioning one week later. At this time, transmission is intact, but will be lost 7–10 days after learning. The mice could acquire the task at this time point in a manner indistinguishable from control animals, yet when long-term memory of the training was assessed 6 weeks later, the mutant animals showed a significant deficit in remote memory (Nakashiba *et al.*, 2009). To establish a link between this consolidation deficit and CA1 physiology, the authors turned once again to in vivo physiology in CA1. Somewhat surprisingly, given the loss of CA3 input, an examination of the CA1 local field potential revealed the presence of ripple oscillations in the mutant animals during periods of quiet wake and slow-wave sleep. However, these ripples were, on average, significantly slower than those observed in control mice, perhaps revealing their alternate source of generation. To more directly assess the correlation between the activity of single neurons and memory consolidation, CA1 place cells were recorded before, during and after exploration of a novel environment. In rats, it had previously been demonstrated that cells which code for nearby locations, and as a result are both spatially and temporally correlated in their firing during exploration, show an experience-dependent increase in their co-activity during ripple events occurring after exploration (Wilson and McNaughton, 1994). Using a similar analysis, the authors replicated this finding in the control mice pairs of cells with spatially similar place fields demonstrated a significant increase in their co-activity in the ripple events that occurred after exploration as compared to before, while pairs of cells with spatially dissimilar fields did not. This increase for overlapping place cells was significantly reduced in the CA3-TeTX mice, providing strong evidence for a link between ripple-associated ensemble reactivation and memory consolidation.

3.3. *Area CA1*

The CA1 region of the hippocampus is perhaps one of the most well-studied regions of the rodent brain. Given its tractability to a variety of approaches; *in vitro* and *in vivo* physiology and imaging, physical, and pharmacological intervention, and robust behavioural phenotypes following its damage, there exists a wealth of data and speculation addressing its role in memory (Rolls and

Kesner, 2006; Silva *et al.*, 2009). Damage to CA1, be it physical or chemical, that results in a loss of output has devastating effects on hippocampal dependent memory, as does pharmacological intervention that inhibits synaptic plasticity (Morris *et al.*, 1982a; Morris, 1981; Morris *et al.*, 1986). However aside from its necessity as an output structure, little still remains known about its precise contribution to hippocampal information processing. The last 15 years has seen an increasing number of studies that combine genetic intervention with *in vivo* physiology. There have now been 16 lines of mutant mice subject to characterisation of changes in CA1 place field activity and/or changes in local field potential following genetic interventions. The molecular targets are numerous, from signalling molecules such as PKA and CaMKII, to receptors like GluR1, GluR2 and D1, to transcription factors such as zif268 and CREB (Buhl *et al.*, 2003; Cacucci *et al.*, 2007; Cacucci *et al.*, 2008; Cho *et al.*, 1998; McHugh *et al.*, 1996; McHugh *et al.*, 2007; Nakashiba *et al.*, 2009; Nakashiba *et al.*, 2008; Nakazawa *et al.*, 2002; Nakazawa *et al.*, 2003; Nolan *et al.*, 2004; Racz *et al.*, 2009; Renaudineau *et al.*, 2009; Rotenberg *et al.*, 2000; Rotenberg *et al.*, 1996; Taverna *et al.*, 2005; Tran *et al.*, 2008; Yan *et al.*, 2002). While limited space prohibits a detailed discussion of all results, the studies involving single unit recordings have been briefly summarised in Table 12.1. The majority of the mice demonstrate plasticity deficits in CA1, and as a whole, have provided strong evidence for a role of CA1 plasticity in place field formation and stability. The remainder of this section will focus on three mice that offer representative phenotypes linking plasticity in CA1 to place coding; the CA1-specific NR1 KO mouse (Tsien *et al.*, 1996) and two transgenic lines with point mutations of CaMKII residue Thr286 which bidirectionally modulate kinase activity and impact synaptic plasticity (Giese *et al.*, 1998; Mayford *et al.*, 1996).

The NMDA receptor is both a ligand- and voltage-gated ion channel and can therefore serve as a molecular coincidence detector ideally suited to trigger synaptic efficacy change (Nowak *et al.*, 1984). Above, I have discussed behavioural and physiological data obtained in mice lacking the receptor,and the plasticity it supports,specifically in the granule cells of the DG or the pyramidal cells of CA3 (McHugh *et al.*, 2007; Nakazawa *et al.*, 2002). These two lines, produced using the Cre/loxP approach, are the intellectual progeny of one of the first conditional mutants in the field of neurobiology, the CA1-NR1 KO mouse (Tsien *et al.*, 1996). In 1996, Tsien *et al.*, created a mouse strain in which the NR1 gene was postnatally knocked out predominantly in CA1 pyramidal cells. This mouse strain grew normally, in contrast to the perinatally lethal NR1 knockout mice produced by conventional knockout technology. In contrast to the DG or CA3-NR1 KO mice, the CA1-NR1 KO mice exhibited a severe deficit in spatial reference memory. When tested in the water maze, the mutants showed profound deficits in task acquisition and a seemingly complete loss of spatial learning (Tsien *et al.*, 1996).

To seek an explanation for this learning deficit, the knockout mice were subjected to *in vivo* recordings of CA1 pyramidal cells as they ran along a familiar linear track.The pyramidal cells in CA1 were found to retain place-specific firing that was stable over repeated exposures in the same day. However, altered properties were observed both on the level of the single-cell and the neuronal ensemble. Individual place fields were larger and less structured, with an average increase in size of about 30%. While the loss of specificity was significant, the authors concluded it could not account for the magnitude of the behavioural deficit observed in the water maze. In further analyses, they took advantage of the ability to record multiple cells simultaneously to examine the coordinate firing of the neural ensemble. It had been previously shown in rats that cells with overlapping place fields reliably fire together during exploration, as you would predict from their average receptive fields (Wilson and McNaughton, 1993). In the mutant mice, this coordinated firing was completely absent, indicating a reliable and coherent spatial map was not being constructed (McHugh *et al.*, 1996). Thus, in the mutant while on average two cells had similar place fields, during any single pass through the region of overlap the firing of one cell did not predict the firing of the other. To date, this remains the strongest and most specific evidence linking synaptic plasticity, spatial learning and hippocampal ensemble activity.

The essential role of Ca^{2+} in neuronal signalling and the modification of synaptic connections has been demonstrated in a variety of ways (Bito *et al.*, 1997; Ghosh and Greenberg, 1995). One of Ca^{2+} important functions is the activation of Ca^{2+}/calmodulin signal transduction enzymes (Bito *et al.*, 1996). α- CaMKII is a neural-specific enzyme which is present both pre- and postsynapticly in the hippocampus (Erondu and Kennedy, 1985; Schulman and Lou, 1989). Upon association with calmodulin-bound Ca^{2+}, CaMKII undergoes an autophosphorylation event at residue threonine-286, converting the kinase to a calcium-independent form (Miller and Kennedy, 1986). This mechanism allows the kinase to remain active, even after the dissipation of the Ca^{2+} signal, a property that makes it an interesting molecule in the events leading to synaptic modification. Using transgenic techniques, two laboratories generated mice with point mutations of CaMKII residue Thr286. Mayford *et al.* constructed a line of mice that carried an α-CaMKII transgene with a point mutation that converted Thr-286 to aspartate (CaMKII-Asp-286), giving the residue a negative charge and mimicking the autophosphorylation event (Mayford *et al.*, 1995). This mutant isoform behaves as a gain-of-function allele, with constitutive activation of the Ca^{2+}-independent kinase. Independently, Giese *et al.* utilised a knockin approach to replace the endogenous α-CaMKII gene with a mutant allele in which Thr286 had been replaced with an alanine residue (CaMKII-T286A). This mutation blocks the autophosphorylation event and creates an allele of the kinase that cannot be converted into a Ca^{2+}/calmodulin-independent state (Giese *et al.*, 1998). Both lines of mice have been extensively analysed and have shed light on the role this kinase plays in the gating of synaptic modification and memory.

Slice electrophysiology performed on the CaMKII-Asp-286 line revelaed normal LTP at high stimulation frequencies (100 Hz), however at lower, more physiological frequencies (1–10 Hz), stimulation resulted in long-term depression of synaptic transmission, as opposed to LTP (Mayford *et al.*, 1995). This shift in response may be particularly significant for hippocampal learning as it is in the theta range, thus alterations in synaptic response at these frequencies may have severe behavioural affects. To assay behaviour, the CaMKII-Asp-286 mice were tested in the Barnes circular maze, a test of spatial learning. These mice were found to be deficient in spatial navigation (Bach *et al.*, 1995). The link between alterations in LTP and behavioural deficits was further investigated using hippocampal place cell recordings.

Rotenberg *et al.* conducted experiments in which they recorded the activity of a total of 52 pyramidal cells from 10 CaMKII-Asp-286 mice (Rotenberg *et al.*, 1996). Using a recording apparatus that consisted of 26 single electrodes placed in the dorsal hippocampus, they monitored the activity of place cells as the mice explored a 49-cm cylindrical environment. The CaMKII-Asp-286 mice were found to have several severe deficiencies in place cell function. The pyramidal cells of the mutant animals were found to encode for place at a significantly lower rate (31% of total mutant cells encoded place vs. 58% in the wild-type controls). Also, a significant decrease in both the average and peak firing rates of mutant place cells was observed. Next, Rotenberg *et al.* looked at the accuracy of the positional firing of the mutant cells and report that the place fields are less precise and more diffuse. However, the magnitude of this deficit could not fully explain the large deficits observed in the behavioural assays. Therefore, Rotenberg *et al.* examined the stability of the mutants' place fields over multiple recording sessions. While assessing the long-term stability of individual place cells can be technically difficult, the authors employed careful experimental controls and were able to detect a loss of place field of stability in the CaMKII-Asp-286 mutant mice. Although the loss of stability was not absolute, the authors concluded that it was the significant enough to account for the behavioural deficits observed in this line of mutant mice.

The second line of CaMKII mutant mice, the CaMKII-T286A line, was first subject to behavioural and *in vitro* electrophysiological characterisation. Giese *et al.* reported that these mice had severe deficits in CA1 LTP, across a range of stimulation frequencies, although LTP in the DG was intact, and were found to lack spatial learning abilities, as assayed by the Morris water maze (Cooke *et al.*, 2006; Giese *et al.*, 1998). This finding suggests that the autophosphorylation event which converts CaMKII into Ca^{2+}/CaM-independent kinase in crucial in synaptic modification and memory formation. To gain a better understanding of these deficits, 14 CA1 place cells across three mutant animals were recorded as the mice explored a four-arm radial maze (Cho *et al.*, 1998). The protocol employed for this study involved multiple recording sessions, interspersed by rotations of the cues in the radial maze environment. Although in some respects this manipulation may make the results somewhat harder to interpret, since even the place cells of wild-type mice can show several possible responses to this change, the authors conducted the experiment in order to assess the mutants' ability to adjust their spatial maps. Regardless, their findings indicate while place fields were observed, some changes, both in the basic physiology of the mutant cells, and their ability to encode for place may account for their behavioural deficits. The pyramidal cells of the CaMKII-T286A mice were found to fire significantly faster while the animal transverses a field that they encoded. In addition, the spatial selectivity of these cells was decreased, with higher firing rates in regions of the environment outside the cell's place field. As with the other mice analysed, these alterations in place cell function are significant, but seem minor to explain the complete loss of spatial learning ability. More severe defects were detected when Cho *et al.* manipulated the environmental cues and then reintroduced the mice to the maze. While a large percentage of wild-type place cells retained their original fields (~66%) and only ~20% developed new spatial representations, the place cells of the mutant mouse showed a tendency to lose their original field (only ~10% stayed the same), and the majority (~50 %) developed entirely new fields. In a final manipulation, the cues were returned to their original positions and the mice were again placed in the maze. The CaMKII-T286A mice once more seemed to largely remap the environment, with little place field stability between the recording sessions. This instability may explain the problems these animals demonstrate in the Morris water maze, as a stable map of the environment is crucial to performing the task.

Recently, a second laboratory recorded CA1 place cells in the same mutant line under a different protocol to address the role of CaMKII in encoding space as mice habituate to a novel environment across days (Cacucci *et al.*, 2007). Cacucci *et al.* examined two slowly developing place cell properties; spatial tuning, related to the field quality, and spatial reproducibility, related to field stability. They report that these properties are dissociable in the CaMKII-T286A mouse, with the mutant mice showing improvement in spatial reproducibility over days of experience in a novel environment, but impairment in the experience-dependent increase in spatial tuning. One possible explanation offered is that much like the CA1-NR1 KO mouse, the tuning deficits in the CaMKII-T286A mouse are related to the loss of CA1-NR–mediated synaptic plasticity. However, the intact plasticity related to field stability may be a consequence of plasticity at another synapse in the circuit, such as the EC-DG pathway. These possibilities remain to be examined using more specific models.

As a whole, these studies agree that genetic interventions that result in a loss of CA1 synaptic plasticity have dire consequences on both spatial learning and on the coding of space on the level of place cells. However, although the rough correlation of LTP deficits, place cell deficits and learning deficits exists, there is not a clear pattern of phenotypes that emerges from the data in Table 12.1. Numerous factors contribute to this, including the lack of regional specificity of many of the genetic manipulations, the variations in recording technique and protocol, and the lack of physiology data acquired during a learning task. Further, even at best these studies all lead to a loss of NMDAR-mediated plasticity at all excitatory synapses to a given cell, and as a result cannot specifically address if or how CA1 integrates the various inputs it receives. Further, as the axons of CA1 pyramidal cells serve as the dominant glutamatergic output of all information flowing through the hippocampus, physical or genetic approaches that completely block or impair CA1 transmission can only conclude that hippocampal output is required for learning, but not what precise computation, if any, this subregion is contributing to information processing in the circuit. One challenge for the field in the coming years is to develop novel genetic tools, as well as novel physiological and behavioural measures, to tackle these difficult problems.

4. Future directions

While the field of hippocampal circuit genetics has made vital contributions to both testing existing hypotheses and generating new ones, many questions remain to be addressed in the coming years. For example, the recent study revealing the unexpected physiology of the inputs to CA2 pyramidal cell in the hippocampal slice suggest that the EC-CA2-CA1 loop may exert a strong influence over hippocampal output, especially when at times when CA3 activity is low (Chevaleyre and Siegelbaum, 2010). It is an intriguing possibility that some of the learning and physiological phenotypes in the CA3-TeTX mutant mice are a direct result of

Table 12.1.

Mouse	Intervention	Synaptic plasticity	Behaviour	*In vivo* physiology	Reference
CA1-NR1 KO	CA1 deletion of the NMDA Receptor	CA1 LTP deficit	No learning in water maze	CA1 place fields are larger	McHugh *et al.*, 1996
			Impaired trace fear conditioning	Cells tuned to similar places less coordinated	
CA3-NR1 KO	CA3 deletion of the NMDA Receptor	CA3 C/A-LTP deficit	Deficits in DMP water maze	CA1 place fields larger in novel environment	Nakazawa *et al.*, 2002
			Impaired water maze recall under partial cue	Cells less active under partial cue conditions	Nakazawa *et al*, 2003
DG-NR1 KO	DG deletion of the NMDA Recptor	DG LTP deficit	Impaired context discrimination conditioning	CA3 place cells show loss of context specific remapping	McHugh *et al.*, 2007
				CA1 place fields larger in novel environment	
CA3-TeTX	Inducible blockade of CA3 output	N/A	Impaired contextual maze fear, normal water	Larger place fields	Nakashiba *et al.*, 2008
			Impaired long-term (6 week) consolidation	Decrease in experience dependent coactivation during ripples	Nakashiba *et al*, 2009

(*Continued*)

Table 12.1. (*Continued*)

Mouse	Intervention	Synaptic plasticity	Behaviour	*In vivo* physiology	Reference
CaMKII T286D	Forebrain expression of mutation that mimics CaMKII autophosphorylation	CA1 LTP deficit in 5–10 Hz Stimulation frequency	Impaired learning in Barnes maze Normal contextual fear conditioning	Place cells less common, less precise and less stable Lower average firing rate	Rotenberg *et al.*, 1996
CamKII T286A	Knock-in of mutation that blocks CaMKII autophosphorylation	CA1 LTP deficit	No learning in water maze	Place cells less selective, less stable Loss-of-experience–dependent spatial tuning	Cho *et al.*, 1998 Cacucci *et al.*, 2007
CREB –/–	Global KO of α/Δ CREB isoforms	CA1 LTp decays to baseline by 90 min	Long-term memory deficit in water maze and contextual fear	Place cells less selective, less stable	Cho *et al.*, 1998
zif268 –/–	Global KO of zif26B	Late phase CA1 LTP deficit	Long-term memory deficit	Exposure to novelty interferes with stored spatial map Impaired long-term (24-h) place cell stability in novel environment	Renaudineau *et al.*, 2009

(*Continued*)

Table 12.1. (*Continued*)

Mouse	Intervention	Synaptic plasticity	Behaviour	*In vivo* physiology	Reference
Tg2576 AD model	Overexpression of human APP mutant APP695SWE	Impaired CA1 & DG LTP in aged mice	Aged mice have deficit in forced-choice T-maze	Aged, but not young, mice have larger fields with lower spatial information	Cacucci *et al.*, 2008
PKA R(AB)	Forebrain expression of R(AB) PKA inhibitory Subunit	Late phase CA1 LTP deficit	24 h deficit in contextual fear	Reduced place field stability at 24 h, but not at 1 h	Rotenberg *et al.*, 2000
NocR –/–	Global KO of Nociceptin receptor	Enhanced CA1 LTP at high stimulation frequencies	Enhanced learning in the water maze	Increased firing rate, fields poorly organized, less responsive to cue rotation Fields unstable at 1 h and 48 h	Taverna *et al.*, 2005
D1R –/–	Global KO of Dopamine D1R receptor	Late phase CA1 LTP deficit in CA1	Impaired learning in water moze	Decreased influence of distal cues on field location Less flexible code for space	Tran *et al.*, 2008
GluR2 –/–	Global KO of GluR2 subunit of AMPAR	Enhanced non-Hebbian LTP	Multiple deficits, including impaired learning in water maze	Place cells less common, less percise and less stable	Yan *et al.*, 2002

this processing loop. Similarly, in terms of genetic intervention, many of the input pathways to the hippocampus from the entorhinal cortex remain untouched. Recent work has suggested divergent roles for the medial and lateral perforant path inputs (Hargreaves *et al.*, 2005; Knierim *et al.*, 2006; Yoganarasimha *et al.*, 2010), as well as possible differential contributions from the various projections from layer II and layer III of the EC.

Another aspect of hippocampal function and information processing much too broad for the scope of this chapter is that of GABAergic inhibition. In CA1 alone, over 30 distinct types of GABAergic inter neurons have been classified, each with a distinct genetic, anatomical and physiological profile (Klausberger and Somogyi, 2008). It is clear that inhibition throughout the entire circuit has a strong role in shaping computation and output and much of this remains to be addressed. Recently, genetic access to a subset of interneurons, the PV+ cells which compromise roughly 25% of all interneurons in CA1, and include the perisomatic targeting basket and axo-axonic cells which exert strong inhibitory control over PC spiking, has been achieved (Klausberger and Somogyi, 2008). Elegant physiological characterisation of mice in which PV+ IN excitability has been drastically reduced via deletion of the GluR1 subunit in these cells revealed augmented ripples in these mice (Racz *et al.*, 2009). Similarly, a mouse with a deletion of the GABA transporter gene, GAT1, displayed significant changes in the theta frequency in CA1 and impaired memory in the water maze, context fear and passive avoidance — all tasks dependent on hippocampal function (Gong *et al.*, 2009). As genetic access to the various classes of INs becomes available, unraveling their unique contribution to hippocampal processing and physiology promises to be a fascinating avenue of research.

Yet another facet of the circuit that has not been fully explored is the role of neuromodulation in hippocampal function. While pharmacological and genetic interventions have suggested that the monoamergic and cholinergic transmission are key to memory formation and plasticity in the hippocampus, the precise role of this signals and how they are recruited and integrated on the level of cells and circuits remains unknown. Cholinergic antagonists such as scopolamine reduce place cell firing and field quality (Brazhnik *et al.*, 2003) and lead to memory encoding deficits in hippocampal-dependent behavioural tasks (Rogers and Kesner, 2003). Likewise, pharmacological silencing of the VTA can cause deficits in hippocampal-dependent behaviour and lead to a remapping of CA1 place fields (Martig and Mizumori, 2010) and drugs targeting the norephinephrine system, which heavily innervates the DG, lead to instability of CA1 firing fields (Tanila, 2001). Finally, some recent studies have identified serotonin as a key modulator of hippocampal function in regards to anxiety. A deletion of the

5-HT1(A) receptor leads to an increase in the magnitude of the hippocampal theta oscillation specifically in anxiety provoking contexts (Gordon *et al.*, 2005) and impacts the communication of the hippocampus with downstream structures such as the prefrontal cortex (Adhikari *et al.*, 2010). Once again, improving genetic access to these systems, combined with behavioural, neurochemical and physiological assays will start to unravel the complex interactions between these complementary systems and hippocampal function.

Finally, the genetic toolbox available to intervene and study circuits in the brain is constantly evolving and expanding. The recent explosion of optogenetic tools that allow exquisite temporal and cell type control of neuronal output will undoubtedly be vital in proving and disproving the current hypotheses regarding hippocampal function (Gradinaru *et al.*, 2010; Gunaydin *et al.*, 2010; Zhang *et al.*, 2010). The ability to control the spiking of discrete classes of cells in the circuit rapidly and repeatedly during behaviour, combined with physiological recording, promises an entirely new level of understanding of how memory is formed, stored and recalled in the hippocampal circuit.

References

Adhikari, A., Topiwala, M.A., and Gordon, J.A. (2010). Synchronized activity between the ventral hippocampus and the medial prefrontal cortex during anxiety. *Neuron* **65**, 257–269.

Alonso, A. and Garcia-Austt, E. (1987a). Neuronal sources of theta rhythm in the entorhinal cortex of the rat. I. Laminar distribution of theta field potentials. *Exp Brain Res* **67**, 493–501.

Alonso, A. and Garcia-Austt, E. (1987b). Neuronal sources of theta rhythm in the entorhinal cortex of the rat. II. Phase relations between unit discharges and theta field potentials. *Exp Brain Res* **67**, 502–509.

Alonso, A., Gaztelu, J.M., Buno, W., Jr., and Garcia-Austt, E. (1987). Cross-correlation analysis of septohippocampal neurons during theta-rhythm. *Brain Res* **413**, 135–146.

Bach, M.E., Hawkins, R.D., Osman, M., Kandel, E.R., and Mayford, M. (1995). Impairment of spatial but not contextual memory in CaMKII mutant mice with a selective loss of hippocampal LTP in the range of the theta frequency. *Cell* **81**, 905–915.

Bannerman, D.M., Deacon, R.M., Seeburg, P.H., and Rawlins, J.N. (2003). GluR-A-Deficient mice display normal acquisition of a hippocampus-dependent spatial reference memory task but are impaired during spatial reversal. *Behav Neurosci* **117**, 866–870.

Barnes, C.A., McNaughton, B.L., Mizumori, S.J., Leonard, B.W., and Lin, L.-H. (1990). Comparison of spatial and temporal characteristics of neuronal activity in sequential stages of hippocampal processing. *Prog Brain Res* **83**, 287–300.

Best, P.J. and White, A.M. (1999). Placing hippocampal single-unit studies in a historical context. *Hippocampus* **9**, 346–351.

Best, P.J., White, A.M. and Minai, A. (2001). Spatial processing in the brain: The activity of hippocampal place cells. *Annu Rev Neurosci* **24**, 459–486.

Bito, H., Deisseeroth, K. and Tsien, R.W. (1996). CREB phosphorylation and dephosphorylation: A Ca^{2+}- and stimulus duration-dependent switch for hippocampal gene expression. *Cell* **87**, 1203–1214.

Bito, H., Deisseroth, K. and Tsien, R.W. (1997). Ca^{2+}-dependent regulation in neuronal gene expression. *Curr Opin Neurobiol* **7**, 419–429.

Bliss, T.V.P. and Collingridge, G.L. (1993). A synaptic model of memory: Long-term potentiation in the hippocampus. *Nature* **361**, 31–39.

Bliss, T.V.P. and Lømo, T. (1973). Long-lasting potentiation of synaptic transmission in the dentate area of the anaesthetized rabbit following stimulation of the perforant path. Journal of Physiology (London) **232**, 331–356.

Bordi, F., Reggiani, A. and Conquet, F. (1997). Regulation of synaptic plasticity by mGluR1 studied in vivo in mGluR1 mutant mice. *Brain Res* **761**, 121–126.

Boss, B.D., Peterson, G.M. and Cowan, W.M. (1985). On the number of neurons in the dentate gyrus of the rat. *Brain Res* **338**, 144–150.

Boss, B.D., Turlejski, K., Stanfield, B.B. and Cowan, W.M. (1987). On the numbers of neurons in fields CA1 and CA3 of the hippocampus of Sprague-Dawley and Wistar rats. *Brain Res* **406**, 280–287.

Brandon, E.P., Zhuo, M., Huang, Y.Y., Qi, M., Gerhold, K.A., Burton, K.A., Kandel, E.R., McKnight, G.S. and Idzerda, R.L. (1995). Hippocampal long-term depression and depotentiation are defective in mice carrying a targeted disruption of the gene encoding the RI beta subunit of cAMP-dependent protein kinase. *Proc Natl Acad Sci USA* **92**, 8851–8855.

Brazhnik, E.S., Muller, R.U. and Fox, S.E. (2003). Muscarinic blockade slows and degrades the location-specific firing of hippocampal pyramidal cells. *J Neurosci* **23**, 611–621.

Brun, V.H., Otnass, M.K., Molden, S., Steffenach, H.A., Witter, M.P., Moser, M.B. and Moser, E.I. (2002). Place cells and place recognition maintained by direct entorhinal-hippocampal circuitry. *Science* **296**, 2243–2246.

Buhl, D.L., Harris, K.D., Hormuzdi, S.G., Monyer, H. and Buzsaki, G. (2003). Selective impairment of hippocampal gamma oscillations in connexin-36 knock-out mouse in vivo. *J Neurosci* **23**, 1013–1018.

Buzsaki, G. (1986). Hippocampal sharp waves: their origin and significance. *Brain Res* **398**, 242–252.

Buzsaki, G. (1989). Two stage model of memory trace formation: A role for "noisy" brain states. *Neurosci* **31**, 551–570.

Buzsaki, G. (1996). The hippocampo-neocortical dialogue. *Cereb Cortex* **6**, 81–92.

Buzsaki, G., Czopf, J., Kondakor, I. and Kellenyi, L. (1986). Laminar distribution of hippocampal rhythmic slow activity (RSA) in the behaving rat: Current-source density analysis, effects of urethane and atropine. *Brain Res* **365**, 125–137.

Buzsaki, G. and Eidelberg, E. (1983). Phase relations of hippocampal projection cells and interneurons to theta activity in the anesthetized rat. *Brain Res* **266**, 334–339.

Buzsaki, G., Grastyan, E., Czopf, J., Kellenyi, L. and Prohaska, O. (1981). Changes in neuronal transmission in the rat hippocampus during behavior. *Brain Res* **225**, 235–247.

Buzsaki, G., Rappelsberger, P. and Kellenyi, L. (1985). Depth profiles of hippocampal rhythmic slow activity ('theta rhythm') depend on behaviour. *Electroencephalogr Clin Neurophysiol* **61**, 77–88.

Cacucci, F., Wills, T.J., Lever, C., Giese, K.P. and O'Keefe, J. (2007). Experience-dependent increase in CA1 place cell spatial information, but not spatial reproducibility, is dependent on the autophosphorylation of the alpha-isoform of the calcium/calmodulin-dependent protein kinase II. *J Neurosci* **27**, 7854–7859.

Cacucci, F., Yi, M., Wills, T.J., Chapman, P. and O'Keefe, J. (2008). Place cell firing correlates with memory deficits and amyloid plaque burden in Tg2576 Alzheimer mouse model. *Proc Natl Acad Sci USA* **105**, 7863–7868.

Chen, C., Magee, J.C., Marcheselli, V., Hardy, M. and Bazan, N.G. (2001). Attenuated LTP in hippocampal dentate gyrus neurons of mice deficient in the PAF receptor. *J Neurophysiol* **85**, 384–390.

Chevaleyre, V. and Siegelbaum, S.A. (2010). Strong CA2 pyramidal neuron synapses define a powerful disynaptic cortico-hippocampal loop. *Neuron* **66**, 560–572.

Cho, Y.H., Giese, K.P., Tanila, H., Silva, A.J. and Eichenbaum, H. (1998). Abnormal hippocampal spatial representations in alphaCaMKIIT286A and CREBalphaDelta-mice. *Science* **279**, 867–869.

Clayton, N.S. and Dickinson, A. (1998). Episodic-like memory during cache recovery by scrub jays. *Nature* **395**, 272–274.

Clelland, C.D., Choi, M., Romberg, C., Clemenson, G.D., Jr., Fragniere, A., Tyers, P., Jessberger, S., Saksida, L.M., Barker, R.A., Gage, F.H., *et al.* (2009). A functional role for adult hippocampal neurogenesis in spatial pattern separation. *Science* **325**, 210–213.

Colgin, L.L., Moser, E.I. and Moser, M.B. (2008). Understanding memory through hippocampal remapping. *Trends Neurosci* **31**, 469–477.

Cooke, S.F., Wu, J., Plattner, F., Errington, M., Rowan, M., Peters, M., Hirano, A., Bradshaw, K.D., Anwyl, R., Bliss, T.V., *et al.* (2006). Autophosphorylation of alphaCaMKII is not a general requirement for NMDA receptor-dependent LTP in the adult mouse. *J Physiol* **574**, 805–818.

Cravens, C.J., Vargas-Pinto, N., Christian, K.M. and Nakazawa, K. (2006). CA3 NMDA receptors are crucial for rapid and automatic representation of context memory. *Eur J Neurosci* **24**, 1771–1780.

Czurko, A., Czeh, B., Seress, L., Nadel, L. and Bures, J. (1997). Severe spatial navigation deficit in the Morris water maze after single high dose of neonatal X-ray irradiation in the rat. *Proc Natl Acad Sci USA* **94**, 2766–2771.

Deng, W., Aimone, J.B. and Gage, F.H. (2010). New neurons and new memories: how does adult hippocampal neurogenesis affect learning and memory? *Nat Rev Neurosci* **11**, 339–350.

Dombeck, D.A., Harvey, C.D., Tian, L., Looger, L.L. and Tank, D.W. (2010). Functional imaging of hippocampal place cells at cellular resolution during virtual navigation. *Nat Neurosci* **13**, 1433–1440.

Dragoi, G. and Buzsaki, G. (2006). Temporal encoding of place sequences by hippocampal cell assemblies. *Neuron* **50**, 145–157.

Dragoi, G., Harris, K.D. and Buzsaki, G. (2003). Place representation within hippocampal networks is modified by long-term potentiation. *Neuron* **39**, 843–853.

Ego-Stengel, V. and Wilson, M.A. (2010). Disruption of ripple-associated hippocampal activity during rest impairs spatial learning in the rat. *Hippocampus* **20**, 1–10.

Eichenbaum, H., Dudchenko, P., Wood, E., Shapiro, M. and Tanila, H. (1999). The hippocampus, memory and place cells: Is it spatial memory or a memory space? *Neuron* **23**, 209–226.

Erondu, N.E. and Kennedy, M.B. (1985). Regional distribution of type II Ca^{2+}/calmodulin dependent protein kinase in rat brain. *J Neurosci* **5**, 3270–3277.

Fanselow, M. (1981). Naloxone and Pavlovian fear conditioning. Learning & Motivation **12**, 389–419.

Fanselow, M. (1990). Factors governing one-trial contextual conditioning. *Anim Learn Behav* **18**, 264–270.

Gabrieli, J.D., Cohen, N.J. and Corkin, S. (1988). The impaired learning of semantic knowledge following bilateral medial temporal-lobe resection. *Brain Cogn* **7**, 157–177.

Ghosh, A. and Greenberg, M.E. (1995). Calcium signaling in neurons: Molecular mechanisms and cellular consequences. *Science* **268**, 239–247.

Giese, K.P., Fedorov, N.B., Filipkowski, R.K. and Silva, A.J. (1998). Autophosphorylation at Thr286 of the alpha calcium-calmodulin kinase II in LTP and learning. *Science* **279**, 870–873.

Gilbert, P.E. and Kesner, R.P. (2006). The role of the dorsal CA3 hippocampal subregion in spatial working memory and pattern separation. *Behav Brain Res* **169**, 142–149.

Gilbert, P.E., Kesner, R.P. and Lee, I. (2001). Dissociating hippocampal subregions: Double dissociation between dentate gyrus and CA1. *Hippocampus* **11**, 626–636.

Girardeau, G., Benchenane, K., Wiener, S.I., Buzsaki, G. and Zugaro, M.B. (2009). Selective suppression of hippocampal ripples impairs spatial memory. *Nat Neurosci* **12**, 1222–1223.

Gold, A.E. and Kesner, R.P. (2005). The role of the CA3 subregion of the dorsal hippocampus in spatial pattern completion in the rat. *Hippocampus* **15**, 808–814.

Gong, N., Li, Y., Cai, G.Q., Niu, R.F., Fang, Q., Wu, K., Chen, Z., Lin, L.N., Xu, L., Fei, J., *et al.* (2009). GABA transporter-1 activity modulates hippocampal theta oscillation and theta burst stimulation-induced long-term potentiation. *J Neurosci* **29**, 15836–15845.

Gordon, J.A., Lacefield, C.O., Kentros, C.G. and Hen, R. (2005). State-dependent alterations in hippocampal oscillations in serotonin 1A receptor-deficient mice. *J Neurosci* **25**, 6509–6519.

Gossen, M., Bonin, A.L., Freundlieb, S. and Bujard, H. (1994). Inducible gene expression systems for higher eukaryotic cells. *Curr Opin Biotechnol* **5**, 516–520.

Gradinaru, V., Zhang, F., Ramakrishnan, C., Mattis, J., Prakash, R., Diester, I., Goshen, I., Thompson, K.R. and Deisseroth, K. (2010). Molecular and cellular approaches for diversifying and extending optogenetics. *Cell* **141**, 154–165.

Gunaydin, L.A., Yizhar, O., Berndt, A., Sohal, V.S., Deisseroth, K. and Hegemann, P. (2010). Ultrafast optogenetic control. *Nat Neurosci* **13**, 387–392.

Hargreaves, E.L., Rao, G., Lee, I., and Knierim, J.J. (2005). Major dissociation between medial and lateral entorhinal input to dorsal hippocampus. *Science* **308**, 1792–1794.

Harvey, C.D., Collman, F., Dombeck, D.A. and Tank, D.W. (2009). Intracellular dynamics of hippocampal place cells during virtual navigation. *Nature* **461**, 941–946.

Hoess, R.H., Ziese, M. and Sternberg, N. (1982). P1 site-specific recombination: Nucleotide sequence of the recombining sites. *Proc Natl Acad Sci USA* **79**, 3398–3402.

Ito, H.T. and Schuman, E.M. (2007). Frequency-dependent gating of synaptic transmission and plasticity by dopamine. *Front Neural Circuits* **1**, 1.

Jarrard, L.E. (1993). On the role of the hippocampus in learning and memory in the rat. *Behav Neural Bio* **60**, 9–26.

Jeffery, K.J. and Hayman, R. (2004). Plasticity of the hippocampal place cell representation. *Rev Neurosci* **15**, 309–331.

Jeltsch, H., Bertrand, F., Lazarus, C. and Cassel, J.C. (2001). Cognitive performances and locomotor activity following dentate granule cell damage in rats: Role of lesion extent and type of memory tested. *Neurobiol Learn Mem* **76**, 81–105.

Ji, D. and Wilson, M.A. (2007). Coordinated memory replay in the visual cortex and hippocampus during sleep. *Nat Neurosci* **10**, 100–107.

Kee, N., Teixeira, C.M., Wang, A.H. and Frankland, P.W. (2007). Preferential incorporation of adult-generated granule cells into spatial memory networks in the dentate gyrus. *Nat Neurosci* **10**, 355–362.

Klausberger, T. and Somogyi, P. (2008). Neuronal diversity and temporal dynamics: The unity of hippocampal circuit operations. *Science* **321**, 53–57.

Knierim, J.J., Lee, I. and Hargreaves, E.L. (2006). Hippocampal place cells: Parallel input streams, subregional processing, and implications for episodic memory. *Hippocampus* **16**, 755–764.

Kudrimoti, H., Barnes, C.A. and McNaughton, B.L. (1999). Reactivation of hippocampal cell assemblies: Effects of behavioral state, experience, and EEG dynamics. *J Neurosci* **19**, 4090–4101.

Lee, A.K., Epsztein, J. and Brecht, M. (2009). Head-anchored whole-cell recordings in freely moving rats. *Nat Protoc* **4**, 385–392.

Lee, A.K., Manns, I.D., Sakmann, B. and Brecht, M. (2006). Whole-cell recordings in freely moving rats. *Neuron* **51**, 399–407.

Lee, A.K. and Wilson, M.A. (2002). Memory of sequential experience in the hippocampus during slow wave sleep. *Neuron* **36**, 1183–1194.

Lee, I. and Kesner, R.P. (2002). Differential contribution of NMDA receptors in hippocampal subregions to spatial working memory. *Nat Neurosci* **5**, 162–168.

Lee, I. and Kesner, R.P. (2004). Differential contributions of dorsal hippocampal subregions to memory acquisition and retrieval in contextual fear-conditioning. *Hippocampus* **14**, 301–310.

Lee, I., Yoganarasimha, D., Rao, G. and Knierim, J.J. (2004). Comparison of population coherence of place cells in hippocampal subfields CA1 and CA3. *Nature* **430**, 456–459.

Lee, M.G., Chrobak, J.J., Sik, A., Wiley, R.G. and Buzsaki, G. (1994). Hippocampal theta activity following selective lesion of the septal cholinergic system. *Neuroscience* **62**, 1033–1047.

Leutgeb, J.K., Leutgeb, S., Treves, A., Meyer, R., Barnes, C.A., McNaughton, B.L., Moser, M.B. and Moser, E.I. (2005a). Progressive transformation of hippocampal neuronal representations in "morphed" environments. *Neuron* **48**, 345–358.

Leutgeb, S., Leutgeb, J.K., Barnes, C.A., Moser, E.I., McNaughton, B.L. and Moser, M.B. (2005b). Independent codes for spatial and episodic memory in hippocampal neuronal ensembles. *Science* **309**, 619–623.

Leutgeb, S., Leutgeb, J.K., Treves, A., Moser, M.B. and Moser, E.I. (2004). Distinct ensemble codes in hippocampal areas CA3 and CA1. *Science* **305**, 1295–1298.

Lever, C., Wills, T., Cacucci, F., Burgess, N. and O'Keefe, J. (2002). Long-term plasticity in hippocampal place-cell representation of environmental geometry. *Nature* **416**, 90–94.

Levy, W.B., Desmond, N.L. and Zhang, D.X. (1998). Perforant path activation modulates the induction of long-term potentiation of the schaffer collateral — hippocampal CA1 response: Theoretical and experimental analyses. *Learn Mem* **4**, 510–518.

Lisman, J.E. and Otmakhova, N.A. (2001). Storage, recall, and novelty detection of sequences by the hippocampus: Elaborating on the SOCRATIC model to account for normal and aberrant effects of dopamine. *Hippocampus* **11**, 551–568.

Louie, K. and Wilson, M.A. (2001). Temporally structured replay of awake hippocampal ensemble activity during rapid eye movement sleep. *Neuron* **29**, 145–156.

Lubenov, E.V. and Siapas, A.G. (2009). Hippocampal theta oscillations are travelling waves. *Nature* **459**, 534–539.

MacVicar, B. and Dudek, F. (1980). Local synaptic circuits in rat hippocampus: Interaction between pyramidal cells. *Brain Res* **184**, 220–223.

Marr, D. (1971). Simple memory: A theory for archicortex. *Philo Trans Royal Soc Lond* **262**, 23–81.

Martig, A.K. and Mizumori, S.J. (2010). Ventral tegmental area disruption selectively affects CA1/CA2 but not CA3 place fields during a differential reward working memory task. *Hippocampus*.

Mayford, M., Bach, M.E., Huang, Y.Y., L., W., Hawkins, R.D. and Kandel, E.R. (1996). Control of memory formation through regulated expression of a CaMKII transgene. *Science* **274**, 1678–1683.

Mayford, M., Wang, J., Kandel, E.R. and O'Dell, T.J. (1995). CaMKII regulates the frequency-response of hippocampal synapses for the production of both LTD and LTP. *Cell* **81**, 891–904.

McClelland, J.L. and Goddard, N.H. (1996). Considerations arising from a complementary learning systems perspective on hippocampus and neocortex. *Hippocampus* **6**, 654–665.

McHugh, T.J., Blum, K.I., Tsien, J.Z., Tonegawa, S. and Wilson, M.A. (1996). Impaired hippocampal representation of space in CA1-specific NMDAR1 knockout mice. *Cell* **87**, 1339–1349.

McHugh, T.J., Jones, M.W., Quinn, J.J., Balthasar, N., Coppari, R., Elmquist, J.K., Lowell, B.B., Fanselow, M.S., Wilson, M.A. and Tonegawa, S. (2007). Dentate gyrus NMDA receptors mediate rapid pattern separation in the hippocampal network. *Science* **317**, 94–99.

McHugh, T.J. and Tonegawa, S. (2009). CA3 NMDA receptors are required for the rapid formation of a salient contextual representation. *Hippocampus* **19**, 1153–1158.

McLamb, R.L., Mundy, W.R. and Tilson, H.A. (1988). Intradentate colchicine disrupts the acquisition and performance of a working memory task in the radial arm maze. *Neurotoxicology* **9**, 521–528.

McNaughton, B.L., Barnes, C.A., Meltzer, J. and Sutherland, R.J. (1989). Hippocampal granule cells are necessary for normal spatial learning but not for spatially-selective pyramidal cell discharge. *Exp Brain Res* **76**, 489–496.

McNaughton, B.L. and Morris, R.G.M. (1987). Hippocampal synaptic enhancement and information storage within a distributed memory system. *Trends Neurosci* **10**, 408–415.

Miles, R. and Wong, R.K. (1986). Excitatory synaptic interactions between CA3 neurones in the guinea-pig hippocampus. *J Physiol* **373**, 397–418.

Miller, S.G. and Kennedy, M.B. (1986). Regulation of brain type II Ca^{2+}/calmodulin-dependent protein kinase by autophosphorylation: A Ca^{2+}-triggered molecular switch. *Cell* **44**, 861–870.

Morris, R.G., Garrud, P., Rawlins, J.N. and O'Keefe, J. (1982a). Place navigation impaired in rats with hippocampal lesions. *Nature* **297**, 681–683.

Morris, R.G.M. (1981). Spatial localization does not require the presence of local cues. *Learning and Motivation* **12**, 239–260.

Morris, R.G.M., Anderson, E., Lynch, G.S. and Baudry, M. (1986). Selective impairment of learning and blockade of long-term potentiation by an N-methyl-D-aspartate receptor antagonist, AP5. *Nature* **319**, 774–776.

Morris, R.G.M., Garrud, P., Rawlins, J.N.P. and O'Keefe, J. (1982b). Place navigation impaired in rats with hippocampal lesions. *Nature* **297**, 681–683.

Moser, E.I., Kropff, E. and Moser, M.B. (2008). Place cells, grid cells, and the brain's spatial representation system. *Annu Rev Neurosci* **31**, 69–89.

Moser, E.I. and Paulson, O. (2001). New excitement in cognitive space: Between place cells and spatial memory. *Curr Opin Neurobiol* **11**, 745–751.

Muller, R. (1996). A quarter of a century of place cells. *Neuron* **17**, 813–822.

Myers, C.E. and Scharfman, H.E. (2010). Pattern separation in the dentate gyrus: A role for the CA3 backprojection. *Hippocampus*.

Nadasdy, Z., Hirase, H., Czurko, A., Csicsvari, J. and Buzsaki, G. (1999). Replay and time compression of recurring spike sequences in the hippocampus. *J Neurosci* **19**, 9497–9507.

Nakashiba, T., Buhl, D.L., McHugh, T.J. and Tonegawa, S. (2009). Hippocampal CA3 output is crucial for ripple-associated reactivation and consolidation of memory. *Neuron* **62**, 781–787.

Nakashiba, T., Young, J.Z., McHugh, T.J., Buhl, D.L. and Tonegawa, S. (2008). Transgenic inhibition of synaptic transmission reveals role of CA3 output in hippocampal learning. *Science* **319**, 1260–1264.

Nakazawa, K., Quirk, M.C., Chitwood, R.A., Watanabe, M., Yeckel, M.F., Sun, L.D., Kato, A., Carr, C.A., Johnston, D., Wilson, M.A., *et al.* (2002). Requirement for hippocampal CA3 NMDA receptors in associative memory recall. *Science* **297**, 211–218.

Nakazawa, K., Sun, L.D., Quirk, M.C., Rondi-Reig, L., Wilson, M.A. and Tonegawa, S. (2003). Hippocampal CA3 NMDA receptors are crucial for memory acquisition of one-time experience. *Neuron* **38**, 305–315.

Niewoehner, B., Single, F.N., Hvalby, O., Jensen, V., Meyer zum Alten Borgloh, S., Seeburg, P.H., Rawlins, J.N., Sprengel, R. and Bannerman, D.M. (2007). Impaired spatial working memory but spared spatial reference memory following functional loss of NMDA receptors in the dentate gyrus. *Eur J Neurosci* **25**, 837–846.

Nolan, M.F., Malleret, G., Dudman, J.T., Buhl, D.L., Santoro, B., Gibbs, E., Vronskaya, S., Buzsaki, G., Siegelbaum, S.A., Kandel, E.R., *et al.* (2004). A behavioral role for dendritic integration: HCN1 channels constrain spatial memory and plasticity at inputs to distal dendrites of CA1 pyramidal neurons. *Cell* **119**, 719–732.

Nowak, L., Bregestovski, P., Ascher, P., Herbert, A. and Prochiantz, A. (1984). Magnesium gates glutamate-activated channels in mouse central neurones. *Nature* **307**, 462–465.

O'Keefe, J. and Dostrovsky, J. (1971). The hippocampus as a spatial map: preliminary evidence from unit activity in the freely-moving rat. *Brain Res* **34**, 171–175.

O'Keefe, J. and Nadel, L. (1978). *The Hippocampus as a Cognitive Map*. (Oxford: Clarendon Press).

O'Keefe, J. and Recce, M.L. (1993). Phase relationship between hippocampal place units and the EEG theta rhythm. *Hippocampus* **3**, 317–330.

O'Reilly, R.C. and McClelland, J.L. (1994). Hippocampal conjunctive encoding, storage, and recall: avoiding a trade-off. *Hippocampus* **4**, 661–682.

O'Reilly, R.C. and Rudy, J.W. (2001). Conjunctive representations in learning and memory: Principles of cortical and hippocampal function. Psychol Rev **108**, 311–345.

Orban, P.C., Chui, D. and Marth, J.D. (1992). Tissue- and site-specific DNA recombination in transgenic mice. *Proc Natl Acad Sci USA* **89**, 6861–6865.

Pastalkova, E., Itskov, V., Amarasingham, A. and Buzsaki, G. (2008). Internally generated cell assembly sequences in the rat hippocampus. *Science* **321**, 1322–1327.

Pavlides, C., Greenstein, Y.J., Grudman, M. and Winson, J. (1988). Long-term potentiation in the dentate gyrus is induced preferentially on the positive phase of theta-rhythm. *Brain Res* **439**, 383–387.

Poucet, B., Save, E. and Lenck-Santini, P.P. (2000). Sensory and memory properties of hippocampal place cells. *Rev Neurosci* **11**, 95–111.

Racz, A., Ponomarenko, A.A., Fuchs, E.C. and Monyer, H. (2009). Augmented hippocampal ripple oscillations in mice with reduced fast excitation onto parvalbumin-positive cells. *J Neurosci* **29**, 2563–2568.

Remondes, M. and Schuman, E.M. (2002). Direct cortical input modulates plasticity and spiking in CA1 pyramidal neurons. *Nature* **416**, 736–740.

Remondes, M. and Schuman, E.M. (2003). Molecular mechanisms contributing to long-lasting synaptic plasticity at the temporoammonic-CA1 synapse. *Learn Mem* **10**, 247–252.

Rempel-Clower, N.L., Zola, S.M., Squire, L.R. and Amaral, D.G. (1996). Three cases of enduring memory impairment after bilateral damage limited to the hippocampal formation. *J Neurosci* **16**, 5233–5255.

Renaudineau, S., Poucet, B., Laroche, S., Davis, S. and Save, E. (2009). Impaired long-term stability of CA1 place cell representation in mice lacking the transcription factor zif268/egr1. *Proc Natl Acad Sci USA* **106**, 11771–11775.

Rogers, J.L. and Kesner, R.P. (2003). Cholinergic modulation of the hippocampus during encoding and retrieval. *Neurobiol Learn Mem* **80**, 332–342.

Rolls, E.T. (1996). A theory of hippocampal function in memory. *Hippocampus* **6**, 601–620.

Rolls, E.T. and Kesner, R.P. (2006). A computational theory of hippocampal function, and empirical tests of the theory. *Prog Neurobiol* **79**, 1–48.

Romberg, C., Raffel, J., Martin, L., Sprengel, R., Seeburg, P.H., Rawlins, J.N., Bannerman, D.M. and Paulsen, O. (2009). Induction and expression of GluA1 (GluR-A)-independent LTP in the hippocampus. *Eur J Neurosci* **29**, 1141–1152.

Rotenberg, A., Abel, T., Hawkins, R.D., Kandel, E.R. and Muller, R.U. (2000). Parallel instabilities of long-term potentiation, place cells, and learning caused by decreased protein kinase A activity. *J Neurosci* **20**, 8096–8102.

Rotenberg, A., Mayford, M., Hawkins, R.D., Kandel, E.R. and Muller, R.U. (1996). Mice expressing activated CaMKII lack low frequency LTP and do not form stable place cells in the CA1 region of the hippocampus. *Cell* **87**, 1351–1361.

Rudell, A.P. and Fox, S.E. (1984). Hippocampal excitability related to the phase of theta rhythm in urethanized rats. *Brain Res* **294**, 350–353.

Rudell, A.P., Fox, S.E. and Ranck, J.B., Jr. (1980). Hippocampal excitability phase-locked to the theta rhythm in walking rats. *Exp Neurol* **68**, 87–96.

Scharfman, H.E. (2007). The CA3 "backprojection" to the dentate gyrus. *Prog Brain Res* **163**, 627–637.

Schmolck, H., Kensinger, E.A., Corkin, S. and Squire, L.R. (2002). Semantic knowledge in patient H.M. and other patients with bilateral medial and lateral temporal lobe lesions. *Hippocampus* **12**, 520–533.

Schulman, H. and Lou, L.L. (1989). Multifunctional Ca^{2+}/calmodulin-dependent protein kinase: Domain structure and regulation. *Trends Biochem Sci* **14**, 62–66.

Scoville, W.B. and Milner, B. (1957). Loss of recent memory after bilateral hippocampal lesions. *J Neuropsychiatry Clin Neurosci* **12**, 103–113.

Siapas, A.G. and Wilson, M.A. (1998). Coordinated interactions between hippocampal ripples and cortical spindles during slow-wave sleep. *Neuron* **21**, 1123–1128.

Silva, A.J., Zhou, Y., Rogerson, T., Shobe, J. and Balaji, J. (2009). Molecular and cellular approaches to memory allocation in neural circuits. *Science* **326**, 391–395.

Sinclair, B.R., Seto, M.G. and Bland, B.H. (1982). theta-Cells in CA1 and dentate layers of hippocampal formation: Relations to slow-wave activity and motor behavior in the freely moving rabbit. *J Neurophysiol* **48**, 1214–1225.

Skaggs, W.E. and McNaughton, B.L. (1996). Replay of neuronal firing sequences in rat hippocampus during sleep following spatial experience. *Science 271*.

Skaggs, W.E., McNaughton, B.L., Wilson, M.A. and Barnes, C.A. (1996). Theta phase precession in hippocampal neuronal populations and the compression of temporal sequences. *Hippocampus* **6**, 149–172.

Squire, L., Shimamura, A.P. and Amaral, D.G. (1989). Memory and the hippocampus. In Byrne, H.A.B., ed. *Neural Models of Plaslicity* (San Diego: Academic Press), pp. 208–239.

Steffenach, H.A., Sloviter, R.S., Moser, E.I. and Moser, M.B. (2002). Impaired retention of spatial memory after transection of longitudinally oriented axons of hippocampal CA3 pyramidal cells. *Proc Natl Acad Sci USA* **99**, 3194–3198.

Stone, S.S., Teixeira, C.M., Zaslavsky, K., Wheeler, A.L., Martinez-Canabal, A., Wang, A.H., Sakaguchi, M., Lozano, A.M. and Frankland, P.W. (2010). Functional convergence of developmentally and adult-generated granule cells in dentate gyrus circuits supporting hippocampus-dependent memory. *Hippocampus*.

Straube, T. and Frey, J.U. (2003). Time-dependent depotentiation in the dentate gyrus of freely moving rats by repeated brief 7 Hz stimulation. *Neurosci Lett* **339**, 82–84.

Straube, T., Korz, V. and Frey, J.U. (2003). Bidirectional modulation of long-term potentiation by novelty-exploration in rat dentate gyrus. *Neurosci Lett* **344**, 5–8.

Sutherland, R.J., Whishaw, I.Q. and Kolb, B. (1983). A behavioural analysis of spatial localization following electrolytic, kainate- or colchicine-induced damage to the hippocampal formation in the rat. *Behav Brain Res* **7**, 133–153.

Tanila, H. (2001). Noradrenergic regulation of hippocampal place cells. *Hippocampus* **11**, 793–808.

Taverna, F.A., Georgiou, J., McDonald, R.J., Hong, N.S., Kraev, A., Salter, M.W., Takeshima, H., Muller, R.U. and Roder, J.C. (2005). Defective place cell activity in nociceptin receptor knockout mice with elevated NMDA receptor-dependent long-term potentiation. *J Physiol* **565**, 579–591.

Tecott, L.H., Logue, S.F., Wehner, J.M. and Kauer, J.A. (1998). Perturbed dentate gyrus function in serotonin 5-HT2C receptor mutant mice. *Proc Natl Acad Sci USA* **95**, 15026–15031.

Thompson, L.T. and Best, P.J. (1990). Long-term stability of the place-field activity of single units recorded from the dorsal hippocampus of freely behaving rats. *Brain Res* **509**, 299–308.

Tran, A.H., Uwano, T., Kimura, T., Hori, E., Katsuki, M., Nishijo, H. and Ono, T. (2008). Dopamine D1 receptor modulates hippocampal representation plasticity to spatial novelty. *J Neurosci* **28**, 13390–13400.

Treves, A. (2004). Computational constraints between retrieving the past and predicting the future and the CA3-CA1 differentiation. *Hippocampus* **14**, 539–556.

Tsien, J.Z., Huerta, P.T. and Tonegawa, S. (1996). The essential role of hippocampal CA1 NMDA receptor-dependent synaptic plasticity in spatial memory. *Cell* **87**, 1327–1338.

Tulving, E. (1972). Episodic and semantic memory. In Tulving, E. and Donaldson, W. eds. *Organization of Memory*, (New York: Academic), pp. 381–403.

Tulving, E. (1983). *Elements of Episodic Memory*. (Oxford: Clarendon Press).

van Praag, H., Schinder, A.F., Christie, B.R., Toni, N., Palmer, T.D. and Gage, F.H. (2002). Functional neurogenesis in the adult hippocampus. *Nature* **415**, 1030–1034.

Vanderwolf, C.H. (1969). Hippocampal electrical activity and voluntary movement in the rat. *Electroencephalogr Clin Neurophysiol* **26**, 407–418.

Vinogradova, O.S. (2001). Hippocampus as comparator: Role of the two input and two output systems of the hippocampus in selection and registration of information. *Hippocampus* **11**, 578–598.

Wills, T.J., Lever, C., Cacucci, F., Burgess, N. and O'Keefe, J. (2005). Attractor dynamics in the hippocampal representation of the local environment. *Science* **308**, 873–876.

Wilson, M.A. and McNaughton, B.L. (1993). Dynamics of the hippocampal ensemble code for space. *Science* **261**, 1055–1058.

Wilson, M.A. and McNaughton, B.L. (1994). Reactivation of hippocampal ensemble memories during sleep. *Science* **265**, 676–679.

Wyble, B.P., Linster, C. and Hasselmo, M.E. (2000). Size of CA1-evoked synaptic potentials is related to theta rhythm phase in rat hippocampus. *J Neurophysiol* **83**, 2138–2144.

Xavier, G.F., Oliveira-Filho, F.J. and Santos, A.M. (1999). Dentate gyrus-selective colchicine lesion and disruption of performance in spatial tasks: Difficulties in "place strategy" because of a lack of flexibility in the use of environmental cues? *Hippocampus* **9**, 668–681.

Xu, L., Anwyl, R. and Rowan, M.J. (1998). Spatial exploration induces a persistent reversal of long-term potentiation in rat hippocampus. *Nature* **394**, 891–894.

Yan, J., Zhang, Y., Jia, Z., Taverna, F.A., McDonald, R.J., Muller, R.U. and Roder, J.C. (2002). Place-cell impairment in glutamate receptor 2 mutant mice. *J Neurosci* **22**, RC204.

Ylinen, A., Bragin, A., Nadasdy, Z., Jando, G., Szabo, I., Sik, A. and Buzsaki, G. (1995a). Sharp wave-associated high-frequency oscillation (200 Hz) in the intact hippocampus: Network and intracellular mechanisms. *J Neurosci* **15**, 30–46.

Ylinen, A., Soltesz, I., Bragin, A., Penttonen, M., Sik, A. and Buzsaki, G. (1995b). Intracellular correlates of hippocampal theta rhythm in identified pyramidal cells, granule cells, and basket cells. *Hippocampus* **5**, 78–90.

Yoganarasimha, D., Rao, G. and Knierim, J.J. (2010). Lateral entorhinal neurons are not spatially selective in cue-rich environments. *Hippocampus*.

Zentall, T.R., Clement, T.S., Bhatt, R.S. and Allen, J. (2001). Episodic-like memory in pigeons. *Psychon Bull Rev* **8**, 685–690.

Zhang, F., Gradinaru, V., Adamantidis, A.R., Durand, R., Airan, R.D., de Lecea, L. and Deisseroth, K. (2010). Optogenetic interrogation of neural circuits: Technology for probing mammalian brain structures. *Nat Protoc* **5**, 439–456.

Zola, S.M. and Squire, L.R. (2001). Relationship between magnitude of damage to the hippocampus and impaired recognition memory in monkeys. *Hippocampus* **11**, 92–98.

Impaired Memory Mechanisms in Mouse Models of Alzheimer's Disease

13

Masuo Ohno*

1. Introduction

Alzheimer's disease (AD) is a progressive neurodegenerative disorder and the most common form of dementia in the elderly. AD is pathologically characterised by two hallmark lesions: (1) senile plaques, extracellular deposits of the amyloid-β (Aβ) peptide derived from amyloid precursor protein (APP), and (2) neurofibrillary tangles (NFTs), intracellular aggregates of the microtubule-associated protein tau that are found in cell bodies and apical dendrites. Although the cause of sporadic AD accounting for the majority of AD cases (inherited forms less than 1% of the total) remains unclear, there is increasing consensus that Aβ accumulation plays a central role in triggering a complex pathogenic cascade that ultimately leads to the emergence of severe memory impairments (Figure 13.1) (Hardy and Selkoe, 2002; Selkoe, 2002). In this regard, the animal model has made substantial contribution to our understanding of molecular mechanisms of AD and memory deficits that are difficult to address in human studies and impossible in cultured cells. Although to date no perfect model has emerged, transgenic animal models of AD, which overexpress the familial AD (FAD) mutant form of human APP, presenilin (PS), mutated human tau genes or combine more than one of these mutations, successfully recapitulate many aspects of AD such as amyloidosis, NFTs, gliosis, neuronal loss (to some extent), synaptic failure, and memory deficits (Eriksen and Janus, 2007; LaFerla and Oddo, 2005; McGowan *et al.*, 2006). Of particular importance, the amyloid cascade hypothesis is supported by a growing body of *in vivo* evidence that genetic, pharmacological, and immunisation-based lowering of cerebral Aβ levels can suppress the neuropathological

*Center for Dementia Research, Nathan Kline Institute, New York University School of Medicine, Orangeburg, New York, USA.

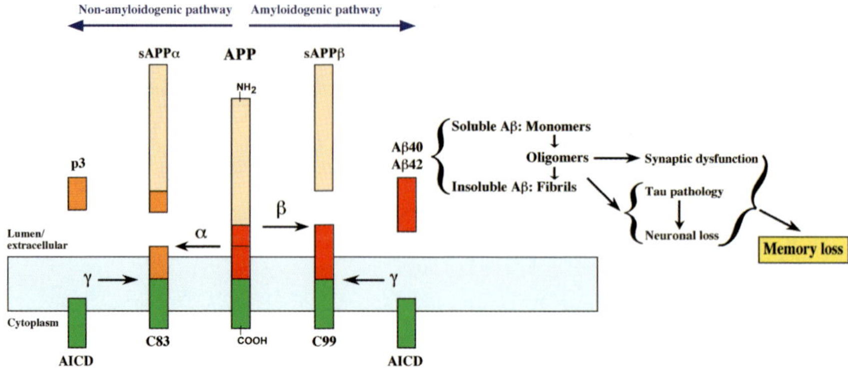

Figure 13.1. APP processing and the amyloid cascade hypothesis. Recent evidence suggests that memory deficits may be mediated by soluble and lower order Aβ oligomers and tau species rather than aggregate forms of amyloid plaques and tangles.

changes associated with AD and improve memory deficits in these animal models (Ohno, 2006; Ohno, 2008). This chapter reviews most recent advances and new insights into the disease mechanisms that the AD mouse models have provided, with special focus on those underlying synaptic and memory dysfunctions. Since increasing evidence indicates soluble oligomeric assemblies are the most neurotoxic forms of Aβ, the emphasis here is on pathogenic roles of Aβ oligomers in causing memory loss. Furthermore, the implications for the development of disease-modifying therapeutic approaches and potential limitations of using transgenic AD animal models are also discussed.

2. APP processing, tau hyperphosphorylation, and transgenic modelling of AD

The major constituent of plaques is a 40–42-amino acid polypeptide that is termed Aβ (Aβ40 or Aβ42) and is generated by the endoproteolysis of APP, a large type I membrane protein. APP is metabolised by at least three enzymes at sites that are referred to as the α-, β- and γ-secretase sites to form a number of peptide fragments (Figure 13.1). The β-secretase (called BACE1 for β-site APP-cleaving enzyme 1) initially cleaves APP to form the N-terminus of Aβ at the Asp + 1 residue of the Aβ sequence (amyloidogenic pathway). The major cleavage products are a secreted APP ectodomain (sAPPβ) and the membrane-bound C-terminal 99 amino acids of APP (C99 or β-CTF: β-secretase-cleaved C-terminal fragment). The α-secretase cleaves APP within the Aβ domain, thus precluding the formation of Aβ (non-amyloidogenic pathway). The resultant products are sAPPα and C83 (or α-CTF). Both the C-terminal fragments C99 and C83 are subsequently cut by the γ-secretase within their transmembrane domain, resulting in the release of Aβ and the 16-residue shorter p3 peptide, respectively.

Under normal conditions α-secretase cleavage is favoured, whereas in early-onset FAD, the mutations in APP are mainly located near the three secretase cleavage sites and alter APP processing to favor the amyloidogenic pathway. FAD mutations that cluster near the β-cleavage site (e.g. the Swedish mutation: K670N/M671L) elevate the production of total Aβ, while those near the γ-site (e.g. the Indiana mutation: V717F) increase the generation of Aβ42 specifically. Similarly, FAD-associated mutations in PS (a catalytic component of the γ-secretase complex) also play a role by favouring excessive generation of pathogenic Aβ42 over Aβ40. The FAD mutations were successfully introduced into the development of transgenic mouse models of AD such as PDAPP and Tg2576 (Table 13.1). Since then, multiple transgenic AD lines have been engineered to overexpress the mutated human APP, PS or both, and consequently overproduce Aβ peptides, especially, longer Aβ42 peptides that are more hydrophobic than Aβ40 and have the propensity to assemble into neurotoxic oligomers and aggregates. These transgenic mice reproduce AD-like pathologies such as amyloid plaques, neuritic dystrophy and gliosis, and exhibit progressive age-related deficits in synaptic plasticity (a cellular basis for learning/memory) and memory function in different behavioural assays (Ashe, 2001; Eriksen and Janus, 2007; Kobayashi and Chen, 2005; McGowan *et al.*, 2006; Ohno, 2006; Rowan *et al.*, 2003). Furthermore, APP/PS transgenic rat models, which reproduce similar AD-like pathology and memory phenotypes, have also been developed, aiming to provide better opportunities for advanced studies that require complex *in vivo* manipulations and/or behavioural testing (Leon *et al.*, 2010; Liu *et al.*, 2008). However, these transgenic rodent models are devoid of tau-tangle pathology and extensive cell loss or gross atrophy as observed in human AD brains (McGowan *et al.*, 2006). Only limited neuronal loss is reported to occur in a subset of early-onset and aggressive amyloid model mice (e.g. APP23, APP/PS1KI and 5XFAD) as disease progresses into severe pathological stages (Calhoun *et al.*, 1998; Casas *et al.*, 2004; Oakley *et al.*, 2006). Somehow, it appears difficult to achieve massive neuronal loss in the mouse possibly because of the short lifespan of mice, less vulnerability of mouse neurons, or the complex nature of Aβ accumulation and other pathological changes. Therefore, the relationship between cognitive dysfunction found in APP/PS transgenic models and memory problems in human AD cases that are accompanied by widespread neuron loss would need to be interpreted with some caution.

Table 13.1. Transgenic mouse models of Alzheimer's disease with APP overexpression.

			Pathology						
Mouse models	**Promoter**	**Gene (Mutations)**	**Plaques**	**Oligomers**	**Intraneuronal Aβ**	**Tangles**	**Neuron death**	**Memory deficits**	**References**
PDAPP	PDGF	APP (V717F)	+	NR	NR	–	–	+	Games *et al.*, 1995
Tg2576	PrP	APP (K670N/M671L)	+	+	NR	–	–	+	Hsiao *et al.*, 1996
APP23	Thy1	APP (K670N/M671L)	+	NR	NR	–	+	+	Sturchler-Pierrat *et al.*, 1997
APP/PS1KI	Thy1	APP (K670N/M671L, PS1KI V717F) (M233T, L235P)	+	+	+	–	+	+	Casas *et al.*, 2004
5XFAD	Thy1	APP (K670N/M671L, I716V, V717I) PS1 (M146L, L286V)	+	+	+	–	+	+	Oakley *et al.*, 2006

(*Continued*)

Table 13.1. (*Continued*)

Mouse models	Promoter	Gene (Mutations)	Pathology: Plaques	Oligomers	Intraneuronal Aβ	Tangles	Neuron death	Memory deficits	References
3xTg-AD	Thy1	APP (K670N/M671L) PS1KI (M146V) MAPT (P301L)	+	+	+	+	–	+	Oddo *et al.*, 2003
TgCRND8	PrP	APP (K670N/M671L, V717F)	+	+	NR	–	–	+	Chishti *et al.*, 2001
APP-E693Δ	PrP	APP (E693Δ)	–	+	+	–	+	+	Tomiyama *et al.*, 2010
ARC6	PDGF	APP (E693G)	+	+ (very low)	NR	–	–	–	Cheng *et al.*, 2007
J20	PDGF	APP (K670N/M671L, V717F)	+	+	NR	–	–	+	Mucke *et al.*, 2000

APP: amyloid precursor protein, PS1: presenilin 1, KI: knock-in, MAPT: microtubule-associated protein tau, NR: not reported, PDGF: platelet-derived growth factor, PrP: prion protein. This table includes only a portion of APP-overexpressing transgenic mouse models that are summarised in this chapter.

Although many APP-overexpressing mice show detectable tau hyperphosphorylation, they do not develop NFTs. To reproduce the features of AD-like neurofibrillary pathology such as aggregated forms of NFTs and neutropil threads in mouse models, they need to express human MAPT in which the tau protein is encoded (Gotz and Ittner, 2008; LaFerla and Oddo, 2005; McGowan *et al.*, 2006). Under physiological conditions, tau is a soluble protein that is mainly localised to the axon and promotes assembly and stabilisation of microtubules and vesicle transport. Under pathological conditions, tau is hyperphosphorylated at the serine and threonine residues, dissociates from microtubules, and is deposited in aggregates forming the core of paired helical filaments of NFTs. The mislocalisation of tau from the axon to the somatodendritic compartment represents a critical event in tauopathies. Although tau mutations are not directly linked to AD, they occur in other neurodegenerative diseases such as frontotemporal dementia with Parkinsonism, indicating that tau dysfunction can cause neurodegeneration and dementia. To better understand the relationship between plaques and NFTs, P301L tau mice were crossed with APP transgenic mice or injected with synthetic Aβ42 fibrils or Aβ-containing brain extracts from aged APP mice (Bolmont *et al.*, 2007; Gotz *et al.*, 2001; Lewis *et al.*, 2001). These studies demonstrated that Aβ dramatically increased tau hyperphosphorylation and NFT numbers in P301L tau mice while plaque formation was not affected in APP/tau doubly transgenic mice compared with APP singly transgenic mice, supporting the view that tangle pathology occurs as a downstream consequence of Aβ accumulation (Figure 13.1). Furthermore, triple-transgenic mice (3xTg-AD) co-expressing mutant forms of human APP, PS1, and tau were recently introduced as one of the most relevant AD models that show temporal- and region-specific Aβ and tangle pathology and are instrumental for studying interactions between both hallmark lesions in AD-like phenotypes including memory deficits (Oddo *et al.*, 2003) (Table 13.1). In spite of equivalent overexpression of human APP and tau transgenes in this model, Aβ deposition develops before the onset of tangle pathology consistent with the amyloid cascade theory. It is reasonable to speculate that the absence of NFT formation in APP/PS models may be linked to their lack of extensive neuronal death. However, this seems unlikely because 3xTg-AD mouse brains fail to mimic significant neuron loss reminiscent of AD. As such, a mouse model that recapitulates all aspects of AD has not yet been produced; however, currently available transgenic models have provided novel and important insights into the molecular and pathophysiological cause of AD. In particular, synaptic dysfunction and the loss of synapses, important early pathological features of AD that are highly relevant to memory deficits (Selkoe, 2002), are well reproduced in these transgenic AD mice and the underlying mechanisms are extensively studied, as summarised in this chapter.

3. Aβ oligomers as a mediator of memory deficits in AD mouse models

Transgenic mouse models of AD exhibit age-related memory decline as assessed by a broad battery of hippocampus-dependent learning paradigms (Ashe, 2001; Eriksen and Janus, 2007; Gotz and Ittner, 2008; Kobayashi and Chen, 2005; Ohno, 2006). It was originally assumed that only Aβ species that had aggregated into large fibrils constituting mature neuritic amyloid plaques would have toxic properties. However, in recent works, small soluble Aβ oligomers, which are also called ADDLs for Aβ-derived diffusible ligands and represent intermediates in the fibril assembly process, have attracted more attention as a pathogenic mediator of memory loss in AD and animal models (Haass and Selkoe, 2007). This idea first comes from the poor correlation between the degree of dementia and amyloid plaque burden in AD (Lue *et al.*, 1999) and the emergence of progressive learning and synaptic deficits at time well before Aβ deposition in transgenic AD mice (Jacobsen *et al.*, 2006; Ohno *et al.*, 2004; Westerman *et al.*, 2002). In fact, Ashe *et al.* identified several kinds of Aβ oligomers in brains of Tg2576 APP transgenic mice at pre-pathological stages (Lesne *et al.*, 2006). Remarkably, levels of 40-kDa (nonamer) and 56-kDa (dodecamer referred to as Aβ*56) Aβ assemblies correlate with the poor spatial memory performance of Tg2576 mice in the Morris water maze. However, it seems unlikely that nonamer and dodecamer are the only oligomeric Aβ assemblies capable of altering brain function. For example, naturally assembled Aβ toxins that are purified from the culture media of APP-transfected cells and human AD brains (mainly dimers and trimers) (Haass and Selkoe, 2007; Shankar *et al.*, 2008), as well as from the Tg2576 brain (Aβ*56/12-mers) (Lesne *et al.*, 2006), disrupt memory and *in vivo* LTP when injected into the brains of normal animals. Taken together, although dominant forms of Aβ assembly that cause neuronal dysfunction may not be necessarily the same through different model systems, the results provide convergent evidence that soluble Aβ oligomers share common synaptotoxic properties that are capable of disrupting memory function in AD and animal models.

Does reducing levels of soluble Aβ oligomers lead to therapeutic improvements in AD-associated memory impairments? First, hippocampal mnemonic and cholinergic dysfunction of Tg2576 APP transgenic mice and their rescue by Aβ reductions with genetic deletion of the β-secretase (BACE1$^{-/-}$) are observed months before the onset of amyloid deposition, clearly indicating that soluble Aβ assemblies are responsible for reversible memory deficits during early phases of AD (Ohno *et al.*, 2004). However, small assemblies and large fibrillar aggregates of Aβ coexist at further advanced stages of disease, rendering it more difficult to evaluate the effects of oligomers and plaques on memory function individually. In this regard, recent studies using AD mouse models have started to dissociate two Aβ species functionally. For example, passive immunisation with monoclonal antibodies, including NAB61 that recognizes a conformation found on oligomeric forms of Aβ, exerts a rapid reversal of memory deficits of APP mice after a single or subchronic treatment without affecting their cortical or hippocampal amyloid load (Dodart *et al.*, 2002; Kotilinek *et al.*, 2002; Lee *et al.*, 2006a), suggesting the importance of sequestration of soluble oligomeric assemblies of Aβ in restoring normal memory function. McLaurin *et al.* (2006) more specifically addressed the question using compounds that inhibit Aβ aggregation into neurotoxic assemblies. They found that inositol stereoisomers (epi-cyclohexanehexol and its more effective enantiomer scyllo-cyclohexanehexol also called AZD-103), which interact with Aβ42 peptides to prevent oligomer formation and toxicity in cell culture, improved spatial memory performance in the water maze as well as synaptic loss in a dose-dependent and stereoisomer-specific manner when they were orally administered to TgCRND8 APP transgenic mice. Importantly, the degree of behavioural improvements following different treatment regimens correlated with reductions in brain levels of soluble Aβ assemblies larger than 40 kDa. The rescue of hippocampal LTP, as well as memory following cyclohexanehexol administration, is also reported in animal models exposed to soluble oligomers of cell-derived human Aβ (Townsend *et al.*, 2006), confirming the beneficial effects of this orally available small-molecule compound on synaptotoxicity.

Consistent with the therapy-oriented findings, novel transgenic mouse lines that harbour a FAD-linked point mutation within Aβ further shed light on the importance of Aβ oligomers in AD pathogenesis (Table 13.1). APP transgenic mice expressing the E693Δ mutation, which promotes Aβ oligomerisation without fibrillisation, develop intraneuronal accumulation of oligomeric Aβ assemblies in the absence of amyloid plaques at >8 months of age (Tomiyama *et al.*, 2010). $APP_{E693\Delta}$-Tg mice reproduce AD-like traits including impaired spatial memory, deficient hippocampal LTP and neuronal loss in the hippocampal CA3 region. Conversely, APP transgenic mice with the "Arctic" mutation (E693G) in the Aβ domain (ARC6), which accelerates Aβ fibrillisation but decreases the abundance of nonfibrillar Aβ assemblies, do not exhibit spatial memory impairments in the water maze (Cheng *et al.*, 2007). Importantly, ARC6 mice have greater hippocampal plaque load and much lower concentrations of Aβ*56 as compared with age-matched J20 APP mice whose spatial memory is significantly impaired. These data demonstrate that soluble Aβ oligomers rather than deposited Aβ fibrils are responsible for memory defects. The results also raise the possibility that efforts to block or reverse the formation of Aβ fibrils at the cost of augmenting pathogenic Aβ oligomers may even worsen memory dysfunction. Meanwhile, promoting fibril formation in a way that bypasses oligomer formation or rapidly sequesters oligomers into more inert fibrils might exert therapeutic benefits. Consistent with this hypothesis, Tg2576 mice at ~12 months of age show a transient reduction in levels of soluble Aβ oligomers (including Aβ*56), which is accompanied by a recovery of spatial memory function in spite of increases in insoluble Aβ fibrils (Lesne *et al.*, 2008). This transient memory rescue coincides with the emergence of mature amyloid plaques and is likely to result from changes in kinetics driving Aβ toward fibrillar Aβ formation from oligomeric intermediates in Tg2576 mice at this particular age. Moreover, it is demonstrated that transgenic overexpression of the Aβ-degrading enzyme neprilysin decreases overall Aβ levels and amyloid plaque burden, whereas it does not reduce levels of Aβ trimers and Aβ*56 or improve spatial memory deficits in J20 APP mice (Meilandt *et al.*, 2009). Taken collectively, these findings provide experimental evidence for a clear dissociation between plaque formation and functional defects and support the idea that soluble oligomeric Aβ assemblies account for synaptic and memory failure in AD. Therefore, Aβ oligomers seem to be an important therapeutic target for treating reversible memory loss during early stages of disease. Nevertheless, we should not conclude that oligomeric assemblies of Aβ represent the sole neurotoxic entity; for example, it is known that fibrillar Aβ deposits are associated with local synaptic abnormalities and breakage of neuronal processes (Tsai *et al.*, 2004).

4. Other APP fragments that may contribute to memory deficits in AD mouse models

In addition to neurotoxic Aβ peptides, C99 (or β-CTF) is well recognised to have the amyloidogenic property. Transgenic overexpression or central administration of C99 has been reported to cause progressive neurodegeneration and impair learning and memory as well as LTP (Choi *et al.*, 2001; Lee *et al.*, 2006b; Nalbantoglu *et al.*, 1997; Song *et al.*, 1998). This issue is quite important from a therapeutic point of view. Both β- and γ-secretases responsible for Aβ production undoubtedly represent excellent therapeutic targets aimed at central lowering of Aβ levels through the inhibition of these enzymes (Figure 13.1). The findings from our laboratory and others have clearly demonstrated that genetic deletion ($BACE1^{-/-}$) or reduction ($BACE1^{+/-}$ and siRNA targeting BACE1) of the β-secretase can suppress the development of AD-like phenotypes such as amyloidosis (including Aβ oligomer formation), glial activation and neuronal loss and rescue synaptic and memory dysfunction in a variety of transgenic mouse models tested by different learning tasks (Devi and Ohno, 2010a; Devi and Ohno, 2010b; Kimura *et al.*, 2010; Laird *et al.*, 2005; Ohno *et al.*, 2006; Ohno *et al.*, 2007; Ohno *et al.*, 2004; Singer *et al.*, 2005). Importantly, these genetic manipulations of BACE1 also suppress build-up of other β-products including C99 (Figure 13.1). Subchronic treatments with the noncompetitive BACE1 inhibitor TAK-070 (9–15 days) similarly reduces Aβ/C99 and ameliorates memory impairments in Tg2576 mice (Fukumoto *et al.*, 2010). Therefore, it is reasonable to conceive that reductions in both neurotoxic species, Aβ peptides and C99 fragments, underlie the functional improvements resulting from β-secretase suppression in APP/PS model mice. Meanwhile, as α- and β-secretases compete for APP as a substrate, the α-cleavage products such as sAPPα are elevated by BACE1 suppression in APP transgenic brains. Given memory-enhancing effects of sAPPα (Meziane *et al.*, 1998), the upregulation of sAPPα may contribute to mechanisms by which β-secretase inhibition benefits memory.

Neuron- or forebrain-specific knockout of PS1 (a catalytic subunit of the γ-secretase complex) reduces Aβ production and prevents amyloid plaque formation in APP mice irrespective of age (Dewachter *et al.*, 2002; Saura *et al.*, 2005). Importantly, although the conditional inactivation of PS1 (PS1 cKO) can successfully rescue impaired hippocampal LTP and memory deficits in relatively younger APP mice as assessed by contextual fear conditioning and water maze tests, these beneficial effects disappear at the advanced age. In contrast with β-secretase inhibition, blocking Aβ generation at the γ-secretase level is accompanied by concurrent acceleration of neurotoxic C99 accumulation (Figure 13.1). In fact, robust increases in C99 are caused by longer-term inhibition of γ-secretase in aged APP mice with PS1 cKO, which seems likely to counteract the cognitive benefits of Aβ reduction (Saura *et al.*, 2005). This mechanistic problem raises potential concerns about therapeutic efforts to improve memory by means of γ-secretase suppression in AD. Comery *et al.* (2005) provided the first demonstration that acute administration of the γ-secretase inhibitor DAPT before training (but not post-training or pre-memory testing treatment) can rapidly reverse impaired contextual fear conditioning concomitant with soluble Aβ reductions in Tg2576 AD model mice at disease stages before and after plaque formation. This study indicates that reducing Aβ during acquisition is important for memory improvements, although it has not been determined whether the ameliorative effects may remain unchanged or decline with chronic treatments with γ-secretase inhibitors especially in relation to C99 accumulation. From a basic science perspective, it will be important to better understand the functional roles of multiple APP metabolites, including but not limited to Aβ peptides, in affecting memory processes (Figure 13.1) so that we can definitively address the molecular cause of memory loss associated with AD. This will enable us to develop truly mechanistic therapeutic approaches that minimize potential adverse effects and achieve specific memory amelioration in clinical settings.

5. Tau-tangle pathology and memory deficits in AD mouse models

Recent tau transgenic model mice have made considerable contribution to our understanding of the relationship between tau abnormalities and memory impairments. First, LaFerla *et al.* developed 3xTg-AD mice that reproduce both plaque and tangle pathologies (Oddo *et al.*, 2003). 3xTg-AD mice show the onset of hippocampal memory impairments (water maze and contextual conditioning) and LTP deficits at 6 months of age, which correlates with intraneuronal Aβ accumulation (including oligomers) but precedes extracellular plaque formation or tangle pathology (Billings *et al.*, 2005; Oddo *et al.*, 2003; Oddo *et al.*, 2006a). Therefore, NFTs are not involved in the early-onset memory decline but may contribute to later stages of cognitive deficits in this mouse model. Consistent with the amyloid cascade hypothesis, it has been demonstrated that the removal of Aβ by immunisation leads to clearance of early somatodendritic accumulation of tau but not late stages of highly hyperphosphorylated and aggregated tau in 3xTg-AD mice (Oddo *et al.*, 2004; Oddo *et al.*, 2006b). Moreover, after the antibody itself is cleared or degraded, amyloid pathology re-emerges prior to tau-tangle pathology. Interestingly, soluble Aβ oligomers seem to mediate neuronal tau hyperphosphorylation (De Felice *et al.*, 2008; Tomiyama *et al.*, 2010) and NFT formation (Oddo *et al.*, 2006a). The immunisation approach was further applied to investigate the mechanisms underlying memory deficits in aged 3xTg-AD mice that have established plaques and NFTs and are thus more relevant to human AD cases. Chronic Aβ immunisation rescues memory deficits of 23-month-old 3xTg-AD mice in alternation T-maze and passive avoidance tasks concomitant with reductions in soluble Aβ (including Aβ*56 dodecamers) and tau, while soluble Aβ reduction alone with acute immunisation is not sufficient to improve memory dysfunction (Oddo *et al.*, 2006b). Aβ immunisation does not significantly affect aggregate forms of amyloid plaques or NFTs in the 3xTg-AD model at the advanced age. Furthermore, administration of lithium, an inhibitor of glycogen synthase kinase-3 (GSK-3), significantly reduces levels of phosphorylated tau but has no effect on Aβ load and is unable to rescue memory deficits in 3xTg-AD mice (Caccamo *et al.*, 2007). Together, these data provide experimental evidence that like soluble Aβ oligomers, intermediate aggregates of abnormal tau molecules rather than NFTs are important for inducing memory impairments and that concomitant reductions in soluble neurotoxic Aβ and tau species, but not either reduction alone, can lead to memory improvements in 3xTg-AD mice. This hypothesis is supported by the observation with an inducible tau transgenic mouse model showing that suppression of tau expression recovers memory dysfunction and ceases neuronal death while NFTs continues to accumulate in brains (Santacruz *et al.*, 2005). Furthermore, knocking out of tau prevents memory decline in APP transgenic mice without altering abundant Aβ levels, implicating endogenous tau as a critical mediator of Aβ-dependent memory deficits (Roberson *et al.*, 2007). Although the tangle pathology does not occur in Aβ-overproducing transgenic mice, soluble Aβ oligomers can produce abnormal tau phosphorylation in these models (Bellucci *et al.*, 2007; Tomiyama *et al.*, 2010). Therefore, it seems likely that tau pathology is not just a marker of dying neurons but that soluble hyperphosphorylated tau molecules (i.e. pre-stages to NFTs) are more dynamically involved in exerting deleterious actions on neuronal function including learning and memory.

6. Impairments of multiple memory systems in AD mouse models

Extensive behavioural studies have analysed the phenotypes of learning and memory defects in a series of AD mouse models using different behavioral paradigms such as the Morris water maze, radial arm maze, Y-maze, contextual or auditory trace fear conditioning, and object or social recognition tests. However, these studies are focused on the acquisition of learning or memories shortly (~24 h) after training, therefore mainly evaluating deficits in hippocampal encoding of various types of memories. The characterisation of many AD transgenic lines in these hippocampus-dependent memory impairments has been described in detail in previous reviews (Ashe, 2001; Eriksen and Janus, 2007; Gotz and Ittner, 2008; Kobayashi and Chen, 2005; Ohno, 2006). Meanwhile, it is important to note that Aβ and abnormal tau accumulation is not restricted to the hippocampus but occur in multiple brain regions and that cognitive declines are observed broadly across almost all aspects of learning and memory including long-term as well as short-term memories during progression of AD. Apparently, previous investigations of transgenic AD mouse models were devoid of the concept of "multiple memory systems" that take place subsequent to or independently of memory processing in the hippocampal circuitry (Figure 13.2). We recently addressed the problems of AD-related multiple memory dysfunctions using 5XFAD transgenic mice, which co-express a total of five FAD mutations

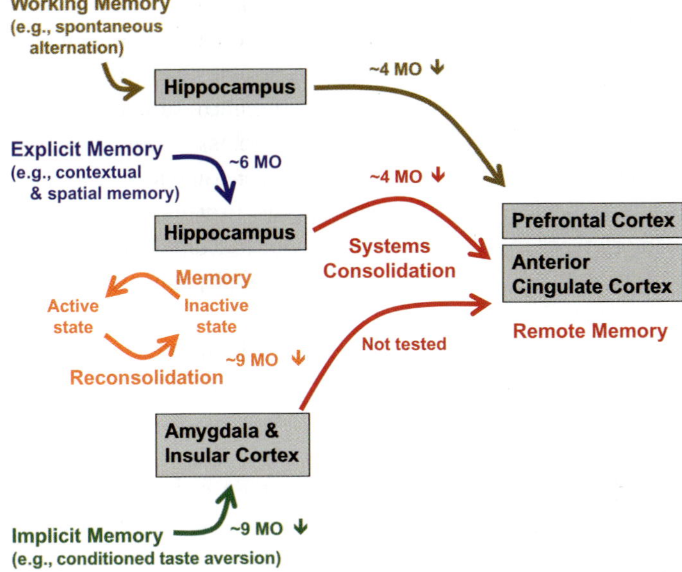

Figure 13.2. Impairments of multiple memory systems in 5XFAD transgenic mice. Multiple aspects of memory function that involve discrete brain regions become impaired differentially during the course of disease progression.

[APP K670N/M671L (Swedish) + I716V (Florida) + V717I (London) and PS1 M146L + L286V] and thus represent one of the most early-onset and aggressive amyloid models concomitant with robust acceleration of Aβ42 production (Table 13.1).

First, two major memory systems are distinguished: the explicit memory refers to a conscious recollection of specific facts and events under the control of the hippocampus while the implicit memory system refers to unconscious acquisition of information and is independent of hippocampal function (Figure 13.2) (Squire, 1992). Our studies of 5XFAD mice during different stages of disease progression have characterised age-dependent impairments in both forms of memory. Deficits in explicit memory (e.g. spatial reference memory in the water maze and contextual memory in the fear conditioning) begin to occur in 5XFAD mice at ~6 months of age consistent with the onset of their hippocampal synaptic dysfunction such as deficient LTP (Devi and Ohno, 2010a; Kimura and Ohno, 2009; Ohno *et al.*, 2006). Furthermore, 5XFAD mice exhibit the earlier onset of impairments in spatial working memory (~4 months of age) as tested by the spontaneous alternation Y-maze assay (short-term memory that depends on interactive neuronal activities of the hippocampus and prefrontal cortex and is categorised as a subset of explicit memory) (Oakley *et al.*, 2006). Meanwhile, implicit memory (e.g. CTA: conditioned taste aversion), which involves the amygdala and insular cortex, starts to decline in 5XFAD mice at ~9 months of age (Devi and Ohno, 2010a). Therefore, it is conceivable that AD model mice show the delayed onset of hippocampus-independent CTA memory deficits as compared to hippocampal synaptic and mnemonic dysfunction. This appears to reflect different sensitivities of both memory types to Aβ levels since amyloid accumulation in the amygdala or insular cortex is not lower than that in the hippocampus. Importantly, our results are in agreement with clinical observations that implicit forms of memory function remain relatively intact or are less severely affected compared with explicit forms of memory in demented patients with AD.

Second, memories are not formed instantaneously but undergo a prolonged period of reorganisation and stabilisation processes that follow initial hippocampal coding (Figure 13.2). After the initial learning and memory consolidation processes, the retrieval or reactivation of acquired memories places them into an additional labile phase during which the memories can be modified or strengthened. This process referred to as reconsolidation requires molecular mechanisms or brain regions that are not necessarily the same as those of initial memory consolidation, representing an important distinctive component of memory processing (Tronson and Taylor, 2007). On the other hand, as memories mature, they become increasingly independent of the hippocampus and memory traces are gradually stabilised and eventually transformed into remote memories in cortical networks such as the anterior cingulate cortex (systems consolidation) (Frankland and Bontempi, 2005). We used the contextual fear conditioning paradigm to examine the phenotypes of 5XFAD mice in these memory reconsolidation and remote memory stabilisation processes (Figure 13.2). Briefly, remote memory during 30 days after training starts to decline in 5XFAD mice at ~4 months of age, which precedes the onset of hippocampal memory and synaptic dysfunctions (~6 months of age) (Kimura and Ohno, 2009). At this stage, it is likely that new memory traces are temporarily stored in hippocampal circuits but they cannot be successfully stabilised into cortical networks. This is consistent with recent data showing that functional connectivity between the hippocampus and cortical areas including the medial prefrontal and anterior cingulate cortices (Wang *et al.*, 2006) as well as neocortical synaptic plasticity like LTP (Battaglia *et al.*, 2007) is disrupted in AD patients and APP transgenic mice. As amyloid pathology further develops, dysfunction extends to memory reconsolidation in 5XFAD mice at ~9 months of age (Ohno, 2009), in which the retrieval-dependent retrograde amnesia due to deficient reconsolidation adds to already weakened hippocampal memory formation. Memory retrieval during reconsolidation in normal subjects conceivably renders memories stronger or more persistent and prevents forgetting (Tronson and Taylor, 2007). In contrast, in 5XFAD mice, retrieval induces memory decline and worsen memory impairments. Although the onset of each type of multiple dysfunctions may depend on the specific neuronal populations selected by the promoter or the distribution pattern of transgene in different AD transgenic models, our studies demonstrate the importance of testing multiple memory systems for better understanding the mechanisms underlying complex AD-associated cognitive disorders. In particular, while the appearance of Aβ*56 dodecamers correlate with the impairment of hippocampus-dependent spatial memory defects in the water maze (Lesne *et al.*, 2006; Lesne *et al.*, 2008), it is possible that other forms of Aβ oligomers may account for impairments of AD model mice in memory reconsolidation and remote memory stabilisation or hippocampus-independent implicit memory. Our recent works indicate that partial reduction of the β-secretase (BACE1$^{+/-}$) rescues the multiple forms of memory dysfunction with reductions in total Aβ accumulation in relevant brain structures of 5XFAD mice (Devi and Ohno, 2010a; Kimura *et al.*, 2010). As such, behavioural evaluations of multiple memory dysfunctions in combination with monitoring changes in the regional levels of specific causal Aβ toxins (e.g. oligomers) in model systems will be crucial for reasoning the optimal design of disease-modifying therapeutic interventions to secure functional benefits in AD.

7. Synaptic and signalling mechanisms underlying memory deficits in AD mouse models

Synaptic loss correlates best with memory decline in AD (Selkoe, 2002). The majority of Aβ is generated at the plasma membrane or in the secretory pathway and is released into the extracellular fluid, while some portion of Aβ is produced via the endocytic pathway within neurons (LaFerla *et al.*, 2007). Currently, it is hypothesised that soluble oligomeric forms of Aβ assemblies, which can diffuse into synaptic clefts or accumulate intraneuronally, cause synaptic dysfunction and influence intracellular signalling pathways, leading to memory loss that is dissociated from neuronal degeneration or permanent neuropathological lesions of the brain (Haass and Selkoe, 2007). In fact, it has been reported that learning- or synaptic plasticity-associated signalling molecules, including receptors, kinases, transcription factors, and immediate-early genes, decline in AD transgenic mouse brains (Dickey *et al.*, 2004; Dickey *et al.*, 2003). Furthermore, AD model mice also exhibit impairments in cellular mechanisms proposed to underlie hippocampal learning and memory such as synaptic plasticity (Jacobsen *et al.*, 2006;

Rowan *et al.*, 2003), place cell firing (Cacucci *et al.*, 2008), plastic changes in intrinsic neuronal excitability (Kaczorowski *et al.*, 2011), and adult neurogenesis (Rodriguez *et al.*, 2008; Verret *et al.*, 2007). Since Aβ peptides target a broad spectrum of synaptic events, signal transductions and neuronal processing, it is difficult to unequivocally determine the mechanistic pathway leading downstream from Aβ to memory deficits. It seems more plausible to conceive that memory failure is brought about as the sum of disrupted multiple synaptic and signalling mechanisms. I focus on only several molecular and cellular mechanisms that appear closely associated with the occurrence of memory decline in AD models and might serve as therapeutic targets; however, the omission from this chapter should not be construed as rejection of other potential mechanisms.

7.1. *Glutamatergic pathways*

The impairment of baseline excitatory transmission (mainly AMPA receptor-mediated synaptic responses) and/or LTP (a model of NMDA receptor-dependent synaptic plasticity) at hippocampal CA1 glutamatergic synapses can be detected before the formation of insoluble Aβ deposits in APP transgenic mouse models, representing the neurophysiological basis by which soluble Aβ species may cause memory defects (Hsia *et al.*, 1999; Jacobsen *et al.*, 2006; Mucke *et al.*, 2000). Notably, we recently found that 5XFAD mice show impaired basal synaptic transmission and LTP at the Schaffer collateral-CA1 pathway in accordance with the onset of hippocampal memory failure (Kimura and Ohno, 2009), while only LTP deficits are restored to normal and the baseline transmission remains significantly reduced in BACE1$^{+/-}$·5XFAD mice (Kimura *et al.*, 2010). Since BACE1$^{+/-}$·5XFAD mice exhibit improvements in the hippocampus-dependent contextual fear conditioning and spontaneous alternation performance concomitant with partial Aβ reductions, Aβ-dependent LTP deficiency at CA1 synapses rather than impaired basal synaptic transmission correlates more closely with memory dysfunction at least in the 5XFAD model. It might be possible that reduced levels of basal transmission can still allow normal memory processing to some extent if it retains the abilities to induce plasticity in response to synaptic stimulation during behavioural learning.

Mounting evidence indicates Aβ oligomers bind to the dendritic processes of specific subtypes of hippocampal neurons with high affinity (Lacor *et al.*, 2004), in particular, to postsynaptic density complexes including NMDA and AMPA receptor subunits (Lacor *et al.*, 2007; Zhao *et al.*, 2010). It has been reported that Aβ and/or Aβ oligomers affect NMDA and AMPA receptor trafficking through the phosphatase 2B (PP2B or calcineurin)-mediated facilitation of endocytosis, resulting in the removal of surface glutamate receptors (Snyder *et al.*, 2005; Zhao *et al.*, 2010). These Aβ-dependent modifications of glutamate receptors not only depress NMDA and AMPA excitatory synaptic transmission (Kamenetz *et al.*, 2003; Ting *et al.*, 2007), but also lead to dendritic spine loss (Hsieh *et al.*, 2006; Shankar *et al.*, 2007). It is proposed that Aβ oligomers shift the activation of NMDA receptor-dependent signalling cascades toward pathways involved in the induction of long-term depression (LTD) accompanied by reductions in the spine volume and number, while they suppress the LTP-inducing mode of signalling that is associated with increased spine volume and new spine formation. This view is confirmed by hippocampal slice electrophysiology showing that the application of soluble Aβ oligomers from different sources (synthetic, cell culture, human AD brain extracts) enhances LTD and inhibits LTP (Li *et al.*, 2009; Shankar *et al.*, 2008). Consistent with these *in vitro* observations, NMDA receptor function and expression levels in the postsynaptic density fraction decline and LTP is impaired in APP transgenic mice (Dewachter *et al.*, 2009), although the phenotypes of LTD has not been examined. It seems reasonable to conceive that partial block of NMDA receptor-mediated Ca^{2+} influx suppresses LTP-mediating signalling (e.g. Ca^{2+}/calmodulin-dependent protein kinase) (Dewachter *et al.*, 2009) while it activates LTD-mediating signalling (e.g. calcineurin) (Dineley *et al.*, 2007), leading to the impaired AMPA receptor trafficking and function (Chang *et al.*, 2006; Gu *et al.*, 2009) and dendritic spine loss (Jacobsen *et al.*, 2006; Lanz *et al.*, 2003) in APP mice that show memory decline before Aβ deposition. Finally, c-Fos induction following fear conditioning is significantly reduced in APP transgenic mice, suggesting deregulated immediate-early genes may be involved in AD-associated memory impairments (Dewachter *et al.*, 2009). It will be important to further investigate changes in learning-induced signalling activation in APP transgenic mice and determine the critical pathways downstream to NMDA signalling that may mediate memory deficits in AD.

7.2. *Cholinergic pathways*

While initially hypothesised to mark early stages of AD, cholinergic neuronal loss is currently thought to be associated with advanced dementia of disease. Similarly, cholinergic degeneration occurs in AD model mice during relatively advanced stages that show massive neuropathology including plaque deposition (Contestabile *et al.*, 2008). Several APP transgenic mouse models recapitulate cholinergic neurodegeneration as evidenced by dystrophic neurites, decreased fiber density in the hippocampus and cortex, and reductions in the basal forebrain neuron volume and/or number (Bellucci *et al.*, 2006; Christensen *et al.*, 2010).

However, cholinergic dysfunction that is electrophysiologically detectable precedes the onset of amyloid pathology and may be more relevant to cognitive decline (Ohno *et al.*, 2004; Zhong *et al.*, 2003). For example, plastic changes in intrinsic neuronal excitability following learning in relevant brain regions represent an important cellular aspect of leaning and memory that may be coordinated with synaptic plasticity (Disterhoft and Oh, 2006; Disterhoft *et al.*, 2004). We found that the cholinergic activation of CA1 pyramidal neuron excitability, as tested by the muscarinic acetylcholine receptor (mAChR) agonist carbachol-induced suppression of postburst afterhyperpolarisation (a measure of cholinergic aspects of learning), is impaired in young Tg2576 mice that show memory deficits but have no visible Aβ deposition (Ohno *et al.*, 2004). Furthermore, trace fear conditioning-induced increases in hippocampal neuronal excitability was impaired specifically in 5XFAD mice that failed to learn the task but not in 5XFAD learners (Kaczorowski *et al.*, 2011). Importantly, the cholinergic deregulation of excitability is restored to normal by $BACE1^{-/-}$ deletion and consequent Aβ reduction concomitant with memory improvements in Tg2576 mice (Ohno *et al.*, 2004). Therefore, the cholinergic dysfunction in regulating CA1 neuron excitability may, at least in part, contribute to the mechanisms by which the overproduction of neurotoxic Aβ (probably, soluble oligomers) causes early-onset memory decline in APP mouse models. In accord, acetylcholinesterase inhibitors, which enhance cholinergic neurotransmission, improve memory impairments in different APP transgenic mice (Dong *et al.*, 2005; Van Dam *et al.*, 2005).

In addition to symptomatic benefits of cholinergic drugs, mAChR activation has been known to stimulate non-amyloidogenic APP processing in cultured cells and brain slices, thus representing some disease-modifying property. To test this hypothesis *in vivo*, LaFerla *et al.* applied AF267B, a selective agonist for the M1 mAChR (the most abundant subtype in the hippocampus and cortex), to 3xTg-AD model mice (Caccamo *et al.*, 2006). Treatments with AF267B (10 weeks) reduced both intracellular and extracellular Aβ immunoreactivity in the cortex and hippocampus of 3xTg-AD mice and significantly decreased soluble/insoluble Aβ42 species. Mechanistically, AF267B lowers BACE1 levels and elevates steady-state levels of the inducible α-secretase ADAM17/TACE, thus shifting the balance of APP processing toward the non-amyloidogenic pathway to yield decreased levels of β-products (e.g. toxic Aβ and C99) and increased levels of α-products such as α-CTF. Furthermore, increased PKC and extracellular signal-regulated kinases (ERK) 1/2 activities seem to underlie the M1 receptor-mediated alterations of APP processing. AF267B also decreased tau pathology in the cortex and hippocampus of 3xTg-AD mice probably by reducing the Aβ-dependent activation of GSK-3β (a major kinase involved in tau hyperphosphorylation). Consequently,

AF267B rescued hippocampus-dependent spatial memory deficits of 3xTg-AD mice in the water maze (Caccamo *et al.*, 2006). In contrast, the same treatments with AF267B did not affect deficient memory of 3xTg-AD mice in the passive avoidance, a behavioural paradigm dependent on both amygdaloid and hippocampal function. Of particular interest, AF267B treatments failed to ameliorate the Aβ or tau pathology in the amygdala, providing a possible explanation for no improvements of impaired passive avoidance performance in 3xTg-AD mice. In this regard, they suspect that the lower expression of ADAM17/TACE in the amygdala as compared with the hippocampus and cortex may account for the lack of effects of AF267B on amygdaloid Aβ pathology.

Conversely, genetic deletion or pharmacological block (dicyclomine for 6 weeks or scopolamine for 14 days) of M1 mAChRs results in the accelerated development of amyloid and tau-tangle pathology in transgenic AD mice (Caccamo *et al.*, 2006; Davis *et al.*, 2010; Liskowsky and Schliebs, 2006). Treatments with dicyclomine significantly elevated BACE1, C99 and Aβ levels in the hippocampus, cortex and amygdala of 3xTg-AD mice and aggravated the impairments in both hippocampus- and amygdala-dependent memory performances of 3xTg-AD mice. Therefore, these animal model studies imply that cholinergic dysfunction may not only contribute to the symptomatic memory decline as a consequence of Aβ accumulation but also play a pathogenic role to a certain extent by favouring the β-cleavage of APP. However, it remains unclear whether strategies reversing only deficient cholinergic pathways (e.g. selective and small-molecule M1 agonists) may be disease-modifying or sufficient to rescue memory defects especially in treating advanced stages of AD in clinical settings.

7.3. *CREB/BDNF pathways*

The cAMP-response element-binding protein (CREB) is a nuclear transcription factor that regulates expression of genes required for long-term memory formation (Alberini, 2009). Sublethal concentrations of Aβ42 peptides suppress glutamate- or brain-derived neurotrophic factor (BDNF)-induced activation of CREB. Meanwhile, one of the CREB target genes is BDNF and Aβ42 treatments also interfere with the CREB-dependent transcription of BDNF gene. Since CREB/BDNF signals in concert with many upstream and downstream molecules play important roles in long-term memory consolidation, synaptic plasticity and neuronal survival (Minichiello, 2009; Pang and Lu, 2004), the relation to Aβ-dependent memory impairments has been investigated using AD model mice. First, CREB dysfunction as evidenced by reductions in the phosphorylation of CREB Ser-133 occurs in AD transgenic mice concomitant with memory deterioration (Gong *et al.*, 2004; Ma *et al.*, 2007). Restoration of CREB signalling with rolipram (phosphodiesterase IV inhibitor) ameliorates deficits of APP/PS1 mice in hippocampal LTP and memory performances on contextual fear conditioning and water maze tasks (Gong *et al.*, 2004). Remarkably, the beneficial effects last for at least 2 months after the cessation of chronic treatments with rolipram (3 weeks). Therefore, it seems likely that the activation of CREB pathways may stabilize synaptic circuitry through changes in gene expression without affecting Aβ levels in APP/PS1 transgenic mice, rendering the synapses less susceptible to the insult inflicted by Aβ toxins. Recent data indicate that rolipram is also efficacious in reversing the progressive loss of spine density in APP/PS1 mice even at advanced disease stages (Smith *et al.*, 2009). Moreover, calcineurin activities are increased in response to Aβ oligomers before plaque formation in Tg2576 APP transgenic mice, while the calcineurin inhibitor FK506 reverses decreases in phospho-CREB (Ser-133) levels (Reese *et al.*, 2008) and improves memory deficits in this model (Dineley *et al.*, 2007; Taglialatela *et al.*, 2009).

It is proposed that a decline in cerebral BDNF levels may be associated with the pathogenesis of AD (Hu and Russek, 2008; Zuccato and Cattaneo, 2009). In accord, APP transgenic mouse models recapitulate the phenotype of BDNF reduction, which correlates with the ratio of Aβ42/Aβ40 and concentrations of larger Aβ oligomers (~115 kDa) (Peng *et al.*, 2009) and emerges concomitant with cognitive decline prior to Aβ deposition (Francis *et al.*, in press). Importantly, BDNF administration or gene delivery in diverse AD model systems including APP transgenic mice exerts substantial neuroprotective effects against Aβ toxicity (Arancibia *et al.*, 2008), reverses synaptic loss, improves cell signalling and aberrant gene expression, and restores learning and memory without altering amyloid plaque load (Nagahara *et al.*, 2009). Moreover, neural stem cell transplantation increases synaptic density and rescues spatial memory decline of 3xTg-AD model mice in the water maze by elevating hippocampal BDNF in the absence of reductions in Aβ or tau-tangle pathology (Blurton-Jones *et al.*, 2009). Meanwhile, we demonstrated that 5XFAD mice with defective CA1 LTP and memory show reductions in levels of hippocampal BDNF and phosphorylated TrkB (an activated form of BDNF receptors), while BDNF/TrkB signalling is recovered to normal by genetic reductions of the β-secretase (BACE1$^{+/-}$) and Aβ, which leads to synaptic and mnemonic improvements in BACE1$^{+/-}$·5XFAD mice (Kimura *et al.*, 2010). Taken collectively, these findings provide experimental evidence that deficient CREB/BDNF pathways resulting from exposure to high concentrations of Aβ oligomers causally mediate synaptic and cognitive dysfunction associated with AD since their restoration is beneficial irrespective of the presence of concomitant Aβ reductions in APP transgenic mice.

7.4. *Insulin signalling pathways*

Increasing data indicate an association between AD and insulin resistance. Experimental induction of diabetes (exposure to high-fat diet or 10% sucrose intake) in AD mouse models exacerbates amyloidosis and memory deficits in the water maze (Cao *et al.*, 2007; Ho *et al.*, 2004; Takeda *et al.*, 2010), consistent with the fact that insulin-resistant type 2 diabetes increases risks for AD. Conversely, rosiglitazone, an insulin-sensitizing drug, reduces Aβ possibly by promoting its clearance and ameliorates spatial and object recognition memories in Tg2576 and J20 AD model mice (Escribano *et al.*, 2010; Pedersen *et al.*, 2006). Recent evidence further shed light on new mechanisms by which insulin signalling (possibly acting through the suppression of GSK-3 activity) provides a protective mechanism against synaptic and memory failure associated with AD. Soluble Aβ oligomers downregulate insulin receptor function and surface expression, thereby causing neuronal insulin resistance (Townsend *et al.*, 2007; Zhao *et al.*, 2008). Interestingly, insulin signalling not only plays a role in learning and memory but also functions to reduce Aβ oligomers to monomers, inhibits pathogenic binding of oligomers to protect synapses from their toxicity, and prevents the oligomer-induced surface insulin receptor loss (De Felice *et al.*, 2009; Zhao *et al.*, 2009). Therefore, although normal insulin receptor activity helps to defend against accumulation of toxic Aβ oligomers, insulin receptors themselves are vulnerable to oligomer-initiated dysfunction. Deficits in insulin function that could arise from multiple factors (e.g. Aβ, diabetes, ageing) would contribute to increased Aβ oligomer formation/accumulation in brains while increasing oligomeric Aβ levels would further suppress insulin receptor activity. Such a detrimental positive feedback loop between insulin insensitivity and Aβ oligomers may contribute to worsening AD-associated memory deficits. However, as opposed to these proposed mechanisms, Killick *et al.* (2009) reported that insulin receptor substrate 2 deletion (IRS2$^{-/-}$), which induces insulin resistance, ameliorates Aβ pathology and contextual memory deficits in Tg2576 mice, suggesting a possibility that some components of insulin signalling may be detrimental rather than protective. Further investigation, especially using *in vivo* model systems, will be needed to establish a coherent picture about the role of insulin signalling in the pathogenesis of AD and memory impairments.

7.5. *p25/cdk5 and GSK-3 pathways*

The p25/cyclin-dependent kinase 5 (cdk5) signalling plays a physiological role in synaptic plasticity and learning and memory, while its deregulated overactivation seems to be implicated in the pathogenesis of AD (Cheung *et al.*, 2006). Interestingly, transient or low levels of p25 expression facilitate hippocampus-dependent memory performances and CA1 LTP and increase the number of dendritic spines and synapses (Fischer *et al.*, 2005; Giese *et al.*, 2005). In contrast, the prolonged or high-level expression leads to memory impairments, which are accompanied by LTP defects and synaptic and neuronal loss. These observations demonstrate the complexity of neuronal function, in which p25/cdk5 signalling may be activated to compensate for memory dysfunction during early AD progression whereas the continued overactivation can cause deleterious impacts on neurons at the advanced stage. The dysregulation of p25/cdk5 pathways also seems to be involved in the pathophysiology of AD by accelerating amyloidogenic processing of APP possibly through STAT3-mideated transcriptional upregulation of BACE1 (Cruz *et al.*, 2006; Wen *et al.*, 2008b). Moreover, although young p25-overexpressing transgenic mice with enhanced cdk5 activity show reductions in GSK-3β activity mediated by the neuregulin receptor ErbB, this inhibitory cross-talk is somehow relieved in old transgenic animals, resulting in GSK-3β activation and increased tau hyperphosphorylation (Plattner *et al.*, 2006; Wen *et al.*, 2008a). In spite of these positive findings with p25 transgenic mice, the p25/cdk5 hypothesis in AD mechanisms is still controversial. In particular, it has not been conclusively determined whether p25 formation is increased in AD brains. Furthermore, the extent to which the p25/cdk5 deregulation may contribute to memory deficits in Aβ-overproducing transgenic models are largely unknown.

Aβ activates GSK-3 (one of the major tau kinases) (Takashima *et al.*, 1996; Terwel *et al.*, 2008) while GSK-3α (but not GSK-3β) regulates APP processing to favor the β-cleavage of APP and increase Aβ production (Phiel *et al.*, 2003). Transgenic overexpression of GSK-3β has been reported to reproduce AD-like phenotypes such as impaired LTP and spatial memory deficits in mice (Hernandez *et al.*, 2002; Hooper *et al.*, 2007). Mechanistically, the activity of GSK-3β seems to favor the induction of LTD, while stimulation of the PI3K-Akt pathway during LTP phosphorylates GSK-3β at Ser-9 to suppress its activity and switch off the LTD mode (Peineau *et al.*, 2007). Furthermore, when CREB is phosphorylated at Ser-133 and activated, GSK-3β further phosphorylates CREB Ser-129 and inhibits the CREB-mediated transcriptional activity that is required for long-term memory and LTP (Bullock and Habener, 1998; Fiol *et al.*, 1994). Therefore, GSK-3 signalling may be a causal mediator of synaptic and memory impairments in AD. This hypothesis is supported by a recent paper that shows that the specific GSK-3 inhibitor can exert beneficial effects against memory defects as well as neuropathologies such as tau phosphorylation, amyloid deposition and neuronal loss in APP/tau transgenic mice (Sereno *et al.*, 2009).

7.6. *BACE1-elevating pathways*

Although the ultimate cause of sporadic AD remains unclear, evidence is accumulating that BACE1 expression and activity levels are significantly elevated in sporadic AD brains (Fukumoto *et al.*, 2002; Li *et al.*, 2004; Yang *et al.*, 2003; Zhao *et al.*, 2007), representing a crucial contributing factor in the progression of AD and Aβ-dependent memory decline. Increased BACE1 in cerebrospinal fluid (CSF) is reported to correlate with decreased hippocampal volume in AD (Ewers *et al.*, 2011) and represents an indicator of mild cognitive impairment (MCI), a population at high risk for AD (Zhong *et al.*, 2007). It is hypothesised that Aβ accumulation induces BACE1 in plaque-surrounding neurons of 5XFAD and AD brains while this BACE1 upregulation further accelerates Aβ generation, forming a pathogenic positive-feedback loop (Zhao *et al.*, 2007). Several mechanisms are suggested to mediate BACE1 elevation associated with AD; for example, caspase-3-dependent cleavage of GGA3, an adaptor protein required for lysosomal degradation of BACE1 (Sarajarvi *et al.*, 2009; Tesco *et al.*, 2007), increased phosphorylation of the translation initiation factor elF2α (Devi and Ohno, 2010b; O'Connor *et al.*, 2008) and changes in microRNA expression profiles (Hebert *et al.*, 2008; Wang *et al.*, 2008). Our recent data indicate that the same degree of β-secretase suppression (e.g. 50% by $BACE1^{+/-}$ deletion) becomes less efficacious and eventually ineffective in rescuing Aβ accumulation, cholinergic neurodegeneration or memory deficits in 5XFAD mice as disease progresses to advanced stages that are accompanied by significant BACE1 upregulation (Devi and Ohno, 2010b). Importantly, levels of phospho-elF2α, which increase age dependently and induce BACE1 elevation in the 5XFAD model, seem to be important for determining the abilities of partial BACE1 reduction to rescue memory impairments. Given some liabilities of complete $BACE1^{-/-}$ deletion such as hypomyelination and altered synaptic functions or neuronal activities (but no adverse phenotypes in $BACE1^{+/-}$ mice) (Cole and Vassar, 2007; Ohno, 2006; Ohno, 2008), strategies combining partial BACE1 suppression with inhibition of the BACE1-elevating pathway (e.g. elF2α phosphorylation) may be useful for increasing the therapeutic benefits while obviating the potential side effects in treating AD and related memory deficits. On the other hand, exposure to stress-like conditions, including hypoxia (Sun *et al.*, 2006), energy inhibition (Velliquette *et al.*, 2005), behavioural constraint (Devi *et al.*, 2010), high cholesterol diet (Ghribi *et al.*, 2006), and stress-level glucocorticoid administration (Green *et al.*, 2006), is known to elevate BACE1 levels and accelerate β-amyloidosis and/or memory decline in various AD animal models. Although the underlying molecular mechanisms may not be necessarily the same, it is interesting to argue that these environmental influences may act together with genetic factors to trigger β-amyloidogenesis through common mechanisms of BACE1 upregulation during the incipient stage of sporadic AD.

8. Concluding remarks

The aim of using AD animal models is to move us closer to the understanding of the ultimate cause of this intractable neurodegenerative disease and related profound memory loss. A variety of transgenic model systems have provided important insights into the relationship between key pathogenic events and the emergence of memory impairments in AD. In particular, it has become increasingly apparent that soluble and diffusible Aβ oligomers cause more damage to synaptic and cognitive functions than amyloid plaques. In concert with advances in basic neurobiology of learning and memory, it will be of crucial importance to address the precise mechanisms by which many different assembly forms of Aβ toxins interfere with the distinctive processing of multiple memory systems. Meanwhile, it is still uncertain whether Aβ accumulation is the sole culprit to trigger disease; for example, the extensive neurodegeneration or almost complete loss of memory function reminiscent of AD has not been reproduced in highly Aβ-overproducing transgenic model mice. Moreover, although complex interactions between the genetic background and various environmental factors are hypothesised to underlie sporadic AD, transgenic models largely fail to encompass the acquired characteristics of disease. As such, it is possible that the current AD models may lack some critical branches of the multifactorial pathogenic cascade. Nevertheless, AD mouse model studies summarised in this chapter have provided a wealth of evidence that is indispensable not only for untangling the molecular causes of AD and memory deficits but also for opening the door to entering an exciting phase, in which many candidate disease-modifying therapeutic approaches against this devastating neurodegenerative disorder are proceeding to clinical testing.

Acknowledgements

This work was supported by Alzheimer's Association grant IIRG-08-91231 (M.O.).

References

Alberini, C. M. (2009). Transcription factors in long-term memory and synaptic plasticity. *Physiol Rev* **89**, 121–145.

Arancibia, S., Silhol, M., Mouliere, F., Meffre, J., Hollinger, I., Maurice, T. and Tapia-Arancibia, L. (2008). Protective effect of BDNF against β-amyloid induced neurotoxicity *in vitro* and *in vivo* in rats. *Neurobiol Dis* **31**, 316–326.

Ashe, K. H. (2001). Learning and memory in transgenic mice modeling Alzheimer's disease. *Learn Mem* **8**, 301–308.

Battaglia, F., Wang, H. Y., Ghilardi, M. F., Gashi, E., Quartarone, A., Friedman, E. and Nixon, R. A. (2007). Cortical plasticity in Alzheimer's disease in humans and rodents. *Biol Psychiatry* **62**, 1405–1412.

Bellucci, A., Luccarini, I., Scali, C., Prosperi, C., Giovannini, M. G., Pepeu, G. and Casamenti, F. (2006). Cholinergic dysfunction, neuronal damage and axonal loss in TgCRND8 mice. *Neurobiol Dis* **23**, 260–272.

Bellucci, A., Rosi, M. C., Grossi, C., Fiorentini, A., Luccarini, I. and Casamenti, F. (2007). Abnormal processing of tau in the brain of aged TgCRND8 mice. *Neurobiol Dis* **27**, 328–338.

Billings, L. M., Oddo, S., Green, K. N., McGaugh, J. L. and LaFerla, F. M. (2005). Intraneuronal Aβ causes the onset of early Alzheimer's disease-related cognitive deficits in transgenic mice. *Neuron* **45**, 675–688.

Blurton-Jones, M., Kitazawa, M., Martinez-Coria, H., Castello, N. A., Muller, F. J., Loring, J. F., Yamasaki, T. R., Poon, W. W., Green, K. N. and LaFerla, F. M. (2009). Neural stem cells improve cognition via BDNF in a transgenic model of Alzheimer disease. *Proc Natl Acad Sci USA* **106**, 13594–13599.

Bolmont, T., Clavaguera, F., Meyer-Luehmann, M., Herzig, M. C., Radde, R., Staufenbiel, M., Lewis, J., Hutton, M., Tolnay, M. and Jucker, M. (2007). Induction of tau pathology by intracerebral infusion of amyloid-β-containing brain extract and by amyloid-β deposition in APP x Tau transgenic mice. *Am J Pathol* **171**, 2012–2020.

Bullock, B. P. and Habener, J. F. (1998). Phosphorylation of the cAMP response element binding protein CREB by cAMP-dependent protein kinase A and glycogen synthase kinase-3 alters DNA-binding affinity, conformation and increases net charge. *Biochemistry* **37**, 3795–3809.

Caccamo, A., Oddo, S., Billings, L. M., Green, K. N., Martinez-Coria, H., Fisher, A. and LaFerla, F. M. (2006). M1 receptors play a central role in modulating AD-like pathology in transgenic mice. *Neuron* **49**, 671–682.

Caccamo, A., Oddo, S., Tran, L. X. and LaFerla, F. M. (2007). Lithium reduces tau phosphorylation but not Aβ or working memory deficits in a transgenic model with both plaques and tangles. *Am J Pathol* **170**, 1669–1675.

Cacucci, F., Yi, M., Wills, T. J., Chapman, P. and O'Keefe, J. (2008). Place cell firing correlates with memory deficits and amyloid plaque burden in Tg2576 Alzheimer mouse model. *Proc Natl Acad Sci USA* **105**, 7863–7868.

Calhoun, M. E., Wiederhold, K. H., Abramowski, D., Phinney, A. L., Probst, A., Sturchler-Pierrat, C., Staufenbiel, M., Sommer, B. and Jucker, M. (1998). *Neuron* loss in APP transgenic mice. *Nature* **395**, 755–756.

Cao, D., Lu, H., Lewis, T. L. and Li, L. (2007). Intake of sucrose-sweetened water induces insulin resistance and exacerbates memory deficits and amyloidosis in a transgenic mouse model of Alzheimer disease. *J Biol Chem* **282**, 36275–36282.

Casas, C., Sergeant, N., Itier, J. M., Blanchard, V., Wirths, O., van der Kolk, N., Vingtdeux, V., van de Steeg, E., Ret, G., Canton, T., *et al.* (2004). Massive CA1/2 neuronal loss with intraneuronal and N-terminal truncated Aβ42 accumulation in a novel Alzheimer transgenic model. *Am J Pathol* **165**, 1289–1300.

Chang, E. H., Savage, M. J., Flood, D. G., Thomas, J. M., Levy, R. B., Mahadomrongkul, V., Shirao, T., Aoki, C. and Huerta, P. T. (2006). AMPA receptor downscaling at the onset of Alzheimer's disease pathology in double knockin mice. *Proc Natl Acad Sci USA* **103**, 3410–3415.

Cheng, I. H., Scearce-Levie, K., Legleiter, J., Palop, J. J., Gerstein, H., Bien-Ly, N., Puolivali, J., Lesne, S., Ashe, K. H., Muchowski, P. J. and Mucke, L. (2007). Accelerating amyloid-β fibrillization reduces oligomer levels and functional deficits in Alzheimer disease mouse models. *J Biol Chem* **282**, 23818–23828.

Cheung, Z. H., Fu, A. K. and Ip, N. Y. (2006). Synaptic roles of Cdk5: implications in higher cognitive functions and neurodegenerative diseases. *Neuron* **50**, 13–18.

Chishti, M. A., Yang, D. S., Janus, C., Phinney, A. L., Horne, P., Pearson, J., Strome, R., Zuker, N., Loukides, J., French, J., *et al.* (2001). Early-onset amyloid deposition and cognitive deficits in transgenic mice expressing a double mutant form of amyloid precursor protein 695. *J Biol Chem* **276**, 21562–21570.

Choi, S. H., Park, C. H., Koo, J. W., Seo, J. H., Kim, H. S., Jeong, S. J., Lee, J. H., Kim, S. S. and Suh, Y. H. (2001). Memory impairment and cholinergic dysfunction by centrally administered Aβ and carboxyl-terminal fragment of Alzheimer's APP in mice. *FASEB* J **15**, 1816–1818.

Christensen, D. Z., Bayer, T. A. and Wirths, O. (2010). Intracellular Aβ triggers neuron loss in the cholinergic system of the APP/PS1KI mouse model of Alzheimer's disease. *Neurobiol Aging* **31**, 1153–1163.

Cole, S. L. and Vassar, R. (2007). The Alzheimer's disease β-secretase enzyme, BACE1. *Mol Neurodegener* **2**, 22.

Comery, T. A., Martone, R. L., Aschmies, S., Atchison, K. P., Diamantidis, G., Gong, X., Zhou, H., Kreft, A. F., Pangalos, M. N., Sonnenberg-Reines, J., *et al.* (2005). Acute γ-secretase inhibition improves contextual fear conditioning in the Tg2576 mouse model of Alzheimer's disease. *J Neurosci* **25**, 8898–8902.

Contestabile, A., Ciani, E. and Contestabile, A. (2008). The place of choline acetyltransferase activity measurement in the "cholinergic hypothesis" of neurodegenerative diseases. *Neurochem Res* **33**, 318–327.

Cruz, J. C., Kim, D., Moy, L. Y., Dobbin, M. M., Sun, X., Bronson, R. T. and Tsai, L. H. (2006). p25/cyclin-dependent kinase 5 induces production and intraneuronal accumulation of amyloid β *in vivo*. *J Neurosci* **26**, 10536–10541.

Davis, A. A., Fritz, J. J., Wess, J., Lah, J. J. and Levey, A. I. (2010). Deletion of M1 muscarinic acetylcholine receptors increases amyloid pathology *in vitro* and *in vivo*. *J Neurosci* **30**, 4190–4196.

De Felice, F. G., Vieira, M. N., Bomfim, T. R., Decker, H., Velasco, P. T., Lambert, M. P., Viola, K. L., Zhao, W. Q., Ferreira, S. T. and Klein, W. L. (2009). Protection of synapses against Alzheimer's-linked toxins: Insulin signaling prevents the pathogenic binding of Aβ oligomers. *Proc Natl Acad Sci USA* **106**, 1971–1976.

De Felice, F. G., Wu, D., Lambert, M. P., Fernandez, S. J., Velasco, P. T., Lacor, P. N., Bigio, E. H., Jerecic, J., Acton, P. J., Shughrue, P. J., *et al.* (2008). Alzheimer's disease-type neuronal tau hyperphosphorylation induced by Aβ oligomers. *Neurobiol Aging* **29**, 1334–1347.

Devi, L., Alldred, M. J., Ginsberg, S. D. and Ohno, M. (2010). Sex- and brain region-specific acceleration of β-amyloidogenesis following behavioral stress in a mouse model of Alzheimer's disease. *Mol Brain*, **3**, 34.

Devi, L. and Ohno, M. (2010a). Genetic reductions of β-site amyloid precursor protein-cleaving enzyme 1 and amyloid-β ameliorate impairment of conditioned taste aversion memory in 5XFAD Alzheimer's disease model mice. *Eur J Neurosci* **31**, 110–118.

Devi, L. and Ohno, M. (2010b). Phospho-eIF2α level is important for determining abilities of BACE1 reduction to rescue cholinergic neurodegeneration and memory defects in 5XFAD mice. *PLoS One* **5**, e12974.

Dewachter, I., Filipkowski, R. K., Priller, C., Ris, L., Neyton, J., Croes, S., Terwel, D., Gysemans, M., Devijver, H., Borghgraef, P., *et al.* (2009). Deregulation of NMDA-receptor function and down-stream signaling in APP[V717I] transgenic mice. *Neurobiol Aging* **30**, 241–256.

Dewachter, I., Reverse, D., Caluwaerts, N., Ris, L., Kuiperi, C., Van den Haute, C., Spittaels, K., Umans, L., Serneels, L., Thiry, E., *et al.* (2002). Neuronal deficiency of presenilin 1 inhibits amyloid plaque formation and corrects hippocampal long-term potentiation but not a cognitive defect of amyloid precursor protein [V717I] transgenic mice. *J Neurosci* **22**, 3445–3453.

Dickey, C. A., Gordon, M. N., Mason, J. E., Wilson, N. J., Diamond, D. M., Guzowski, J. F. and Morgan, D. (2004). Amyloid suppresses induction of genes critical for memory consolidation in APP + PS1 transgenic mice. *J Neurochem* **88**, 434–442.

Dickey, C. A., Loring, J. F., Montgomery, J., Gordon, M. N., Eastman, P. S. and Morgan, D. (2003). Selectively reduced expression of synaptic plasticity-related genes in amyloid precursor protein + presenilin-1 transgenic mice. *J Neurosci* **23**, 5219–5226.

Dineley, K. T., Hogan, D., Zhang, W. R. and Taglialatela, G. (2007). Acute inhibition of calcineurin restores associative learning and memory in Tg2576 APP transgenic mice. *Neurobiol Learn Mem* **88**, 217–224.

Disterhoft, J. F. and Oh, M. M. (2006). Learning, aging and intrinsic neuronal plasticity. Trends Neurosci **29**, 587–599.

Disterhoft, J. F., Wu, W. W. and Ohno, M. (2004). Biophysical alterations of hippocampal pyramidal neurons in learning, ageing and Alzheimer's disease. *Ageing Res* Rev **3**, 383–406.

Dodart, J. C., Bales, K. R., Gannon, K. S., Greene, S. J., DeMattos, R. B., Mathis, C., DeLong, C. A., Wu, S., Wu, X., Holtzman, D. M. and Paul, S. M. (2002).

Immunization reverses memory deficits without reducing brain Aβ burden in Alzheimer's disease model. *Nat Neurosci* **5**, 452–457.

Dong, H., Csernansky, C. A., Martin, M. V., Bertchume, A., Vallera, D. and Csernansky, J. G. (2005). Acetylcholinesterase inhibitors ameliorate behavioral deficits in the Tg2576 mouse model of Alzheimer's disease. Psychopharmacology (Berl) **181**, 145–152.

Eriksen, J. L. and Janus, C. G. (2007). Plaques, tangles, and memory loss in mouse models of neurodegeneration. Behav Genet **37**, 79–100.

Escribano, L., Simon, A. M., Gimeno, E., Cuadrado-Tejedor, M., Lopez de Maturana, R., Garcia-Osta, A., Ricobaraza, A., Perez-Mediavilla, A., Del Rio, J. and Frechilla, D. (2010). Rosiglitazone rescues memory impairment in Alzheimer's transgenic mice: Mechanisms involving a reduced amyloid and tau pathology. Neuropsychopharmacology **35**, 1593–1604.

Ewers, M., Cheng, X., Nural, H. F., Walsh, C., Meindl, T., Teipel, S. J., Buerger, K., Shen, Y. and Hampel, H. (2011). Increased CSF-BACE1 activity associated with decreased hippocampus volume in Alzheimer's disease. *J Alzheimers Dis* **25**, 378–381.

Fiol, C. J., Williams, J. S., Chou, C. H., Wang, Q. M., Roach, P. J. and Andrisani, O. M. (1994). A secondary phosphorylation of $CREB^{341}$ at Ser^{129} is required for the cAMP-mediated control of gene expression. A role for glycogen synthase kinase-3 in the control of gene expression. *J Biol Chem* **269**, 32187–32193.

Fischer, A., Sananbenesi, F., Pang, P. T., Lu, B. and Tsai, L. H. (2005). Opposing roles of transient and prolonged expression of p25 in synaptic plasticity and hippocampus-dependent memory. *Neuron* **48**, 825–838.

Francis, B. M., Kim, J., Barakat, M. E., Fraenkl, S., Yucel, Y. H., Peng, S., Michalski, B., Fahnestock, M., McLaurin, J. and Mount, H. T. (in press). Object recognition memory and BDNF expression are reduced in young TgCRND8 mice. *Neurobiol Aging*, doi:10.1016/j.neurobiolaging.2010.1004.1003.

Frankland, P. W. and Bontempi, B. (2005). The organization of recent and remote memories. *Nat Rev Neurosci* **6**, 119–130.

Fukumoto, H., Cheung, B. S., Hyman, B. T. and Irizarry, M. C. (2002). β-Secretase protein and activity are increased in the neocortex in Alzheimer disease. Arch Neurol **59**, 1381–1389.

Fukumoto, H., Takahashi, H., Tarui, N., Matsui, J., Tomita, T., Hirode, M., Sagayama, M., Maeda, R., Kawamoto, M., Hirai, K., *et al.* (2010). A noncompetitive BACE1 inhibitor TAK-070 ameliorates Aβ pathology and behavioral deficits in a mouse model of Alzheimer's disease. *J Neurosci* **30**, 11157–11166.

Games, D., Adams, D., Alessandrini, R., Barbour, R., Berthelette, P., Blackwell, C., Carr, T., Clemens, J., Donaldson, T., Gillespie, F., *et al.* (1995). Alzheimer-type neuropathology in transgenic mice overexpressing V717F β-amyloid precursor protein. *Nature* **373**, 523–527.

Ghribi, O., Larsen, B., Schrag, M. and Herman, M. M. (2006). High cholesterol content in neurons increases BACE, β-amyloid, and phosphorylated tau levels in rabbit hippocampus. *Exp Neurol* **200**, 460–467.

Giese, K. P., Ris, L. and Plattner, F. (2005). Is there a role of the cyclin-dependent kinase 5 activator p25 in Alzheimer's disease? Neuroreport **16**, 1725–1730.

Gong, B., Vitolo, O. V., Trinchese, F., Liu, S., Shelanski, M. and Arancio, O. (2004). Persistent improvement in synaptic and cognitive functions in an Alzheimer mouse model after rolipram treatment. J Clin Invest **114**, 1624–1634.

Gotz, J., Chen, F., van Dorpe, J. and Nitsch, R. M. (2001). Formation of neurofibrillary tangles in P301L tau transgenic mice induced by Aβ 42 fibrils. *Science* **293**, 1491–1495.

Gotz, J. and Ittner, L. M. (2008). Animal models of Alzheimer's disease and frontotemporal dementia. *Nat Rev Neurosci* **9**, 532–544.

Green, K. N., Billings, L. M., Roozendaal, B., McGaugh, J. L. and LaFerla, F. M. (2006). Glucocorticoids increase amyloid-β and tau pathology in a mouse model of Alzheimer's disease. *J Neurosci* **26**, 9047–9056.

Gu, Z., Liu, W. and Yan, Z. (2009). β-Amyloid impairs AMPA receptor trafficking and function by reducing Ca^{2+}/calmodulin-dependent protein kinase II synaptic distribution. *J Biol Chem* **284**, 10639–10649.

Haass, C. and Selkoe, D. J. (2007). Soluble protein oligomers in neurodegeneration: lessons from the Alzheimer's amyloid β-peptide. *Nat Rev Mol Cell Biol* **8**, 101–112.

Hardy, J. and Selkoe, D. J. (2002). The amyloid hypothesis of Alzheimer's disease: Progress and problems on the road to therapeutics. *Science* **297**, 353–356.

Hebert, S. S., Horre, K., Nicolai, L., Papadopoulou, A. S., Mandemakers, W., Silahtaroglu, A. N., Kauppinen, S., Delacourte, A. and De Strooper, B. (2008). Loss of microRNA cluster miR-29a/b-1 in sporadic Alzheimer's disease correlates with increased BACE1/β-secretase expression. *Proc Natl Acad Sci USA* **105**, 6415–6420.

Hernandez, F., Borrell, J., Guaza, C., Avila, J. and Lucas, J. J. (2002). Spatial learning deficit in transgenic mice that conditionally over-express GSK-3β in the brain but do not form tau filaments. *J Neurochem* **83**, 1529–1533.

Ho, L., Qin, W., Pompl, P. N., Xiang, Z., Wang, J., Zhao, Z., Peng, Y., Cambareri, G., Rocher, A., Mobbs, C. V., *et al.* (2004). Diet-induced insulin resistance promotes amyloidosis in a transgenic mouse model of Alzheimer's disease. *FASEB J* **18**, 902–904.

Hooper, C., Markevich, V., Plattner, F., Killick, R., Schofield, E., Engel, T., Hernandez, F., Anderton, B., Rosenblum, K., Bliss, T., *et al.* (2007). Glycogen synthase kinase-3 inhibition is integral to long-term potentiation. *Eur J Neurosci* **25**, 81–86.

Hsia, A. Y., Masliah, E., McConlogue, L., Yu, G. Q., Tatsuno, G., Hu, K., Kholodenko, D., Malenka, R. C., Nicoll, R. A. and Mucke, L. (1999). Plaque-independent disruption of neural circuits in Alzheimer's disease mouse models. *Proc Natl Acad Sci USA* **96**, 3228–3233.

Hsiao, K., Chapman, P., Nilsen, S., Eckman, C., Harigaya, Y., Younkin, S., Yang, F. and Cole, G. (1996). Correlative memory deficits, Aβ elevation, and amyloid plaques in transgenic mice. *Science* **274**, 99–102.

Hsieh, H., Boehm, J., Sato, C., Iwatsubo, T., Tomita, T., Sisodia, S. and Malinow, R. (2006). AMPAR removal underlies Aβ-induced synaptic depression and dendritic spine loss. *Neuron* **52**, 831–843.

Hu, Y. and Russek, S. J. (2008). BDNF and the diseased nervous system: A delicate balance between adaptive and pathological processes of gene regulation. *J Neurochem* **105**, 1–17.

Jacobsen, J. S., Wu, C. C., Redwine, J. M., Comery, T. A., Arias, R., Bowlby, M., Martone, R., Morrison, J. H., Pangalos, M. N., Reinhart, P. H. and Bloom, F. E. (2006). Early-onset behavioral and synaptic deficits in a mouse model of Alzheimer's disease. *Proc Natl Acad Sci USA* **103**, 5161–5166.

Kaczorowski, C. C., Sametsky, E., Shah, S., Vassar, R. and Disterhoft, J. F. (2011). Mechanisms underlying basal and learning-related intrinsic excitability in a mouse model of Alzheimer's disease. *Neurobiol Aging* **32**, 1452–1465.

Kamenetz, F., Tomita, T., Hsieh, H., Seabrook, G., Borchelt, D., Iwatsubo, T., Sisodia, S. and Malinow, R. (2003). APP processing and synaptic function. *Neuron* **37**, 925–937.

Killick, R., Scales, G., Leroy, K., Causevic, M., Hooper, C., Irvine, E. E., Choudhury, A. I., Drinkwater, L., Kerr, F., Al-Qassab, H., *et al.* (2009). Deletion of Irs2 reduces amyloid deposition and rescues behavioural deficits in APP transgenic mice. Biochem Biophys Res Commun **386**, 257–262.

Kimura, R., Devi, L. and Ohno, M. (2010). Partial reduction of BACE1 improves synaptic plasticity, recent and remote memories in Alzheimer's disease transgenic mice. *J Neurochem* **113**, 248–261

Kimura, R. and Ohno, M. (2009). Impairments in remote memory stabilization precede hippocampal synaptic and cognitive failures in 5XFAD Alzheimer mouse model. *Neurobiol Dis* **33**, 229–235.

Kobayashi, D. T. and Chen, K. S. (2005). Behavioral phenotypes of amyloid-based genetically modified mouse models of Alzheimer's disease. Genes Brain Behav **4**, 173–196.

Kotilinek, L. A., Bacskai, B., Westerman, M., Kawarabayashi, T., Younkin, L., Hyman, B. T., Younkin, S. and Ashe, K. H. (2002). Reversible memory loss in a mouse transgenic model of Alzheimer's disease. *J Neurosci* **22**, 6331–6335.

Lacor, P. N., Buniel, M. C., Chang, L., Fernandez, S. J., Gong, Y., Viola, K. L., Lambert, M. P., Velasco, P. T., Bigio, E. H., Finch, C. E., *et al.* (2004). Synaptic targeting by Alzheimer's-related amyloid β oligomers. *J Neurosci* **24**, 10191–10200.

Lacor, P. N., Buniel, M. C., Furlow, P. W., Clemente, A. S., Velasco, P. T., Wood, M., Viola, K. L. and Klein, W. L. (2007). Aβ oligomer-induced aberrations in synapse composition, shape and density provide a molecular basis for loss of connectivity in Alzheimer's disease. *J Neurosci* **27**, 796–807.

LaFerla, F. M., Green, K. N. and Oddo, S. (2007). Intracellular amyloid-β in Alzheimer's disease. *Nat Rev Neurosci* **8**, 499–509.

LaFerla, F. M. and Oddo, S. (2005). Alzheimer's disease: Aβ, tau and synaptic dysfunction. Trends Mol Med **11**, 170–176.

Laird, F. M., Cai, H., Savonenko, A. V., Farah, M. H., He, K., Melnikova, T., Wen, H., Chiang, H. C., Xu, G., Koliatsos, V. E., *et al.* (2005). BACE**1**, a major determinant of selective vulnerability of the brain to amyloid-β amyloidogenesis, is essential for cognitive, emotional and synaptic functions. *J Neurosci* **25**, 11693–11709.

Lanz, T. A., Carter, D. B. and Merchant, K. M. (2003). Dendritic spine loss in the hippocampus of young PDAPP and Tg2576 mice and its prevention by the ApoE2 genotype. *Neurobiol Dis* **13**, 246–253.

Lee, E. B., Leng, L. Z., Zhang, B., Kwong, L., Trojanowski, J. Q., Abel, T. and Lee, V. M. (2006a). Targeting amyloid-β peptide (Aβ) oligomers by passive immunization with a conformation-selective monoclonal antibody improves learning and memory in Aβ precursor protein (APP) transgenic mice. *J Biol Chem* **281**, 4292–4299.

Lee, K. W., Im, J. Y., Song, J. S., Lee, S. H., Lee, H. J., Ha, H. Y., Koh, J. Y., Gwag, B. J., Yang, S. D., Paik, S. G. and Han, P. L. (2006b). Progressive neuronal loss and behavioral impairments of transgenic C57BL/6 inbred mice expressing the carboxy terminus of amyloid precursor protein. *Neurobiol Dis* **22**, 10–24.

Leon, W. C., Canneva, F., Partridge, V., Allard, S., Ferretti, M. T., DeWilde, A., Vercauteren, F., Atifeh, R., Ducatenzeiler, A., Klein, W., *et al.* (2010). A novel transgenic rat model with a full Alzheimer's-like amyloid pathology displays pre-plaque intracellular amyloid-β-associated cognitive impairment. J Alzheimers Dis **20**, 113–126.

Lesne, S., Koh, M. T., Kotilinek, L., Kayed, R., Glabe, C. G., Yang, A., Gallagher, M. and Ashe, K. H. (2006). A specific amyloid-β protein assembly in the brain impairs memory. *Nature* **440**, 352–357.

Lesne, S., Kotilinek, L. and Ashe, K. H. (2008). Plaque-bearing mice with reduced levels of oligomeric amyloid-β assemblies have intact memory function. Neuroscience **151**, 745–749.

Lewis, J., Dickson, D. W., Lin, W. L., Chisholm, L., Corral, A., Jones, G., Yen, S. H., Sahara, N., Skipper, L., Yager, D., *et al.* (2001). Enhanced neurofibrillary degeneration in transgenic mice expressing mutant tau and APP. *Science* **293**, 1487–1491.

Li, R., Lindholm, K., Yang, L. B., Yue, X., Citron, M., Yan, R., Beach, T., Sue, L., Sabbagh, M., Cai, H., *et al.* (2004). Amyloid β peptide load is correlated with increased β-secretase activity in sporadic Alzheimer's disease patients. *Proc Natl Acad Sci USA* **101**, 3632–3637.

Li, S., Hong, S., Shepardson, N. E., Walsh, D. M., Shankar, G. M. and Selkoe, D. (2009). Soluble oligomers of amyloid β protein facilitate hippocampal long-term depression by disrupting neuronal glutamate uptake. *Neuron* **62**, 788–801.

Liskowsky, W. and Schliebs, R. (2006). Muscarinic acetylcholine receptor inhibition in transgenic Alzheimer-like Tg2576 mice by scopolamine favours the amyloidogenic route of processing of amyloid precursor protein. Int J Dev Neurosci **24**, 149–156.

Liu, L., Orozco, I. J., Planel, E., Wen, Y., Bretteville, A., Krishnamurthy, P., Wang, L., Herman, M., Figueroa, H., Yu, W. H., *et al.* (2008). A transgenic rat that develops Alzheimer's disease-like amyloid pathology, deficits in synaptic plasticity and cognitive impairment. *Neurobiol Dis* **31**, 46–57.

Lue, L. F., Kuo, Y. M., Roher, A. E., Brachova, L., Shen, Y., Sue, L., Beach, T., Kurth, J. H., Rydel, R. E. and Rogers, J. (1999). Soluble amyloid β peptide concentration as a predictor of synaptic change in Alzheimer's disease. *Am J Pathol* **155**, 853–862.

Ma, Q. L., Harris-White, M. E., Ubeda, O. J., Simmons, M., Beech, W., Lim, G. P., Teter, B., Frautschy, S. A. and Cole, G. M. (2007). Evidence of Aβ- and transgene-dependent defects in ERK-CREB signaling in Alzheimer's models. *J Neurochem* **103**, 1594–1607.

McGowan, E., Eriksen, J. and Hutton, M. (2006). A decade of modeling Alzheimer's disease in transgenic mice. Trends Genet **22**, 281–289.

McLaurin, J., Kierstead, M. E., Brown, M. E., Hawkes, C. A., Lambermon, M. H., Phinney, A. L., Darabie, A. A., Cousins, J. E., French, J. E., Lan, M. F., *et al.* (2006). Cyclohexanehexol inhibitors of Aβ aggregation prevent and reverse Alzheimer phenotype in a mouse model. *Nat Med* **12**, 801–808.

Meilandt, W. J., Cisse, M., Ho, K., Wu, T., Esposito, L. A., Scearce-Levie, K., Cheng, I. H., Yu, G. Q. and Mucke, L. (2009). Neprilysin overexpression inhibits plaque formation but fails to reduce pathogenic Aβ oligomers and associated cognitive deficits in human amyloid precursor protein transgenic mice. *J Neurosci* **29**, 1977–1986.

Meziane, H., Dodart, J. C., Mathis, C., Little, S., Clemens, J., Paul, S. M. and Ungerer, A. (1998). Memory-enhancing effects of secreted forms of the β-amyloid precursor protein in normal and amnestic mice. *Proc Natl Acad Sci USA* **95**, 12683–12688.

Minichiello, L. (2009). TrkB signalling pathways in LTP and learning. *Nat Rev Neurosci* **10**, 850–860.

Mucke, L., Masliah, E., Yu, G. Q., Mallory, M., Rockenstein, E. M., Tatsuno, G., Hu, K., Kholodenko, D., Johnson-Wood, K. and McConlogue, L. (2000). High-level neuronal expression of Aβ1–42 in wild-type human amyloid protein precursor transgenic mice: Synaptotoxicity without plaque formation. *J Neurosci* **20**, 4050–4058.

Nagahara, A. H., Merrill, D. A., Coppola, G., Tsukada, S., Schroeder, B. E., Shaked, G. M., Wang, L., Blesch, A., Kim, A., Conner, J. M., *et al.* (2009). Neuroprotective effects of brain-derived neurotrophic factor in rodent and primate models of Alzheimer's disease. *Nat Med* **15**, 331–337.

Nalbantoglu, J., Tirado-Santiago, G., Lahsaini, A., Poirier, J., Goncalves, O., Verge, G., Momoli, F., Welner, S. A., Massicotte, G., Julien, J. P. and Shapiro, M. L. (1997). Impaired learning and LTP in mice expressing the carboxy terminus of the Alzheimer amyloid precursor protein. *Nature* **387**, 500–505.

O'Connor, T., Sadleir, K. R., Maus, E., Velliquette, R. A., Zhao, J., Cole, S. L., Eimer, W. A., Hitt, B., Bembinster, L. A., Lammich, S., *et al.* (2008). Phosphorylation of the translation initiation factor eIF2α increases BACE1 levels and promotes amyloidogenesis. *Neuron* **60**, 988–1009.

Oakley, H., Cole, S. L., Logan, S., Maus, E., Shao, P., Craft, J., Guillozet-Bongaarts, A., Ohno, M., Disterhoft, J., Van Eldik, L., *et al.* (2006). Intraneuronal β-amyloid aggregates, neurodegeneration and neuron loss in transgenic mice with five familial

Alzheimer's disease mutations: Potential factors in amyloid plaque formation. *J Neurosci* **26**, 10129–10140.

Oddo, S., Billings, L., Kesslak, J. P., Cribbs, D. H. and LaFerla, F. M. (2004). Aβ immunotherapy leads to clearance of early, but not late, hyperphosphorylated tau aggregates via the proteasome. *Neuron* **43**, 321–332.

Oddo, S., Caccamo, A., Shepherd, J. D., Murphy, M. P., Golde, T. E., Kayed, R., Metherate, R., Mattson, M. P., Akbari, Y. and LaFerla, F. M. (2003). Triple-transgenic model of Alzheimer's disease with plaques and tangles: Intracellular Aβ and synaptic dysfunction. *Neuron* **39**, 409–421.

Oddo, S., Caccamo, A., Tran, L., Lambert, M. P., Glabe, C. G., Klein, W. L. and LaFerla, F. M. (2006a). Temporal profile of amyloid-β (Aβ) oligomerization in an *in vivo* model of Alzheimer disease. A link between Aβ and tau pathology. *J Biol Chem* **281**, 1599–1604.

Oddo, S., Vasilevko, V., Caccamo, A., Kitazawa, M., Cribbs, D. H. and LaFerla, F. M. (2006b). Reduction of soluble Aβ and tau, but not soluble Aβ alone, ameliorates cognitive decline in transgenic mice with plaques and tangles. *J Biol Chem* **281**, 39413–39423.

Ohno, M. (2006). Genetic and pharmacological basis for therapeutic inhibition of β- and γ-secretases in mouse models of Alzheimer's memory deficits. Rev Neurosci **17**, 429–454.

Ohno, M. (2008). β-Secretase as a prime therapeutic target for Alzheimer's disease: A perspective from mouse model studies. In Araki, W. ed. *Recent Advances in the Biology of Secretases, Key Proteases* in *Alzheimer's Disease*, (Kerala: Research Signpost), pp. 1–25.

Ohno, M. (2009). Failures to reconsolidate memory in a mouse model of Alzheimer's disease. *Neurobiol Learn Mem* **92**, 455–459.

Ohno, M., Chang, L., Tseng, W., Oakley, H., Citron, M., Klein, W. L., Vassar, R. and Disterhoft, J. F. (2006). Temporal memory deficits in Alzheimer's mouse models: Rescue by genetic deletion of BACE1. *Eur J Neurosci* **23**, 251–260.

Ohno, M., Cole, S. L., Yasvoina, M., Zhao, J., Citron, M., Berry, R., Disterhoft, J. F. and Vassar, R. (2007). BACE1 gene deletion prevents neuron loss and memory deficits in 5XFAD APP/PS1 transgenic mice. *Neurobiol Dis* **26**, 134–145.

Ohno, M., Sametsky, E. A., Younkin, L. H., Oakley, H., Younkin, S. G., Citron, M., Vassar, R. and Disterhoft, J. F. (2004). BACE1 deficiency rescues memory deficits and cholinergic dysfunction in a mouse model of Alzheimer's disease. *Neuron* **41**, 27–33.

Pang, P. T. and Lu, B. (2004). Regulation of late-phase LTP and long-term memory in normal and aging hippocampus: Role of secreted proteins tPA and BDNF. *Ageing Res Rev* **3**, 407–430.

Pedersen, W. A., McMillan, P. J., Kulstad, J. J., Leverenz, J. B., Craft, S. and Haynatzki, G. R. (2006). Rosiglitazone attenuates learning and memory deficits in Tg2576 Alzheimer mice. *Exp Neurol* **199**, 265–273.

Peineau, S., Taghibiglou, C., Bradley, C., Wong, T. P., Liu, L., Lu, J., Lo, E., Wu, D., Saule, E., Bouschet, T., *et al.* (2007). LTP inhibits LTD in the hippocampus via regulation of GSK3β. *Neuron* **53**, 703–717.

Peng, S., Garzon, D. J., Marchese, M., Klein, W., Ginsberg, S. D., Francis, B. M., Mount, H. T., Mufson, E. J., Salehi, A. and Fahnestock, M. (2009). Decreased brain-derived neurotrophic factor depends on amyloid aggregation state in transgenic mouse models of Alzheimer's disease. *J Neurosci* **29**, 9321–9329.

Phiel, C. J., Wilson, C. A., Lee, V. M. and Klein, P. S. (2003). GSK-3α regulates production of Alzheimer's disease amyloid-β peptides. *Nature* **423**, 435–439.

Plattner, F., Angelo, M. and Giese, K. P. (2006). The roles of cyclin-dependent kinase 5 and glycogen synthase kinase 3 in tau hyperphosphorylation. *J Biol Chem* **281**, 25457–25465.

Reese, L. C., Zhang, W., Dineley, K. T., Kayed, R. and Taglialatela, G. (2008). Selective induction of calcineurin activity and signaling by oligomeric amyloid beta. *Aging Cell* **7**, 824–835.

Roberson, E. D., Scearce-Levie, K., Palop, J. J., Yan, F., Cheng, I. H., Wu, T., Gerstein, H., Yu, G. Q. and Mucke, L. (2007). Reducing endogenous tau ameliorates amyloid β-induced deficits in an Alzheimer's disease mouse model. *Science* **316**, 750–754.

Rodriguez, J. J., Jones, V. C., Tabuchi, M., Allan, S. M., Knight, E. M., LaFerla, F. M., Oddo, S. and Verkhratsky, A. (2008). Impaired adult neurogenesis in the dentate gyrus of a triple transgenic mouse model of Alzheimer's disease. *PLoS One* **3**, e2935.

Rowan, M., Klyubin, I., Cullen, W. and Anwy, l. R. (2003). Synaptic plasticity in animal models of early Alzheimer's disease. *Philos Trans R Soc Lond B Biol Sci* **358**, 821–828.

Santacruz, K., Lewis, J., Spires, T., Paulson, J., Kotilinek, L., Ingelsson, M., Guimaraes, A., DeTure, M., Ramsden, M., McGowan, E., *et al.* (2005). Tau suppression in a neurodegenerative mouse model improves memory function. *Science* **309**, 476–481.

Sarajarvi, T., Haapasalo, A., Viswanathan, J., Makinen, P., Laitinen, M., Soininen, H. and Hiltunen, M. (2009). Down-regulation of seladin-1 increases BACE1 levels and activity through enhanced GGA3 depletion during apoptosis. *J Biol Chem* **284**, 34433–34443.

Saura, C. A., Chen, G., Malkani, S., Choi, S. Y., Takahashi, R. H., Zhang, D., Gouras, G. K., Kirkwood, A., Morris, R. G. and Shen, J. (2005). Conditional inactivation of presenilin 1 prevents amyloid accumulation and temporarily rescues contextual and spatial working memory impairments in amyloid precursor protein transgenic mice. *J Neurosci* **25**, 6755–6764.

Selkoe, D. J. (2002). Alzheimer's disease is a synaptic failure. *Science* **298**, 789–791.

Sereno, L., Coma, M., Rodriguez, M., Sanchez-Ferrer, P., Sanchez, M. B., Gich, I., Agullo, J. M., Perez, M., Avila, J., Guardia-Laguarta, C., *et al.* (2009). A novel GSK-3β inhibitor reduces Alzheimer's pathology and rescues neuronal loss *in vivo*. *Neurobiol Dis* **35**, 359–367.

Shankar, G. M., Bloodgood, B. L., Townsend, M., Walsh, D. M., Selkoe, D. J. and Sabatini, B. L. (2007). Natural oligomers of the Alzheimer amyloid-β protein induce reversible synapse loss by modulating an NMDA-type glutamate receptor-dependent signaling pathway. *J Neurosci* **27**, 2866–2875.

Shankar, G. M., Li, S., Mehta, T. H., Garcia-Munoz, A., Shepardson, N. E., Smith, I., Brett, F. M., Farrell, M. A., Rowan, M. J., Lemere, C. A., *et al.* (2008). Amyloid-β protein dimers isolated directly from Alzheimer's brains impair synaptic plasticity and memory. *Nat Med* **14**, 837–842.

Singer, O., Marr, R. A., Rockenstein, E., Crews, L., Coufal, N. G., Gage, F. H., Verma, I. M. and Masliah, E. (2005). Targeting BACE1 with siRNAs ameliorates Alzheimer disease neuropathology in a transgenic model. *Nat Neurosci* **8**, 1343–1349.

Smith, D. L., Pozueta, J., Gong, B., Arancio, O. and Shelanski, M. (2009). Reversal of long-term dendritic spine alterations in Alzheimer disease models. *Proc Natl Acad Sci USA* **106**, 16877–16882.

Snyder, E. M., Nong, Y., Almeida, C. G., Paul, S., Moran, T., Choi, E. Y., Nairn, A. C., Salter, M. W., Lombroso, P. J., Gouras, G. K. and Greengard, P. (2005). Regulation of NMDA receptor trafficking by amyloid-β. *Nat Neurosci* **8**, 1051–1058.

Song, D. K., Won, M. H., Jung, J. S., Lee, J. C., Kang, T. C., Suh, H. W., Huh, S. O., Paek, S. H., Kim, Y. H., Kim, S. H. and Suh, Y. H. (1998). Behavioral and neuropathologic changes induced by central injection of carboxyl-terminal fragment of β-amyloid precursor protein in mice. *J Neurochem* **71**, 875–878.

Squire, L. R. (1992). Memory and the hippocampus: A synthesis from findings with rats, monkeys, and humans. *Psychol Rev* **99**, 195–231.

Sturchler-Pierrat, C., Abramowski, D., Duke, M., Wiederhold, K., Mistl, C., Rothacher, S., Ledermann, B., Burki, K., Frey, P., Paganetti, P., *et al.* (1997). Two amyloid precursor protein transgenic mouse models with Alzheimer disease-like pathology. *Proc Natl Acad Sci USA* **94**, 13287–13292.

Sun, X., He, G., Qing, H., Zhou, W., Dobie, F., Cai, F., Staufenbiel, M., Huang, L. E., and Song, W. (2006). Hypoxia facilitates Alzheimer's disease pathogenesis by up-regulating BACE1 gene expression. *Proc Natl Acad Sci USA* **103**, 18727–18732.

Taglialatela, G., Hogan, D., Zhang, W. R. and Dineley, K. T. (2009). Intermediate- and long-term recognition memory deficits in Tg2576 mice are reversed with acute calcineurin inhibition. *Behav Brain Res* **200**, 95–99.

Takashima, A., Noguchi, K., Michel, G., Mercken, M., Hoshi, M., Ishiguro, K. and Imahori, K. (1996). Exposure of rat hippocampal neurons to amyloid β peptide (25–35) induces the inactivation of phosphatidyl inositol-3 kinase and the activation of tau protein kinase I/glycogen synthase kinase-3β. *Neurosci Lett* **203**, 33–36.

Takeda, S., Sato, N., Uchio-Yamada, K., Sawada, K., Kunieda, T., Takeuchi, D., Kurinami, H., Shinohara, M., Rakugi, H. and Morishita, R. (2010). Diabetes-accelerated memory dysfunction via cerebrovascular inflammation and Aβ deposition in an Alzheimer mouse model with diabetes. *Proc Natl Acad Sci USA* **107**, 7036–7041.

Terwel, D., Muyllaert, D., Dewachter, I., Borghgraef, P., Croes, S., Devijver, H. and Van Leuven, F. (2008). Amyloid activates GSK-3β to aggravate neuronal tauopathy in bigenic mice. *Am J Pathol* **172**, 786–798.

Tesco, G., Koh, Y. H., Kang, E. L., Cameron, A. N., Das, S., Sena-Esteves, M., Hiltunen, M., Yang, S. H., Zhong, Z., Shen, Y., *et al.* (2007). Depletion of GGA3 stabilizes BACE and enhances β-secretase activity. *Neuron* **54**, 721–737.

Ting, J. T., Kelley, B. G., Lambert, T. J., Cook, D. G. and Sullivan, J. M. (2007). Amyloid precursor protein overexpression depresses excitatory transmission through both presynaptic and postsynaptic mechanisms. *Proc Natl Acad Sci USA* **104**, 353–358.

Tomiyama, T., Matsuyama, S., Iso, H., Umeda, T., Takuma, H., Ohnishi, K., Ishibashi, K., Teraoka, R., Sakama, N., Yamashita, T., *et al.* (2010). A mouse model of amyloid β oligomers: Their contribution to synaptic alteration, abnormal tau phosphorylation, glial activation, and neuronal loss *in vivo*. *J Neurosci* **30**, 4845–4856.

Townsend, M., Cleary, J. P., Mehta, T., Hofmeister, J., Lesne, S., O'Hare, E., Walsh, D. M. and Selkoe, D. J. (2006). Orally available compound prevents deficits in memory caused by the Alzheimer amyloid-β oligomers. *Ann Neurol* **60**, 668–676.

Townsend, M., Mehta, T. and Selkoe, D. J. (2007). Soluble Aβ inhibits specific signal transduction cascades common to the insulin receptor pathway. *J Biol Chem* **282**, 33305–33312.

Tronson, N. C. and Taylor, J. R. (2007). Molecular mechanisms of memory reconsolidation. *Nat Rev Neurosci* **8**, 262–275.

Tsai, J., Grutzendler, J., Duff, K. and Gan, W. B. (2004). Fibrillar amyloid deposition leads to local synaptic abnormalities and breakage of neuronal branches. *Nat Neurosci* **7**, 1181–1183.

Van Dam, D., Abramowski, D., Staufenbiel, M. and De Deyn, P. P. (2005). Symptomatic effect of donepezil, rivastigmine, galantamine and memantine on cognitive deficits in the APP23 model. Psychopharmacology (Berl) **180**, 177–190.

Velliquette, R. A., O'Connor, T. and Vassar, R. (2005). Energy inhibition elevates β-secretase levels and activity and is potentially amyloidogenic in APP transgenic mice: Possible early events in Alzheimer's disease pathogenesis. *J Neurosci* **25**, 10874–10883.

Verret, L., Jankowsky, J. L., Xu, G. M., Borchelt, D. R. and Rampon, C. (2007). Alzheimer's-type amyloidosis in transgenic mice impairs survival of newborn neurons derived from adult hippocampal neurogenesis. *J Neurosci* **27**, 6771–6780.

Wang, L., Zang, Y., He, Y., Liang, M., Zhang, X., Tian, L., Wu, T., Jiang, T. and Li, K. (2006). Changes in hippocampal connectivity in the early stages of Alzheimer's disease: Evidence from resting state fMRI. *Neuroimage* **31**, 496–504.

Wang, W. X., Rajeev, B. W., Stromberg, A. J., Ren, N., Tang, G., Huang, Q., Rigoutsos, I. and Nelson, P. T. (2008). The expression of microRNA miR-107 decreases early in Alzheimer's disease and may accelerate disease progression through regulation of β-site amyloid precursor protein-cleaving enzyme 1. *J Neurosci* **28**, 1213–1223.

Wen, Y., Planel, E., Herman, M., Figueroa, H. Y., Wang, L., Liu, L., Lau, L. F., Yu, W. H. and Duff, K. E. (2008a). Interplay between cyclin-dependent kinase 5 and glycogen synthase kinase 3β mediated by neuregulin signaling leads to differential effects on tau phosphorylation and amyloid precursor protein processing. *J Neurosci* **28**, 2624–2632.

Wen, Y., Yu, W. H., Maloney, B., Bailey, J., Ma, J., Marie, I., Maurin, T., Wang, L., Figueroa, H., Herman, M., *et al.* (2008b). Transcriptional regulation of β-secretase by p25/cdk5 leads to enhanced amyloidogenic processing. *Neuron* **57**, 680–690.

Westerman, M. A., Cooper-Blacketer, D., Mariash, A., Kotilinek, L., Kawarabayashi, T., Younkin, L. H., Carlson, G. A., Younkin, S. G. and Ashe, K. H. (2002). The relationship between Aβ and memory in the Tg2576 mouse model of Alzheimer's disease. *J Neurosci* **22**, 1858–1867.

Yang, L. B., Lindholm, K., Yan, R., Citron, M., Xia, W., Yang, X. L., Beach, T., Sue, L., Wong, P., Price, D., *et al.* (2003). Elevated β-secretase expression and enzymatic activity detected in sporadic Alzheimer disease. *Nat Med* **9**, 3–4.

Zhao, J., Fu, Y., Yasvoina, M., Shao, P., Hitt, B., O'Connor, T., Logan, S., Maus, E., Citron, M., Berry, R., *et al.* (2007). β-Site amyloid precursor protein cleaving enzyme 1 levels become elevated in neurons around amyloid plaques: Implications for Alzheimer's disease pathogenesis. *J Neurosci* **27**, 3639–3649.

Zhao, W. Q., De Felice, F. G., Fernandez, S., Chen, H., Lambert, M. P., Quon, M. J., Krafft, G. A. and Klein, W. L. (2008). Amyloid β oligomers induce impairment of neuronal insulin receptors. *FASEB* J **22**, 246–260.

Zhao, W. Q., Lacor, P. N., Chen, H., Lambert, M. P., Quon, M. J., Krafft, G. A. and Klein, W. L. (2009). Insulin receptor dysfunction impairs cellular clearance of neurotoxic oligomeric Aβ. *J Biol Chem* **284**, 18742–18753.

Zhao, W. Q., Santini, F., Breese, R., Ross, D., Zhang, X. D., Stone, D. J., Ferrer, M., Townsend, M., Wolfe, A. L., Seager, M. A., *et al.* (2010). Inhibition of calcineurin-mediated endocytosis and α-amino-3-hydroxy-5-methyl-4-isoxazolepropionic acid (AMPA) receptors prevents amyloid β oligomer-induced synaptic disruption. *J Biol Chem* **285**, 7619–7632.

Zhong, P., Gu, Z., Wang, X., Jiang, H., Feng, J. and Yan, Z. (2003). Impaired modulation of GABAergic transmission by muscarinic receptors in a mouse transgenic model of Alzheimer's disease. *J Biol Chem* **278**, 26888–26896.

Zhong, Z., Ewers, M., Teipel, S., Burger, K., Wallin, A., Blennow, K., He, P., McAllister, C., Hampel, H. and Shen, Y. (2007). Levels of beta-secretase (BACE1) in cerebrospinal fluid as a predictor of risk in mild cognitive impairment. *Arch Gen Psychiatry* **64**, 718–726.

Zuccato, C. and Cattaneo, E. (2009). Brain-derived neurotrophic factor in neurodegenerative diseases. *Nat Rev Neurol* **5**, 311–322.

Restoring Memory Deficits in Cognitive Diseases

14

Andre Fischer*

1. Introduction

The main function of our brain is to processes stimuli, thereby allowing us to adapt our behaviour in a constantly changing environment. To understand the molecular and cellular mechanisms that underlie such processes is one of the most fascinating, and still unresolved questions in neuroscience. In order to better understand a system, it is sometimes useful to have a closer look at the mechanisms that cause its dysfunction. In the case of brain research, this is also of extreme importance because dysfunction of the brain is linked to the pathogenesis of severe neurological disorders. As such, many neuropathological conditions eventually manifest as learning and memory deficits.

For example, the famous patient Henry G. Molaison, better knows as H.M., suffered for severe epileptic seizures when doctors decided to bilaterally remove parts of the mediotemporal lobe, including the hippocampus. While the seizures were gone, it became soon clear that H.M. suffered from a retrograde amnesia. He could perfectly remember everything until 4 weeks before the surgery but was unable to acquire new long-term memories (Scoville and Milner, 1957). It is therefore important to reiterate that memory formation can be distinguished into different phases such as acquisition, consolidation, storage, and retrieval of information. The learning impairments of H.M. are also observed in rodents. To this end, rodents that are trained in the Pavlovian fear conditioning paradigm (Figure 14.1) are able to recall this event already 24 h later. Notably, this form of associative fear memory can also be readily retrieved 4 or more weeks after the training (Kim and Fanselow, 1992; Fischer *et al.*, 2007). Animals that suffer from a hippocampal lesion are unable to learn this task. This also holds true when the hippocampus was destroyed immediately after the training. However, animals in

*Laboratory for Aging and Cognitive Diseases, European Neuroscience Institute, Grisebach Str. 5, D-37077 Goettingen, Germany.

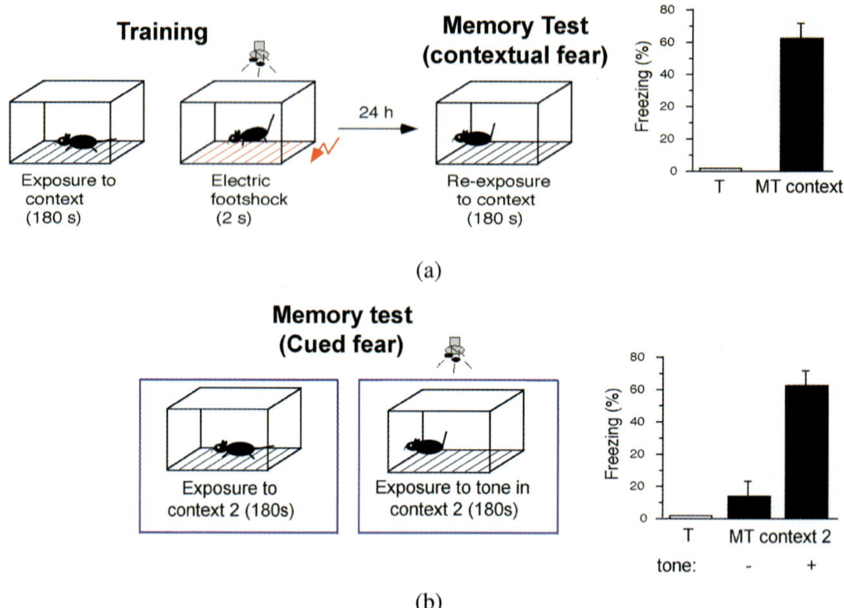

Figure 14.1. The fear-conditioning paradigm: A model to study associative learning. (A) The 'fear conditioning paradigm' helped to elucidate the mechanisms underlying memory formation. In the Pavlovian fear conditioning paradigm, the training procedure consists of rodents which are exposed to a novel context (context 1), consisting of an observation chamber with defined lightning, odor, background noise and geometry. After exploring this novel environment for 180 s followed by a mild electric foot shock for 2 s. Additionally, the mice can be exposed to a tone for 30 s. On the basis of associative learning, mice display aversive freezing behaviour during a memory test consisting of re-exposure to the conditioning context 24 h after the training. Freezing is an inborn behaviour that rodents express in threatening situations. It is defined as the complete absence of movement, except respiration and heartbeat. Whereas mice display no freezing behaviour during training, a significant increase of freezing behaviour is observed upon re-exposure to the conditioned context only. Therefore, freezing behaviour is commonly used to quantify the amount of learning. (B) To test cued learning (e.g. freezing to a tone) 24 h after training mice are exposed to a novel context (context2) that differs significantly from context 1. The animals show no freezing behavior in this novel context demonstrating the specificity of the associated fear memory. Upon presentation of the tone, mice will show freezing indicating cued fear. The fear conditioning paradigm is particularly suitable to study the molecular mechanisms of memory consolidation since it allows to take a molecular snap shot of memory formation. This is possible because a single training session is sufficient to induce long-term memory consolidation that can be analysed already 24 h later. As such, a typical experiment often involves the exposure of rodents to the training procedure followed by the molecular analysis 1 h later. ***$P < 0.0001$ freezing on E1 vs. E4/5. T, training; MT, memory test.

which the hippocampus was inactivated 4 weeks after a fear memory was acquired could perfectly well recall this memory trace but were unable to form new hippocampus-dependent memories (Kim and Fanselow, 1992). This data shows that the cellular substrates that are required for the formation of memories are likely to be different form the structures by which the memory is eventually maintained (Frankland *et al.*, 2004). Although the mechanisms by which memories are maintained for a prolonged time period are not well understood, it is interesting to note that this process is very dynamic. For example, inhibition of the action of NMDA receptors in the adult forebrain, appears to be not only important for the acquisition of memories but also for their maintenance. To this end, deletion of the essential NMDA receptor subunit NR1 from the forebrain of mice impairs the acquisition of contextual fear memories (Tsien *et al.*, 1996). Notably, when NR1 was removed from excitatory forebrain neurons after a stable memory has formed, animals were unable to retrieve this memory (Cui *et al.*, 2004). Most remarkably, restoration of NR1 expression allowed the recovery of formerly lost long-term memories (Cui *et al.*, 2004). The most likely explanation of this data is that NMDA receptor function in the adult forebrain is required for memory retrieval. As such, loss of NR1 would not erase the original memory but rather render it inaccessible. This data is in line with reports on dementia patients that under some circumstances can temporarily regain access to formally lost memories (Bradshaw *et al.*, 2004; Voss *et al.*, 2006). Such data raises the hope that a better understanding of the mechanisms that underlie memory function may eventually help to restore learning and memory processes in the diseased brain.

2. Restoring memory deficits in Alzheimer's disease

Disorders that are associated with a decline of cognitive function have become a severe social and economical problem for our societies. This is most obvious for the case of dementia. Dementia is characterized by the loss of social, emotional and cognitive abilities. The major reason for dementia is Alzheimer's disease (AD) the most common of all neurodegenerative diseases.

Because ageing is the most significant risk factor for AD, the number of patients is expected to double by the year 2025 causing a huge economical and emotional burden to our societies. Patients that suffer from AD exhibit difficulties to form new memories and fail to efficiently process novel information (Mesulam, 1999). At the later stages of the disease, patients also show a severe memory loss and are unable to remember important life events or to recognize close relatives. I will therefore discuss Alzheimer's disease as a suitable example to illustrate current approaches to restore memory deficits in neurological diseases.

Alzheimer's disease arises on the pathological background of amyloid-β (Aβ) plaques and neurofibrillary tangels (NFTs). Aβ plaques are extracellular aggregates of Aβ-peptides that are cleaved from the amyloid-precursor protein (APP) (Haass and Selkoe, 2007). NFTs are intracellular aggregates of hyperphosphorylated Tau protein, a microtuble-associated protein (Mandelkow, 1999). The commonly accepted view argues that toxic Aβ-peptides cause synaptic dysfunction (Palop and Mucke, 2010). Most recent data suggest Tau-protein plays an important role in this process and indeed deletion of Tau was able to rescue disease pathogenesis and memory impairment in a mouse model for Aβ-pathology (Roberson *et al.*, 2007; Ittner *et al.*, 2010). Although those pathological hallmarks were already described by Alois Alzheimer at the beginning of the last century, the precise mechanism by which Aβ- and Tau pathology lead to neuronal cell death and memory dysfunction are still not understood. As a consequence, there is still no effective treatment available and currently used drugs only act symptomatically. Yet there has been substantial progress and numerous reports identified treatments that are neuroprotective or neuroregenerative in various animal models for Alzheimer's disease. Nevertheless, when tested in clincial settings, most promising approaches including strategies to reduce Aβ-pathology via vaccination (Schenk *et al.*, 1999; Holmes *et al.*, 2008) or the usage of gamma-secretase inhibitors (Cummings, 2010) failed to improve cognition. One argument, which is often used to explain such lack of effectiveness and which is indeed very valid, states that most therapeutic interventions in humans simply start too late to show any effect. In line with this, similar observations were made in studies that tested for particular life-style such as the consumption of fish that includes the high levels of omega-3 fatty acids such as docosahexaenocid acids (DHA) (Morris, 2010). While epidemiological data clearly suggest that consumption of DHA lowers the risk of cognitive decline at older age, a recent clinical study failed to see any effect when patients that were diagnosed with mild cognitive impairment (MCI) were treated with DHA (Quinn *et al.*, 2010). Similarly, epidemilogical data clearly suggested that certain activities are neuroprotective and lower the risk of cognitive decline and dementia at older age (Verghese *et al.*, 2003). Such activities are usually a combination of cognitive and physical exercise such as dancing or playing a musical instrument. It has been argued that such activities build up a "cognitive reserve" that delay decline of cognitive function during old age or Alzheimer's disease (Nithianantharajah, 2009). On the other hand, experimental data from animal models show that a combination of cognitive and physical exercise act neuroprotectively but also neuroregeneratively (Jankowsky *et al.*, 2005; Lazarov *et al.*, 2005; Fischer *et al.*, 2007; Nithianantharajah and Hannan, 2006). In rodents, the most commonly used paradigm is environmental

(A) (B)

Figure 14.2. Environmental enrichment. (A) Cage commonly used for standard housing of rodents. (B) Mice in an enriched environment.

enrichment (Nithianantharajah and Hannan, 2006). While control animals live in a home cage that allows free access to food and water, enriched rodents are placed in a more challenging environment that includes free access to running wheels, tunnels etc. and may involve the completion of certain cognitive task to access food and water (Figure 14.2). While many variations of the environmental enrichment protocol exist, a unanimous finding across experiments is that EE facilitates synaptic plasticity and cognitive function in wild type rodents (Nithianantharajah and Hannan, 2006). Most importantly, enrichment was also found to be neuroprotective in various mice models for neurodegenerative diseases, including Alzheimer's disease (Lazarov *et al.*, 2005; Jankowsky *et al.*, 2005). In addition to the epidemiological data and findings from animal studies, EE can enhance brain plasticity in humans suffering form memory impairment (Singer *et al.*, 2005). For example, cognitive impairment is commonly observed in individuals suffering from schizophrenia and in fact cognitive impairment is predictive for the unfavorable course of pathology (Andreasen *et al.*, 2005). Moreover, a reduced hippocampal volume is often observed in schizophrenia patients (Maller *et al.*, 2010). It is therefore interesting to note that 3 months of regular exercise was sufficient to induce a relative increase of hippocampal volume in schizophrenia patients. This was accompanied by increased aerobic fitness and facilitated short-term memory (Pajonk *et al.*, 2010). Similarly, several studies suggest that MCI patients may benefit from exercise and cognitive rehabilitation programs (Kurz *et al.*, 2009; Geda *et al.*, 2010). This data suggest that a better understanding of the molecular mechanisms that underlie EE could help to develop novel therapeutic strategies to treat memory impairment. Taking into account the wealth of studies in which EE was employed to assess its neuroprotective and neuroregenerative potential in preclinical settings (Nithianantharajah and Hannan, 2006; Laviola *et al.*, 2008), is its quite surprising that still little is known on the mechanisms that underlie the beneficial effect of EE (Need and Giese, 2003; Tang *et al.*, 2001; van Praag *et al.*, 1999). Cellular substrates such as enhanced neurogenesis, synaptogenesis as well as changes in gene expression and protein production and activity have been observed in response to EE and it appears likely that EE enhances memory function by facilitating the induction of learning-regulated processes (Laviola *et al.*, 2008; Need and Giese, 2003; Tang *et al.*, 2001; Fischer *et al.*, 2007; Arai *et al.*, 2009). A fascinating aspect of the EE training protocol is that its beneficial effect on memory function can be transmitted from one generation to another. For example, exposure of 15-day-old mice to a 2-week enrichment training protocol resulted in long-lasting enhancement of hippocampal long-term potentiation (LTP) in adult life (Arai *et al.*, 2009). Strikingly, the offspring of those enriched mice also exhibited enhanced LTP and facilitated hippocampal cAMP/p38 MAP kinase signaling, although those animals were never exposed to EE training. Such transgenerational effects are not only observed for "positive" stimuli such as facilitated cognitive function after EE training but also in response to "negative" environmental stimuli. It was, for example, shown that in rodents maternal separation or stress can influences emotional behaviour in subsequent generations (Weaver *et al.*, 2004; Franklin *et al.*, 2010).

It is now generally accepted that epigenetic processes are important for such effects (Sananbenesi and Fischer, 2009). The term "epigenetics" was initially coined by Conrad H. Waddington. He added the greek prefix "epi" (over; above) to the word "genetics" to illustrate that its not simply the DNA sequence that defines a cellular phenotype (Waddington, 1953). Since then, the definition of epigenetics has undergone an evolution by itself. The most common way to define epigenetics nowadays is "*the study of heritable changes in gene expression that cannot be explained by changes in DNA sequence*" (Holliday, 1994).

3. Epigenetics: The role of genome–environment interactions during recovery of memory loss

Epigenetic mechanisms are key processes to regulate genome–environment interactions. It is therefore interesting to consider that health or disease of an individual critically depends on the interaction of genes and environment. A number of recent studies suggested that epigenetic mechanisms may play an important role during memory function and EE (Fischer *et al.*, 2007; Levenson and Sweatt, 2005; Fischer, 2010). In a broader sense, the term "epigenetics" is now widely used to refer to changes in gene expression that involve classical epigenetic mechanisms, namely, DNA methylation and histone modifications (Berger *et al.*, 2006; Bird, 2002). Recent data suggest that changes in DNA methylation are critically involved in cognitive function (Koshibu *et al.*, 2009). For example, inhibiton of DNA methylation by intrahippocampal injections of 5-azacytidine or zebularine (drugs that prevent DNA methylation) right before the training severely impair memory consolidation in rodents. This correlated with transient changes of DNA methylation in the promoter regions of the Reelin and protein phosphatase 1 genes, both implicated with neuronal plasticity, and corresponding gene expression (Miller *et al.*, 2008; Day and Sweatt, 2010). This data indicates that DNA methyaltion might be critically involved in learning and memory processes in the adult brain. In line with this, it was shown that exposing animals to a fear conditioning training protocol induces very selective and reversible changes in the promoter methylation of the BDNF gene (Lubin *et al.*, 2008) and that the deletion of DNMT1 and DNMT3a from the adult forebrain of mice impairs memory function (Feng *et al.*, 2010). This data indicates that DNA methylation is involved in the plastic response of neurons during memory formation.

Similarly, histone acetylation, a transient epigenetic modification that is usually associated with active gene expression and enables chromatin plasticity, has been associated with learning and memory processes in the adult brain (Fischer *et al.*, 2010).

The fundamental unit of the chromatin is the nucleosome, an octamer consisting of two molecules of each core histone (H) H2A, H2B, H3, and H4, which is wrapped around 147 bp of DNA (Sananbenesi and Fischer, 2009). An additional histone, H1, acts as a linker molecule that regulates the packaging of nucleosomes into higher-order structures. Histones are highly conserved basic proteins, containing a flexible N-terminus that protrudes from the surface of the nucleosome and is often named the "histone tail." Histone tails are subjected to multiple posttranslational modifications such as acetylation, phosphorylation, methylation, ubiquitination, sumoylation, ADP-ribosylation, and biotinylation (Vaquero *et al.*, 2003), which are targeted to specific residues. For example, histone acetylation occurs primarily upon lysine residues (K) (Figure 14.3).

Histone acetylation is regulated by the opposing activities of histone acetyltransferases (HATs) and histone deacetylases (HDACs). HAT proteins are subdivided into five families that have high sequence similarity and related substrate

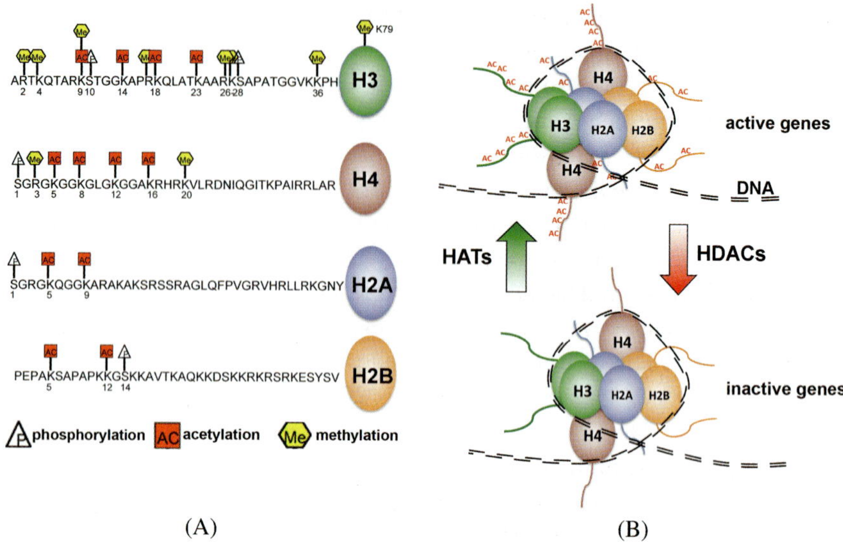

Figure 14.3. Histone acetylation regulates gene expression. (A) The N-terminus of histones protrudes from the surface of the nucleosome and is subjected to posttranslational modifications that affect the availability of the DNA for transcription. Therefore, histone modification is a key mechanism to regulate gene expression. (B) Histone acetylation is regulated by the counteracting activity of histone acetyltransferases (HATs) and histone deacetylases (HDAC) that add or remove acetyl groups from lysine residues. As such, HAT activity generally leads to active gene expression while HDAC activity induced gene silencing. As such, inhibitors of HDAC activity can induce gene expression.

specificity. The best-characterized HAT families are the Gcn5/PCAF, the MYST, and the CBP/p300 families (Allis *et al.*, 2007).

The human and the rodent genome encode 11 HDAC proteins. Class I HDACs consist of HDAC1, HDAC2, HDAC3 and HDAC8, and share sequence homology with the yeast Rpd3 protein. The class I HDACs all contain a nuclear localization signal (NLS) and, with the exception of HDAC 3 (Gregoretti *et al.*, 2004), lack a nuclear export signal (NES). Thus, the class I HDACs are primarily found within the nucleus, where they regulate histone acetylation. The class II HDACs are generally larger proteins than the class I HDACs since they contain additional regulatory domains. Class II HDACs contain both a NLS and a NES, and are known to shuttle between the cytoplasm and nucleus in a manner regulated by phosphorylation via calcium/calmodulin-dependent kinase II (CaMKII) (Gregoretti *et al.*, 2004; Chawla *et al.*, 2003). HDAC11 shares sequence similarities with both class I and class II HDACs and, from phylogenetic analysis, HDAC11 seems to be most closely related to HDAC3 and HDAC8, but is considered to be the sole class IV HDAC member (de Ruijter *et al.*, 2003).

Interestingly, it was observed that rodents display a transient increase in hippocampal histone acetylation 1 h after exposure to the fear conditioning paradigm (see Figure 14.1) that initiates associative learning (Levenson *et al.*, 2004; Peleg *et al.*, 2010). The finding that memory consolidation correlates with increased histone acetylation could be reproduced in other species and is not limited to fear conditioning but was also observed in the novel object recognition paradigm (Fontán-Lozano *et al.*, 2008). The transient increase in hippocampal H3 acetylation after fear conditioning is, at least in part, mediated via NMDA receptor-dependent activation of MAPK signalling (Chwang *et al.*, 2006; Chwang *et al.*, 2007). Consistent with the finding that learning increased hippocampal histone acetylation, mouse models for altered HAT activity display memory impairments. For example, loss of CREB (cAMP response element-binding protein)-binding protein (CBP) impairs hippocampal histone acetylation and leads to impairments in various memory tests such as novel object recognition and spatial memory (Korzus *et al.*, 2004; Alarcon *et al.*, 2004; Wood *et al.*, 2005; Wood *et al.*, 2005; Vecsey *et al.*, 2007; Valor *et al.*, 2010; Chen *et al.*, 2010). Notably, even if the CREB binding site of CBP is mutated severe learning and histone acetylation deficits were observed in those mice (Wood *et al.*, 2005). Interestingly, learning impairment in mutant CBP mice has been linked to the inducible expression of a limited number of genes (Vecsey *et al.*, 2007). In addition to a role of CBP in learning and memory, recent data suggest that HATs such as P300 and PCAF also play an important role in memory consolidation (Oliveira *et al.*, 2007; Maurice *et al.*, 2008). Importantly, administration of HDAC inhibitors increase histone acetylation and were able to restore memory function in mice lacking proper CBP activity (Alarcon *et al.*, 2004). Learning was also facilitated in wild-type mice and rats after treatment with HDAC inhibitor, suggesting that increased histone acetylation is an important mechanisms in memory formation (Fischer *et al.*, 2007; Vecsey *et al.*, 2007; Levenson *et al.*, 2004).

In conclusion, there is now substantial evidence that epigenetic processes regulate gene expression programs that are essential for memory function.

In turn, recent data indicate that deregulation of such mechanisms is causatively involved in the onset of memory disturbances. To this end, a number of studies indicated that neurodegenerative diseases, such as AD, correlate with altered epigenetic gene expression in those brain regions that are first affected. It was observed that particularly genes involved with synaptic plasticity were deregulated during AD pathogenesis (Berchtold *et al.*, 2008; Lu *et al.*, 2004; Scheff, 2003). Taken into account that ageing is the major risk factor for AD, it is interesting to note that a number of studies using post-mortem human brain samples showed that age itself correlates with a deregulation of "plasticity genes". The cause of this deregulation of gene expression is not understood but it can be speculated that epigenetic processes might play a role. In line with this hypothesis, recent findings suggest that altered histone acetylation is indeed causally linked to the onset of memory disturbances (Peleg *et al.*, 2010). Consistent with previous data (Levenson *et al.*, 2004), our laboratory found that learning transiently increased hippocampal bulk histone acetylation in young mice. In aged mice, however, the onset of memory disturbances correlated with a lack of learning-induced acetylation of histone 4 at lysine 12 (H4K12), while other histone modifications were not affected. Notably, a lack of H4K12 acetylation severely impaired a hippocampal gene expression program required for memory formation.

By analysing the distribution of H4K12 acetylation in the young and old brains during learning, we found that impaired H4K12 acetylation was selectively associated with the coding regions of genes that were normally upregulated during learning. While histone acetylation proximal to the transcription start site of genes has been linked to transcriptional initiation, histone modifications that persist throughout the coding region of genes are associated with proper transcriptional elongation. Interestingly, a role of H4K12 acetylation in transcriptional elongation was also found in T cells, where H4K12 was one of the few histone acetylations that was elevated throughout the entire coding region of highly expressed genes (Wang *et al.*, 2008). Importantly, acute administration of the pan-HDAC inhibitor, SAHA, was able to rescue learning-induced H4K12 acetylation in aged mice and reinstate memory function, while the HDAC1-specific inhibitor, MS-275, failed to do so. These data suggest that H4K12 acetylation is an early biomarker for an impaired genome environment interaction in the aging brain. Moreover, such data led to the current view that HDAC inhibitors could be suitable strategies to treat cognitive dysfunction. A major breakthrough in this field was the finding that the memory enhancing effect of EE can be recapitulated by HDAC inhibitors and that changes in epigenetic gene-expression seems to be an essential step during EE training. In a recent study from our laboratory, we found that exposure of CK-p25 mice to EE improved memory formation and retrieval even when up to 25% of forebrain neurons were degenerated (Fischer *et al.*, 2007). The CK-p25 mouse model allows the inducible expression of the p25 protein, a risk factor for AD, and animals develop severe neuronal loss within 6 weeks of induction (Cruz *et al.*, 2003). EE was able to restore memory function but also enable animal to access memories that were formally inaccessible. Importantly, the memory restoring action of EE training in this mouse model correlated with changes in hippocampal and cortical histone acetylation and the upregualtion of genes involved with synaptic plasticity (Fischer *et al.*, 2007). As such, we hypothesised that a master regulatory program involving epigenetic mechanisms might be crucial to coordinate the response to EE. In line with this it was show that when HDAC inhibitors were administered to CK-p25 mice histone acetylation was increased and memory formation and retrieval was much improved (Fischer *et al.*, 2007). Although more data is needed to better understand those mechanisms, HDAC inhibitors were able to replicate the effect of EE and recent studies could confirm those results in mouse models for amyloid pathology (Ricobaraza *et al.*, 2009; Kilgore *et al.*, 2010).

Taken together, those data suggest that histone acetylation is an important mechanism of memory formation and that HDAC inhibitors, by at least partly recapitulating the induction of gene expression programs initiated by EE, could be suitable therapeutic strategies to treat cognitive diseases. However, we have to acknowledge that a lot of work is still needed to further elucidate the role of epigenetic mechanisms in memory formation.

4. Conclusion

Great progress has been made towards the understanding of the molecular mechanisms by which the brain processes, acquires and stores information. Although the precise molecular clockwork that regulates memory function is far from being understood, such research is of great clinical relevance since memory disturbances are key features of various neurological disorders. In this chapter, I have discussed novel therapeutic approaches that aim to treat memory impairments in neurodegenerative disorders and specifically address the role of epigenetic mechanisms to reinstate learning and memory function in the diseases brain.

References

Alarcon, J.M., Malleret, G., Touzani, K., Vronskaya, S., Ishii, S., Kandel, E.R. and Barco, A. (2004). Chromatin acetylation, memory, and LTP are impaired in CBP+/– mice: A model for the cognitive deficit in Rubinstein-Taybi syndrome and its amelioration. *Neuron* **42**, 947–959.

Allis, C.D., Berger, S.L., Cote, J., Dent, S., Jenuwein, T., Kouzarides, T., Pillus, L., Reinberg, D., Shi, Y., Shiekhattar, R., *et al.* (2007). New nomenclature for chromatin-modifying enzymes. *Cell* **131**, 633–636.

Andreasen, N.C., Carpenter, W.T., Kane, J.M., Lasser, R.A., Marder, S.R. and Weinberger, D.R. (2005). Remission in schizophrenia: Proposed criteria and rationale for consensus. *Am J Psychiatry* **162**, 441–449.

Arai, J.A., Li, S., Hartley, D.M. and Feig, L.A. (2009). Transgenerational rescue of a genetic defect in long-term potentiation and memory formation by juvenile enrichment. *J Neurosci* **29**, 1496–1502.

Berchtold, N.C., Cribbs, D.H., Coleman, P.D., Rogers, J., Head, E., Kim, R., Beach, T., Miller, C., Troncoso, J., Trojanowski, J.Q., *et al.* (2008). Gene expression changes in the course of normal brain aging are sexually dimorphic. *Proc Natl Acad Sci USA* **105**, 15605–15610.

Berger, I., Stahl, S., Rychkova, N. and Felbor, U. (2006). VEGF receptors on PC12 cells mediate transient activation of ERK1/2 and Akt: Comparison of nerve growth factor and vascular endothelial growth factor. *J Negat Results Biomed* **5**, 8.

Bird, A. (2002). DNA methylation patterns and epigenetic memory. *Genes Dev* **16**, 6–21.

Bradshaw, J., Saling, M., Hopwood, M., Anderson, V. and Brodtmann, A. (2004). Fluctuating cognition in dementia with Lewy bodies and Alzheimer's disease is qualitatively distinct. *J Neurol Neurosurg Psychiatry* **75**, 382–387.

Chawla, S., Vanhoutte, P., Arnold, F.J., Huang, C.L. and Bading, H. (2003). Neuronal activity-dependent nucleocytoplasmic shuttling of HDAC4 and HDAC5. *J Neurochem* **85**, 151–159.

Chen, G., Zou, X., Watanabe, H., van Deursen, J.M. and Shen, J. (2010). CREB binding protein is required for both short-term and long-term memory formation. *J Neurosci* **30**, 13066–13070.

Chwang, W.B., Arthur, J.S., Schumacher, A. and Sweatt, J.D. (2007). The nuclear kinase mitogen- and stress-activated protein kinase 1 regulates hippocampal chromatin remodeling in memory formation. *J Neurosci* **27**, 12732–12742.

Chwang, W.B., O'Riordan, K.J., Levenson, J.M. and Sweatt, J.D. (2006). ERK/MAPK regulates hippocampal histone phosphorylation following contextual fear conditioning. *Learn Mem* **13**, 322–328.

Cruz, J.C., Tseng, H.C., Goldman, J.A., Shih, H. and Tsai, L.H. (2003). Aberrant Cdk5 activation by p25 triggers pathological events leading to neurodegeneration and neurofibrillary tangles. *Neuron* **40**, 471–483.

Cui, Z., Wang, H., Tan, Y., Zaia, K.A., Zhang, S. and Tsien, J.Z. (2004). Inducible and reversible NR1 knockout reveals crucial role of the NMDA receptor in preserving remote memories in the brain. *Neuron* **41**, 781–793.

Cummings, J. (2010). What can be inferred from the interruption of the semagacestat trial for treatment of Alzheimer's disease? *Biol Psychiatry* **68**, 876–878.

Day, J.J. and Sweatt, J.D. (2010). DNA methylation and memory formation. *Nat Neurosci* **13**, 1319–1323.

de Ruijter, A.J., van Gennip, A.H., Caron, H.N., Kemp, S. and van Kuilenburg, A.B. (2003). Histone deacetylases (HDACs): Characterization of the classical HDAC family. *Biochem J* **370**, 737–749.

Feng, J., Zhou, Y., Campbell, S.L., Le, T., Li, E., Sweatt, J.D., Silva, A.J. and Fan, G. (2010). Dnmt1 and Dnmt3a maintain DNA methylation and regulate synaptic function in adult forebrain neurons. *Nat Neurosci* **13**, 423–430.

Fischer, A. (2010). HDAC Inhibitoren als Therapie für Neuronale Erkrankungen. *Pharm Unserer Zeit* **39**, 204–209.

Fischer, A., Sananbenesi, F., Mungenast, A. and Tsai, L.H. (2010). Targeting the right HDAC(s) to treat cognitive diseases. *Trends Pharmacol Sci* **31**, 605–617.

Fischer, A., Sananbenesi, F., Wang, X., Dobbin, M. and Tsai, L.H. (2007). Recovery of learning & memory after neuronal loss is associated with chromatin remodeling. *Nature* **447**, 178–182.

Fontán-Lozano, A., Romero-Granados, R., Troncoso, J., Múnera, A., Delgado-García, J.M. and Carrión, A.M. (2008). Histone deacetylase inhibitors improve learning consolidation in young and in KA-induced-neurodegeneration and SAMP-8-mutant mice. *Mol Cell Neurosci* **39**, 193–201.

Frankland, P.W., Bontempi, B., Talton, L.E., Kaczmarek, L. and Silva, A.J. (2004). The involvement of the anterior cingulate cortex in remote contextual fear memory. *Science* **304**, 881–883.

Franklin, T.B., Russig, H., Weiss, I.C., Gräff, J., Linder, N., Michalon, A., Vizi, S. and Mansuy, I.M. (2010). Epigenetic transmission of the impact of early stress across generations. *Biol Psychiatry* **68**, 408–415.

Geda, Y.E., Roberts, R.O., Knopman, D.S., Christianson, T.J., Pankratz, V.S., Ivnik, R.J., Boeve, B.F., Tangalos, E.G., Petersen, R.C. and Rocca, W.A. (2010). Physical exercise, aging, and mild cognitive impairment: A population-based study. *Arch Neurol* **67**, 80–86.

Gregoretti, I.V., Lee, Y.M. and Goodson, H.V. (2004). Molecular evolution of the histone deacetylase family: Functional implications of phylogenetic analysis. *J Mol Biol* **338**, 17–31.

Haass, C. and Selkoe, D.J. (2007). Abstract Soluble protein oligomers in neurodegeneration: Lessons from the Alzheimer's amyloid beta-peptide. *Nat Rev Mol Cell Biol* **8**, 112–116.

Holliday, R. (1994). Epigenetics: An overview. *Dev Genet* **15**, 453–457.

Holmes, C., Boche, D., Wilkinson, D., Yadegarfar, G., Hopkins, V., Bayer, A., Jones, R.W., Bullock, R., Love, S., Neal, J.W., *et al.* (2008). Long-term effects of Abeta42 immunisation in Alzheimer's disease: Follow-up of a randomised, placebo-controlled phase I trial. *Lancet* **372**, 216–232.

Ittner, L.M., Ke, Y.D., Delerue, F., Bi, M., Gladbach, A., van Eersel, J., Wölfing, H., Chieng, B.C., Christie, M.J., Napier, I.A., *et al.* (2010). Dendritic function of tau mediates amyloid-beta toxicity in Alzheimer's disease mouse models. *Cell* **142**, 387–397.

Jankowsky, J.L., Melnikova, T., Fadale, D.J., Xu, G.M., Slunt, H.H., Gonzales, V., Younkin, L.H., Younkin, S.G., Borchelt, D.R. and Savonenko, A.V. (2005). Environmental enrichment mitigates cognitive deficits in a mouse model of Alzheimer's disease. *J Neurosci* **25**, 5217–5224.

Kilgore, M., Miller, C.A., Fass, D.M., Hennig, K.M., Haggarty, S.J., Sweatt, J.D. and Rumbaugh, G. (2010). Inhibitors of class 1 histone deacetylases reverse contextual memory deficits in a mouse model of Alzheimer's disease. *Neuropsychopharmacology* **35**, 870–880.

Kim, J.J. and Fanselow, M.S. (1992). Modality-specific retrograde amnesia of fear. *Science* **256**, 675–677.

Korzus, E., Rosenfeld, M.G. and Mayford, M. (2004). CBP histone acetyltransferase activity is a critical component of memory consolidation. *Neuron* **42**, 961–972.

Koshibu, K., Gräff, J., Beullens, M., Heitz, F.D., Berchtold, D., Russig, H., Farinelli, M., Bollen, M. and Mansuy, I.M. (2009). Protein phosphatase 1 regulates the histone code for long-term memory. *J Neurosci* **29**, 12079–12089.

Kurz, A., Pohl, C., Ramsenthaler, M. and Sorg, C. (2009). Cognitive rehabilitation in patients with mild cognitive impairment. *Int J Geriatr Psychiatry* **24**, 163–168.

Laviola, G., Hannan, A.J., Macrì, S., Solinas, M. and Jaber, M. (2008). Effects of enriched environment on animal models of neurodegenerative diseases and psychiatric disorders. *Neurobiol Dis* **31**, 159–168.

Lazarov, O., Robinson, J., Tang, Y.P., Hairston, I.S., Korade-Mirnics, Z., Lee, V.M., Hersh, L.B., Sapolsky, R.M., Mirnics, K. and Sisodia, S. (2005). Environmental enrichment reduces Abeta levels and amyloid deposition in transgenic mice. *Cell* **120**, 572–574.

Levenson, J.M., O'Riordan, K.J., Brown, K.D., Trinh, M.A., Molfese, D.L. and Sweatt, J.D. (2004). Regulation of histone acetylation during memory formation in the hippocampus. *J Biol Chem* **279**, 40545–40559.

Levenson, J.M. and Sweatt, J.D. (2005). Epigenetic mechanisms in memory formation. *Nat Rev* **6**, 108–119.

Lu, T., Pan, Y., Kao, S.Y., Li, C., Kohane, I., Chan, J. and Yankner, B.A. (2004). Gene regulation and DNA damage in the ageing human brain. *Nature* **429**, 883–891.

Lubin, F.D., Roth, T.L. and Sweatt, J.D. (2008). Epigenetic regulation of BDNF gene transcription in the consolidation of fear memory. *J Neurosci* **28**, 10576–10586.

Maller, J.J., Daskalakis, Z.J., Thomson, R.H., Daigle, M., Barr, M.S. and Fitzgerald, P.B. (2010). Hippocampal volumetrics in treatment-resistant depression and schizophrenia: The devil's in De-Tail. Hippocampus, *in press*.

Mandelkow, E. (1999). Alzheimer's disease. The tangled tale of tau. *Nature* **402**, 588–589.

Maurice, T., Duclot, F., Meunier, J., Naert, G., Givalois, L., Meffre, J., Célérier, A., Jacquet, C., Copois, V., Mechti, N., *et al.* (2008). Altered memory capacities and response to stress in p300/CBP-associated factor (PCAF) histone acetylase knockout mice. *Neuropsychopharmacology* **33**, 1584–1602.

Mesulam, M.M. (1999). Neuroplasticity failure in Alzheimer's disease: Bridging the gap between plaques and tangles. *Neuron* **24**, 521–529.

Miller, C.A., Campbell, S.L. and Sweatt, J.D. (2008). DNA methylation and histone acetylation work in concert to regulate memory formation and synaptic plasticity. *Neurobiol Learn Mem* **89**, 599–603.

Morris, M.C. (2010). The role of nutrition in Alzheimer's disease: Epidemiological evidence. *Eur J Neurol* **16**, 1–7.

Need, A.C. and Giese, K.P. (2003). Handling and environmental enrichment do not rescue learning and memory impairments in alphaCamKII(T286A) mutant mice. *Genes Brain Behav.* **2**, 132–139.

Nithianantharajah, J. and Hannan, A.J. (2006). Enriched environments, experience-dependent plasticity and disorders of the nervous system. *Nat Rev Neurosci* **7**, 697–709.

Oliveira, A.M., Wood, M.A., McDonough, C.B. and Abel, T. (2007). Transgenic mice expressing an inhibitory truncated form of p300 exhibit long-term memory deficits. *Learn Mem* **14**, 564–572.

Pajonk, F.G., Wobrock, T., Gruber, O., Scherk, H., Berner, D., Kaizl, I., Kierer, A., Mueller, M., Oest, M., Meyer, T., *et al.* (2010). Hippocampal plasticity in response to exercise in schizophrenia. *Arch Gen Psychiatry* **67**, 133–143.

Palop, J.J. and Mucke, L. (2010). Amyloid-beta-induced neuronal dysfunction in Alzheimer's disease: From synapses toward neural networks. *Nat Neurosci* **13**, 812–818.

Peleg, S., Sananbenesi, F., Zovoilis, A., Burkhardt, S., Bahari-Java, S., Agis-Balboa, R.C., Cota, P., Wittnam, J., Gogul-Doering, A., Opitz, L., *et al.* (2010). Altered histone acetylation is associated with age-dependent memory impairment in mice. *Science* **328**, 753–756.

Quinn, J., Raman, R., Thomas, R.G., Yurko-Mauro, K., Nelson, E.B., Van Dyck, C., J.E., G., Emond, J., Jack, C.R., Weiner, M., *et al.* (2010). Docosahexaenoic acid supplementation and cognitive decline in Alzheimer disease: A randomized trial. *JAMA* **304**, 1903–1911.

Ricobaraza, A., Cuadrado-Tejedor, M., Pérez-Mediavilla, A., Frechilla, D., Del Río, J. and García-Osta, A. (2009). Phenylbutyrate ameliorates cognitive deficit and reduces tau pathology in an Alzheimer's disease mouse model. *Neuropsychopharmacology* **34**, 1721–1732.

Roberson, E.D., Scearce-Levie, K., Palop, J.J., Yan, F., Cheng, I.H., Wu, T., Gerstein, H., Yu, G.Q. and Mucke, L. (2007). Reducing endogenous tau ameliorates amyloid beta-induced deficits in an Alzheimer's disease mouse model. *Science* **316**, 750–754.

Sananbenesi, F. and Fischer, A. (2009). The epigenetic bottleneck of neurodegenerative and psychiatric diseases. *Biol Chem* **390**, 1145–1153.

Scheff, S. (2003). Reactive synaptogenesis in aging and Alzheimer's disease: Lessons learned in the Cotman laboratory. *Neurochem Res* **11**, 1625–1630.

Schenk, D., Barbour, R., Dunn, W., Gordon, G., Grajeda, H., Guido, T., Hu, K., Huang, J., Johnson-Wood, K., Khan, K., *et al.* (1999). Immunization with amyloid-beta attenuates Alzheimer-disease-like pathology in the PDAPP mouse. *Nature* **400**, 173–177.

Scoville, W.B. and Milner, B. (1957). Loss of recent memory after bilateral hippocampal lesions. *Neuropsychiatry Clin Neurosci 2000 (classical article)* **1**, 103–113.

Singer, B., Friedman, E., Seeman, T., Fava, G.A. and Ryff, C.D. (2005). Protective environments and health status: cross-talk between human and animal studies. *Neurobiol Ageing* **26**, 113–118.

Tang, Y.P., Wang, H.S., Feng, M., Kyin, Y.Z. and Tsien, J.Z. (2001). Differential effects of enrichment on learning and memory function in NR2B transgenic mice. *Neuropharmacology* **41**, 779–790.

Tsien, J.Z., Huerta, P.T. and Tonegawa, S. (1996). The essential role of hippocampal CA1 NMDA receptor-dependent synaptic plasticity in spatial memory. *Cell* **87**, 1327–1338.

Valor, L.M., Pulopulos, M.M., Jimenez-Minchan, M., Olivares, R., Lutz, B. and Barco, A. (2011). Neuronal-specific loss of CBP causes strong reduction of histone acetylation and memory impairment, but does not affect neuronal viability and the induction of immediate early genes. *J Neurosci* **31**, 1652–1663.

van Praag, H., Kempermann, G. and Gage, F.H. (1999). Running increases cell proliferation and neurogenesis in the adult mouse dentate gyrus. *Nat Neurosci* **2**, 266–270.

Vaquero, A., Loyola, A. and Reinberg, D. (2003). The constantly changing face of chromatin. *Sci Aging Knowledge Environ* **14**, RE4.

Vecsey, C.G., Hawk, J.D., Lattal, K.M., Stein, J.M., Fabian, S.A., Attner, M.A., Cabrera, S.M., McDonough, C.B., Brindle, P.K., Abel, T. and Wood, M.A. (2007). Histone deacetylase inhibitors enhance memory and synaptic plasticity via CREB:CBP-dependent transcriptional activation. *J Neurosci* **27**, 6128–6140.

Verghese, J., Lipton, R.B., Katz, M.J., Hall, C.B., Derby, C.A., Kuslansky, G., Ambrose, A.F., Sliwinski, M. and H., B. (2003). Leisure activities and the risk of dementia in the elderly. *N Engl J Med* **348**, 2508–2516.

Voss, H.U., Uluc, A.M., Dyke, J.P., Watts, R., Kobylarz, E.J., McCandliss, B.D., Heier, L.A., Beattie, B.J., Hamacher, K.A., Vallabhajosula, S., *et al.* (2006). Possible axonal regrowth in late recovery from the minimally conscious state. *J Clin Invest* **116**, 2005–2011.

Waddington, C.H. (1953). Epigenetics and evolution. *Symp Soc Exp Biol* **7**, 186–199.

Wang, Z., Zang, C., Rosenfeld, J.A., Schones, D.E., Barski, A., Cuddapah, S., Cui, K., Roh, T.Y., Peng, W., Zhang, M.Q. and K., Z. (2008). Combinatorial patterns of histone acetylations and methylations in the human genome. *Nat Genet* **40**, 897–903.

Weaver, I.C., Cervoni, N., Champagne, F.A., D'Alessio, A.C., Sharma, S., Seckl, J.R., Dymov, S., Szyf, M. and Meaney, M.J. (2004). Epigenetic programming by maternal behavior. *Nat Neurosci* **7**, 847–854.

Wood, M.A., Kaplan, M.P., Park, A., Blanchard, E.J., Oliveira, A.M., Lombardi, T.L. and Abel, T. (2005). Transgenic mice expressing a truncated form of CREB-binding protein (CBP) exhibit deficits in hippocampal synaptic plasticity and memory storage. *Learn Mem* **12**, 111–119.

The Impact of Studies of Memory Mechanisms for Psychiatry

15

Shahid H. Zaman*

1. Introduction

What role does memory play in behaviour, cognition and emotions in the field of psychiatry? I propose that understanding the mechanisms of memory function will help us to, (1) identify factors that are key to the development of certain mental disorders, (2) understand the phenomenology of some of the symptoms that may present to psychiatry, such as delusions, confabulations and phobias, and (3) provide opportunities for managing behavioural and mental disorders.

I will first introduce clinical psychiatry and neuropsychology and then outline the relevance of the function of memory to psychiatry. I shall highlight relevant concepts that will help provide a framework for understanding the underlying mechanisms of memory and how they are relevant to psychiatry. I will provide examples from research to illustrate my points.

2. Psychiatry — A brief general introduction

Psychiatry attempts to understand disorders of the mind whether in the presence or the apparent absence of primary brain dysfunction. It aims to treat mental health problems and to alleviate suffering. Like other branches of medicine, it uses a systematic approach, which involves taking a detailed history (a subjective account) from the patient, making observations of the patient's behaviours and undertaking a detailed assessment of the mental state of the person. Often, a corroborative account from a witness is crucial as patients may have an altered sense of reality and believe that they are not ill. A psychiatrist will have to perform a thorough relevant physical examination (with an especial focus on the central

* Cambridge Intellectual & Developmental Disabilities Research Group, Department of Developmental Psychiatry, University of Cambridge, Cambridge CB2 8AH, UK.

nervous system) of the patient to take into account any medical diagnosis that may have an impact on the diagnosis or the psychiatric management of the patient (Gelder *et al.*, 2006; MacKinnon 2010).

The focus of the history and mental state is on a person's appearance, behaviours, emotions, thoughts (including beliefs), perceptions, cognition, and self-appraisal (or insight). Unusual perceptual experiences, such as perceiving in the absence of external stimuli (visual or auditory hallucinations) or having false thoughts or bizarre beliefs that are impervious to contradictory evidence (delusions), e.g. thinking that the Central Intelligence Agency are involved in surveillance of the patient should be sought and described. A bedside assessment of cognition (attention, orientation, registration of information, memory, visuo-spatial skills and language) is also undertaken (see Table 15.1).

Table 15.1. Formal mental state examination.

Feature	Examples of findings and their associations
Appearance and behaviour	Unkempt (a state of neglect which may indicate, e.g. depression, psychosis, or abuse) Expressionless "unemotional" face (autistic, negative symptoms of schizophrenia, Parkinsonism, schizotypal personality disorder) Over-familiar or disinhibited (mania, dissocial personality disorder, intoxicated (alcohol or other drugs) Restless or agitated (anxiety, depression, mania, hyperactivity syndrome, substance misuse) Involuntary movements such as tics (Gilles de la Tourrette's, akathisia, medication adverse effects) Psychomotor retardation (depression, hypothyroidism)
Speech or talk	Fast rate, high volume (large number of words spoken) and loud and uninterruptible (mania or drugs) Monotonic and slow rate (depression)
Affect	Non-reactive to appropriate cues to social communication Cheerful or sad, irritable or angry etc.
Mood	How does the patient feel now in their mood? Are there any "biological symptoms" namely: appetite and weight changes, early morning wakening, diurnal variation of mood, poor concentration, loss of energy, loss of libido.

(*Continued*)

Table 15.1. (*Continued*)

Feature	Examples of findings and their associations
	Note the severity of symptoms and the presence of any psychotic symptoms, irritability, mania, incongruent mood (i.e. mood that is inconsistent with the prevailing difficulties or delusion or hallucination, e.g. laughing when talking about a recent bereavement).
Thoughts	Contents-including delusions persecutory (paranoid), grandiose, nihilistic (commoner in psychotic depression e.g. "my insides are rotting") Ideas of reference: when the patient experiences a special connection to an experience or event to which he or she places an extraordinary meaning or significance, e.g. "The newsreader on the T.V. is talking about things about me". Form (structure of thought, how it is constructed or organised and the form of its conveyance to another) Flow rate of thoughts (fast in mania, slow in depression) Passivity: experiences of one's body, thoughts (see below) or emotions being controlled by outside forces, and not by the patient. Thought withdrawal, insertion or broadcast passivity phenomena where the patient's thoughts are being taken away from them, being implanted from elsewhere or are available for all to see (telepathy). Any ruminations or obsessions or compulsions or rituals
Perception	Perceptual abnormalities in all modalities of perceptions including auditory (e.g. in the second or third person, or are of any voices giving a running commentary on the patients thoughts or activities in patients with schizophrenia), gustatory or olfactory. The latter two modalities are more likely to be due to an organic cause, e.g. focal temporal epilepsy. Any illusions?
Cognition	Degree of alertness (in acute confusional states attention is abnormal), orientation, working memory test, memory test (immediate and delayed recall), recognition memory, verbal and visual memory, visuo-perceptual testing; language; praxis.
Insight	Does the person acknowledge that he or she is unwell? Why? Are the symptoms described considered abnormal and why? Would the person consider being treated or admitted to hospital?

Once all the information is gathered, a differential diagnosis is made. To help evaluate specific factors in the aetiology (the cause of the disorder) and the reason for the clinical presentation, one attempts to identify the predisposing, precipitating and perpetuating factors. A framework with biological, psychological and social factors in mind is constructed (Engel, 1977, 1981). Such a formulation helps to identify appropriate management strategies such as psychological or drug therapies (Gelder *et al.*, 2006).

BOX 1: Formulating a Case

History

A 35-year-old married woman, presented with a 6-month history of feeling increasingly tired, lacking in energy and taking more time off her part-time work as a personal assistant for a legal firm and in danger of losing her job. She had noted that she was less hungry as she was skipping meals and she had lost about 7% of her body weight in 10 months. Her husband (who had lost his job) noted that she was less interested in socialising and that she was awake at night for least 1–2 h. Her husband also had noted that she had started consuming more alcohol than before, often being embarrassingly drunk and inappropriate with it. She had never had such a constellation of problems before. She did recollect that in that past she had had times of up to two to three days when she would not be able to sleep at all and be full of energy and overwhelming enthusiasm to plan and do things, even if she had had limited resources or abilities. There was no history of misuse of substance prior to this, although during the nights of no sleep, she did also recall that she was taking unnecessary risks such as trying illicit substances and having unprotected casual sex. There was a family history on her mother's side of a great-aunt who was described as "eccentric" and she and another grandparent who had committed suicide. As a person, she has been described as a gregarious and chatty person, who deals with stress by "ignoring it" and diverting attention to another activity. She is not religious and has not been involved with the police. Her birth, and development were unremarkable and at school she was bullied (following the divorce of her parents when she was 11 years old). At school, she achieved average academic grades. She had a first sexual relationship at the age of 15 years and she has had 16 partners before she met her current partner. She has been married for five years and has no children, which she would like. She has attempted to get pregnant but is finding it difficult; she is consulting a fertility specialist. She is on no medication and there is no significant medication history. Her physical examination was unremarkable, except that the weight loss was confirmed. Her mental state revealed a woman who was

(*Continued*)

BOX 1 (*Continued*)

casually dressed, with poor eye-to eye contact as her gaze was mostly downward. Her speech was slow, monosyllabic, and she produced answers to questions that were very brief and monotonic. Her affect was blunted and her mood was moderately low with some biological symptoms of depression (disrupted sleep, anorexia, lack of concentration, lack of libido, weight loss). She felt "depressed". Her thoughts were slow in rate and she was quite pessimistic of the future; she had thoughts of harming herself, but there were no plans for suicide. She thought very negatively of herself, and she felt that it was her fault for feeling like this. She did not feel that she deserved to get better. Her concentration was poor, and she struggled to complete the bedside cognitive test, scoring only just within the normal limits. She was willing to engage with any treatments offered but felt that it was a hopeless undertaking.

Below is a deconstruction of this hypothetical case that helps a psychiatrist to understand the aetiology and help with the management of the case.

	Predisposing factors	Precipitating factors	Perpetuating factors
Biological	Grandparents have possible affective disorder (depression/ bipolar). There may be a genetic predisposition	She increased the use of alcohol and this may have an adverse effect on her mood	Any ongoing alcohol misuse or substances of misuse
Psychological	Was bullied at school & divorced parents; this may have had an impact on her personality or coping skills	Undergoing treatment for sub-fertility	Less socialising; strained relationship with husband
Social	Husband's business performing badly in recent years	Financial worries as had lost job	Ongoing financial worries; possibility of losing her job

There is no doubt that she has severe depression. It is also likely that it is part of a bipolar affective disorder, given the history of what appears to be at least one manic episode and a family history.

2.1. *Classification of mental health disorders*

The ICD-10 or International Classification of Diseases-10 (World Health Organization, 1993) and the Diagnostic and Statistical Manual of Mental Disorder-IV (American Psychiatric Association, 1994) — which was undergoing revision at the time of writing describe operational criteria to help to classify psychiatric disorders. A cluster of symptoms, which have developed characteristically over time and have affected a sufferer's quality of life, defines the psychiatric disorder or syndrome. Although different disorders may have the same symptoms, additional details such as age, gender, family history, chronology or progression of symptoms, personality traits, and history of substance misuse, medical conditions (epilepsy, head injuries) and social circumstances need to be taken into account to arrive at a diagnosis and formulation. For example, schizophrenia and cannabis misuse can present similarly, however it is not characteristic of schizophrenia to present to a clinician in a middle-aged man. It usually presents in adolescence or early adulthood and it is slightly commoner in males. See Appendix for the highlights of the ICD-10 classification.

The "organic" psychiatric disorders (or neuropsychiatric conditions) refer to constellations of psychiatric symptoms, e.g. agitation, rituals, obsessions, anxiety mania, depression and psychosis that have an identifiable medical or neurological cause. For example, following a traumatic head injury from a road traffic accident, a person may present with a mood or psychotic disorder directly related to the brain injury. Psychiatric disorders that are not organic are termed, "functional" psychiatric disorders and have no measureable structural pathology identifiable using current routine clinical tools, such as magnetic resonance imaging (MRI). A subset of functional disorders present with memory complaints, an example is psychogenic amnesia. More often, organic disorders such as dementia or brain injury will present with a memory complaint (see examples of such memory disorders in Table. 15.2). Memory problems may only be transient (e.g. transient global amnesia), fixed (e.g. limbic encephalitis) or progressive (dementia of Alzheimer's disease).

3. An outline of the neuropsychology of memory

Memory is a construct used to explain certain kinds of behavioural and experiential phenomena. In human neuropsychological studies of memory the "behavioural responses that subjects make in cognitive memory tasks in the laboratory serve merely as reports of cognitive processes", and "The ultimate output of cognitive memory systems is expressed in conscious awareness, which can, but

Table 15.2. Some examples of conditions presenting with memory function deficits.

Disorder or disease	Comment
Dementia including, Alzheimer's disease, vascular dementia, frontotemporal dementia, dementia with Lewy bodies	Declarative memory is affected and not working memory. In some fronto-temporal dementias, a semantic memory problem may be evident with episodic memory relatively intact. There are many other causes of dementia not listed here.
Traumatic brain injury	In a closed head injury, (i.e. where there is no penetrative injury), due to diffuse tissue damage, attention and concentration is often affected that can have an indirect impact on memory. If there is mesial temporal lobe damage, then declarative memory is usually defective.
Stroke	Ischaemic damage to mesial temporal lobes can occur because of hypoxaemia from narrowing of blood vessels or emboli (dislodged microclots) in vessels supplying the area.
Transient global amnesia	This lasts for several hours. There is both declarative amnesia and inability to learn new material during the attack. It has been shown to be due to mesial temporal lobe dysfunction. Following recovery there is amnesia for the period of the attack.
Limbic encephalitis	Associated with viral infections (especially due to Herpes simplex) or autoimmune mediated diseases (such as autoantibodies to potassium channels) or in the context of an occult tumour, i.e. paraneoplastic limbic encephalitis
Korsakoff's psychosis	Due to thiamine deficiency in alcoholism and malnutrition. Anterograde and retrograde memory is poor, but working memory is intact. There may be dysexecutive neuropsychological problems.

need not, be converted into overt behaviour such as verbal expression" (Tulving, 1995). For example, the attentiveness of the tested participant or the emotional content of the memory tested will have an impact on the memory evaluation.

The way in which the memory construct has been broken down varies greatly with many overlapping but differing cognitive models proposed (e.g. working memory-see below). However, a common framework proposes distinguishing memory functions in terms of modalities (i.e. verbal, visual, olfactory), time (i.e. very short term, long term and remote), process (e.g. encoding, storage, and retrieval) and the nature of the to-be-remembered material (i.e. semantic knowledge, personal experience, patterns of behaviour).

Memory entails the acquisition of information (or encoding), consolidation, storage and retrieval. The memory store can be viewed as a repository of information that has accumulated over the lifetime of an individual through his or her experiences. Not all of these experiences will be accessible to the person, as the inaccessible ones will be under "implicit" or "automatic" control. The integrity of memory function will have an impact on the perception, thoughts, self-knowledge, feelings, language and the meaning of experiences to the individual.

A popular view is that memory is not a unitary concept (Hirsh, 1974; Nadel and O′Keefe, 1974; Tulving, 1972) but that it consists of a number of systems that exist in parallel (Gabrieli, 1998; Squire *et al.*, 1993; Tulving, 1995). The process of memory for each individual system involves the acquisition and retrieval of information. The storage of information can be either short term ("working memory" is discussed below) or long term, if there is consolidation or stabilization of the memory trace (Squire and Zola, 1996). An important aspect of memory after its acquisition concerns the nature of the stability of long-term storage or consolidation. A consolidated memory once retrieved becomes temporarily labile requiring *de novo* protein synthesis for re-establishing it as a long-term storage memory trace. The term "reconsolidation" (Dudai, 2004; Nader, 2003; Nader *et al.*, 2000) is used to describe memory that has destabilised after retrieval that subsequently restabilises (see Chapter 10). Reconsolidation is not a mere recapitulation of memory and the underlying mechanisms may differ from those that bring about consolidation. So once memory is consolidated, it is more labile than was once thought; it is a mutable memory, which once retrieved after initial consolidation requires additional processing for reconsolidation (Tronson and Taylor, 2007). The function of reconsolidation remains unclear, but it is possibly to update memories or enable continued learning.

The model divides long-term memory into declarative (or explicit) which is accessible to conscious control, and non-declarative (or implicit; see Figure 15.1) which is not accessible to conscious effort (Cohen and Squire, 1980; Corkin, 2002; Squire, 1992; Squire *et al.*, 2004; Squire and Zola, 1996). Declarative memory can be further subdivided into episodic memory (the ability to remember autobiographical events — the "what" and "where") and semantic memory (the store of general knowledge-facts, ideas and concepts; (Tulving, 2002). Non-declarative memory is further subdivided into, procedural (habits and skills), priming, classical conditioning and habituation or sensitisation. The evidence for this model is largely from lesion studies of the medial temporal lobe (the

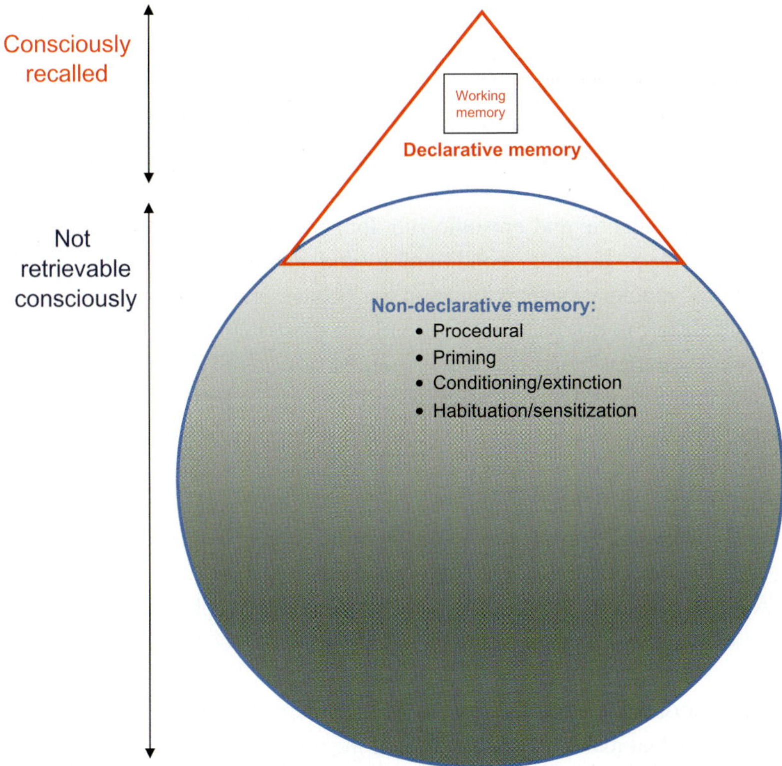

Figure 15.1. Illustration to help visualise the relationship between different forms of memory. Note that declarative memory is further subdivided into episodic and semantic memory.

celebrated case of HM is a well-studied example (Scoville and Milner, 1957), where declarative but not non-declarative memory is affected.

Autobiographical memory is "reconstructed" (Bartlett, 1932; Holland and Kensinger, 2010; Schacter *et al.*, 1998) as it is the result of what is remembered, what is not recollected and upon one's cognitive schemata. Memory is more akin to a perception of the past and gaps in recall are filled unwittingly by non-veridical recollections (Koriat *et al.*, 2000). Thus, it is not a precise chronologically accurate recall of the learning experience or event. Sometimes this may lead to false memories (Newman and Lindsay, 2009). Emotional and attentive states together with psychiatric disorder influence the representations in memory stores and the efficiency of retrieval (Dere *et al.*, 2010; Haas and Canli, 2008; Hamann, 2001). For example, in depressive disorders (Bradley *et al.*, 1995; Castaneda *et al.*,

2008; Gotlib and Joormann, 2010), there is a bias to recollecting negative aspects of experiences and the generic aspects of autobiographical memory more than details (Mark *et al.*, 2007).

Tulving (1995) places an emphasis on the role that each memory system described has on cognition, thought and behaviour. He employs the terms, "procedural" and "non-procedural" to classify the above memory systems. *Apart from* the working, semantic and episodic sub-domains, all memory sub-domains are "procedural" (e.g. priming, conditioning, and habituation). He proposes that "procedural" memory systems function as "behavioural or cognitive *action* systems", whereas, "non-procedural" systems are those that, "…mediate changes in cognition or thought.", and the output from these systems guide overt behaviour, and that this is an "optional postretrieval process" (Tulving, 1995); here he invokes the idea of volition and control. The difference between the procedural and the other types of memory is of the "feasibility of characterisation of the changes that result from learning or acquisition in a propositional or symbolic form". By including "propositional or symbolic forms" in this model, the role of language becomes central to memory systems in humans.

Animal studies of hippocampal-based memories emphasise the processing aspects of memory whereas human based studies make consciousness-based distinctions (Henke, 2010). Henke (2010) points out that conscious recall is a poor differentiator of declarative and non-declarative memory systems. She instead proposes a model that makes the distinctions of the various memory types that focuses upon the encoding or learning stage regardless of consciousness. She argues that hippocampal episodic learning (as opposed to classical fear conditioning, for example) is attuned to rapid encoding and flexible associations of the experience even without requiring conscious awareness. In her scheme, the slow encoding of rigid associations that are characteristic of procedural memory, classical conditioning and semantic memory are supported by basal ganglia, cerebellum and neocortex, whereas the hippocampus and neocortex undertake rapid encoding of flexible associations. The parahippocampus and neocortex undertake rapid encoding of single or unitised items, which results in priming and familiarity memory.

3.1. *Working memory*

Working memory (Baddeley, 2003) is the term used to try to explain the systems supporting the holding and manipulation of information over short time periods. The working memory model includes "slave" systems responsible for holding verbal (articulatory loop) or visual (visuo-spatial sketchpad) information in mind.

They have a limited capacity. For example, it is thought our working memory "span" is about seven plus or minus two items. If held for a sufficiently long period, representations in these temporary stores consolidate into long-term storage. The episodic buffer is a more recent addition to the model. It serves the purpose of holding much larger, conceptually integrated, amounts of information (as in the case of an episodic memory, recounting a story, etc.). The central executive has a regulatory role in managing the interactions between these systems, and in manipulating the content of working memory. As explained by Baddeley (2003), working memory is a limited capacity system that "supports human thought process by providing an interface between perception, long-term memory and action" (Baddeley, 2003).

3.2. *A conceptual framework for understanding the role of memory in psychiatric disorders*

A conceptual framework to understand the role of memory in psychiatric disorders should include (1) a direct output of the memory stores to behaviour, language, thought (including, self-awareness or self-knowledge), mood and perception, and (2) a dynamic reciprocal relationship between the memory store and perception, mood, etc. that allows one to mutate the other and *vice versa*. The information or representations stored can be processed further for cognitive tasks before they influence behavioural and emotional outputs.

I present a model that is similar to the one described in the previous section consisting of a system of memory structures and their respective components with their underlying mechanisms. The various components of the systems may have mechanisms that either are the same or shared with the other components. The processes particular to a given memory system would include reconsolidation to allow long-term memories to be strengthened or weakened and so provide opportunities for understanding mechanisms underlying the causes and the treatments of certain psychiatric disorders (Tronson and Taylor, 2007).

The model also includes "moderator" elements of the memory stores and memory inputs and outputs. The moderator elements may be the main part of the memory systems or neuromodulatory pathways such as dopamine. Dopamine as a moderator element would have a role in behaviours involving reward and novelty (Berke and Hyman, 2000; Bromberg-Martin *et al.*, 2010; Fadok *et al.*, 2010; Fields *et al.*, 2007). Such elements feed into memory components and function to monitor, control, evaluate, appraise and interpret (e.g. decision-making; (Britton *et al.*, 2011; Glimcher, 2002; Schall, 2001) the memory representations.

Executive functioning is the ability to plan, organise, set goals, initiate thought or action, inhibit responses or shift attention, so providing cognitive flexibility (Badre and Wagner, 2007; Buckner, 2003; Fletcher and Henson, 2001; Stuss and Levine, 2002). It is also proposed that executive functioning, which is dysfunctional in frontal lobe damage, provides a moderator element, with reciprocal dynamic interactions between the memory components and executive functioning components. Many psychiatric disorders manifest executive functioning neuropsychological abnormalities (Bishop, 2007; Palmer *et al.*, 2009; Quraishi and Frangou, 2002) which are likely to have an effect on the encoding, storage and retrieval aspects of memory. The frontal lobe may also exercise control in the form of "repression" of painful or traumatic (declarative) memories (Anderson and Green, 2001; Axmacher *et al.*, 2010) that would prevent distress or maladaptive behaviours.

An important aspect of the model presented here is that its fundamental feature is that of a representational system (Nadel and Hardt, 2011). Nadel and Hardt (2011) advocate a model where the conventional view of memory as a system, where perception and the storage of experience are separated (with the latter being designated the function of "memory") is replaced with the view that information inputs (via perception or internal inputs) are stored as a given set of "representations of knowledge". The types of information acquired may be emotional and the response of a person to the stored information may includes emotional outputs (Cahill *et al.*, 1994; Cahill *et al.*, 2002; Kensinger, 2009; Kensinger and Corkin, 2004). The key is that both processing and the storage of information takes place during a memory task and "…there is neither perceptual nor [separate] memory systems *per se*" (Nadel and Hardt, 2011). Importantly, such a representational model constitutes the "contents" (i.e. what is the information about) of the representation, *and* the "processing function" or what this representation can do.

A model presented by Britton *et al.* (2011) is similar to the one presented here. It is a model of anxiety disorders in humans that incorporates the elements of appraisal, attention orienting and cognitive appraisal in the context of fear learning. In conclusion, this sort of model in the context of psychiatric disorders does not consider memory systems in isolation but proposes that the control of the processing, the memory representations and their inputs and the outputs dynamically and reciprocally interact.

4. Memory complaints at a psychiatric clinic

Amnesia is defined as a deficit in memory functioning (Kopelman, 2002). As a primary memory complaint, amnesia does not usually present to a general psychiatrist. Instead, a neuropsychiatrist or a neurologist with expertise in behavioural and cognitive disorders (see Table 15.2) is more likely to have such a clinical encounter.

Patients suffering from disorders of memory functioning can present with complaints of forgetfulness or lapses in memory. Often, the patient may not be fully aware of it, and the partner, relative or friend will provide a more accurate picture of the deficits. In the clinical setting, the inability to learn new material is termed anterograde amnesia and the inability to recollect the past (declarative memory) is termed retrograde amnesia. Such amnesia may occur following a pathological event such as a head injury.

As part of the clinical assessment, the clinician undertakes a detailed bedside assessment of the ability to register information, recall new information immediately (to test working memory capacity) and after a delay of 5 and 30 or more min (long-term memory). Asking the patient to recall recently acquired test information, answer questions about personal historical details and general knowledge allows the testing of autobiographical memory, semantic and recognition memory systems. The pattern of deficits helps define the type of disease or disorder (see Table 15.2).

A memory deficit may present in some clinical conditions where the underlying deficit is not due to a primary memory dysfunction but due to problems with the attention system (e.g. in delirium) or working memory, which can have a major impact on acquisition as information cannot be held long enough in this ultra-short term type of memory to be encoded. As executive functioning is involved in retrieval strategies and monitoring output, frontal lobe dysfunction can impact strategic rather than associative aspects of memory functioning (Davidson *et al.*, 2006; Pannu and Kaszniak, 2005; Stuss and Levine, 2002).

A memory problem may be due to qualitative deficits rather than amnesia, e.g. déjà vu, false memories, confabulations, flashbulb and flashback memories (Berrios and Hodges, 2000), these have been termed "paramnesias" (Berrios, 1995). Such memory symptoms are associated with certain psychiatric and neuropsychiatric syndromes, although some memory symptoms such as dissociative (psychogenic) amnesia are recognised as disease categories in their own right according to the ICD-10 and DSM-IV.

5. Using knowledge of the mechanisms of memory to improve outcomes in psychiatric disorders

Although cognitive impairment is associated with the major psychiatric disorders such as schizophrenia (see below), major depression and bipolar affective disorder (e.g. (Malhi *et al.*, 2007; Quraishi and Frangou, 2002; Ragland *et al.*, 2009; Ranganath *et al.*, 2008), these do not primarily present with memory problems.

These clinical syndromes have been defined largely by non-memory cognitive function categories and the focus of treatment has been upon reducing the defining symptoms with medication and psychosocial strategies. When neuropsychological measurements are undertaken they can confirm the clinical suspicion of cognitive deficits in such disorders. Cognitive impairment that includes dysfunction of episodic and working memory have a big impact on daily functioning of the patient and therefore clinical outcomes (Green, 1996; Green *et al.*, 2000). One might argue though that it is at the strategic or executive level that memory problems arise as the retrieval process stops at the level of generic representations, and some might argue as an attempt to avoid recall of potentially unpleasant memories or associated negative affect. Major efforts are now underway to address means of specifically treating such cognitive deficits in the hope of improving clinical outcome that is often poor in chronic cases treated conventionally.

In order to understand especially the genetic significance of cognitive impairments in these psychiatric disorders, many researchers employ the concept of "endophenotype" (Gottesman and Gould, 2003). A gene may result in more than one phenotype, or a given phenotype may be as a consequence of different genes, so a given phenotype is not always a specific marker of the corresponding gene (Gottesman and Gould, 2003). The endophenotype concept assumes that a psychiatric condition, such as schizophrenia, is constructed of a number of intermediate phenotypes, which correspond more closely to a given gene, or a set of genes. For example, in schizophrenia, working memory has been shown to be impaired in both index cases and their family members (Cannon *et al.*, 2000; Park *et al.*, 1995). With this in mind, I now give examples of psychiatric disorders and treatments (such as psychotherapies) where knowledge of the underlying mechanisms of memory impairment could lead to improved outcomes.

5.1. *Schizophrenia and memory impairment*

The operational clinical definition of schizophrenia is dominated by "positive symptoms" (hallucinations, delusions and disordered thinking), which tend to fluctuate throughout the course of the condition. Clinical studies, however, point to more pervasive deficits in aspects of cognition that have a greater impact on daily functioning than positive symptoms (Green, 1996; Green *et al.*, 2000; Palmer *et al.*, 2009). Usually in chronic cases of schizophrenia, "negative symptoms" (e.g. apathy, poor engagement with others, lack of enjoyment and emotional flatness) predominate. Negative symptoms correlate strongly with cognitive impairment and with a poor quality of life (Tamminga *et al.*, 1998).

As cognitive deficits present before the core features of the classical symptoms of schizophrenia, some consider cognitive deficits to be the core feature of the disorder. Indeed delusions or perceptual abnormalities may be understood from a cognitive dysfunctioning perspective (David *et al.*, 1997; Elvevag and Goldberg, 2000; Jones *et al.*, 1994; Lepage *et al.*, 2007; Murray, 2011; Saykin *et al.*, 1994). Although not all studies agree with the exact range and degree of deficits, a number of cognitive domains have been shown to be affected including attention, working memory, episodic memory, executive functioning and intelligence (Cannon *et al.*, 2000; Heinrichs and Zakzanis, 1998; McKenna, 1991; Palmer *et al.*, 2009; Park *et al.*, 1995; Weinberger *et al.*, 1986). It is for this reason that much research effort is currently focused on developing means of cognitive remediation for instance with the development of pharmacological cognitive enhancers.

In schizophrenia, the dopaminergic pathways play a critical role in explaining the therapeutic effects of antipsychotics, the drugs used to treat schizophrenia. The first generation antipsychotic medications (which are still widely used) have a very high binding affinity for the dopamine-type 2 (D2) receptor subtype. The second generation or "atypical antipsychotics" (e.g. risperidone, olanzapine and clozapine) have a higher affinity for 5-hydroxytryptamine-type-2 receptors than D2 receptors, and some studies using atypical antipsychotics have been shown benefit in schizophrenia-related cognitive impairment (Harvey and Keefe, 2001; Keefe *et al.*, 2004; Mishara and Goldberg, 2004). Using the cognitive impairment framework, drugs being developed for schizophrenia are those likely to target not only DA receptors, but also cholinergic, noradrenergic and glutamatergic pathways (Breier, 2005) as these neuromodulators also impact memory functioning.

5.2. *Anxiety and related disorders*

From Shakespeare's Macbeth, Act V Scene III

Macbeth is referring to his wife who is suffering from guilt of being complicit in the murder of the King.

Macbeth. Cure her of that.

Canst thou not minister to a mind diseas'd,

Pluck from the memory a rooted sorrow,

Raze out the written troubles of the brain,

And with some sweet oblivious antidote

Cleanse the stuff'd bosom of that perilous stuff

Which weights upon the heart?

A simple idea from Shakespeare's play presents the notion of being able to "Pluck from memory a rooted sorrow". The idea of memories that are "pathological" and therefore causative to the suffering of the patient is not new. If simple phobias (e.g. of heights, open spaces or spiders) and other anxiety disorders are a consequence of "bad learning", then it would be an attractive idea if one could eliminate a given memory trace responsible for a particular maladaptive behaviour, thought or emotion in order to treat such disorders (Kindt *et al.*, 2009; Otto, 2002). In psychotherapy, early experiences of trauma (emotional and physical) or abandonment may have culminated in "bad memories" that dominate the patient's emotional life (Kandel, 1999). The role of psychotherapy would be to either erase these, to modify the contents or to align them to a different meaning. Tyron notes that all successful psychotherapies involve learning and memory as a pre-requisite to a change in behaviour, cognition and emotion (Tryon, 2000; Tryon, 2005). Some models of anxiety disorders assume they are learned behaviours (through fear conditioning or procedural learning) where associations have formed with stimuli that would normally be innocuous. Anxiety treatments can be viewed as treatments that alter a "situational" memory or create new ones (Brewin, 1989).

How can one alter memory? There are at least two different ways. One is through reconsolidation and the other through extinction. After a memory is retrieved or reactivated it becomes labile, and it requires a further process of consolidation to commit it to long-term memory. This process of reconsolidation (see Chapter 10) provides a window of opportunity to remove unwanted or bad memories for therapeutic purposes (Schiller *et al.*, 2010). For example, if the β-adrenoreceptor antagonist, propanolol is given after the reactivation of a given fear memory, it causes amnesia for that memory (Cahill *et al.*, 2000; Pitman *et al.*, 2002; Przybyslawski *et al.*, 1999). The β-adrenergic pathways have been shown in human studies to be involved in the formation of memory with emotional valence (Cahill *et al.*, 2002). Pharmacological manipulations using γ-amino-butyric acid receptor type-A ($GABA_AR$) modulators such as the benzodiazepine, midazolam also disrupted reconsolidation (Bustos *et al.*, 2006, 2009).

Extinction learning is thought to operate in exposure behavioural psychotherapy (Chambless and Ollendick, 2001; Phelps *et al.*, 2004) and is also influenced by manipulations of the $GABA_AR$ (Ehrlich *et al.*, 2009; Makkar *et al.*, 2010) and glucocorticoids (Bentz *et al.*, 2010). One complication of the reconsolidation notion is that the experimental means by which it is usual evoked can also evoke the process of extinction (Debiec *et al.*, 2002; Myers and Davis, 2002).

Posttraumatic stress disorder (PTSD) can also be viewed as a disorder of memory and learning (Brewin, 2011; Ferreri *et al.*, 2011; McCleery and Harvey,

2004; Rothbaum and Davis, 2003). It is triggered by a traumatic (usually perceived as a life-threatening) event that results in symptoms that persist where the sufferer repeatedly re-experiences the trauma. The patient suffers with nightmares, intrusive vivid memories of the event and an exaggerated fear and anxiety response to reminders or cues to the event (World Health Organization, 1993)-ICD-10). The fear response in PTSD is exaggerated and/or fear inhibition is deficient when compared with healthy people. One proposal is that the decreased ability to inhibit fear is a phenotype that increases the likelihood of developing PTSD and that when there is PTSD there is decreased extinction retention (Jovanovic and Ressler, 2010).

Better models of anxiety disorders include the notion of an automatic threat bias to real or imagined dangers, and attentional processes and cognitive appraisal that moderate the output of the fear response (Britton *et al.*, 2011). Some support for such models is borne out by meta-analysis of clinical studies where fear conditioning in patients and healthy controls showed modest exaggerated acquisition and extinction of fear responses in patients (Lissek *et al.*, 2005).

5.3. *Psychotherapies*

Cognitive behavioural therapy (CBT) is a widely used technique for the treatment of many neuroses such as panic disorder, phobic disorders, depression and social phobias. According to one popular form of therapy, as proposed by Beck (Beck *et al.*, 1985; Beck *et al.*, 1979), some mental health problems are due to a bias in the way information is processed by the brain (Beck and Clark, 1997). For example, in a patient with depression, there is a bias to think or evaluate the environmental interactions as negative. In anxiety disorders, the bias is towards being over attentive to dangers and threats. The therapy attempts to address these biases (termed cognitive distortions or thinking errors) with cognitive and behavioural exercises that challenge these assumptions by "checking the reality" and altering emotional learning. By using the psychological notion that one's thoughts, feelings (affect) and behaviours (actions) are interdependent, and what happens in one modality directly influences the others, biases are challenged. For example, if there is a negative thought or fear, this will result in a negative mood and negative behaviour. Conversely, if one's mood is upbeat, this will have a positive impact on one's thoughts and behaviour. In therapy, target problems are identified and reformulated in terms of the associated thoughts, moods and actions. The therapy focuses on the thoughts and teaches one to challenge any negative or false beliefs by evaluating the evidence.

CBT also encompasses behavioural techniques of exposure-based therapies. The central idea is that the process of exposure allows the patient and therapist to engage in cognitive change; changing the meaning associated with the emotional reaction or with the stimulus is crucial. If the technique of exposure is used, the patient may have to undergo exercises such as exposure to the anxiety-provoking situation or object (e.g. a spider), either in one's mind or for real. Upon exposure, the person will associate an altered meaning to the object of fear and the arousal would not manifest any longer. For example, the usual treatment offered for simple phobias is of exposure in the context of CBT where the patient is gradually exposed to the offending object or situation after some mental preparation. After several attempts, the patient is able to "desensitise" to the anxiety-provoking situation, and the symptoms abate.

Exposure therapy presumably facilitates fear extinction. There is evidence from animal studies that fear conditioning is mediated by *N*-methyl-*D*-aspartate (NMDA) receptors and the basolateral amygdala plays a crucial role (Cox and Westbrook, 1994; Davis, 2002; Fanselow and LeDoux, 1999; Lee and Kim, 1998). By targeting the NMDA receptor, it was proposed that it might enhance exposure therapy when the consolidation or the reconsolidation of memory is manipulated (Ressler *et al.*, 2004). Based on this premise, *D*-cyloserine, an N-methyl-D-aspartate (NMDA) receptor partial agonist at its glycine binding site, has been used to facilitate fear extinction (Norberg *et al.*, 2008) by facilitating the consolidation of extinction learning (Davis *et al.*, 2006).

Clinical trials using D-cylcoserine have been undertaken in suffers of panic disorder (Otto *et al.*, 2010), social anxiety and obsessive compulsive disorder (Hofmann *et al.*, 2006; Kushner *et al.*, 2007; Wilhelm *et al.*, 2008) that have shown some success. It is however not certain if the mechanism of action of D-cylcoserine in humans is the same as in rodents. As Grillon (2009) points out, fear conditioning occurs at different levels and may involve higher order cognition. He proposes that there may be two levels operating in humans and only one "low-level" (fast and reflexive) in lower species. Fear conditioning is influenced by the past memories, expectation and it is context dependent. Unlike in humans, extinction is speeded up in rodents with *D*-cyloserine. He proposes that in humans higher cognitive processes of danger and expectation (available for conscious access) play a role. Fear learning can involve not just fear conditioning, but other learning pathways i.e. vicarious learning (social learning — where there is no encounter with an aversive stimulus) or verbal information acquisition (Grillon, 2009). For the best-targeted therapies, regard will have to be given to the types of learning and memories evoked at the time of the experience. People with anxiety disorders have a memory bias of tending to assign personal events as threatening (Mitte, 2008) and this provides a target for therapy.

In the light of contemporary cognition-emotion research, CBT has evolved considerably and now the focus has shifted from contents to cognitive processes. This is true in the newer techniques of mindfulness meditation based methods, attention training and bias modification strategies (Kuyken *et al.*, 2010).

Other forms of psychotherapies may work via memory mechanisms. As mentioned by Kandel (1999), psychodynamic psychotherapy is a form of talk therapy that deals with the development of the maladaptive behaviour or emotional difficulties. Although there is little clinical evidence supporting the use of psychodynamic psychotherapy in schizophrenia and other major mental illness, it is still used for patients with e.g. anxiety disorders and personality disorders (Malmberg and Fenton, 2001). Even though psychodynamic psychotherapy is not strictly based upon the method of natural science because it proposes many abstractions that are experimentally not testable, it has provided a framework for understanding behaviour and is still religiously preached by some practitioners (Kandel, 1998).

Psychodynamic psychotherapy is based on the notion that early childhood experiences influence the trajectory of personality development, the degree to which a person is able to deal with emotional adversity and the degree of resilience to coping with psychological stress. The quality of early relationships, any early emotional or physical adversities, attachments and abandonments have a direct impact on personality. Psychological defence (or coping) mechanisms develop that help reduce distress and anxiety. The psychological defence mechanisms that manifest in a patient and the "transference" (see below) that occurs during psychotherapy sessions help to unearth the "unconscious" mind and so provide an opportunity to resolve emotional difficulties. According to psychodynamic theory [see Tamietto and de Gelder (2010) and Kandel (1999)] information is suppressed and rendered unavailable to the "conscious" mind by the "unconscious" mind.

Transference is a technique employed by psychotherapists which relies upon monitoring the emotional responses experienced by the therapist during a session, which are believed to reflect the underlying unconscious feelings and thoughts of the patient; the unconscious feelings are unavailable to the patient (Bateman and Fonagy, 2007; Hoglend *et al.*, 2008). Many of the unconscious processes are hypothesised to rely upon non-declarative memory and being able to manipulate these through the understanding of the mechanism may be a means of facilitating the therapeutic process (Gabbard, 2000; Grosjean, 2005).

It is useful in psychiatry (and especially in psychotherapy) to view a person as one whose make-up is explicitly influenced by the meanings and narratives, whether these are consciously accessible or not, that are held in memory. The idea of meaning or the narrative helps one to explain given behaviours in a given person in a given situation (Bolton, 2003; Kandel, 1999). Meanings are constructed through learning and the storage of experience throughout the lifetime of a person, but there will also be interpersonal, societal and cultural influences that impinge upon the making of meanings. As memory systems are not fully developed until later in childhood (Bauer, 2006), it is likely that early experiences will have a different impact on adult behaviour than those acquired later in life. For example, declarative memory systems take longer to mature than procedural memory systems probably because the dentate of the hippocampus and the frontal lobe development takes longer (Bauer, 2006). By exploring the narrative and understanding the meaning of an act or behaviour one has a good basis for understanding many forms of behaviour (Bolton and Hill, 2004). Some aspects of psychiatric disorders are better understood by exploring a narrative explanation as opposed to a mechanistic explanation. A narrative captures the salient facts and reveals the meaning they hold for the patient (MacKinnon 2010).

It may be straightforward to accept the notion of the processing of information in the brain, but how meaning is represented and used by memory systems is harder to understand; a possibility is that it is encoded in neuronal assemblies and modulated through synaptic plasticity (Eichenbaum, 2004; Gabbard, 2000; Wang and Morris, 2010). Connectionist models propose that memory is encoded in the efficacy of synaptic connections and with new learning the neuronal network connectivity changes with the change in the synaptic weights. Long-term potentiation and depression are forms of activity-dependent synaptic plasticity which are proposed mechanisms for memory (Martin *et al.*, 2000; Wang and Morris, 2010).

As pointed out by MacKinnon (2010), we have in some ways moved very little from Ancient Times when melancholia was considered to result from too much black bile, to our modern account of a serotonin imbalance problem. What is missing in psychiatry is a model of the mind (Kendler, 2005). Many aspects of psychiatric disorder are understood piecemeal such that a model used to explain a given clinical scenario might change in a given patient depending upon the circumstances. Bridging this chasm needs to be an aim of scientists in the field of memory research when applied to mental health.

6. Consequences of treatments

When treating psychiatric disorders that target memory processes there may be unintended consequences. If, as is likely, early childhood experiences have a major influence in constructing a person's character, beliefs and attitudes, then by altering the contents or the quality of a memory (whether implicit or explicit) it is likely to have an impact on the personality and cognitive abilities of the person. Thus, the erasure of a bad memory with treatment may also eliminate another memory that was important to the person as it carried an important meaning for them. It is an ethical question whether altering the persona by such means is acceptable to the individual and society. In terms of who is the best to make that decision, by the principle of autonomy, many would argue that it should be up to the individual as long as there is the capacity for the individual to make the decision. Like with any medical intervention, there are likely to be unintended consequences in the treatment of psychiatric disorders through this route and it is important to be aware of any (de Jongh *et al.*, 2008). Notions of autonomy and

BOX 2: A summary of some key points

"Memory" is fundamentally a representation of information

Learning/adaptation occurs with experience and environmental exposure (resulting in neuroplasticity); it is dependent upon the neurodevelopmental stage (e.g. "critical periods")

Most psychiatric patients do not present with amnesia

Memory disorders may be transient, progressive or fixed

It is clinically useful to divide memory into, declarative, non-declarative and working memory.

Some psychopathology is best understood, if related to underlying memory mechanisms

Some molecular mechanisms of behavioural phenotypes (e.g. addiction) share molecular mechanisms with memory

Lesions that result in memory dysfunction may be, acquired, congenital or genetic

Memory dysfunction may not have a "molecular", but a systems or network dysfunction explanation.

Certain neuromodulators that do not play a primary role in memory formation can modulate memory functioning

Treatment or management of patients with memory dysfunction may involve developing novel learning strategies that invokes the non-diseased pathways because (a) the defective parts may be irreversibly damaged or (b) remediation may be difficult because aspects of memory gained during the lifetime of the patient are lost or (c) the treatment itself may have unintended consequences doing no harm to the individual and society are desirable aims (Liao and Sandberg, 2008).

Exposure and reprocessing interventions like the CBT approach for PTSD (Ehlers and Clark, 2000, 2008) involves engaging with the avoided memory or situation and 'reprocessing' the meanings or content of the memory such that it becomes integrated into the autobiographical memory store, rendering it less "toxic". However, in brain injured patients especially with executive impairment, the elevated affect associated with this process could potentially have undesirable effects. King (2002) presents a case of "perseverative re-experiencing of trauma" in such a case. It is important to understand and work on affect regulation and coping skills thoroughly before undertaking such psychotherapy work. It raises questions about the nature of relationships between cognition and emotion at both physiological and neuropsychological levels, from which sense might be made of complex presentations where there are either temporary or permanent disruptions to these systems through mental disorder or acquired cognitive impairment.

7. Conclusions

The notion of implicit versus explicit (or declarative versus non-declarative) memory remains a useful concept in helping to understand some psychiatric disorders. Basic research in the neurosciences of memory is providing new perspectives to the nature of psychiatric disorders and helping to carve out new avenues to manipulate memory to treat disorders. To understand memory, an elucidation of mechanisms is needed at molecular, cellular and systems levels within a context provided by an understanding of psychiatric disorders. Conversely, clinical psychiatric research needs to be aware of the new findings of the basic science of memory research and apply these to clinical practice.

8. Acknowledgements

I am very grateful to Drs Fergus Gracey, Gurprit Pannu and Howard Ring for their time in critically reading and appraising drafts of this Chapter. I take responsibility for any errors.

References

American Psychiatric Association (1994). *Diagnostic and Statistical Manual of Mental Disorders: DSM-IV* (Washington, DC: American Psychiatric Association).

Anderson, M.C. and Green, C. (2001). Suppressing unwanted memories by executive control. *Nature* **410**, 366–369.

Axmacher, N., Do Lam, A.T.A., Kessler, H. and Fell, J. (2010). Natural memory beyond the storage model: Repression, trauma, and the construction of a personal past. *Front Human Neurosci* **4**, 12.

Baddeley, A. (2003). Working memory: Looking back and looking forward. *Nat Rev Neurosci* **4**, 829–839.

Badre, D. and Wagner, A.D. (2007). Left ventrolateral prefrontal cortex and the cognitive control of memory. *Neuropsychologia* **45**, 2883–2901.

Bartlett, F.C. (1932). *Remembering*. (Cambridge University Press).

Bateman, A. and Fonagy, P. (2007). The use of transference in dynamic psychotherapy. *Am J Psychiatry* **164**, 680.

Bauer, P.J. (2006). Constructing a past in infancy: A neuro-developmental account. *Trends Cogn Sci* **10**, 175–181.

Beck, A.T. and Clark, D.A. (1997). An information processing model of anxiety: Automatic and strategic processes. *Behaviour Research and Therapy* **35**, 49–58.

Beck, A.T., Emery, G. and Greenberg, R. (1985). *Anxiety Disorders and Phobias: A Cognitive Perspective*. (New York: Basic Books).

Beck, A.T., Rush, A.J., Shaw, B.F. and Emery, G. (1979). *Cognitive Therapy of Depression*. (New York: Guilford Press).

Bentz, D., Michael, T., de Quervain, D.J. and Wilhelm, F.H. (2010). Enhancing exposure therapy for anxiety disorders with glucocorticoids: From basic mechanisms of emotional learning to clinical applications. *J Anxiety Disord* **24**, 223–230.

Berke, J.D. and Hyman, S.E. (2000). Addiction, dopamine, and the molecular mechanisms of memory. *Neuron* **25**, 515–532.

Berrios, G.E. (1995). Deja vu in France during the 19th century: A conceptual history. *Compr Psychiatry* **36**, 123–129.

Berrios, G.E. and Hodges, J.R. (2000). *Memory Disorders in Psychiatric Practice*. (Cambridge, UK: Cambridge University Press).

Bishop, S.J. (2007). Neurocognitive mechanisms of anxiety: An integrative account. *Trends Cogn Sci* **11**, 307–316.

Bolton, D. (2003). Chapter 7. In Nature and narative, an introduction to the new philosophy of psychiatry, B. Fulford, K. Morris, J. Sadler and G. Stanghellini, eds. (OUP, Oxford), pp. 113–126.

Bolton, D. and Hill, J. (2004). *Mind, Meaning and Mental Disorder: The Nature of Causal Explanation in Psychology and Psychiatry*. 2nd ed. (OUP, Oxford).

Bradley, B.P., Mogg, K. and Williams, R. (1995). Implicit and explicit memory for emotion-congruent information in clinical depression and anxiety. *Behav Res Ther* **33**, 755–770.

Breier, A. (2005). Developing drugs for cognitive impairment in schizophrenia. *Schizophr Bull* **31**, 816–822.

Brewin, C.R. (1989). Cognitive change processes in psychotherapy. *Psychol Rev* **96**, 379–394.

Brewin, C.R. (2011). The nature and significance of memory disturbance in posttraumatic stress disorder. *Annu Rev Clin Psychol.*

Britton, J.C., Lissek, S., Grillon, C., Norcross, M.A. and Pine, D.S. (2011). Development of anxiety: The role of threat appraisal and fear learning. *Depress Anxiety* **28**, 5–17.

Bromberg-Martin, E.S., Matsumoto, M. and Hikosaka, O. (2010). Dopamine in motivational control: Rewarding, aversive, and alerting. *Neuron* **68**, 815–834.

Buckner, R.L. (2003). Functional-anatomic correlates of control processes in memory. *J Neurosci* **23**, 3999–4004.

Bustos, S.G., Maldonado, H. and Molina, V.A. (2006). Midazolam disrupts fear memory reconsolidation. *Neuroscience* **139**, 831–842.

Bustos, S.G., Maldonado, H. and Molina, V.A. (2009). Disruptive effect of midazolam on fear memory reconsolidation: Decisive influence of reactivation time span and memory age. *Neuropsychopharmacology* **34**, 446–457.

Cahill, L., Pham, C.A. and Setlow, B. (2000). Impaired memory consolidation in rats produced with beta-adrenergic blockade. *Neurobiol Learn Mem* **74**, 259–266.

Cahill, L., Prins, B., Weber, M. and McGaugh, J.L. (1994). [beta]-Adrenergic activation and memory for emotional events. *Nature* **371**, 702–704.

Cahill, L., Prins, B., Weber, M. and McGaugh, J.L. (2002). beta-adrenergic activation and memory for emotional events. *Nature* **371**, 702–704.

Cannon, T.D., Huttunen, M.O., Lonnqvist, J., Tuulio-Henriksson, A., Pirkola, T., Glahn, D., Finkelstein, J., Hietanen, M., Kaprio, J. and Koskenvuo, M. (2000). The inheritance of neuropsychological dysfunction in twins discordant for schizophrenia. *Am J Hum Genet* **67**, 369–382.

Castaneda, A.E., Tuulio-Henriksson, A., Marttunen, M., Suvisaari, J. and Lonnqvist, J. (2008). A review on cognitive impairments in depressive and anxiety disorders with a focus on young adults. *J Affect Disord* **106**, 1–27.

Chambless, D.L. and Ollendick, T.H. (2001). Empirically supported psychological interventions: Controversies and evidence. *Annu Rev Psychol* **52**, 685–716.

Cohen, N.J. and Squire, L.R. (1980). Preserved learning and retention of pattern-analyzing skill in amnesia: Dissociation of knowing how and knowing that. *Science* **210**, 207–210.

Corkin, S. (2002). What's new with amnesic patient H.M.? *Nat Rev Neurosci* **3**, 153–160.

Cox, J. and Westbrook, R.F. (1994). The NMDA receptor antagonist MK-801 blocks acquisition and extinction of conditioned hypoalgesic responses in the rat. *Q J Exp Psychol B* **47**, 187–210.

David, A.S., Malmberg, A., Brandt, L., Allebeck, P. and Lewis, G. (1997). IQ and risk for schizophrenia: A population-based cohort study. *Psychol Med* **27**, 1311–1323.

Davidson, P.S., Troyer, A.K. and Moscovitch, M. (2006). Frontal lobe contributions to recognition and recall: Linking basic research with clinical evaluation and remediation. *J Int Neuropsychol Soc* **12**, 210–223.

Davis, M. (2002). Role of NMDA receptors and MAP kinase in the amygdala in extinction of fear: Clinical implications for exposure therapy. *Eur J Neurosci* **16**, 395–398.

Davis, M., Ressler, K., Rothbaum, B.O. and Richardson, R. (2006). Effects of D-cycloserine on extinction: Translation from preclinical to clinical work. *Biol Psychiatry* **60**, 369–375.

de Jongh, R., Bolt, I., Schermer, M. and Olivier, B. (2008). Botox for the brain: Enhancement of cognition, mood and pro-social behavior and blunting of unwanted memories. *Neurosci Biobehav Rev* **32**, 760–776.

Debiec, J., LeDoux, J.E. and Nader, K. (2002). Cellular and systems reconsolidation in the hippocampus. *Neuron* **36**, 527–538.

Dere, E., Pause, B.M. and Pietrowsky, R. (2010). Emotion and episodic memory in neuropsychiatric disorders. *Behav Brain Res* **215**, 162–171.

Dudai, Y. (2004). The neurobiology of consolidations, or, how stable is the engram? *Annu Rev Psychol* **55**, 51–86.

Ehlers, A. and Clark, D.M. (2000). A cognitive model of posttraumatic stress disorder. *Behav Res Ther* **38**, 319–345.

Ehlers, A. and Clark, D.M. (2008). Post-traumatic stress disorder: The development of effective psychological treatments. *Nord J Psychiatry* 62 Suppl **47**, 11–18.

Ehrlich, I., Humeau, Y., Grenier, F., Ciocchi, S., Herry, C. and Luthi, A. (2009). Amygdala inhibitory circuits and the control of fear memory. *Neuron* **62**, 757–771.

Eichenbaum, H. (2004). Hippocampus: Cognitive processes and neural representations that underlie declarative memory. *Neuron* **44**, 109–120.

Elvevag, B. and Goldberg, T.E. (2000). Cognitive impairment in schizophrenia is the core of the disorder. *Crit Rev Neurobiol* **14**, 1–21.

Engel, G.L. (1977). The need for a new medical model: A challenge for biomedicine. *Science* **196**, 129–136.

Engel, G.L. (1981). The clinical application of the biopsychosocial model. *J Med Philos* **6**, 101–123.

Fadok, J.P., Darvas, M., Dickerson, T.M.K. and Palmiter, R.D. (2010). Long-term memory for pavlovian fear conditioning requires dopamine in the nucleus accumbens and basolateral amygdala. *PLoS ONE* **5**, 1–6.

Fanselow, M.S. and LeDoux, J.E. (1999). Why we think plasticity underlying Pavlovian fear conditioning occurs in the basolateral amygdala. *Neuron* **23**, 229–232.

Ferreri, F., Lapp, L.K. and Peretti, C.S. (2011). Current research on cognitive aspects of anxiety disorders. *Curr Opin Psychiatry* **24**, 49–54.

Fields, H.L., Hjelmstad, G.O., Margolis, E.B. and Nicola, S.M. (2007). Ventral tegmental area neurons in learned appetitive behavior and positive reinforcement. *Annu Rev Neurosci* **30**, 289–316.

Fletcher, P.C. and Henson, R.N. (2001). Frontal lobes and human memory: Insights from functional neuroimaging. *Brain* **124**, 849–881.

Gabbard, G.O. (2000). A neurobiologically informed perspective on psychotherapy. *Br J Psychiatry* **177**, 117–122.

Gabrieli, J.D. (1998). Cognitive neuroscience of human memory. *Annu Rev Psychol* **49**, 87–115.

Gelder, M., Harrison, P. and Cowen, P. (2006). *Shorter Oxford Textbook of Psychiatry.* 5th ed. (OUP: Oxford).

Glimcher, P. (2002). Decisions, decisions, decisions: Choosing a biological science of choice. *Neuron* **36**, 323–332.

Gotlib, I.H. and Joormann, J. (2010). Cognition and depression: Current status and future directions. *Annu Rev Clin Psychol* **6**, 285–312.

Gottesman, II. and Gould, T.D. (2003). The endophenotype concept in psychiatry: Etymology and strategic intentions. *Am J Psychiatry* **160**, 636–645.

Green, M.F. (1996). What are the functional consequences of neurocognitive deficits in schizophrenia? *Am J Psychiatry* **153**, 321–330.

Green, M.F., Kern, R.S., Braff, D.L. and Mintz, J. (2000). Neurocognitive deficits and functional outcome in schizophrenia: Are we measuring the "right stuff"? *Schizophr Bull* **26**, 119–136.

Grillon, C. (2009). D-cycloserine facilitation of fear extinction and exposure-based therapy might rely on lower-level, automatic mechanisms. *Biol Psychiatry* **66**, 636–641.

Grosjean, B. (2005). From synapse to psychotherapy: The fascinating evolution of neuroscience. *Am J Psychother* **59**, 181–197.

Haas, B.W. and Canli, T. (2008). Emotional memory function, personality structure and psychopathology: A neural system approach to the identification of vulnerability markers. *Brain Res Rev* **58**, 71–84.

Hamann, S. (2001). Cognitive and neural mechanisms of emotional memory. *Trends Cogn Sci* **5**, 394–400.

Harvey, P.D. and Keefe, R.S. (2001). Studies of cognitive change in patients with schizophrenia following novel antipsychotic treatment. *Am J Psychiatry* **158**, 176–184.

Heinrichs, R.W. and Zakzanis, K.K. (1998). Neurocognitive deficit in schizophrenia: A quantitative review of the evidence. *Neuropsychology* **12**, 426–445.

Henke, K. (2010). A model for memory systems based on processing modes rather than consciousness. *Nat Rev Neurosci* **11**, 523–532.

Hirsh, R. (1974). The hippocampus and contextual retrieval of information from memory: A theory. *Behav Biol* **12**, 421–444.

Hofmann, S.G., Meuret, A.E., Smits, J.A., Simon, N.M., Pollack, M.H., Eisenmenger, K., Shiekh, M. and Otto, M.W. (2006). Augmentation of exposure therapy with D-cycloserine for social anxiety disorder. *Arch Gen Psychiatry* **63**, 298–304.

Hoglend, P., Bogwald, K.-P., Amlo, S., Marble, A., Ulberg, R., Sjaastad, M.C., Sorbye, O., Heyerdahl, O. and Johansson, P. (2008). Transference interpretations in dynamic psychotherapy: Do they really yield sustained effects? *Am J Psychiatry* **165**, 763–771.

Holland, A.C. and Kensinger, E.A. (2010). Emotion and autobiographical memory. *Phys Life Rev* **7**, 88–131.

Jones, P., Rodgers, B., Murray, R. and Marmot, M. (1994). Child development risk factors for adult schizophrenia in the British 1946 birth cohort. *Lancet* **344**, 1398–1402.

Jovanovic, T. and Ressler, K.J. (2010). How the neurocircuitry and genetics of fear inhibition may inform our understanding of PTSD. *Am J Psychiatry* **167**, 648–662.

Kandel, E.R. (1998). A new intellectual framework for psychiatry. *Am J Psychiatry* **155**, 457–469.

Kandel, E.R. (1999). Biology and the future of psychoanalysis: A new intellectual framework for psychiatry revisited. *Am J Psychiatry* **156**, 505–524.

Keefe, R.S., Seidman, L.J., Christensen, B.K., Hamer, R.M., Sharma, T., Sitskoorn, M.M., Lewine, R.R., Yurgelun-Todd, D.A., Gur, R.C., Tohen, M., *et al.* (2004). Comparative effect of atypical and conventional antipsychotic drugs on neurocognition in first-episode psychosis: A randomized, double-blind trial of olanzapine versus low doses of haloperidol. *Am J Psychiatry* **161**, 985–995.

Kendler, K.S. (2005). Toward a philosophical structure for psychiatry. *Am J Psychiatry* **162**, 433–440.

Kensinger, E.A. (2009). Remembering the details: Effects of emotion. *Emotion Rev* **1**, 99–113.

Kensinger, E.A. and Corkin, S. (2004). Two routes to emotional memory: Distinct neural processes for valence and arousal. *Proc Natl Acad Sci USA* **101**, 3310–3315.

Kindt, M., Soeter, M. and Vervliet, B. (2009). Beyond extinction: Erasing human fear responses and preventing the return of fear. *Nat Neurosci* **12**, 256–258.

King, N.S. (2002). Perseveration of traumatic re-experiencing in PTSD; a cautionary note regarding exposure based psychological treatments for PTSD when head injury and dysexecutive impairment are also present. *Brain Injury* **16**, 65–74.

Kopelman, M.D. (2002). Disorders of memory. *Brain* **125**, 2152–2190.

Koriat, A., Goldsmith, M. and Pansky, A. (2000). Toward a psychology of memory accuracy. *Annu Rev Psychol* **51**, 481–537.

Kushner, M.G., Kim, S.W., Donahue, C., Thuras, P., Adson, D., Kotlyar, M., McCabe, J., Peterson, J. and Foa, E.B. (2007). D-cycloserine augmented exposure therapy for obsessive-compulsive disorder. *Biol Psychiatry* **62**, 835–838.

Kuyken, W., Watkins, E., Holden, E., White, K., Taylor, R.S., Byford, S., Evans, A., Radford, S., Teasdale, J.D. and Dalgleish, T. (2010). How does mindfulness-based cognitive therapy work? *Behav Res Ther* **48**, 1105–1112.

Lee, H. and Kim, J.J. (1998). Amygdalar NMDA receptors are critical for new fear learning in previously fear-conditioned rats. *J Neurosci* **18**, 8444–8454.

Lepage, M., Sergerie, K., Pelletier, M. and Harvey, P.O. (2007). Episodic memory bias and the symptoms of schizophrenia. *Can J Psychiat-Rev Can Psychiat* **52**, 702–709.

Liao, S.M. and Sandberg, A. (2008). The normativity of memory modification. *Neuroethics* **1**, 85–99.

Lissek, S., Powers, A.S., McClure, E.B., Phelps, E.A., Woldehawariat, G., Grillon, C. and Pine, D.S. (2005). Classical fear conditioning in the anxiety disorders: A meta-analysis. *Behav Res Ther* **43**, 1391–1424.

MacKinnon , D.F. (2010). *Trouble in Mind: An Unorthodox Introduction to Psychiatry*. (Baltimore, Maryland: The Johns Hopkins University Press).

Makkar, S.R., Zhang, S.Q. and Cranney, J. (2010). Behavioral and neural analysis of GABA in the acquisition, consolidation, reconsolidation and extinction of fear memory. *Neuropsychopharmacology* **35**, 1625–1652.

Malhi, G.S., Ivanovski, B., Hadzi-Pavlovic, D., Mitchell, P.B., Vieta, E. and Sachdev, P. (2007). Neuropsychological deficits and functional impairment in bipolar depression, hypomania and euthymia. *Bipolar Disord* **9**, 114–125.

Malmberg, L. and Fenton, M. (2001). Individual psychodynamic psychotherapy and psychoanalysis for schizophrenia and severe mental illness. *Cochrane Database Syst Rev*, CD001360.

Mark, J., Williams, G., Barnhofer, T., Crane, C., Hermans, D., Raes, F., Watkins, E. and Dalgleish, T. (2007). Autobiographical memory specificity and emotional disorder. *Psychol Bull* **133**, 122–148.

Martin, S.J., Grimwood, P.D. and Morris, R.G. (2000). Synaptic plasticity and memory: An evaluation of the hypothesis. *Annu Rev Neurosci* **23**, 649–711.

McCleery, J.M. and Harvey, A.G. (2004). Integration of psychological and biological approaches to trauma memory: Implications for pharmacological prevention of PTSD. *J Trauma Stress* **17**, 485–496.

McKenna, P.J. (1991). Memory, knowledge and delusions. *Br J Psychiatry Suppl*, 36–41.

Mishara, A.L. and Goldberg, T.E. (2004). A meta-analysis and critical review of the effects of conventional neuroleptic treatment on cognition in schizophrenia: Opening a closed book. *Biol Psychiatry* **55**, 1013–1022.

Mitte, K. (2008). Memory bias for threatening information in anxiety and anxiety disorders: A meta-analytic review. *Psychol Bull* **134**, 886–911.

Murray, G.K. (2011). The emerging biology of delusions. *Psychological Medicine* **41**, 7–13.

Myers, K.M. and Davis, M. (2002). Systems-level reconsolidation: Reengagement of the hippocampus with memory reactivation. *Neuron* **36**, 340–343.

Nadel, L. and Hardt, O. (2011). Update on memory systems and processes. *Neuropsychopharmacology* **36**, 251–273.

Nadel, L. and O'Keefe, J. (1974). The hippocampus in pieces and patches: An essay on modes of explanation in physiological psychology. In Bellairs, R. and Gray, E.G. eds. *Essays on the Nervous System*, (Oxford, The Clarendon Press).

Nader, K. (2003). Memory traces unbound. *Trends Neurosci* **26**, 65–72.

Nader, K., Schafe, G.E. and Le Doux, J.E. (2000). Fear memories require protein synthesis in the amygdala for reconsolidation after retrieval. *Nature* **406**, 722–726.

Newman, E.J. and Lindsay, D.S. (2009). False memories: What the hell are they for? *Appl Cogn Psychol* **23**, 1105–1121.

Norberg, M.M., Krystal, J.H. and Tolin, D.F. (2008). A meta-analysis of D-cycloserine and the facilitation of fear extinction and exposure therapy. *Biol Psychiatry* **63**, 1118–1126.

Otto, M.W. (2002). Learning and "unlearning" fears: Preparedness, neural pathways and patients. *Biol Psychiatry 52*, 917–920.

Palmer, B.W., Dawes, S.E. and Heaton, R.K. (2009). What do we know about neuropsychological aspects of schizophrenia? *Neuropsychol Rev* **19**, 365–384.

Pannu, J.K. and Kaszniak, A.W. (2005). Metamemory experiments in neurological populations: A review. *Neuropsychol Rev* **15**, 105–130.

Park, S., Holzman, P.S. and Goldman-Rakic, P.S. (1995). Spatial working memory deficits in the relatives of schizophrenic patients. *Arch Gen Psychiatry* **52**, 821–828.

Phelps, E.A., Delgado, M.R., Nearing, K.I. and LeDoux, J.E. (2004). Extinction learning in humans: Role of the amygdala and vmPFC. *Neuron* **43**, 897–905.

Pitman, R.K., Sanders, K.M., Zusman, R.M., Healy, A.R., Cheema, F., Lasko, N.B., Cahill, L. and Orr, S.P. (2002). Pilot study of secondary prevention of posttraumatic stress disorder with propranolol. *Biol Psychiatry* **51**, 189–192.

Przybyslawski, J., Roullet, P. and Sara, S.J. (1999). Attenuation of emotional and nonemotional memories after their reactivation: Role of beta adrenergic receptors. *J Neurosci* **19**, 6623–6628.

Quraishi, S. and Frangou, S. (2002). Neuropsychology of bipolar disorder: A review. *J Affect Disord* **72**, 209–226.

Ragland, J.D., Laird, A.R., Ranganath, C., Blumenfeld, R.S., Gonzales, S.M. and Glahn, D.C. (2009). Prefrontal activation deficits during episodic memory in schizophrenia. *Am J Psychiatry* **166**, 863–874.

Ranganath, C., Minzenberg, M.J. and Ragland, J.D. (2008). The cognitive neuroscience of memory function and dysfunction in schizophrenia. *Biol Psychiatry* **64**, 18–25.

Ressler, K.J., Rothbaum, B.O., Tannenbaum, L., Anderson, P., Graap, K., Zimand, E., Hodges, L. and Davis, M. (2004). Cognitive enhancers as adjuncts to psychotherapy: Use of D-cycloserine in phobic individuals to facilitate extinction of fear. *Arch Gen Psychiatry* **61**, 1136–1144.

Rothbaum, B.O. and Davis, M. (2003). Applying learning principles to the treatment of post-trauma reactions. *Ann N Y Acad Sci* **1008**, 112–121.

Saykin, A.J., Shtasel, D.L., Gur, R.E., Kester, D.B., Mozley, L.H., Stafiniak, P. and Gur, R.C. (1994). Neuropsychological deficits in neuroleptic naive patients with first-episode schizophrenia. *Arch Gen Psychiatry* **51**, 124–131.

Schacter, D.L., Norman, K.A. and Koutstaal, W. (1998). The cognitive neuroscience of constructive memory. *Annu Rev Psychol* **49**, 289–318.

Schall, J.D. (2001). Neural basis of deciding, choosing and acting. *Nat Rev Neurosci* **2**, 33–42.

Schiller, D., Monfils, M.H., Raio, C.M., Johnson, D.C., Ledoux, J.E. and Phelps, E.A. (2010). Preventing the return of fear in humans using reconsolidation update mechanisms. *Nature* **463**, 49–53.

Scoville, W.B. and Milner, B. (1957). Loss of recent memory after bilateral hippocampal lesions. *J Neurol Neurosurg Psychiatry* **20**, 11–21.

Squire, L.R. (1992). Memory and the hippocampus: A synthesis from findings with rats, monkeys, and humans. *Psychol Rev* **99**, 195–231.

Squire, L.R., Knowlton, B. and Musen, G. (1993). The structure and organization of memory. *Annu Rev Psychol* **44**, 453–495.

Squire, L.R., Stark, C.E. and Clark, R.E. (2004). The medial temporal lobe. *Annu Rev Neurosci* **27**, 279–306.

Squire, L.R. and Zola, S.M. (1996). Structure and function of declarative and nondeclarative memory systems. *Proc Natl Acad Sci USA* **93**, 13515–13522.

Stuss, D.T. and Levine, B. (2002). Adult clinical neuropsychology: Lessons from studies of the frontal lobes. *Annu Rev Psychol* **53**, 401–433.

Tamietto, M. and de Gelder, B. (2010). Neural bases of the non-conscious perception of emotional signals. *Nat Rev Neurosci* **11**, 697–709.

Tamminga, C.A., Buchanan, R.W. and Gold, J.M. (1998). The role of negative symptoms and cognitive dysfunction in schizophrenia outcome. *Int Clin Psychopharmacol 13 Suppl* **3**, S21–26.

Tronson, N.C. and Taylor, J.R. (2007). Molecular mechanisms of memory reconsolidation. *Nat Rev Neurosci* **8**, 262–275.

Tryon, W.W. (2000). Behavior therapy as applied learning theory. *The Behavior Therapist* **23**, 131–134.

Tryon, W.W. (2005). Possible mechanisms for why desensitization and exposure therapy work. *Clin Psychol Rev* **25**, 67–95.

Tulving, E. (1972). Episodic and semantic memory. In Tulving E., and Donaldson W., eds. *Organisation of Memory*, (New York: Academic Press).

Tulving, E. (1995). Organization of memory: Quo vadis? In G. M. S., ed. *The Cognitive Neurosciences*, (Cambridge: MIT Press), pp. 839–847.

Tulving, E. (2002). Episodic memory: From mind to brain. *Annu Rev Psychol* **53**, 1–25.

Wang, S.H. and Morris, R.G. (2010). Hippocampal-neocortical interactions in memory formation, consolidation, and reconsolidation. *Annu Rev Psychol* **61**, 49–79, C41–44.

Weinberger, D.R., Berman, K.F. and Zec, R.F. (1986). Physiologic dysfunction of dorsolateral prefrontal cortex in schizophrenia. I. Regional cerebral blood flow evidence. *Arch Gen Psychiatry* **43**, 114–124.

Wilhelm, S., Buhlmann, U., Tolin, D.F., Meunier, S.A., Pearlson, G.D., Reese, H.E., Cannistraro, P., Jenike, M.A. and Rauch, S.L. (2008). Augmentation of behavior therapy with D-cycloserine for obsessive-compulsive disorder. *Am J Psychiatry* **165**, 335–341; quiz 409.

World Health Organization. (1993). International Statistical Classification of Diseases and Related Health Problems — 10th version (World Health Organization).

APPENDIX

International Classification of Disease-10th Edition

Chapter 5 has the following list of conditions categorised (a briefer version is presented for non-organic conditions):

ICD-10 Code-Range of Disorders	Examples of disorders
(F00–F09) Organic, including symptomatic, mental disorders	(F00) Dementia in Alzheimer's disease; (F01) Vascular dementia; (F01.1) Multi-infarct dementia; (F02.2) Dementia in Huntington's disease; (F02.3) Dementia in Parkinson's disease; (F02.4) Dementia in human immunodeficiency virus (HIV) disease; (F04.) Organic amnesic syndrome, not induced by alcohol and other psychoactive substances; (F05.) Delirium, not induced by alcohol and other psychoactive substances; (F06.1) Organic catatonic disorder; (F06.0) Organic hallucinosis; (F06.2) Organic delusional (schizophrenia-like) disorder; (F06.3) Organic mood (affective) disorders; (F06.4) Organic anxiety disorder; (F06.7) Mild cognitive disorder; (F06.8) Other specified mental disorders due to brain damage and dysfunction and to physical disease; (F07.) Personality and behavioural disorders due to brain disease, damage and dysfunction; (F07.0) Organic personality disorder; (F07.2) Postconcussional syndrome.
(F10–F19) Mental and behavioural disorders due to psychoactive substance use	Note: the following conditions are subtypes of each code from F10–19: (F10–F19) Mental and behavioural disorders due to psychoactive substance use; (F1x.0) acute intoxication; (F1x.1) harmful use; (F1x.2) dependence syndrome; (F1x.3) withdrawal state; (F1x.4) withdrawal state with delirium; (F1x.5) psychotic disorder; (F1x.6) amnesic syndrome; (F1x.7) Residual and late-onset psychotic disorder; (F1x.8) other mental and behavioural disorder; (F1x.9) unspecified mental and behavioural disorder.
(F20–F29) Schizophrenia, schizotypal and delusional disorders	(F20) Schizophrenia; (F20.0) Paranoid schizophrenia; (F20.1) Hebephrenic schizophrenia (Disorganized schizophrenia); (F20.2) Catatonic schizophrenia; (F21.) Schizotypal disorder; (F22.) Persistent delusional disorders; (F22.0) Delusional disorder; (F22.8) Other persistent delusional disorders, e.g. Delusional dysmorphophobia; (F23.) Acute and transient

(*Continued*)

(*Continued*)

	psychotic disorders; (F23.0) Acute polymorphic psychotic disorder without symptoms of schizophrenia; (F23.1) Acute polymorphic psychotic disorder with symptoms of schizophrenia; (F23.2) Acute schizophrenia-like psychotic disorder; (F24.) Induced delusional disorder; (F25.) Schizoaffective disorders ; (F25.0) Schizoaffective disorder, manic type; (F25.1) Schizoaffective disorder, depressive type; (F25.2) Schizoaffective disorder, mixed type;
(F30–F39) Mood (affective) disorders	(F30) Manic episode; (F30.0) Hypomania; (F30.1) Mania without psychotic symptoms; (F30.2) Mania with psychotic symptoms; (F31.) Bipolar affective disorder; (F31.0) Bipolar affective disorder, current episode hypomanic; (F31.1) Bipolar affective disorder, current episode manic without psychotic symptoms; (F31.2) Bipolar affective disorder, current episode manic with psychotic symptoms; (F31.3) Bipolar affective disorder, current episode mild or moderate depression; (F31.4) Bipolar affective disorder, current episode severe depression without psychotic symptoms; (F31.5) Bipolar affective disorder, current episode severe depression with psychotic symptoms; (F31.6) Bipolar affective disorder, current episode mixed; (F31.7) Bipolar affective disorder, currently in remission; (F32) Depressive episode; (F32.0) Mild depressive episode; (F32.1) Moderate depressive episode; (F32.2) Severe depressive episode without psychotic symptoms; (F32.3) Severe depressive episode with psychotic symptoms; (F32.8) Other depressive episodes, e.g. Atypical depression; (F33) Recurrent depressive disorder; (F33.0) Recurrent depressive disorder, current episode mild; (F33.1) Recurrent depressive disorder, current episode moderate; (F33.2) Recurrent depressive disorder, current episode severe without psychotic symptoms; (F33.3) Recurrent depressive disorder, current episode severe with psychotic symptoms; (F33.4) Recurrent depressive disorder, currently in remission; (F34.) Persistent mood (affective) disorders; (F34.0) Cyclothymia; (F34.1) Dysthymia;
(F40–F48) Neurotic, stress-related and somatoform disorders	(F40) Phobic anxiety disorders; (F40.0) Agoraphobia; (F40.1) Social phobias; (F40.2) Specific (isolated) phobias, e.g. Acrophobia, Animal phobias, Claustrophobia; (F41.0) Panic disorder (episodic paroxysmal anxiety); (F41.1) Generalized anxiety disorder; (F42) Obsessive-compulsive disorder;

(*Continued*)

(*Continued*)

	(F43) Reaction to severe stress, and adjustment disorders; (F43.0) Acute stress reaction; (F43.1) Post-traumatic stress disorder; (F43.2) Adjustment disorder; (F44) Dissociative (conversion) disorders; (F44.0) Dissociative amnesia; (F44.1) Dissociative fugue; (F44.2) Dissociative stupor; (F44.3) Trance and possession disorders; (F44.4) Dissociative motor disorders; (F44.5) Dissociative convulsions; (F44.6) Dissociative anaesthesia and sensory loss; (F44.7) Mixed dissociative (conversion) disorders; (F44.8) Other dissociative (conversion) disorders, e.g. Ganser's syndrome, Multiple personality; (F45) Somatoform disorders; (F45.0) Somatisation disorder, e.g. Briquet's disorder; (F45.2) Hypochondriacal disorder, e.g. Body dysmorphic disorder Dysmorphophobia (nondelusional), Hypochondriacal neurosis, Hypochondriasis; (F45.3) Somatoform autonomic dysfunction, e.g. Cardiac neurosis; (F45.4) Persistent Somatoform Pain Disorder; (F48.0) Neurasthenia; (F48.1) Depersonalisation-derealisation syndrome.
(F50–F59) Behavioural syndromes associated with physiological disturbances and physical factors	(F50) Eating disorders; (F50.0) Anorexia nervosa; (F50.2) Bulimia nervosa; (F51) Nonorganic sleep disorders; (F52) Sexual dysfunction, not caused by organic disorder or disease; (F53) Mental and behavioural disorders associated with the puerperium, not elsewhere classified;
(F60–F69) Disorders of adult personality and behaviour	(F60) Specific personality disorders; (F60–F69) Disorders of adult personality and behaviour; (F60.0) Paranoid personality disorder; (F60.1) Schizoid personality disorder; (F60.2) Dissocial personality disorder; (F60.3) Emotionally unstable personality disorder (Borderline personality disorder; (F60.4) Histrionic personality disorder); (F60.5) Anankastic personality disorder (Obsessive-compulsive personality disorder); (F60.6) Anxious (avoidant) personality disorder; (F60.7) Dependent personality disorder; (F63) Habit and impulse disorders; (F63.0) Pathological gambling; (F63.1) Pathological fire-setting (pyromania); (F63.2) Pathological stealing (kleptomania); (F63.3) Trichotillomania; (F64) Gender identity disorders; (F64.0) Transsexualism; (F65) Disorders of sexual preference; (F65.0) Sexual fetishism; (F65.1) Fetishistic transvestism; (F65.2) Exhibitionism; (F65.3) Voyeurism;

(*Continued*)

(*Continued*)

	(F65.4) Paedophilia; (F65.5) Sadomasochism; (F65.8) Other disorders of sexual preference (Necrophilia, Zoophilia); (F68) Other disorders of adult personality and behaviour; (F68.0) Elaboration of physical symptoms for psychological reasons; (F68.1) Intentional production or feigning of symptoms or disabilities, either physical or psychological (factitious disorder), i.e. Munchausen syndrome;
(F70–F79) Mental retardation	(F70) Mild mental retardation; (F71) Moderate mental retardation; (F72) Severe mental retardation; (F73) Profound mental retardation
(F80–F89) Disorders of psychological development	(F80) Specific developmental disorders of speech and language (F81); Specific developmental disorders of scholastic skills; (F81.0) Specific reading disorder (Developmental dyslexia); (F81.1) Specific spelling disorder; (F81.2) Specific disorder of arithmetical skills; (F84) Pervasive developmental disorders (F84.0) Childhood autism; (F84.1) Atypical autism; (F84.2) Rett's syndrome; (F84.3) Other childhood disintegrative disorder; (F84.4) Overactive disorder associated with mental retardation and stereotyped movements; (F84.5) Asperger syndrome;
(F90–F98) Behavioural and emotional disorders with onset usually occurring in childhood and adolescence	(F90) Hyperkinetic disorders; (F90.0) Disturbance of activity and attention (Attention-deficit hyperactivity disorder; Attention deficit syndrome with hyperactivity); (F90.1) Hyperkinetic conduct disorder; (F91) Conduct disorders; (F92) Mixed disorders of conduct and emotions; (F93) Emotional disorders with onset specific to childhood; (F93.0) Separation anxiety disorder of childhood; (F94); Disorders of social functioning with onset specific to childhood and adolescence; (F94.0) Elective mutism; (F94.1) Reactive attachment disorder of childhood; (F94.2) Disinhibited attachment disorder of childhood; (F95.) Tic disorders; (F95.2) Combined vocal and multiple motor tic disorder (de la Tourette); (F98) Other behavioural and emotional disorders with onset usually occurring in childhood and adolescence; (F98.0) Non-organic enuresis; (F98.3) Pica of infancy and childhood.

Index